ADVANCES IN DISORDERED SYSTEMS, RANDOM PROCESSES AND SOME APPLICATIONS

A unified perspective on the study of complex systems is offered to scholars of various disciplines including mathematics, physics, computer science, biology, economics and social science.

The contributions, written by leading scientists, cover a broad set of topics: new approaches to data science, the connection between scaling limits and conformal field theories, new ideas on the Legendre duality approach in statistical mechanics of disordered systems. The volume moreover explores results on extreme values of correlated random variables and their connection with the Riemann zeta functions, the relation between diffusion phenomena and complex systems and the Brownian web, which appears as the universal scaling limit of several probabilistic models. Written for researchers from a broad range of scientific fields, this text examines a selection of recent developments in complex systems from a rigorous perspective.

PIERLUIGI CONTUCCI is Full Professor of Mathematical Physics at the Alma Mater Studiorum and his research interests are mathematical physics, statistical mechanics and their applications. He is a member of the Istituto Cattaneo, Bologna and in 2000 he received the Schloessman Award from the Max Planck Society.

CRISTIAN GIARDINÀ is Professor in Mathematical Physics at the University of Modena and Reggio Emilia and his research interests are mathematical statistical physics and probability theory. Currently he is also visiting full professor in probability at Delft University.

ADVANCES IN DISORDERED SYSTEMS, RANDOM PROCESSES AND SOME APPLICATIONS

PIERLUIGI CONTUCCI
University of Bologna, Italy

CRISTIAN GIARDINÀ
University of Modena and Reggio Emilia, Italy

CAMBRIDGE
UNIVERSITY PRESS

University Printing House, Cambridge CB2 8BS, United Kingdom

One Liberty Plaza, 20th Floor, New York, NY 10006, USA

477 Williamstown Road, Port Melbourne, VIC 3207, Australia

4843/24, 2nd Floor, Ansari Road, Daryaganj, Delhi - 110002, India

79 Anson Road, #06-04/06, Singapore 079906

Cambridge University Press is part of the University of Cambridge.

It furthers the University's mission by disseminating knowledge in the pursuit of education, learning, and research at the highest international levels of excellence.

www.cambridge.org
Information on this title: www.cambridge.org/9781107124103
10.1017/9781316403877

© Pierluigi Contucci and Cristian Giardinà 2017

This publication is in copyright. Subject to statutory exception and to the provisions of relevant collective licensing agreements, no reproduction of any part may take place without the written permission of Cambridge University Press.

First published 2017

Printed in the United States of America by Sheridan Books, Inc.

A catalogue record for this publication is available from the British Library.

Library of Congress Cataloging-in-Publication Data
Names: Contucci, Pierluigi, 1964– | Giardinà, Cristian.
Title: Advances in disordered systems, random processes, and some applications / Pierluigi Contucci, Università di Bologna, Cristian Giardinà, Università degli Studi di Modena, Italy, [editors].
Description: Cambridge : Cambridge University Press, [2017] | Includes bibliographical references and index.
Identifiers: LCCN 2016036260 | ISBN 9781107124103 (hardback : alk. paper)
Subjects: LCSH: System analysis. | Chaotic behavior in systems.
Classification: LCC QA402.A2975 2017 | DDC 003/.857–dc23
LC record available at https://lccn.loc.gov/2016036260

ISBN 978-1-107-12410-3 Hardback

Cambridge University Press has no responsibility for the persistence or accuracy of URLs for external or third-party Internet Web sites referred to in this publication and does not guarantee that any content on such Web sites is, or will remain, accurate or appropriate.

Contents

List of Contributors		*page* vi
Preface		viii
1	Topological Field Theory of Data: Mining Data Beyond Complex Networks *Mario Rasetti and Emanuela Merelli*	1
2	A Random Walk in Diffusion Phenomena and Statistical Mechanics *Elena Agliari*	43
3	Legendre Structures in Statistical Mechanics for Ordered and Disordered Systems *Francesco Guerra*	142
4	Extrema of Log-correlated Random Variables: Principles and Examples *Louis-Pierre Arguin*	166
5	Scaling Limits, Brownian Loops, and Conformal Fields *Federico Camia*	205
6	The Brownian Web, the Brownian Net, and their Universality *Emmanuel Schertzer, Rongfeng Sun and Jan M. Swart*	270
Index		369

Contributors

Mario Rasetti
ISI Foundation, Via Alassio 11-C; 10126 Torino, Italy

Emanuela Merelli
School of Science and Technology, University of Camerino, Via del Bastione 1; 62032 Camerino, Italy

Elena Agliari
Dipartimento di Matematica, Sapienza Universita di Roma, Piazzale Aldo Moro 5, 00185 Roma, Italy

Francesco Guerra
Dipartimento di Fisica, Sapienza Universita di Roma and Istituto Nazionale di Fisica Nucleare, Sezione di Roma Piazzale A. Moro 5, 00185 Roma, Italy

Louis-Pierre Arguin
Dép. de Mathématique et Statistique, Université de Montréal, C.P. 6128, succ. Centre-Ville, Montréal, Québec, Canada

Federico Camia
VU University Amsterdam, the Netherlands and New York University, Abu Dhabi, United Arab Emirates

Emmanuel Schertzer
UPMC University Paris 6, Laboratoire de Probabilités et Modèles Aléatoires, CNRS UMR 7599, Paris, France

Rongfeng Sun
Department of Mathematics, National University of Singapore, 10 Lower Kent Ridge Road, 119076 Singapore

Jan M. Swart
Institute of Information Theory and Automation of the ASCR (UTIA), Pod Vodárenskou věží 4, 18208 Praha 8, Czech Republic

Preface

Modern science is witnessing a peak of intense activity toward the study of complex systems. This new topic is a very heterogeneous variety of approaches, methods and perspectives that share the common attempt to understand the collective behavior of a very large number of units correlated by simple competitive and cooperative interactions. These types of investigations have indeed appeared several times in the past. Recently though, the availability of large databases and the advent of unprecedented computer facilities have created a fertile ground for the current boost. Moreover, the progress made in the studies of non-homogenous, disordered systems of the last two decades has produced new promising approaches and technical tools for applied research topics. The analysis of such objects has led to a fruitful dialog involving mathematics and physics as founding and guiding disciplines, with an increasingly growing contribution of specific problems from the socio-economic, biological and other sciences. The present book is a collection of selected contributions by leading world experts toward the process of building up a rigorous conceptual framework within these new ideas that have irrigated the exact sciences, revealing a host of new questions, strategies and solutions.

The volume opens with the contribution by Mario Rasetti and Emanuela Merelli that focuses on a new approach to Data Science that challenges the traditional ones. The authors, after a broad introduction that encompasses motivational arguments and epistemological reasonings, build up a general framework identifying it with a topological field theory of data. The theoretical physics scheme of field theory is proposed as a guide where the relevant information is encoded into topological features as it happens, for instance, in general relativity and its geometrical curvature theory. The content is grounded on a rich and articulated scientific culture and presents a wide set of novel and brilliant ideas. The strategy and vision introduced have the potential to widely impact the field by providing a unified, pre-axiomatic framework, that is able to guide instances and specific case studies.

The contribution by Elena Agliari deals with stochastic processes for diffusion phenomena and statistical mechanics techniques for complex systems, showing the strong analogies between the two theories from a methodological point of view. In particular, after diffusion is initially analyzed and approached from different perspectives, the author provides the basic elements of the theory of random walks on graphs as paradigmatic models of diffusion phenomena. The quantum analogue of classical random walk is also addressed. The description of basic statistical mechanics models defined on infinite graphs is briefly reviewed in the third chapter in order to show that the main concepts characterizing random walks, such as recurrence, transience and spectral dimension, also determine the properties of these models. Finally, taking the Curie-Weiss model as an example, the problem of the explicit evaluation of its free energy is mapped into a random walk framework. This dictionary allows exploitation of some of the methodologies originally used in one field to apply them efficiently to the other. The clear and well organized exposition make this work a suitable starting point for scholars wanting to investigate topics at the interplay between diffusion phenomena and statistical mechanics.

The contribution by Francesco Guerra is, as typical of the author, towards a quest for a mathematically simple and physically deep understanding of the mean field spin glass phase. It starts with a brief summary of the classical Legendre dual (energy-entropy) structure that occupies a central role in the study of statistical mechanics models. The author therefore introduces a new kind of Legendre duality that occurs in disordered models. The latter is inspired by the interpolation methods and convexity arguments, of which the author is inventor and master, successfully used in the analysis of mean field disordered models (REM, GREM, SK and p-spin). The main features of this new Legendre duality are the facts that it appears as an inequality in the opposite direction with respect to the usual one and involves the covariance of the random interaction instead of the energy. Taking as a pedagogical example the REM model, the author shows how to construct in an explicit way the Legendre functional considering a variational problem involving the square of the Hamiltonian function. The author suggests moreover that this procedure works also in other cases and may provide a general framework to investigate the properties of disordered models from a new perspective. The general concepts introduced in this work, despite their deep meaning and consequences, are developed in a simple and clear way and can be useful to obtain new insights in the field.

The contribution by Louis-Pierre Arguin also finds its roots in the field of disordered systems. The work deals with the extreme value of correlated random variables, such as the energy levels of a spin glass model. The topic of Arguin's contribution is to show general arguments for obtaining the leading order behavior of the extrema of log-correlated random variables, as well as subleading corrections and fluctuations. The main ideas that Arguin emphasizes are the three steps that lead to the results: multiscale decomposition, dichotomy of scales and self-similarity of scales. He introduces the ideas with the Gaussian branching random walk for pedagogical reasons. But he also shows some key ideas of how to apply it to the two-dimensional Gaussian free field. Potential applications to seemingly unrelated problems are also discussed. For instance, the relation between the maxima of the characteristic polynomials of random matrices and the maxima of the Riemann zeta function on an interval of the critical line is analyzed.

The main topic of the contribution by Federico Camia is the continuum scaling limit of planar lattice models. This topic is at the interface between statistical physics and Euclidean field theory. The link between the two areas emerges via the scaling limit that turns a lattice model into a continuum model as the mesh of the lattice is sent to zero. The author chooses to focus on a concrete example, namely the random walk loop soup, that plays the role of the ideal gas (being made of non-interacting lattice loops) and that is also related to the discrete Gaussian free field. In the scaling limit the Brownian loop soup emerges, which is a model already appearing within the Schramm-Loewner Evolution (SLE) context. The model is also used to explain the deep connection between scaling limits and conformal field theory of two-dimensional critical systems.

The volume closes with the contribution by Emmanuel Schertzer, Rongfeng Sun and Jan Swart on the Brownian web and the Brownian net. The Brownian web is formally a collection of one-dimensional coalescing Brownian motions starting from everywhere in space and time, and the Brownian net is one in which the above is also branching. Brownian web and net arise as the diffusive scaling limit of many one-dimensional interacting particle systems with branching and coalescence. Here the authors focus on coalescing random walks and random walks in i.i.d. space-time random environments. The prominent role of Brownian web/net is due to their appearance as universal scaling limits of several other models. The authors argue why this is the case by considering several models: the biased voter model, true self-avoiding walks, true self-repelling motion, drainage networks and others. This contribution is clearly a great source of the current state of research on Brownian web and Brownian net, and will prove to be very handy for researchers who wish to be introduced to the area and work on it.

Pierluigi Contucci and Cristian Giardinà

1

Topological Field Theory of Data: Mining Data Beyond Complex Networks

MARIO RASETTI AND EMANUELA MERELLI

1.1 A Philosophical Introduction

It has become increasingly obvious that very detailed, intricate interactions and interdependencies among and within large systems are often central to most of the important problems that science and society face. Distributed information technologies, neuroscience and genomics are just a few examples of rapidly emerging areas where very complex large-scale system interactions are viewed more and more as central to understanding, as well as to practical advances. Decision makers in these environments increasingly use computer models, simulation and data access resources to try to integrate and make sense of information and courses of action. There is also mounting concern that, in spite of the extended use of these simulations and models, we are repeatedly experiencing unexpected cascading systemic failures in society. We feel that, without resolving the issue of learning how to cope with complex situations, we also do not know enough about our methods of modeling complex systems to make effective decisions.

In the late eighties Saunders Mac Lane started a philosophical debate which, over thirty years later, is still going on with varying interest in the outcomes. This paper stems partly out of the crucial fundamental question that debate gave life to in contemporary science. This deep long-standing philosophical question, that can be formulated in several different ways, concerns mathematics. Are the formalisms of mathematics based on or derived from the facts and, if not, how are they derived? Alternatively, if mathematics is a purely formal game – an elaborate and tightly connected network of formal structures, axiom systems and connections – why do the formal conclusions in most of the cases fit the facts? Or, is mathematics invented or discovered? In the language of Karl Popper, statements of a science should be falsifiable by factual data; those of mathematics are not. Thus mathematics is not a science, it is something else. Yet the mathematical network is tied to numberless sources in human activities, to crucial parts of human knowledge and, most especially, to the various sciences.

What is intriguing is not only the number of connections between mathematics and science, but the fact that they often bear on subjects which are at the very core of the mathematical network, not just on the basic topics at the edge of the network. The external connections of mathematics are numerous and tight, but they do not fully describe or determine the mathematical subjects. Basic mathematical concepts may be derived from human activity, but they are not themselves such activity; nor are they the phenomena involved as the background of such activity: the axiomatic method is a declaration of independence for mathematics.

Even though science has an inherent, natural tendency toward specialization, contemporary mathematics is more and more pursuing a general theory of structures. One such theory is category theory. "Category theory has come to occupy a central position in contemporary mathematics and theoretical computer science, and has also successfully entered physics. Roughly, it is a general mathematical theory of structures and of systems of structures. As category theory is still evolving, its functions are correspondingly developing, expanding and multiplying."[1] It is first of all a powerful language, a conceptual framework allowing us to see the universal components of a family of structures of a given kind, and how structures of different kinds are interrelated.

The message emerging is: the subjects of mathematics are extracted from the environment, that is from activities or phenomena of science and society. This notion of *extraction* is close to the more familiar term *abstraction*, with the intent that the mathematical subject resulting from an extraction is indeed abstract. Mathematics is not *about* human activity or phenomena, it is about the extraction and formalization of ideas and their manifold consequences. The formalization of such ideas in certain cases took centuries, but then it often opened the way to deep unexpected interconnections that in turn opened the way to looking at certain human activities in a completely new and diverse fashion.

The forces driving the development of the mathematical framework are manifold: for instance, generalization from specific cases, by analogy or by modification, and abstraction (once more) by analogy, or by deletion or yet by shift of focus; the appearance of novel problems; or simply, just plain curiosity. But questions arising from the variety of human and scientific activity have been and can be the most important sources of novel mathematics. Computer science also brings up new mathematical ideas. There is a wealth of new algorithms which bear on decisive conceptual aspects, such as the subtle question of computational complexity.

---

[1] Stanford Encyclopedia of Philosophy

On the other hand probably the most important fact in modern science is that dramatic change in paradigms that has seen reductionism challenged by holism. This is the story: an integrated set of methods and concepts have emerged in science since the mid-eighties under several designations, of which complexity science is the simplest and most comprehensive. Complex systems can be simply defined as systems composed of many non-identical elements, entangled in loops of nonlinear interactions. A typical example is neurons in the brain cortex. The challenge is to describe the collective properties of these systems, getting from the mere description of their components to the global properties of the whole system – in the example, from the description of neurons to the cognitive properties of the brain.

A difficult issue arises here, for when the composing elements and their interactions are highly simplified, the global properties are typically very hard to predict. The global description in terms of attractors of system model dynamics can be a strong and insightful simplification with respect to a full description of the *microscopic* components; this is exactly the same in thermodynamics, where global properties of a system can be described independently from the complete description of its microscopic elements, which is partially done, instead, by statistical mechanics. Yet, a real theory of complex systems, relating to the wide phenomenology of complex phenomena and data in the way in which statistical mechanics is related to thermodynamics, is still missing.

There is an overwhelming evidence that the current emphasis of numerous sciences, not only sciences of nature but sciences of society as well, on this novel paradigm of complexity (holism versus reductionism) sorely requires a rigorous scientific framing of its methodologies, which is not yet available. If it is true that wide classes of systems and problems from various disciplines share universal features that lead us to imagine the existence of common structures directing their dynamics, it is equally true that the simplified schemes whereby they are handled, once reduced to the conventional form of decision problems, can often be approached and solved only by resorting to very drastic, generally ad hoc simplifications. All problems dealt with in the framework of multi-agent complex systems, usually approached by network theory, belong to this latter family, which includes a huge number of applications, from bio- and eco-systems to economic and sociological decision making issues. Such simplifications are typically dictated by the utter lack of mathematical tools that are powerful or flexible enough to lead to a true theory.

A typical feature of complex systems is the *emergence* of nontrivial superstructures that cannot be reconstructed by a reductionist approach. Not only do higher emergent features of complex systems arise out of the lower level interactions, but the patterns that they create act back on those lower levels. This ensures that complex systems possess a characteristic robustness with respect to large scale or multi-dimensional perturbations or disruptions, whereby they are endowed with an inherent ability to adapt or persist in a stable way. Because of their inherent structure, which requires analysis at many scales of space and time, complex systems face science with unprecedented challenges of observation, description and control. Complex systems do not have a *blueprint* and are perceived only through very large amounts of data. Therefore a typical task scientists are required to face is to simulate, model and control them, and mostly to develop theories for their behavior, control, management or prediction.

In science, methods generally come before theory; theory is the synthesis of knowledge gained by the application of systematic or heuristic methods. Although wide classes of systems from various disciplines share universal features that lead us to imagine the existence of common structures, their analysis is often based on drastic, generally ad hoc, simplifications, and their description resorts to the specific language proper to the most affine discipline, losing the richness of universality. On the other hand, a full theoretical understanding, for example, of the mechanism linking individual and collective behavior, along with the possibility of exploring the related systems with sufficiently powerful reliable simulations, cannot but lead to profound new insight in various areas. Metaphors should be avoided: metaphors are dangerous, because a metaphor is not a theory nor does it give much indication on specific applications.

To bridge the extraction of mathematical structures out of the phenomenology of complexity science and to give life to an efficient and complete collection of concepts and methods of mathematics appropriate for complexity theory is the challenge, and the most universal, potential setting frame for this is category theory: namely the construction of categorical structures for system modeling. Born with the aim of reorganizing algebra, looking not only at the objects (sets, groups, or rings) but also at the mapping between them (functions between sets, homomorphisms between groups or rings), category theory provides an elegant conceptual tool for expressing relationships across many branches of mathematics. It considers mathematical relations as *arrows* between *objects*. This approach fits in our case not only algebra but topology, where the arrows are continuous maps and objects are spaces, and geometry, with arrows that are smooth maps and objects which are manifolds. Category theory is a powerful, far-reaching formal tool for the investigation of concepts such as space, system, and even truth. It can be applied to the study of logical systems at the syntactic, proof-theoretic, and semantic levels. It is an alternative to set theory, with a foundational role for mathematics and computer science that answers many questions about mathematical ontology and epistemology.

Clearly, the choice to use the language of categories should not be made a priori, but should naturally impose itself due to the need to translate the seemingly purely mathematical objectives related to basic complexity science questions into theoretical computer science issues, and to establish a number of conceptual paradigms and technical instruments.

In complex systems, reconstruction is searching for a model that can be represented as a computer simulation program able to reproduce the observed data *reasonably well*. In this sense, reconstruction is the inverse problem of simulation. The statistics community addresses two closely related questions, namely, *what is a statistical model?* and *what is a parameter?* These questions, that are deeply ingrained in applied statistical work and reasonably well understood at an intuitive level as they are, are absent from most formal theories of modeling and inference. Whilst using category theory, these concepts can be well defined in algebraic terms, proving that a given model is a functor between appropriate categories. The objective that will guide us here is to construct an articulated and extended pathway connecting globally many apparently isolated (sub-)structures – those belonging to the functional (language) and behavioral (dynamics) features of complex systems; i.e., not simply gluing together a collection of local maps. This will be done by resorting to the language of category theory.

The novel approach to the problems of data-based complexity science described in this paper consists in the setting up of a new methodology, which is a sort of algebraic (in the sense of algebraic topology) complex systems theory, that pursues the idea that there exist suitable categories $\mathfrak{A}$ and $\mathfrak{B}$, functors $\mathcal{F} : \mathfrak{A} \to \mathfrak{B}$ and $\mathcal{G} : \mathfrak{B} \to \mathfrak{A}$, and a natural equivalence between them $\mathfrak{h} : \mathcal{F} \sim \mathcal{G}$, such that: $\mathcal{F}$ is a *simulation* and $\mathcal{G}$ is a *reconstruction*. In such schemes, systems of systems may be represented by n-categories, i.e., categories whose objects are arrows, arrows between arrows, and so on. Emergence may happen in any graph representing relationships between agents or multi-agents, in which spaces (or objects of some category) are attached to the vertices, and maps (or morphisms) are attached to the edges. As will be discussed in detail below, one can build out of such a graph an associated simplicial complex, whose *persistent homology* is the way to study its *shape* in a functorial way. Adaptivity arises in this way. Notice that no limitation is imposed in this perspective on the topology of the underlying graph, i.e., loops and self-loops are allowed, implying that systems with feedback can be included.

The categories to be involved in the conceptual scheme – why they emerge and how they can be linked together up to the completion of a global picture – come naturally out of the rationale of going beyond the traditional point of view and paradigms (networks, predicates, multi-agent schemes) by introducing in the framework of complex system theory the study of spaces in place of agents, connecting them by morphisms instead of functions. Then one shall be able, from the study of the homology of the simplicial complexes generated by data clouds, to turn the data environment into a space of random variables connected by conditional probability distributions.

Categorification, the process of finding category-theoretic analogues of set-theoretic concepts by replacing sets with categories, functions with functors, and equations between functions with natural isomorphisms between functors satisfying the required *coherence laws*, can be iterated. This leads to n-categories, algebraic structures having objects, morphisms between objects, and also 2-morphisms between morphisms and so on up to n-morphisms. The morphisms of the old category *preserve* the additional structure.

This can be achieved through the description of the *process algebras* involved in terms of *quivers* and *path algebras*, and their representations. A quiver Q is a directed graph, possibly with self loops and multiple edges between two vertices. A representation of Q in a given category $\mathfrak{C}$ is obtained by attaching an object $o \in \mathfrak{C}$ to each vertex of Q and labeling each arrow of Q by a morphism between the objects sitting on its vertices. Given Q and $\mathfrak{C}$ there exists an algebra, $\mathcal{P}_Q$, such that a representation of Q in $\mathfrak{C}$ is the same representation that would be obtained from $\mathcal{P}_Q$ in $\mathfrak{C}$. Oriented paths in Q can be multiplied by concatenation and form a basis of $\mathcal{P}_Q$. This gives an equivalence of categories and allows us to study the local properties of the quiver globally by means of its path algebra in a new scheme that is a very rich algebraic structure.

Graphical models, i.e., probabilistic models in which a graph describes the conditional independence structure between random variables, are commonly used in probability theory, statistics (particularly Bayesian statistics) and machine learning. The rules of discrete probability express the observed probabilities as polynomials in the parameters, parameterizing the graphical model as an algebraic variety. *Belief propagation*, Judea Pearl's algorithm, and all *message passing* methods of this kind are rooted in an environment of this sort. This work aims to overcome the limitations of these methods by importing the analysis tools from algebra, algebraic topology and quiver theory.

Homology is the mathematical device that converts information about a topological space into an algebraic structure in a functorial way. This implies that topologically equivalent (homotopic) spaces have algebraically equivalent (isomorphic) homology groups, and that topological maps between spaces induce algebraic maps (homomorphisms) on homology groups. Different homology theories have been developed for different spaces and needs; here we are interested in a special kind of homology which is called *persistent homology*. Given a discrete set in a higher dimensional space, persistent homology will allow us to attach to it a homological complex, which in turn will allow us to study the *shape* of the data set.

Long-lived topological features can thus be distinguished from short-lived ones in data sets, resorting to the simplicial complexes one can construct out of complex networks. The persistent homology of the complex identifies a graded module over a polynomial ring.

Most algebraic and combinatorial/configurational properties of the representation methods, such as structural isomorphism classes over graphs, maps and orders of local state evaluation, give rise to moduli over multi-graded vector spaces which are quiver representations. However, nearly all the usual homogeneity, symmetry and approximately infinite sizes that are essential for conventional statistical mechanics and other simplifications such as those necessary for the pursuit of network scaling and scale-free properties, are simply **not** present in meaningful treatments of interaction-based systems. The world of complex systems data is a much stranger, richer and more beautiful world than that. The challenge of understanding the collective emergent properties of these systems, from knowledge of components to global behavior is this: will Wigner's notion of "unreasonable effectiveness of mathematics" hold for complex systems as well?

Another deep philosophical question behind our work is an important one that was recently brought up by Vint Cerf [1]: whether or not there is any real *science* in computer science, namely if all the well posed questions can be approached by a truly scientific methodology: universal and self-contained. Of course, whenever computing implies the use of formal methods, i.e., mathematical techniques of some kind, it is reasonable to say that there is a rigorous element of science in the field. Computability, complexity analysis, theorem proving, correctness and completeness analysis, etc., are all abilities that fall into the category *scientific*. Since computing is a dynamical process rather than a static process, there is a need for stronger scientific tools that allow us to predict behaviors in computational processes. The challenge lies in being able to manage the explosive state space that arises from the interaction of the processes themselves with inputs, outputs, and with each other. In computer science, the need to constrain the unprecedented width of the state space range is often dealt with through the use of abstraction. Modeling is a form of abstraction, adequate to represent systems with fidelity, i.e., well defined in the abstract representation and suitable to be rigorously analyzed. Judea Pearl's *causal reasoning* in conditional probabilities is grounded on graphical models, linking the various conditional statements in chains of cause-effect: this introduces a sort of inherent time variable (reminding us of the *arrow of time* proper to statistical physics – the link being provided by entropy) and hence the ground for true dynamics.

Such a scenario is represented by diagrams analogous to those of Feynman's representation of quantum field interactions, that make it possible to construct analytic equations that not only characterize the problem, but make its solution computable. Both are abstractions of complex processes, which aid our ability to analyze and make predictions about the system's behavior. Abstraction is a powerful tool: it eliminates unimportant details while revealing structure; a way of dealing with the problem that recalls statistical mechanics (smoothing out fluctuations, interaction-induced noise, renormalization) and chaos theory (the dynamical disorder effect of nonlinearity), where patterns emerge despite the apparent randomness of the processes. Our ability to understand and make predictions on data-represented complex processes rests on our cleverness in creating more efficient high-level query languages that allow unnecessary details to be suppressed and *theories* to emerge.

Information technology is facing its *fifth revolution*: the era of Big Data Science is challenged to handle information at unprecedented scales and needs to do so under diverse perspectives which share the common objective of selecting meaningful information from data. This means to be able to identify, within the space of data, the existing, typically hidden, correlation patterns, and formalize a consistent description of the data space structure that thus emerges. Such a structure contains the inherent, explicit representation of the organized information that data encode. Big Data Science needs to treat this massive corpus as a laboratory of the human condition. The challenge that arises is different, not only because it is much harder, but because – as the motto of complexity science asserts – *more is different*.

In this context, a 2008 editorial of *Wired* magazine with the provocative title "The End of Theory" prospected the idea that computers, algorithms and Big Data may generate more insightful, useful, accurate, true results than scientific theories, which traditionally rely on carefully crafted, targeted hypotheses and research strategies. This provocative notion has indeed entered not just the popular imagination, but also the research practices of corporations, governments and also academics. The idea is that data, shadow of information trails, can reveal secrets that we were once unable to extract, but that we now have the prowess to uncover, with no need of resorting to any underlying or pre-existing conceptual model.

Present work grows out of the conviction that, at today's scale, information is no longer a matter of simple low-dimensional taxonomy statistics and order, but rather of dimensionally agnostic pattern individuation. It calls for an entirely different approach; one that requires us to renounce the tether of data as something that can be embraced in its entirety. It instead forces us to view data mathematically, so as to be able to extract from it such rigorous information that will permit establishing its context. We claim that, contrary to the *Wired* magazine prophecy, this can be done and must be done, which establishes a well-defined theoretical context for a complex process that is unprecedentedly hard to handle. In other words, it is not true that we no longer need to speculate and hypothesize, while simply we have to let machines lead us to patterns, trends, and relationships. We need to have a conceptual frame for handling the impending data deluge if we want to understand and control its implications, and construct a fully innovative theoretical conceptual structure that is a consistent stage for all plays.

On the other hand, a characteristic feature of complex systems is the *emergence* of nontrivial superstructures that cannot be reconstructed by a reductionist approach. Our goal is to build a tool for discovering directly from the observation of data those mathematical relations (patterns) that emerge as correlations among events at a global level, or alternatively, as local interactions among systemic components. Not only do higher emergent features of complex systems arise out of such lower level interactions, the patterns they create may also react back, implying the capacity to develop tools to support a learning process as well.

We develop here a topological field theory for data space, a concrete (though conceptual) objective that is itself proof-of-concept of its breakthrough capacity. The problem at stake can be seen as a far-reaching evolution/generalization of *data mining*, which is the analysis step of knowledge generation in data sets, and focuses on the discovery of unknown features that data can conceal. Data mining uses typically artificial intelligence methods (such as *machine learning*), but often with different goals. Machine learning employs *unsupervised learning* to improve the learner accuracy in the design of algorithms, allowing computers to evolve its major focus: to recognize complex patterns in data and make intelligent decisions based on it. The difficulty here is that the set of all possible behaviors, given all possible inputs, is too large to be covered by the set of observed examples (training data). Predictions are based on known properties learned from the training data: the true task of data mining is then the automatic analysis of large quantities of data, aimed at extracting interesting patterns to be used in predictive analytics. We argue that the data tsunami we are facing can be dealt with only by mathematical tools that are able to incorporate data in a topological setting, enabling us to explore the space of data **globally**, so as to be able to control its structure and hidden information.

In spite of their robustness – namely the capacity they are endowed with to adapt and persist in stable forms – and the emphasis of science on the paradigm of complexity, complex systems are hard to represent and harder to predict. One of the reasons for this is that complex systems knowledge is mostly based not on a shared, well-defined phenomenology, but on data. Yet there are clear elements of universality in the dynamical features of such systems. A real theory of complex systems having a direct bearing on complex phenomena and data in the same way as statistical mechanics bears on thermodynamics, is still not available. A deeper question is thus: can it ever be available? Gödel's theorem and Cantor's set theory appear to forbid it, implying as they do that an infinite multiplicity of conceptual models should exist, but the challenge of a *statistical dynamics* with no background ergodic hypothesis, no thermodynamic limit, no identical *particles* (agents), and above all, not based on repeatable experiments but data driven, is certainly there and needs to be faced. The latter reason is what makes us focus our attention first on the Big Data issue.

Data collection, maintenance and access are central to all crucial issues of society, because the increasingly large influx of data bears not only on science but on a correct governance of all societal processes as well. Large integrated data sets can potentially provide a much deeper understanding of nature but they are also critical for addressing key problems of society. We claim that the data tsunami we are facing can be dealt with only with mathematical tools that are able to incorporate data information in a geometric/topological way, based on a space of data thought of as a collection of finite samples taken from (possibly noisy) geometric objects.

Our work rests on three pillars, interlaced in such a way as to reach the specific objective of devising a new method to recognize structural patterns in large data sets, which allows us to perform data mining in a more efficient way and to extract more easily valuable effectual information. Such pillars are: i) topological data analysis (homology driven), and the related geometric/algebraic/combinatorial architecture; ii) topological field theory for data space as generated by the (simplicial complex) data structure, the construction of a measure over data space, and the identification of a gauge group; iii) formal language (semantic) representation of the transformations presiding the field evolution.

1.2 The Reference Landscape

Complex Systems are ubiquitous: they are complex, multi-level, multi-scale systems and are found everywhere in nature and also in the Internet, the brain, the climate, the spread of pandemics, in economy and finance; in other words, in society. Here we intend to address the deep, intriguing question that has been raised in a previous section about complex systems: can we envisage the construction of a bona fide *Complexity Science Theory*? In other words, does it make sense to think of a conceptual construct playing for complex systems the same role that Statistical Mechanics played for Thermodynamics?

As it has already been mentioned, the challenge is indeed enormous. In statistical mechanics a number of assumptions play a crucial constraining role: i) *ergodicity*, ensuring that all accessible states of the system considered are visited with equal probability; ii) the so called *thermodynamic limit*, $N \to \infty$, requiring that the number of degrees of freedom N (proportional to the number of particles, measured essentially by the Avogadro number), could be assumed as essentially infinite; iii) particles are identical (or possibly indistinguishable): particles of the same species are identical and interact with each other pairwise all in the same way, that is, obeying the same interaction law – in the quantum case they are indistinguishable; iv) an analytical structure is definable for the underlying dynamics, namely equations of motion exist at the micro-scale – analyticity breaking and singularities only appear as a signal of the macro-phenomenon of phase transition; v) experiment-based – phenomenology, implying that phenomena are repeatable, as in reductionist science: under the same initial and boundary conditions the same experiment must give the same outcome.

In contrast, typically complex systems, in particular those representing societal phenomena, have the following hallmarks: i) they are NOT ergodic; ii) their number of agents, N, is ordinarily finite, even though it can be large on a social scale; iii) their agents are NOT identical – they are quite distinguishable complex systems themselves, with their strategies and autonomous behaviors; iv) they are NEVER representable by analytic, perhaps in certain cases not even by recursive, functions; v) above all, they are DATA-based, usually NO repeatable experiment is possible under external control.

The world we live in is no doubt complex and dramatically data-based. More than 4 billion people (more than half of the world's population) own a mobile phone (which makes this the first device in human history owned by more than a half of the world's inhabitants); every day over 300 billion e-mails and 25 billion SMSs are exchanged, 500 million pictures are uploaded on Facebook, etc. The information created and exchanged in a year added up in 2013 to 4 zettabytes (1 zettabyte = 10^{21} bytes) and every year it grows 40% (in 4 years it will reach a yottabyte, 10^{24}, a number larger than Avogadro's number!). For this reason we concentrate first on the last item of the above list: *data*, indeed Big Data. The challenge is to extract all the information, as a norm hidden within, from the huge collection of data flowing in and around complex systems.

Big Data have a variety of diverse features. They have always been present in science where they have played a central role (though today even science has difficulties in dealing with the immense quantity of data made available by measures and experiments; see, e.g., the Hubble and Genome projects, the CERN data archive, etc.): typically scientific data is well organized in high quality data-bases. Today Big Data also plays a role in society, where it may for the first time allow for a true societal tomography, making possible predictions not envisioned before (see, e.g., the H1N1 pandemics of 2009), or for unprecedented targets and strategies. Big Data pose a demanding hardware challenge, (high performance computing), and also a strenuous data manipulation challenge, both in computer science (new computing paradigms; interaction-based computing; beyond the Turing machine) and data analytics (new approach to data mining; nonlinear causal inference). Also, the ever-more blurred boundaries between the digital and physical worlds that characterize our digitalized global world are bound to progressively fade away as IT becomes an integral part of the fabric of nature and society.

A parallel goal is to endow IT with innovation and to use more and more efficient tools to play its role in the hard process of turning *data* into *information*, information into *knowledge*, and eventually knowledge into *wisdom*; in other words, to give life to a new paradigm for data manipulation capable of managing the complex dialectic relations between structural and functional properties of systems, in a way analogous to that with which the human brain interacts with information and behaves as a set of embodied computers. An exercise in *artificial intelligence*.

We explore the possibility of taming Big Data with topology (the geometry of *shapes*), building on a fundamental notion from computer science when dealing with data: the concept of *space of data*. It is the latter that provides the structure (represented geometrically) within which information is encoded, such as the frameworks for algorithmic (digital) thinking, and the lode in which to perform data mining, i.e., to extract patterns of correlated information. It is the very notion of data space that engenders the objective: finding new ways – based on its geometrical (topological) and combinatorial features – to extract (*mine*) information from data.

The ideas proposed by Carlsson, Edelsbrunner and others will now be expanded upon. They all argued that geometry and topology are the natural tools to handle large, high-dimensional, complex spaces of data in this process. *Why*? Because *global*, though *qualitative*, information is relevant; data users aim to obtain maximum knowledge, i.e., to understand how data is organized on a large, global scale rather than locally. *Metrics are not theoretically justified*: while in physics, most phenomena naturally lead to elegant, clear-cut theories which imply – as an outcome of the theory itself – the metrics to be used; in the life or social sciences this is either less cogent or it is simply not there. *Coordinates are not natural*: data is typically conveyed and received in the form of strings of symbols, typically numbers in some field, and vector-like objects whose *components* have no meaning as such and whose linear combinations are not objects in data space. In other words, the space of data is *not* a vector space. Thus those properties of data space that depend on a specific choice of coordinates cannot be considered relevant. *Summaries only are valuable*. The conventional method of handling data is based on the construction of a graph (*network*) whose vertex set is the collection of points in data space (each point possibly itself a collection of data) where two vertices are connected by an edge if their *proximity measure* is, say, less than a given threshold η; followed by the attempt to find (determine) the optimal choice of η. The complete diagram that illustrates the arrangement produced by data hierarchical clustering is, however, much more informative. It is able to capture at once the summary of all relevant features with all possible values of η. The difficulty is to get to know how the global features of data space vary upon varying η.

For all these reasons the methods to be adopted should be inspired by topology, because: *topology* is the branch of mathematics that deals with both *local* and *global qualitative* geometric information in a (topological) ambient space, specifically connectivity, classification of loops and higher dimensional manifolds, and invariants, which are properties that are preserved under homeomorphisms of the background space. *Topology* studies geometric properties in a way that is *insensitive to metrics*; it ignores the notion of distance and replaces it just with the concept of proximity ($\eta \approx$ connective *nearness*, in the sense of Grothendieck topology). *Topology* deals with those properties of geometric objects that *do not depend on coordinates* but only on intrinsic geometric features; it is *coordinate-free*.

Besides, and perhaps even more importantly, in topology relationships involve maps between objects; thus they are naturally a manifestation of *functoriality*. Also, the invariants are related not just to objects, but to maps between objects as well. Thus functoriality reflects an inherent *categorical* structure, allowing for computation of global invariants from local information.

Finally, the whole information about topological spaces is typically faithfully contained in their *simplicial complex* representation, that is itself a piece-wise linear (PL), combinatorially complete, discrete realization of functoriality. As already mentioned, the conventional way to convert a collection of points in data space into a global object is to use the vertex set of a network, whose edges are determined by proximity. However, while such a graph is able to capture data connectivity (a local property of the network), it ignores a wealth of higher order global features, which are instead well discerned by a higher-dimensional object, the *simplicial complex*, that can be thought of as the scaffold (the 1-skeleton) of the graph. The latter is a PL space built by gluing together simple pieces (the simplices) identified combinatorially along their faces, which are obtained by the completion of the graph.

1.3 The Challenge

Current thinking in Information Technology is at the crossroads of different evolution pathways. On the one hand, is what is now universally referred to as the

Big Data question [2], which urges fully innovative methodologies to approach data analytics – in particular data mining – to be able to extract information from data with the required efficiency and reliability. On the other hand, is the ever increasing number of real world instances, in science as well as in society, of problems that ask us to go computationally *beyond Turing* [3], which touches on such basic issues as *decidability*, *computability*, and even *embodied computation*.

Following Cerf's view, we assume modeling of computational processes as the most inspiring candidate for the construction of a true *theory* that is able to lead to credible predictions about complex processes through the analysis of the large data sets that represent them. We also keep Pearl's diagrams [4], whose surprising analogy with Feynman's representation of interactions in quantum field theory, with cause-effect relations replacing time flow direction, which was already previously mentioned, in mind as a reference paradigm for the construction of analytic equations that not only fully characterize this type of problem, but make their solution accessible. A crucial issue here is indeed to eliminate unimportant details while revealing the relevant underlying structure: a method well known to statistical mechanics (this is exactly what the renormalization group method does), dealing with fluctuations and noise induced by interactions, and to chaos theory (one of the dramatic dynamical effects of nonlinearity), where patterns emerge despite the apparent randomness of the process.

This paper intends to describe a long-term program designed to generate a novel pathway to face some of the challenges posed above – in particular, the issue of sustaining predictions about the dynamics of complex processes through the analysis of Big Data sets – by paving the way for the creation of new high-level query languages that allow insignificant details to be suppressed and meaningful information to emerge as *mined out* correlations. The main goal of this paper is to find the definition of a theoretical framework, described essentially as a nonlinear topological field theory, as a possible alternative to conventional machine learning or other artificial intelligence data mining techniques, allowing for an efficient analysis of and extraction of information from large sets of data.

The approach proposed differs from all previous ones in its deep roots in the inference of globally rather than locally coded data features. Its focus is on the integration of the pre-eminent constructive elements of topological data analysis (*facts* as *forms*) into a topological field theory for the data space (which becomes in this way the logical space of forms), relying on the structural and syntactical features generated by the formal language whereby the transformation properties of the space of data are faithfully represented. The latter is a sort of *language of forms* recognized by an automaton naturally associated with the field theory.

This perspective has a profound, far-reaching philosophical meaning. As Wittgenstein beautifully phrased it [5]: "The world is the totality of facts, not things.... The facts in logical space are the world.... A logical picture of the facts is a thought." and "To imagine a language means to imagine a form of life.... The meaning of a word is its use in the language game."

The decisive outcome of the approach proposed will be a way to extract directly from the space of observations (the collection of data) those relations that encode – by means of this novel language – the emergent features of the complex systems represented by data; *patterns* that data themselves describe as correlations among events at the global level, the result of interactions among systemic components at local level. The complex system's global properties are hard to represent and even harder to predict, just because – contrary to what happens in traditional reductionist science – complex systems knowledge in general does not bear on repeatable experiments and phenomenology that, incidentally, provide the necessary shared information leading to the statistical characteristics of the system properties, but on *data* or on virtual artificial representations of real systems built out of data.

There are three bodies of knowledge that constitute the three pillars our scheme rests on, which need to operate synergically: i) *homology theory*; ii) *topological field theory* and iii) *formal language theory*. The *singular homology methods* (i) furnishes the necessary tools for the efficient (re-)construction of the (simplicial) topological structures in the space of data which encode patterns. It enables us to make topological data analysis homology driven and coherently consistent with the global topological, algebraic and combinatorial architectural features of the space of data, when equipped with an appropriate *measure*. The *topological field theory* (ii) provides the construct, mimicking physical field theories (as connected to statistical field theories), for extracting the necessary information to characterize the patterns in a way that might generate, in view of the field nonlinearity and self-interaction, the reorganization of the data set itself, as feedback. The construction of the *statistical/topological field theory of data space*, is generated by the simplicial structure underlying data space, by an action and the corresponding fiber (block) bundle. An action depends on the topology of the space of data and on the nature of the data, as they characterized by the properties of the processes whereby they can be manipulated, a *gauge group* that embodies these same two features: data space topology and process algebra structure. The *formal language theory* (FLT) (iii) offers the way to study the syntactical aspects of languages generated by the field theory through its algebraic structure, i.e., the inner configuration of its patterns, and to reason and understand how they behave. It allows us to map the semantics of the transformations implied by the nonlinear field dynamics into automated self-organized learning processes. These three pillars are interlaced in such a way as to allow us to identify structural patterns in large data sets, and efficiently perform data mining. The outcome is a new pattern discovery method, based on extracting information from field correlations that produces an automaton as a recognizer of the data language.

1.4 Step One: Topological Data Analysis

The main pillar of the construction of our theory is the notion of data space, whose crucial feature is that it is neither a metric space nor a vector space – a property that is unfortunately still uncritically assumed even by the most distinguished authors (see, e.g., Hopcroft and Kannan [6]) – but it is a topological space. This is at the root of most aspects of the scheme proposed: whether the higher dimensional, global structures encoding relevant information can be efficiently inferred from lower dimensional, local representations; whether the reduction process performed (filtration; the progressive finer and finer simplicial complex representation of the data space) may be implemented in such a way as to preserve maximal information about the global structure of data space; whether the process can be carried over in a truly metric-free way [7]; whether from such global topological information *knowledge* can be extracted, as well as correlated information, in the form of patterns in the data set.

The basic principles of this approach stem from the seminal work of a number of authors: G. Carlsson [8], H. Edelsbrunner and J. Harer [9], A. J. Zomorodian [10], and others. Its fundamental goal is to overcome the conventional method of converting the collection of points in data space into a *network* – a graph $\mathcal{G}$ encompassing all relevant *local* topological features, whose edges are determined by the given notion of *proximity*, characterized by parameter η that fixes a coordinate-free metric for *distance*. Indeed, while $\mathcal{G}$ captures pretty well *local* connectivity data, it ignores an abundance of higher-order features, most of which have a *global* nature, and misses its rich and complex combinatorial structure. All these can instead be accurately perceived and captured by focusing on a different object than $\mathcal{G}$, say $\mathcal{S}$. $\mathcal{S}$ is a higher-dimensional, discrete object, of which $\mathcal{G}$ is the 1-skeleton, generated by combinatorially completing the graph $\mathcal{G}$ to a *simplicial complex*. $\mathcal{S}$ is constructed from higher and higher-dimensional simple pieces (simplices) identified combinatorially along their faces. It is this recursive and combinatorially exhaustive way of construction that makes the subtlest features of the data set, seen as a topological space $X \sim \mathcal{S}$, manifest and accessible.

In this representation, X has an *hypergraph* structure whose hyperedges generate, for a given η, the set of relations induced by η itself as a measure of proximity. In other words, each hyperedge is a *many-body relational* simplex, namely a simplicial complex built by gluing together lower-dimensional relational simplices that satisfy the η property. This makes η effectively metric independent:

in fact an *n*-relation here is nothing but a subset of *n* related data points, satisfying the property represented by η. Dealing with the simplicial complex representation of *X* by the methods of algebraic topology, specifically the theory of persistent homology that explores it at various proximity levels by varying η, i.e., filtering relations by their robustness with respect to η, allows for the construction of a parameterized ensemble of inequivalent representations of *X*. The filtration process identifies those topological features which persist over a significant parameter range, making them eligible as candidates to be thought about as *signal*, whereas those that are short-lived can be assumed to characterize *noise*. Moreover, it implicitly defines the notion of an η-parametrized semigroup connecting spaces in the ensemble.

Key ingredients of this form of analysis are the homology groups, $H_i(X)$, $i = 0, 1, \ldots$, of *X* and in particular the associated *Betti numbers* b_i, the *i*-th Betti number, $b_i = b_i(X)$, being the rank of $H_i(X)$ – a basic set of topological invariants of *X*. Intuitively, homology groups are functional algebraic tools that are easy to deal with (as they are abelian) to pick up the qualitative features of a topological space represented by a simplicial complex. They are connected with the existence of *i-holes* (holes in *i* dimensions) in *X*. Holes simply mean *i*-dimensional cycles which don't arise as boundaries of $(i+1)$ or higher-dimensional objects. Indeed, the number of *i*-dimensional holes is b_i, the dimension of $H_i(X)$, because $H_i(X)$ is realized as the quotient vector space of the group of *i*-cycles with the group of *i*-boundaries. In the torsion-free case, knowing the b_i's is equivalent to knowing the full space homology and the b_i are suffcient to fully identify *X* as topological space.

Efficient algorithms are known for the computation of homology groups [11]. Indeed, for $\mathcal{S}$, a simplicial complex of vertex-set $\{v_0, \ldots, v_N\}$, a simplicial *k*-chain is a finite formal sum $\sum_{i=1}^{N} c_i \sigma_i$, where each c_i is an integer and σ_i is an oriented *k*-simplex $\in \mathcal{S}$. One can define on $\mathcal{S}$ the group of *k*-chains $\mathcal{C}_k$ as the free abelian group which has a basis in one-to-one correspondence with the set of *k*-simplices in $\mathcal{S}$. The boundary operator

$$\partial_k : \mathcal{C}_k \to \mathcal{C}_{k-1} \tag{1.1}$$

is the homomorphism defined by:

$$\partial_k \sigma = \sum_{i=0}^{k} (-1)^i (v_0, \ldots, \widehat{v_i}, \ldots, v_k), \tag{1.2}$$

where the oriented simplex $(v_0, \ldots, \widehat{v_i}, \ldots, v_k)$ is the *i*-th face of σ obtained by deleting its *i*-th vertex.

In $\mathcal{C}_k$ elements of the subgroup

$$Z_k = \ker(\partial_k) \qquad (1.3)$$

are referred to as *cycles*, whereas those of the subgroup

$$B_k = \text{im}(\partial_{k+1}) \qquad (1.4)$$

are called *boundaries*.

Direct computation shows that $\partial^2 = 0$, simply meaning that the boundary of anything has no boundary. The abelian groups $(\mathcal{C}_k, \partial_k)$ form a *chain complex* in which both B_k and Z_k are contained; B_k is included in Z_k.

The k-th homology group H_k of $\mathcal{S}$ is defined to be the quotient abelian group

$$H_k(\mathcal{S}) = Z_k/B_k. \qquad (1.5)$$

There follows that the homology group $H_k(\mathcal{S})$ is non-zero exactly when there are k-cycles on $\mathcal{S}$ which are not boundaries, meaning that there are k-dimensional holes in the complex.

Holes can be of different dimensions. The rank of the k-th homology group, the number

$$b_k = \text{rank}(H_k(\mathcal{S})), \qquad (1.6)$$

the k-th Betti number of $\mathcal{S}$, gives just a measure of the number of k-dimensional holes in $\mathcal{S}$.

Persistent homology is generated recursively, starting with a specific complex $\mathcal{S}_0$, characterized by a given $\eta = \eta_0$ and constructing from it the succession of chain complexes $\mathcal{S}_\eta$ and chain maps for an increasing sequence of values of η, say $\eta_0 \leq \eta \leq \eta_0 + \Lambda$, for some Λ. The size of the $\mathcal{S}_\eta$ grows monotonically with η, thus the chain maps generated by the filtration process can be naturally identified with a sequence of successive inclusions.

In algebraic topology most invariants are difficult to compute efficiently, but homology is not: it is actually somewhat exceptional not only because – as we have seen – its invariants arise as quotients of finite-dimensional spaces but also because some of its properties can sometimes be derived from *physical* models. In standard topology, invariants were historically *constructed* out of geometric properties and manifestly able to distinguish between objects of different shape but homeomorphically invariant globally. Other invariants were instead obtained in physics, and these were in fact *discovered*, based, e.g., on topological quantum field theory technology [12]. These invariants provide information about properties that are purely topological but one cannot detect, not even guess, based only on geometric representation.

It is this perspective that we adopt here, namely the idea of constructing a reliable *physical* scenario for data spaces, where no structure is visible. *Physical* should of course be interpreted metaphorically: we aim to construct a coherent formal framework in the abstract space of data, where no equation is available giving the information it encodes as an outcome, that is capable to describe through its topology the hidden correlation patterns that link data into information. This is metaphorically analogous to what one has, say, in general relativity, when a given distribution of masses returns the full geometry of space-time. Here we expect that a given amount of information hidden in data would return the full topology of data space. Of course we don't have a priori equations to rely on, yet we argue that a topological, nonlinear field theory can be designed over data space whereby global, topology-related pattern structures can indeed be reconstructed, providing a key to the information they encode.

All this bears of course on how patterns must be interpreted, as it deals rather with pattern *discovery* than pattern *recognition*. This requires at least a remark. In logic there are approaches to the notion of pattern that, drawing on abstract algebra and on the theory of relations in formal languages – as opposed to others that deal with patterns via the theory of algorithms and effective constructive procedures – define a pattern as that kind of structural regularity, namely organization of configurations or regularity, that one identifies with the notion of *correlations* in (statistical) physics [13]. These logical paradigms will guide our strategy.

A subtle and delicate issue here is that simplicial complex S (typically but not automatically a finite Constantine Whitehead (CW) complex whose cellular chain complex is endowed with Poincaré duality) is not necessarily a manifold; it is only if the links of all vertices are simplicial spheres, which is, indeed, the very definition of manifold in a piecewise linear context. The difficulty resides in the feature that n-spheres are straightforwardly identifiable only for $n = 1, 2$. The problem is tractable for $n = 3$ and possibly 4 only with exponential resources, and it is *undecidable* for $n \geq 5$ [14]. However, given a singular chain complex S, a *normal* map endows it with the homotopy-theoretic global structure of a closed manifold. Sergei P. Novikov proved that for $\dim S \geq 5$ only the *surgery* obstruction prevents S from being homotopy equivalent to a closed manifold. The meaning of this is the following: if S is homotopy equivalent to a manifold then the complex behaves as the base space of a unique Spivak normal fibration, because a manifold has a unique tangent bundle and a unique stable normal bundle. A finite Poincaré complex does not possess such a unique bundle; nevertheless, it possesses an affine fibration that is unique, which is just the Spivak normal fibration. This implies that if S is homotopy equivalent to a manifold then the spherical fibration associated to the pullback of the normal bundle of that manifold is isomorphic to the Spivak normal fibration; but the latter has fiber a that is homotopically equivalent to a sphere. This finally entails that all finite simplicial complexes have at least the homotopy type of manifolds with boundary.

We further observe that all available algorithms to compute persistent homology groups are based on the notion of filtered simplicial complex, consisting of pairs: i) the simplex generated at each given step in the recursive construction, and ii) the order-number of the step, a time-like discrete parameter that orders (labels) the collection of complexes by the step at which that simplex appeared in the filtration. The emerging picture can be naturally interpreted as the representation of a *process*, which is endowed with inherent characteristic dynamics that remind us of a discrete-time renormalization group flow [15]. One may then expect that, as it happens with dynamical triangulations of simplicial gravity, the combinatorially different ways in which one may realize the sampling of (*inequivalent*) structures in the persistence construction process, varying the complex shape, give rise to a *natural* probability measure. The measure thus generated is constrained by and must be consistent with the data space invariants and transformation properties.

1.5 Step Two: from Data Topology to Data Field

Besides the customary filtrations due to Vietoris-Rips [16], whose k-simplices are the unordered $(k+1)$-tuples of points pairwise within distance η, and to Čech [17], where k-simplices are instead unordered $(k+1)$-tuples of points whose $\frac{1}{2}\eta$-ball neighborhoods intersect, or other complexes such as the witness complex [18], which provide natural settings to implement persistence, another filtration, Morse filtration, needs to be considered, that enters here naturally into play.

In the case of those simplicial complexes that are manifold, Morse filtration is a filtration by excursion sets, in terms of what for differentiable manifolds would be curvature-like data. It is indeed a non-smooth, discretized, intrinsic, metric-free version thereof, which is appropriate for the wild simplicial complex that is data space, that can be thought of as the simplicial, combinatorial analogue of the Hodge construction.

It is worth pointing out that, even though it apparently deals with metric-dependent features, in fact Morse filtration is purely topological, namely it is independent on both the Morse function and the pseudo-metric adopted. Also, Morse theory generates a set of inequalities for alternating sums of Betti numbers in terms of corresponding alternating sums of the numbers of critical points of the Morse function for each given index. The analogy with the Hodge scheme is far reaching: simplicial Morse theory generates notions of intrinsic, discrete gradient vector field and gradient flow, associated to any given Morse function f_M. The latter played a particularly significant role – which has been interpreted in the framework of discrete differential calculus – in applications to classical field theory over arbitrary discrete sets [19], which is well described in a non-commutative geometry setting [20].

A Morse complex, built out of the critical points of (any) Morse function with support on the vertex set of $\mathcal{S}$, has the same homology as the underlying structure. This assumes particular importance because the Morse stratification induced [21] is essentially the same as the Harder-Narasimhan [22] stratification of algebraic geometry: one can construct the PL analogue of local *co-ordinates* at the Morse critical points and provide a viable representation of the normal bundle to the critical sets. It helps recalling that the relation between Morse and homology theory is generated by the property that the number of critical points of index i of a given function f_M is equal to the number of i cells in the simplicial complex obtained *climbing* f_M, that manifestly bears on b_i. Morse homology is isomorphic to the singular homology; Morse and Betti numbers encode the same information, yet Morse numbers allow us to think of the underlying true *manifold*.

Inspired by what happens in the simpler context of gravity, we select the Gromov-Hausdorff (GH) topology [23, 24] to construct a self-consistent measure over $\mathcal{S}$. Gromov's spaces of bounded geometries in fact provide the natural framework to address the measure-theoretical questions posed by simplicial geometry in higher-dimensions. Specifically, it allows us to establish tight entropy estimates that characterize the distribution of combinatorially inequivalent simplicial configurations. In gravity theory the latter problem was solved [25]; however we should keep in mind that one deals with an underlying metric vector space that gives rise, under triangulation, to a simplicial complex, which is a Lorentz manifold.

The GH topology leads naturally to the construction of a statistical field theory of data, as its statistical features are fully determined by the *homotopy* types of the space of data [26]. Complexity and randomness of spaces of bounded geometry can be quite large in the case of Big Data, since the number of *coverings* of a simplicial complex of bounded geometry grows exponentially with the volume. A sort of *thermodynamic limit* then needs to be realized over the more and more random growing filtrations of simplicial complexes. To explain this within the present context, a well defined statistical field theory is required to deal with the extension of the statistical notion of Gibbs field to the case where the substrate is not simply a graph but a simplicial complex, which amounts to proving the property that the substrate underlying the Gibbs field may itself be in some way random. This can be done by resorting to Gibbs *families* [27], so that the ensuing ensemble of geometric systems – a sort of phase space endowed with a natural measure – behaves as a statistical mechanics object. There ensues the possibility of finding a critical behavior as diversified phase structures emerge – entailing a sort of phase transition when the system passes from one homotopy type to another. The final message is: the deep connection between the simplicial complex structure of data space and the information that such space hides, which is encoded at its deepest levels, resides in the property that data can be partitioned in a variety of equivalence classes and classified by their homotopy type, all elements of each of which encode similar information. In our metaphor, in X information behaves as a sort of *order parameter*.

1.6 The Topological Field Theory of Data

A single mathematical object encompasses most of the information about the global topological structure of the data space: the Hilbert-Poincaré series $\mathcal{P}(z)$ (in fact a polynomial in some indeterminate z), generating function for the Betti numbers of the related simplicial complex. $\mathcal{P}(z) = \sum_{i \geq 0} b_i z^i$ can be generated through a field theory, as it turns out to be nothing but one of the functors of the theory itself for an appropriate choice of the field action.

The best known analogy to refer to for this formal setup – naturally keeping in mind not only the analogies but mostly the deep structural differences: continuous vs. discrete, tame vs. wild, finite vs. infinite gauge group – is Yang-Mills field theory (YMFT) [28]. In YMFT the variables are a connection field over a manifold M (in this case, a Riemann surface), and the gauge group G is SU(N) (a Lie group of $n \times n$ unitary matrices), under which the Chern-Simons (CS) action (i.e., the $(2k-1)$-form defined in such a way that its exterior derivative equals the trace of the k-th power of the curvature) is invariant.

Paraphrasing Terry Tao [29], one may think of a *gauge* as simply a *global coordinate system* that varies depending on one's location over the reference (ambient) space. A gauge transformation is nothing but a change of coordinates *consistently* performed at *each* such location, and a gauge theory is the model for a system whose dynamics is left unchanged if a gauge transformation is performed on it. A global coordinate system is an isomorphism between some geometric or combinatorial objects in a given class and a standard reference object in that same class. Within a gauge-invariant perspective – as all geometric quantities must be converted to the values they assume in that specific representation – it ensues that every geometric statement has to be invariant under coordinate changes. When this can be done, the theory can be cast into a coordinate-free form. Given the coordinate system and an isomorphism of the standard object, a new coordinate system is simply obtained by composing the global coordinate system and the standard object isomorphism, namely operating with the group of all transformations that leave the gauge invariant. Every coordinate system arises in this manner. The space of coordinate systems can then be fully identified with the isomorphism group G of the standard object. This group is the *gauge group* for the class of objects considered. This very general and simple definition of gauge group allows us to introduce in our scheme a general notion of *coordinates*. These can be straightforwardly identified by the existing intrinsic way to identify mutual relations between objects introduced by the data space topology and by the proximity criterion adopted. It is worth noticing how different such a notion is from the customary notion of coordinates in a vector space.

Let us continue the YMFT analogy. The base-space for YMFT is a smooth manifold, M, over which the connection field is well defined and allows for a consistent definition of the action, since the curvature, which is simply the exterior derivative of the connection plus the wedge product of the connection by itself, is well defined everywhere. Field equations in this case are nothing but a *variational machinery* that takes a symmetry constraint as input, expressed as invariance with respect to G, and gives as output a field satisfying that constraint. In YMFT, connections allow us to do calculus with the appropriate type of field attaching to each point p of M a vector space – a *fiber* over that point: the field at p is simply an element of such a fiber. The resulting collection of objects (manifold M plus a fiber at every point $p \in M$) is a *vector bundle*. In the presence of a gauge symmetry, every fiber must be a representation (not necessarily different) of the gauge group, G. The field structure is that of a G-bundle. Atiyah and Bott [30], via an infinite-dimensional Morse theory with the CS action functional as Morse function, in addition to Harder and Narasimhan [22], via a purely combinatorial approach, have both established a formula that expresses the Hilbert-Poincaré series as a functor of the YMFT, in terms of the partition functions corresponding to all Levi subgroups of G, a form that is reminiscent of the relation between grand-canonical and canonical partition functions in statistical mechanics.

For the space of data the picture is obviously more complex, because of the more complex underlying structure. Vector bundles of the differential category have a PL category analogue, referred to as *block bundles* [31]. These allow us to reduce geometric and transformation problems characteristic of manifolds to homotopy theory for the groups and the complexes involved. This leads in a natural way to the reconstruction of the G-bundle moduli space in a discretized setting. For simplicial complexes that, as already noticed, may not be manifolds, Novikov's lesson is that this can be done just in homotopy terms [14]. Since the homotopy class of a map fully determines its homology class, the simplicial block-bundle construction furnishes all necessary tools to compute the Poincaré series. Also, in spite of its topological complexity, data space offers a natural, simple choice for the action. Indeed an obvious candidate to start with the exponentiated action is the Heat Kernel $\mathcal{K}$, because the Heat Kernel's trace is precisely proportional to the Poincaré series [32]. $\mathcal{K}$ can be obtained by constructing an intrinsic (metric-free) combinatorial Laplacian over the simplicial complex [33]. This is done by the ad hoc construction of the Hodge decomposition over $\mathcal{S}$ and the related Dirac operator.

An oriented simplicial complex is one in which all simplices in the complex, except for the vertices and empty simplex, are oriented. For any finite simplicial complex K and any nonnegative integer d, the collection of d-chains of K, $\mathcal{C}_d$, is a vector space over $\mathbb{R}$ (nevertheless, the chains still form a group; we refer to the set of chains of a given dimension as the chain group of that dimension). A basis for $\mathcal{C}_d$ is given by the elementary chains associated with the d-simplices of K, so $\mathcal{C}_d$ has finite dimension $D_d(K)$. If the elements of $\mathcal{C}_d$ are looked at as coordinates relative to this basis of elementary chains, we have the standard inner product on these coordinate vectors, and this basis of elementary chains is orthonormal. The d-th *boundary* operator is a linear transformation $\partial_d : \mathcal{C}_d \to \mathcal{C}_{d-1}$.

Each boundary operator $\partial_d : \mathcal{C}_d \to \mathcal{C}_{d-1}$ of K relative to the standard bases for $\mathcal{C}_d$ and $\mathcal{C}_{d-1}$ with some given orderings has a matrix representation $\mathbf{B}_d$. The number of rows in $\mathbf{B}_d$ is the number of $(d-1)$-simplices in K, and the number of columns is the number of d-simplices. Associated with the boundary operator ∂_d is its adjoint ∂^*, $\partial^* : \mathcal{C}_{d-1} \to \mathcal{C}_d$.

It is known that the transpose of the matrix for the d-th boundary operator relative to the standard orthonormal basis of elementary chains with the given ordering, $\mathbf{B}_d^t$, is the matrix representation of the d-th adjoint boundary operator, ∂^*, with respect to this same ordered basis. It is worth recalling that the d-th adjoint boundary operator of a finite oriented simplicial complex K is in fact the same as the d-th coboundary operator $\delta_d : \mathcal{C}^{d-1}(K, \mathbb{R}) \to \mathcal{C}^d(K, \mathbb{R})$ under the isomorphism $\mathcal{C}^d(K, \mathbb{R}) = \text{Hom}(\mathcal{C}_d(K, \mathbb{R}) \simeq \mathcal{C}_d(K)$.

For K a finite oriented simplicial complex, and $d \geq 0$ an integer, the d-th *combinatorial Laplacian* is the linear operator $\Delta_d : \mathcal{C}_d \to \mathcal{C}_d$ given by

$$\Delta_d = \partial_{d+1} \circ \partial_{d+1}^* + \partial_d^* \circ \partial_d. \tag{1.7}$$

As for the group G, notice that the space of data has a deep, far-reaching property: it is fully characterized only by its topological properties, neither metric nor geometric, thus – as the objects of the theory have no internal degrees of freedom, they are constrained by the manipulation processes they can be submitted to – there is only one natural symmetry it needs to satisfy, which is the invariance under all those transformations of data space into itself that do not change its topology and are consistent with the constraints.

This requires a more thorough discussion of homomorphisms of topological spaces. Let $\mathfrak{X}$ be a topological space like the space of data, i.e., a space endowed with some notion of *nearness* between its points. The set $\mathcal{H} = \{\mathfrak{h}\}$ of all homeomorphisms $\mathfrak{h} : \mathfrak{X} \mapsto \mathfrak{X}$ representable as continuous, invertible functions can be thought of itself as a space. $\mathcal{H} = \{\mathfrak{h}\}$ is also a group under functional composition. One can define a topology also on $\mathcal{H}$, space of homeomorphisms $\mathfrak{h}(\mathfrak{X})$. The open sets of $\mathcal{H}$ are made up of sets of functions that map compact subsets $\mathcal{K} \subset \mathfrak{X}$ into open subsets $\mathcal{U} \subset \mathfrak{h}(\mathfrak{X})$ as $\mathcal{K}$ ranges throughout $\mathfrak{X}$, and $\mathcal{U}$ ranges through the images of $\mathfrak{X}$ under all allowed homeomorphisms $\mathfrak{h}$ (completed with their finite intersections – which must be open by definition of topology – and arbitrary unions, that once more must be open). This gives a notion of continuity on the space of functions, so that one can consider continuous deformation of the homeomorphisms themselves: the *homotopies*. The *Mapping Class Group* $\mathfrak{G}_{\mathfrak{MC}}$ is defined by taking *homotopy classes of homeomorphisms*, and inducing the group structure from the functional composition group structure – which is already present on the space of homeomorphisms. This general definition allows us to export the notion of mapping class group to the PL case when $\mathfrak{X}$ is a simplicial complex.

The notion of mapping class group is typically used in the context of manifolds. Indeed, for a given manifold $\mathcal{M}$, $\mathfrak{G}_{\mathfrak{MC}}(\mathcal{M})$ can be interpreted as the group of isotopy classes of automorphisms of $\mathcal{M}$. Thus, if $\mathcal{M}$ is a topological manifold, its mapping class group is the group of isotopy-classes of homeomorphisms of $\mathcal{M}$. In the metric case, if $\mathcal{M}$ is smooth $\mathfrak{G}_{\mathfrak{MC}}(\mathcal{M})$ is the group of isotopy-classes of the *diffeomorphisms* of $\mathcal{M}$. Whenever the group of automorphisms of an object $\mathfrak{X}$ has a natural topology, $\mathcal{M}$, $\mathfrak{G}_{\mathfrak{MC}}(\mathfrak{X})$ is defined as $\text{Aut}(\mathfrak{X})/\text{Aut}_0(\mathfrak{X})$ where $\text{Aut}_0(\mathfrak{X})$ is the *path component* of the identity in $\text{Aut}(\mathfrak{X})$ (in the compact-open topology, path components and isotopy classes coincide); so that there is a short-exact sequence of groups:

$$1 \to \text{Aut}_0(\mathfrak{X}) \to \text{Aut}(\mathfrak{X}) \to \mathfrak{G}_{\mathfrak{MC}}(\mathfrak{X}) \to 1. \tag{1.8}$$

All this implies that the gauge group should be chosen as the semidirect product $\mathcal{G} \wedge \mathfrak{G}_{\mathfrak{MC}}$ of the group $\mathcal{G} \sim \mathcal{P}_\mathcal{Q}$ of the *path algebra* associated with the process algebra characteristic of the data set in the representation defined by quiver $\mathcal{Q}$, and the (simplicial analog) $\mathfrak{G}_{\mathfrak{MC}}$ of the *mapping class group* [34] for the space of data.

Recall that *process algebra* refers to the *behavior* of a *system* [35]. A system is indeed anything able to exhibit a behavior, which is the entire collection of events or actions that it can perform, together with the order in which they are executed and other relevant aspects of this execution, such as timing or probabilities, that define the process. The term algebra refers to the fact that the language used to represent the behavior is algebraic and axiomatic. For this reason operations on processes can be defined in terms of quivers, and their effects can be formally represented in terms of the universal algebra associated with the path algebra.

In analogy with the definition of a group, a *process algebra* is any mathematical structure satisfying the axioms given for its operators. A process is then an element of the universe of the process algebra. The axioms allow calculations with processes. Even if process algebras have their roots in universal algebra, it often goes beyond the bounds of universal algebra: for example, the restriction to a single universe of elements can be relaxed and different types of elements can be used, sometimes, also binding operators. The structure is capable of supporting mathematical reasoning about behavioral equivalences, meaning that whatever the specific approach followed for their definition, these are congruences with respect to behavioral operators.

On the other hand, a process can be modeled as an automaton: an abstract machine with a discrete number of *states* (among which the initial state, not necessarily unique, and the final state) and of *transitions*, i.e., all possible ways of going from a state to its *neighbor* states through the execution of elementary actions, the basic units of a behavior. Then, a generic behavior is an execution path of a number of elementary actions that leads from some initial state to its final state, and an automaton is characterized by the complete set of execution paths. Considering these actions as elements of an alphabet, an automaton is the finite representation of a *formal language*. The important issue of deciding when two automata can be considered equal is in this view expressed by a notion of *semantic equivalence*, specifically of *language equivalence*: two automata are equal when they have the same set of execution paths, or – differently stated – they accept/recognize the same language. In this context, an algebra that allows reasoning about automata is the algebra of regular expressions [36].

Since in automata theory the notion of *interaction* is missing, in order to model a system that interacts with other similar systems, *concurrency theory* is typically used: the theory of interacting, parallel, distributed or reactive systems that provides a process algebra with parallel composition among its basic operators. In this case, the abstract, universal model is the *transition systems* in which the notion of equivalence is not necessarily restricted to language equivalence, but rather to *bisimilarity*. Two transition systems are bisimilar if, and only if, they can mimic each other's behavior in any state they may reach.

Finally, we must take into account that any algebra with a finite number of generators and a finite number of relations can be written as a quiver with relations (though not necessarily in a unique way) by thinking of the set of execution paths of the automaton's actions as the basis of a *k-path algebra* with composition law induced by the structure of the combinatorial data of a suitable *k*-Quiver (*kQ*). For a given quiver *kQ*, a relation is simply a *k*-linear combination of paths in *kQ*. Given a finite number of relations, one can form their two sided ideal $\mathcal{R}$ in the path algebra, and thus define the algebra $\mathcal{A} \sim kQ/\mathcal{R}$ as a *quiver with relations*. Process algebras can always be assumed to be representable by a quiver with relations. $\mathcal{G}$ is the group associated with $\mathcal{A}$.

A few technicalities are needed here to better define the notion of *process algebra* adopted here. Any finite-dimensional algebra which is *basic* (i.e., all of its irreducible modules are one-dimensional) is isomorphic to a quotient of the path algebra $\mathfrak{P}_Q$ of its quiver Q modulo an admissible ideal $\mathfrak{I}$.

An analogous, but more subtle, result holds at the basic coalgebra level, through the notion of *path coalgebra* $\mathfrak{C}$ of a *quiver with relations* $(Q, \mathcal{R})$, where $\mathcal{R}$ is the set of relations induced by $\mathfrak{I}$.

Fix a field, say $\mathcal{K}$. A $\mathcal{K}$-coalgebra – that we shall simply denote as $\mathfrak{C}$ – is a triple $(\mathfrak{C}_\mathcal{K}, \Delta, \epsilon)$ consisting of a $\mathcal{K}$-vector space $\mathfrak{C}_\mathcal{K}$ and two $\mathcal{K}$-linear maps: the *coproduct* $\Delta : \mathfrak{C}_\mathcal{K} \to \mathfrak{C}_\mathcal{K} \otimes \mathfrak{C}_\mathcal{K}$ and the *co-unit* $\epsilon : \mathfrak{C}_\mathcal{K} \to \mathcal{K}$, such that the two equalities hold:

$$(\Delta \otimes \mathbb{I}) \Delta = \Delta (\mathbb{I} \otimes \Delta), \quad (\epsilon \otimes \mathbb{I}) \Delta = (\mathbb{I} \otimes \epsilon) \Delta = \mathbb{I}, \tag{1.9}$$

$\mathbb{I}$ denoting the identity map in $\mathfrak{C}$.

A sub-coalgebra $\mathfrak{A}$ of $\mathfrak{C}$, if it exists, is a $\mathcal{K}$-vector subspace $\mathfrak{A}_\mathcal{K}$ of $\mathfrak{C}_\mathcal{K}$ such that $\Delta(\mathfrak{A}) \subseteq \mathfrak{A} \otimes \mathfrak{A}$. Henceforth we shall drop index $\mathcal{K}$ whenever it is not necessary.

In this setting *quiver* Q is a actually a quadruple (Q_0, Q_1, s, e), where: Q_0 is a set of vertices and Q_1 a set of *arrows* (oriented edges) in some given ambient space, and for each $\mathfrak{a} \in Q_1$ the vertices $s(\mathfrak{a})$ and $e(\mathfrak{a})$ in Q_0 are respectively the *source* (start point) and the *sink* (end point) of $\mathfrak{a}$. When $e(\mathfrak{a}) \equiv s(\mathfrak{a})$, arrow $\mathfrak{a}$ is said to be a *loop*.

For κ and ℓ vertices ($\kappa, \ell \in Q_0$) an oriented path $\mathfrak{p}_L$ of length L in Q from κ to ℓ is the formal ordered composition of arrows

$$\mathfrak{p}_L = \mathfrak{a}_L \circ \mathfrak{a}_{L-1} \circ \cdots \circ \mathfrak{a}_2 \circ \mathfrak{a}_1, \tag{1.10}$$

where $s(\mathfrak{a}_1) \equiv \kappa$, $e(\mathfrak{a}_L) \equiv \ell$, and, for $j = 2, \ldots, L$, $e(\mathfrak{a}_{j-1}) \equiv s(\mathfrak{a}_j)$. Also, to any vertex $\kappa \in Q_0$ one formally attaches a trivial path of length 0, $\mathfrak{p}_0$, starting and ending at κ, such that for any arrow $\mathfrak{a} \in Q_1$ such that $s(\mathfrak{a}) = \kappa$, or $\mathfrak{b} \in Q_1$ such that $e(\mathfrak{b}) = \kappa$, one has – respectively – $\mathfrak{a} \circ \mathfrak{p}_0 = \mathfrak{a}$, $\mathfrak{p}_0 \circ \mathfrak{b} = \mathfrak{b}$. The set of trivial paths can be identified with the set of vertices Q_0. A path $\mathfrak{c}$ that starts and ends at the same vertex is a *cycle*. Loops are cycles.

Let $\mathcal{H}_{\mathcal{K}Q}$ be the $\mathcal{K}$-vector space generated by the set of all paths in Q. Endow $\mathcal{H}_{\mathcal{K}Q}$ with the structure of a $\mathcal{K}$ algebra (note, not necessarily unitary) by defining the algebra composition law (we may call it *multiplication*) as that induced by simple concatenation of paths: for $\mathfrak{p}_L = \mathfrak{a}_L \circ \cdots \circ \mathfrak{a}_1$, $\mathfrak{q}_M = \mathfrak{b}_M \circ \cdots \circ \mathfrak{b}_1$,

$$\mathfrak{p}_L \bullet \mathfrak{q}_M \doteq \begin{cases} \mathfrak{a}_L \circ \cdots \circ \mathfrak{a}_1 \circ \mathfrak{b}_M \circ \cdots \circ \mathfrak{b}_1, & \text{if } e(\mathfrak{b}_M) \equiv s(\mathfrak{a}_1), \\ \emptyset, & \text{otherwise.} \end{cases} \quad (1.11)$$

The algebra $\mathfrak{P}_{\mathcal{KQ}}$ thus generated is the *path algebra* of the quiver $\mathcal{Q}$.

$\mathfrak{P}_{\mathcal{KQ}}$ has a natural *grading*:

$$\mathfrak{P}_{\mathcal{Q}} \equiv \mathfrak{P}_{\mathcal{KQ}} = \mathfrak{P}_{\mathcal{Q}_0} \oplus \mathfrak{P}_{\mathcal{Q}_1} \oplus \cdots \oplus \mathfrak{P}_{\mathcal{Q}_m} \oplus \cdots, \quad (1.12)$$

where $\mathcal{Q}_m$ denotes the set of all paths of length m, $\mathcal{Q}_m = \{\mathfrak{p}_m \mid m \in \mathbb{N}\}$, that form a complete set of primitive, orthogonal idempotents of $\mathfrak{P}_{\mathcal{Q}}$.

$\mathfrak{P}_{\mathcal{Q}}$ is unitary if $\mathcal{Q}_0$ is finite; $\mathfrak{P}_{\mathcal{Q}}$ is finite-dimensional if and only if $\mathcal{Q}$ is finite and has no cycles.

An ideal $\mathfrak{I} \subseteq \mathfrak{P}_{\mathcal{Q}}$ is called *ideal of relations* if $\mathfrak{I} \subseteq \mathfrak{P}_{\mathcal{Q}_2} \oplus \mathfrak{P}_{\mathcal{Q}_3} \oplus \cdots \doteq \mathfrak{P}_{\mathcal{Q}_{\geq 2}}$.

For $\mathcal{Q}$ finite, ideal $\mathfrak{I}$ of $\mathfrak{P}_{\mathcal{Q}}$ is *admissible* if and only if there exists an integer $n \geq 2$ such that, denoting by $\mathfrak{P}_{\mathcal{Q}_{\geq n}}$ the ideal $\mathfrak{P}_{\mathcal{Q}_{\geq n}} \doteq \mathfrak{P}_{\mathcal{Q}_n} \oplus \mathfrak{P}_{\mathcal{Q}_{n+1}} \oplus \cdots$, one has $\mathfrak{P}_{\mathcal{Q}_{\geq n}} \subseteq \mathfrak{I} \subseteq \mathfrak{P}_{\mathcal{Q}_{\geq 2}}$.

Finally, a *quiver with relations* $\mathcal{Q}_{\mathcal{R}}$ is a pair $(\mathcal{Q}, \mathcal{R})$, namely a quiver $\mathcal{Q}$ endowed with the ideal generated by the relations $\mathcal{R}$ induced by $\mathfrak{I}$. If $\mathfrak{I}$ is admissible then $\mathcal{Q}_{\mathcal{R}}$ is a *bound quiver*.

If for $\mathfrak{p}_L = \mathfrak{a}_L \circ \mathfrak{a}_{L-1} \circ \cdots \circ \mathfrak{a}_2 \circ \mathfrak{a}_1$ a path of length L in $\mathcal{Q}$ from vertex κ to vertex ℓ one defines:

$$\Delta(\mathfrak{p}_L) \doteq \mathfrak{p}_0^{(\ell)} \otimes \mathfrak{p}_L + \mathfrak{p}_L \otimes \mathfrak{p}_0^{(\kappa)} + \sum_{j=1}^{L-1} \mathfrak{a}_L \circ \cdots \circ \mathfrak{a}_{j+1} \otimes \mathfrak{a}_j \circ \cdots \circ \mathfrak{a}_1$$

$$\doteq \sum_{\substack{\mathfrak{r}, \mathfrak{s} \\ \mathfrak{r} \bullet \mathfrak{s} = \mathfrak{p}_L}} \mathfrak{r} \otimes \mathfrak{s}, \quad (1.13)$$

whereas, for any trivial path $\mathfrak{p}_0$, $\Delta(\mathfrak{p}_0) = \mathfrak{p}_0 \otimes \mathfrak{p}_0$; and

$$\epsilon(\mathfrak{p}) \doteq \begin{cases} 1, & \text{if } \mathfrak{p} \in \mathcal{Q}_0, \\ 0, & \text{if } \mathfrak{p} \text{ is a path of length } \geq 1, \end{cases} \quad (1.14)$$

then $(\mathfrak{P}_{\mathcal{KQ}}, \Delta, \epsilon)$ is the path coalgebra of quiver $\mathcal{Q}$ (or $\mathcal{Q}_{\mathcal{R}}$ if $\mathcal{Q}$ is endowed with relations $\mathcal{R}$). This completes the toolkit necessary for the construction of the factor $\mathcal{G} \sim \mathcal{P}_{\mathcal{Q}}$ of the gauge group.

As for $\mathfrak{G}_{\mathfrak{M}\mathfrak{C}}$, a few extra comments are needed to clarify how one has to proceed to coherently and practically construct its simplicial complex representation. A key aspect here is the set of actions of the mapping class groups on spaces of different sorts, encoding characteristic geometric and topological features. Among these homotopy classes, foliations, conformal structures have all been extensively studied [37]. All these actions are induced by corresponding actions of the homeomorphisms of the base space on the objects selected. Moreover, the spaces on which the mapping class groups act can be equipped with different structures, e.g., groups, simplicial complexes, or manifolds, and the mapping class groups are embedded accordingly into groups of algebraic isomorphisms, simplicial automorphisms, isometries of the related metrics – if any. For most of these actions, the natural homomorphism from the mapping class group to the automorphism group of the given structure is an isomorphism. Among these, particularly interesting in the present context are the actions by simplicial automorphisms on the abstract simplicial complexes associated to X; namely, actions by piecewise linear automorphisms of the associated measured foliations space, equipped, for example, with the train-track piecewise linear structure introduced by Thurston [38] or with the set of self-preserving intersection functions.

The latter structure is related with the braid group, whose central extension – not unexpectedly – is the group of permutations. $\mathfrak{G}_{\mathfrak{M}\mathcal{C}}$ is finite and finitely presented; its presentation, as well as its representations, can be completely constructed once one knows the full homotopy of the simplicial complex. Recently, a complete representation of $\mathfrak{G}_{\mathfrak{M}\mathcal{C}}$ realized in terms of the group $SU(1,1)$ of hyperbolic rotations has been obtained by the authors (and is reported in [39]).

We claim that, in spite of the formal difficulties, mimicking the block bundle approach for the appropriate simplicial complex structure and given G, the data space topological invariants (among which Betti numbers) can be computed in the context of the proposed field theory through the (recursively computable) subsets of symmetries of $\mathcal{G} \wedge \mathfrak{G}_{\mathfrak{M}\mathcal{C}}$. The benefit is twofold. On the one hand the cosets of $\mathcal{G} \wedge \mathfrak{G}_{\mathfrak{M}\mathcal{C}}$ order data in equivalence classes with respect to isotopy, leading to a canonical system in the related process algebras. On the other hand, one can make a unique choice among the several possible theories – the multiplicity being related with the plurality of topological structures due to the passage through Morse numbers (Morse and Betti numbers are related through inequalities, not equalities) – in the following way. One begins by constructing, for all manifolds in the family generated by the collection of Morse numbers, the *free* field theories whose exponentiated action is simply the Heat Kernel, for which the partition function is the generating function of the manifold Betti numbers. By self-consistency, i.e., simply comparing the coefficients of $\mathcal{P}(z)$ with the Betti numbers outcome of the *phenomenological* persistent homology one identifies which is the effective data manifold.

In this way, not only do we fully recover through the construction proposed the whole data space topology (for example, the set of Betti numbers), but we are able to continue to construct an autonomous, self-consistent topological data field theory (TDFT) on the space of data: once more *the fascination of unexpected links in mathematics* [40].

As a final remark, notice that the resulting picture comprises a surprising amount of information on the associated moduli spaces as well; markedly, the quiver representation for the path algebra $\mathcal{A}$ (see also [41, 42]), basic tools for the description of processes involving maps and transformations of data sets.

1.7 The Formal Language Theory Facet

The construction outlined so far naturally brings to light a new facet: formal language theory, which conveys a dimension as much unexpected as elegant in its form.

A preliminary question to raise at this point is whether the adopted topological landscape is inherently coherent with the structure of Formal Language Theory (FLT). As we know, a central issue in the theory of computation is to determine classes of languages whose representation has finite specification [36]. A formal language defined over a finite alphabet $\mathfrak{A}$ of symbols is a subset of the set $\mathfrak{A}^*$ of all strings of any length that can be represented by that alphabet. As a consequence, the number of possible representations is countably infinite and the set of all possible languages over a given alphabet $\mathfrak{A}$ is uncountably infinite. Under these conditions we are obviously unable to represent all languages. Coupled with this issue there is the limit posed by well-known Gold's theorem for which the *minimum automaton identification from given data is NP-Complete* [43]. In the TDFT context, the challenge is to construct a finite representation of the language defined over the alphabet whose symbols are the generators of gauge group G, and whose cosets partition the data space X in equivalence classes of finitely presented objects. Such languages can be finite or infinite; what is interesting here is that their presentation can always be finitely given in $\mathcal{G} \wedge \mathfrak{G}_{\mathfrak{M}\mathfrak{C}}$. In other words, such languages are each a collection of discrete spaces containing a finite number of homeomorphic objects; by the TDFT we construct a language of data, the language proper to topological shape $\mathcal{S}$.

Interpreting the gauge group G as topological shape language requires a resort to a notion of duality somehow similar to that entering the construction of Langland's dual group, yet designed to represent the relationship between structure and function of a behavior: a *mirror* symmetry that allows each to affect the other in the same way. As a consequence, we can characterize the data language as the process algebra whose processes are well-behaved with respect to *modulo bisimulation* [44], by attributing them the same, unique (bi-)algebra induced by the gauge group $\mathcal{G} \wedge \mathfrak{G}_{\mathfrak{M}\mathfrak{C}}$, with $\mathcal{G}$, as mentioned, the group of $\mathcal{A}$.

The role of $\mathfrak{G}_{\mathfrak{M}\mathfrak{C}}$ in the discrete case can be naturally traced back to *Automatic Groups* [45], i.e., finitely generated groups equipped with several finite-state automata that are able to distinguish whether or not a given word – a representation of a group element – is in *canonical form*, and hence if two elements in canonical form differ, and if they do, by which generators. It may be worth recalling that automatic groups were originally introduced in connection with topology, in particular with the study of the fundamental group, and of the homotopy (3-manifolds), because the class of automatic groups can be extended to include the fundamental group of every compact 3-manifold, thus satisfying Thurston's geometrization [38]. In the topological structure we are dealing with here – where we consider collections of *relational* simplexes, built by combinatorially gluing together relational simplices – the task is much more complex. However, as the basic structure is fully controlled by homotopy types, turning the generation of a family of parametrized simplicial complexes into a classification problem in FLT is natural and straightforward in its statement, if not in its solution. One should be aware, however, that issues of uncontrollable algorithmic complexity or even of undecidability may possibly arise.

Moreover, the syntax of a language in FLT is traditionally described by using the notion of grammar, defined by the relations that are necessary to build correct syntax constructs from atomic entities (symbols). This is what allows us to describe the syntax of a formal language universally, in spite of the representation of its texts. In addition, the syntax constructs are typically described as resorting to the notion of syntax diagram, $\mathbb{D}$, that is the connected multigraph with nodes labeled in terms of the formal language's alphabet $\mathfrak{A}$ and connections – in our representation not only edges or links, but also higher-dimensional simplices – that represent the syntax relations. The multigraph of a syntax diagram may be directed or not, and in view of its combinatorial structure, inherited from the simplicial structure of data space and accounted for in the FTL vision, it is itself to all effects a simplicial complex. It is possible to select specific syntax diagrams (referred to as *correct*, as defined below) out of the set of all syntax diagrams on $\mathfrak{A}$ to construct different grammars. The formalism used to do this requires a fundamental notion: that of neighbor grammars [46], whose meaning is the following. Define for each $\mathbb{D}$, the collection of subdiagrams labeled by the set of pairs $(\mathbb{D}', \mathfrak{s})$ where $\mathbb{D}' \subseteq \mathbb{D}$ is another syntax diagram and $\mathfrak{s}$ is the inclusion map of $\mathbb{D}'$ into $\mathbb{D}$. The neighborhood of a symbol of $\mathfrak{A}$ is a syntax diagram that contains the node singled out by this symbol. The neighbor grammar of the given grammar consists of the finite family of neighborhoods defined for each symbol of $\mathfrak{A}$. A given syntax diagram is said to be *correct* if for each of its nodes, labeled by some symbol of $\mathfrak{A}$, it includes some a neighborhood of this symbol. Such a neighborhood should contain all simplices adjoining to its center. There is therefore at least one cover consisting of neighborhoods for each correct syntax diagram in the given neighbor grammar. Such cover is the *syntax*. Furthermore, the category $\mathfrak{D}$ of syntax diagrams over the given alphabet can be introduced, based on the neighboring grammar. It is known [46] that the category of correct syntax diagrams, defined as $\mathfrak{D}$ but limited to correct syntax diagrams, admits a Grothendieck topology [47].

It is the formal language generated by the field theory through its gauge group that makes the TDFT consistent with a formal language architecture. This comes exactly from the property of having a Grothendieck topology at our disposal. Indeed, the Grothendieck topology is a structure on a category $\mathcal{C}$ which makes the objects of $\mathcal{C}$ behave like the open sets of a topological space $\mathcal{X}$. Its characteristic is that it replaces the notion of a collection of open subsets of $\mathcal{U} \subseteq \mathcal{X}$ which is stable under inclusion by the notion of a *sieve*. If c is an object in $\mathcal{C}$, a sieve $\mathfrak{S}$ on c is a *subfunctor* of the functor $\mathrm{Hom}(-,c)$ – i.e., for all objects $c \in \mathcal{C}$, $\mathfrak{S}(c) \subseteq \mathrm{Hom}(c,c)$, and for all arrows $f : c \to c$, $\mathfrak{S}(f)$ is the restriction of $\mathrm{Hom}(f,c)$, *pullback* by f to $\mathrm{Hom}(c,c)(c)$: the Yoneda embedding applied to c. In the case of $\mathcal{O}(\mathcal{X})$ [the category whose objects are the open subsets $\mathcal{U} \subseteq \mathcal{X}$ and whose morphisms are the inclusion maps $\mathcal{V} \to \mathcal{U}$ of open sets $\mathcal{U}$ and $\mathcal{V}$ of $\mathcal{X}$], a sieve $\mathfrak{S}$ on an open set $\mathcal{U}$ just selects the collection of open subsets of $\mathcal{U}$ which is stable under inclusion. If $\mathcal{W} \subset \mathcal{V}$, then there is a morphism $\mathfrak{S}(\mathcal{V}) \to \mathfrak{S}(\mathcal{W})$ given by composition with the inclusion $\mathcal{W} \to \mathcal{V}$. If $\mathfrak{S}(\mathcal{V})$ is non-empty, there follows that $\mathfrak{S}(\mathcal{W})$ is also non-empty. The pullback of $\mathfrak{S}$ along f, that we denote by $f * \mathfrak{S}$, is – for $\mathfrak{S}$ a sieve on $\mathcal{X}$ and $f : \mathcal{Y} \to \mathcal{X}$ a morphism, left composition by f – the sieve on $\mathcal{Y}$ defined as the fibered product $\mathfrak{S} \times_{\mathrm{Hom}(-,\mathcal{X})} \mathrm{Hom}(-,\mathcal{Y})$ together with its natural embedding in $\mathrm{Hom}(-,\mathcal{Y})$. More concretely, for each object $\mathcal{Z}$ of $\mathcal{C}$, $f * \mathfrak{S}(\mathcal{Z}) = \{g : \mathcal{Z} \to \mathcal{Y} | f g \in \mathfrak{S}(\mathcal{Z})\}$, and $f * \mathfrak{S}$ inherits its action on morphisms by being a subfunctor of $\mathrm{Hom}(-,\mathcal{Y})$. Finally, a Grothendieck topology $\mathfrak{G}_\mathcal{C}$ on a category $\mathcal{C}$ is a collection, for each object $c \in \mathcal{C}$, of distinguished sieves on c, say $\mathfrak{G}_\mathcal{C}(c)$, called covering sieves of c. The selection process, whereby such collection is generated, will be subjected to a number of axioms. A sieve $\mathfrak{S}$ on an open set $\mathcal{U} \in \mathcal{O}(\mathcal{X})$ will be a covering sieve if, and only if, the union of all the open sets $\mathcal{V}$ for which $\mathfrak{S}(\mathcal{V})$ is non-empty and equals $\mathcal{U}$; in other words, if and only if $\mathfrak{S}$ gives us a collection of open sets which cover $\mathcal{U}$ in the customary sense.

1.8 Language, Structure and Behavior, Automata

The TDFT construct has crucial consequences in terms of theoretical computer science. In particular, the three basic identifications it implies have a far reaching interpretation: i) the *architectural structure* of the dataset seen as a *G-fiber bundle*, consisting of a base space, the space of data X, dealt with as a topological space, a fiber attached to each point of X, the set of *fibers*, each as a representation of the *gauge group* $G = \mathcal{G} \wedge \mathfrak{G}_{\mathfrak{M}\mathfrak{C}}$; ii) the *field* as an element of the *fiber* at each point of the data space; iii) an *action* – in the simplest non-interacting case is the combinatorial Laplacian – able to describe the processes over data as transformations of the global topological landscape.

This architecture is indeed what allows us to touch the final goal: the definition of a universal methodology whereby, starting from the exploration of (large) data sets, we may construct a *language* capable of describing processes over data as a unified operational system of structure and behavior. This new object can be interpreted as *true* (effective, extended) *data space*, which includes, besides the topological features inherent in the data set, the set of all possible transformations allowed on data, which are generated by the group of all its possible topology-preserving transformations as well as by the related process algebra and reflected in the resulting equivalence classes. In such perspective, the system becomes itself a *self-organizing program*, whose identifiers are the interactions that characterize the field action. Such interactions correlate parts of potential processes (embedded programs) of real life applications: a feature typically caught in the $S[B]$ paradigm [48].

The principle of *self-organization* has long entered as a fundamental feature in the theory of nonlinear, possibly discrete, dynamical systems. It provides the clue to obtain diverse representations of the relation between lower-level elements and higher-order structures in (multi-level) complex systems. Its basic idea is that the interactions among low-level elements, in which each element adjusts to the others, is local because it does not make reference to patterns that are global. It is however this latter feature that leads to the emergence of highly coherent structures and complex behavior over the system as a whole. Such structures, in turn, are able to provide correlations for the lower-level elements with no need of higher-order agents to induce their emergence [50, 51]. In other words, rather than being imposed from above or from outside, the higher-order structures emerge from the interactions internal to the system or between the system and its environment.

From an algebraic perspective, it is the language signature that becomes a measure of the interactions which generate the environment associated with the data set. This is exactly what happens in the $S[B]$ model when one establishes which states connected to B (i.e., which *behavior*) satisfy the constraints imposed by the set of states S (i.e., the states defining the system's *structure*). In TFTD this is equivalent to the process of determining which is the global fiber bundle to which a given element in a fiber attached to a specific point of the topological space belongs.

In this perspective, the structure of $S[B]$ can be naturally identified as a fiber bundle: $S[B] = (B, S, \pi, \mathcal{B} \doteq \{B_j | j \in \mathcal{J}\})$, with total space B, base space $S \sim X$ (the different notation S is to remind us the we are dealing with a simplicial complex), projection map $\pi : B \to S$ and fiber set, $\mathcal{B}$. $\mathcal{J}$ is a label set tagging points $x_j \in S, j \in \mathcal{J}$. In $\mathcal{B}$ each single fiber B_j specifies the global topological constraints conditioning all the correlations of the x_j. It should be recalled here that S is a higher-dimensional *standard* object that provides the frame for the data space X. This defines the internal homeomorphisms within the equivalence classes on the fiber, for any subset of constraints corresponding to a given choice of the global invariants.

Fiber B_j is the topological space of *computations* induced by those constraints in S compatible with the fiber structure, whose subset $S_j = \pi^{-1}(B_j)$ is itself a subspace of S. It is the commutativity of diagram $\mathfrak{D}_j$:

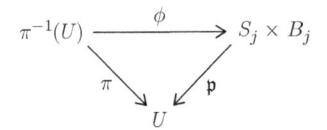

$$\pi^{-1}(U) \xrightarrow{\phi} S_j \times B_j$$
$$\pi \searrow \swarrow \mathfrak{p}$$
$$U$$

that allows us to identify the homeomorphism $\mathfrak{p}$ as the projection map that establishes the one-to-one relationship restricting S_j to the subfiber U of $S[B]$ that we can finally denote as $S_j[B_j]$. The latter has the same topological invariants as $S_j \leftrightarrow x_j$, so that $\mathfrak{p}$ actually *entangles* computation and its context (i.e., the objects living in $S_j \times B_j, \forall j \in \mathcal{J}$).

Our TDFT can in this way be viewed as generated by symmetric monoidal functors from the monoidal pseudo n-fold category to a monoidal n-fold category of spans of sets. The possible resulting degeneracy (more than a single automaton associated with the same language; i.e., strongly connected oriented graphs) reflects the non-uniqueness at the simplicial complex level of the correspondence *Betti numbers to Morse numbers* at the field theoretical level. In the present scheme it is resolved by self-consistency.

Before proceeding, let us recall a few definitions. First, a *Tensor Category* (TC) is a sextuple $(\mathcal{C}; \otimes; a_{...}; \mathbf{1}; \ell_{.}; r_{.})$, where $\mathcal{C}$ is a category; operation $\otimes : \mathcal{C} \times \mathcal{C} \to \mathcal{C}$ is a (bi)functor; $a_{XYZ} : (X \otimes Y) \otimes Z \simeq X \otimes (Y \otimes Z)$ is a (functorial) associativity constraint; $\mathbf{1}$ is the unit object; while $\ell_X : \mathbf{1} \otimes X \simeq X$ and $r_X : X \otimes \mathbf{1} \simeq X$, subject to a number of axioms. Considering only $\mathbb{C}$-linear abelian tensor categories (with bilinear tensor product), the TC must satisfty two sets of basic (defining) axioms:

the pentagon axiom:

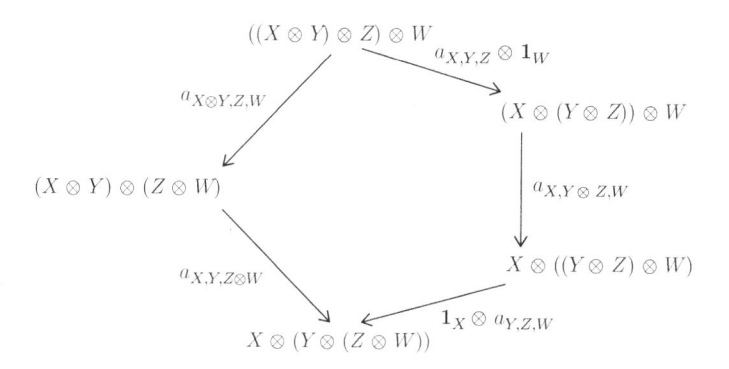

$$\begin{array}{c}
((X \otimes Y) \otimes Z) \otimes W \\
\end{array}$$

with morphisms $a_{X \otimes Y, Z, W}$, $a_{X,Y,Z} \otimes 1_W$, $a_{X, Y \otimes Z, W}$, $a_{X, Y, Z \otimes W}$, $1_X \otimes a_{Y,Z,W}$ going around to $(X \otimes Y) \otimes (Z \otimes W)$, $(X \otimes (Y \otimes Z)) \otimes W$, $X \otimes ((Y \otimes Z) \otimes W)$, $X \otimes (Y \otimes (Z \otimes W))$.

and the triangle axiom:

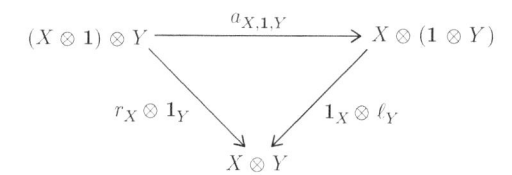

$$(X \otimes 1) \otimes Y \xrightarrow{a_{X,1,Y}} X \otimes (1 \otimes Y)$$

with $r_X \otimes 1_Y$ and $1_X \otimes \ell_Y$ to $X \otimes Y$.

Two categories $\mathcal{C}_1, \mathcal{C}_2$ are said to be *tensor equivalent* if there exists a functor $\mathcal{F}: \mathcal{C}_1 \longrightarrow \mathcal{C}_2$, together with an isomorphism $\mathcal{F}(\mathbf{1}) \simeq \mathbf{1}$ and a functorial isomorphism $\iota_{X,Y}: \mathcal{F}(X \otimes Y) \longrightarrow \mathcal{F}(X) \otimes \mathcal{F}(Y)$, such that

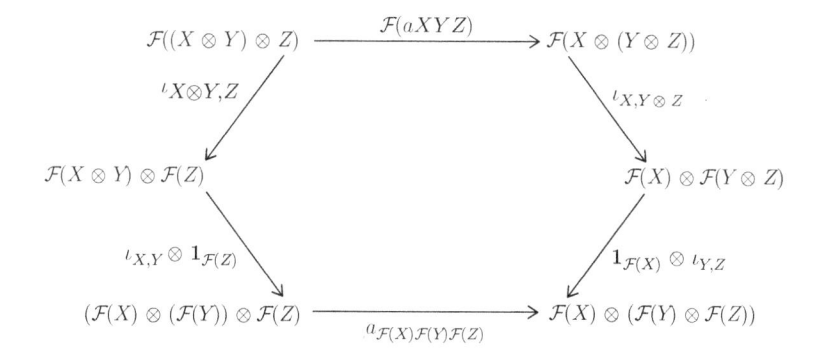

$$\begin{array}{ccc}
\mathcal{F}((X \otimes Y) \otimes Z) & \xrightarrow{\mathcal{F}(a_{XYZ})} & \mathcal{F}(X \otimes (Y \otimes Z)) \\
\downarrow \iota_{X \otimes Y, Z} & & \downarrow \iota_{X, Y \otimes Z} \\
\mathcal{F}(X \otimes Y) \otimes \mathcal{F}(Z) & & \mathcal{F}(X) \otimes \mathcal{F}(Y \otimes Z) \\
\downarrow \iota_{X,Y} \otimes 1_{\mathcal{F}(Z)} & & \downarrow 1_{\mathcal{F}(X)} \otimes \iota_{Y,Z} \\
(\mathcal{F}(X) \otimes \mathcal{F}(Y)) \otimes \mathcal{F}(Z) & \xrightarrow{a_{\mathcal{F}(X)\mathcal{F}(Y)\mathcal{F}(Z)}} & \mathcal{F}(X) \otimes (\mathcal{F}(Y) \otimes \mathcal{F}(Z))
\end{array}$$

If moreover a functional isomorphism $c_{XY} : X \otimes Y \simeq Y \otimes X$ exists, satisfying – both for c_{XY} and $c_{XY}^{(rev)} \equiv c_{YX}^{-1}$ – the hexagon axiom:

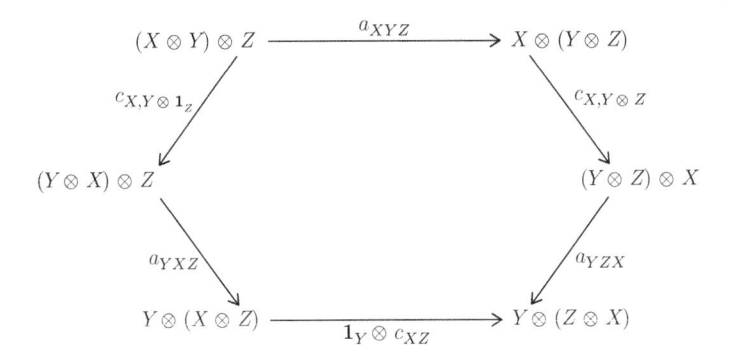

$$\begin{array}{ccc}
(X \otimes Y) \otimes Z & \xrightarrow{a_{XYZ}} & X \otimes (Y \otimes Z) \\
{}_{c_{X,Y} \otimes 1_Z} \swarrow & & \searrow {}^{c_{X,Y \otimes Z}} \\
(Y \otimes X) \otimes Z & & (Y \otimes Z) \otimes X \\
{}_{a_{YXZ}} \searrow & & \swarrow {}^{a_{YZX}} \\
Y \otimes (X \otimes Z) & \xrightarrow{1_Y \otimes c_{XZ}} & Y \otimes (Z \otimes X)
\end{array}$$

then $\mathcal{C}$ is a called a *braided tensor category*, and the pure braid group $\mathcal{PB}_n$ acts on $X_1 \otimes X_n$, whereas the braid group $\mathcal{B}_n$ acts on $X^{\otimes n}$. A braided tensor category is symmetric if $c_{XY} \circ c_{YX} = id$ (i.e., $c^{(rev)} \equiv c$), $\forall X, Y$.

For $\mathcal{C}$ a TC, a *module category* over $\mathcal{C}$ is a quadruple $(\mathcal{K}, \otimes, a_{...}, \ell_{.})$, $\mathcal{K}$ being a $\mathbb{C}$-linear category and the (exact) bifunctor $\otimes$ denoting now the operation $\otimes$: $\mathcal{C} \times \mathcal{K} \longrightarrow \mathcal{K}$, satisfying the pentagon and triangle axioms.

An important example of this construction comes from conformal field theory (CFT). In statistical field theory a *conformal theory* is fully determined by its correlation functions, exactly like it happens in TDFT and in S[B]. In CFT, correlation functions are bilinear combinations of conformal blocks (sets of correlators that implement the identities and constraints that follow from the global gauge symmetries of the theory), and a monodromy for conformal blocks arises that is encoded into a modular tensor category $\mathfrak{T}$. Given conformal blocks with monodromy described by $\mathfrak{T}$, specifying the correlation functions is equivalent to selecting another category, the *module category* $\mathfrak{M}$ over $\mathfrak{T}$. Also, in a CFT conformal blocks are controlled by a *vertex* algebra $\mathcal{V}$ [52]. A deep theorem [53] states that for $\mathfrak{M}$ to be indecomposable over the representations of $\mathcal{V}$ one can combine conformal blocks of $\mathcal{V}$ into a globally consistent system of correlation functions.

In this complex construction a crucial notion emerges: that of the (asynchronously) $\mathfrak{L}$-combable group [54]. The latter is a group to each element of which we can associate a word in some free group within an arbitrary, abstract family of languages $\mathfrak{L}$. The nature of $\mathfrak{L}$ is rather flexible: it can be the family of regular languages, context-free languages, or indexed languages. Words representing group elements in some of these languages [55] describe (flow-like) transformations over the data set. The class of combable regular languages consists of precisely those groups that are asynchronously automatic. Recalling the Atiyah-Bott and Harder-Narashiman results for manifolds, it is relevant to try and classify the (normal) sub-groups of G, and this can be done in the group and language theoretical setting.

In the algebraic theory of languages, a regular language is fully represented by its syntactic monoid (meaning that the properties of that language, e.g., the expressive power of its first-order logic, are fully contained in the structure of the monoid), which is typically finite. In this framework regular languages are referred to as *languages of data words*. A rigorous, but simple construction, of data words consists in identifying first the alphabet, say $\mathfrak{A}$, and focusing then the attention on words and languages over $\mathfrak{A}$ and on the algebraic theory they generate. The field theoretical construction of the Betti number generating function for data sets is an instance of representation of the complex language of data words associated with the simplicial realization of $\mathfrak{G}_{\mathfrak{M}\mathfrak{C}}$, which is known to be combable (though in some cases possibly not automatic) [56].

Automata models can be developed for languages of data words, whose basic feature is that they provide a trade-off among three crucial properties: strong *expressivity*, good *closure* properties and decidable (or efficiently decidable) *emptiness*. It strikes an acceptable balance in the trade-off. Logics have been developed to establish the properties of data words: in particular a language of data words is definable in first-order logic *if* its syntactic monoid is aperiodic; a statement that links the feature of definability in first-order logic to a property that in our framework is dynamical.

In the topological setting, the relevant emerging relationships naturally involve invariants that are related not only to objects but maps between pairs of objects as well. Once again we find here an explicit manifestation of *functoriality*. This is the way in which the theories of automata and formal languages merge with the field theoretical picture, because the field theory generates sequences of symbols that enter into play in the simplicial construction of the G-bundle associated with the gauge group G, as well as the relations among them. In turn, this bears on the enumerative combinatorics content of the theory (because G is reduced essentially to homotopy braids) that provides the language recognized by the automata. Also, combinatorics on words pertains to the wide set of natural operations on languages, in particular to the property – crucial for the final step of pattern discovery in data space – that the orbit of any language in $\mathfrak{L}$ under the *monoid* generated by such a set is finite and bounded, independently of what $\mathfrak{L}$ is.

The use of formal languages leads to the recognition of automatically generated domain-specific languages. The latter are languages appropriate to single out specific topological objects (*concepts*) and their mutual relations, hidden in the noisy landscape of the large data space, and to manage, query and reason over those concepts so as to infer new knowledge. This recalls Codd's theory of database management with its basic tool, *relational algebra* – derived from the algebra of sets and first-order logic when dealing with finite relations closed under specific operations. Codd's approach tackles the problem top-down, first defining the conceptual model, then classifying data through relations, and finally manipulating such relations through their schemas. The approach based on the topology of data space, on the contrary, tackles the problem bottom-up. The two approaches can thus be associated to two different, complementary ways of thinking: the former, based on the assumption that the agent knows a priori, at least in part, the properties of data (characteristic, e.g., of artificial intelligence approaches to data mining, such as *machine learning*); the latter aimed at inferring new knowledge for the agent, extracting from data (*ontological emergence*) those relations that define hidden structural knowledge-generating patterns, but with no a priori information on what data is about.

The dialectical question about the nature of *patterns*, grounded in the antithesis between pattern *recognition* and pattern *discovery*, has guided us naturally – in the field theoretical context – to search for a way to describe patterns at the same time algebraic, computational, intrinsically probabilistic, yet causal. In TDFT, patterns can be collected in ensembles resorting to equivalence classes of histories, or of sets of states. The strength of such patterns (e.g., their predictive, i.e., information retrieval, capability) and their statistical complexity (via state entropy, or the amount of information retained) provide, for each particular process, a measure of the forecasting ability of the theory over the entire data space.

1.9 Emergence of Patterns

We need now to finally merge all the above ingredients into a unique field-theoretical picture, consistent not only with the representation of the space of data equivariant with respect to the transformation properties induced by the simplicial topological scheme itself and by the processes the system may undergo, but also on the full set of characteristic patterns within the data set – via the field correlation functions. The weights depend on the notion of proximity adopted, on the formal language on which the theory is based, on the field action functional selected and on the Morse stratification corresponding to it, as well as on the set of transformations of the data space into itself that preserve its topology. The choice of correlations to represent patterns is crucial: it enables us to make predictions without violating the unavoidable restriction (a mixture of the second law of thermodynamics with the principle of relativity) that predictions can only be based on the process's past, not on any outside source of information except the data in X. In a perspective of this sort, patterns belong to the intrinsic structure of the process, not to the rest of the universe; aggregated pieces of information that share a common structure, and say little about what that pattern is. This is just what correlations are about.

Patterns as represented by field correlations are: *robust*, because they are derived from persistent homology (mediated, if necessary, by the statistical mechanics manipulation process, e.g., smoothing out the role of very high order topological invariants) and hence free, to any desired accuracy, of irrelevant noisy components; *global*, as they describe deep lying correlations dictated by the non-local features of the space topology inherited by the field; *optimal*, based as they are on the variational principle proper to the field theory; *flexible*, due to the vast diversity inherent in their language theoretic structure. This is why they provide essential strategic directions as how to search data space. Whilst several details of the theory remain to be exhaustively worked out, its grand design does not. Of course several of its subtle technicalities need to be completed. A number of applications have started to confirm its potential reach and validity. Among these we mention in particular two: the formulation of a novel *many body* approach to the construction of an effective immune system model [48], and the analysis of the nature of altered consciousness in the *psychoactive drug controlled state* based on functional magnetic resonance imaging data [7, 57].

1.10 Conclusions

To conclude, we have outlined the construction of a topological gauge field theory for data space when these data encode information. Such a theory is capable of acting as a machine whose inputs are a space of data and the symmetry group generated by its simplicial complex approximation as resulting from persistent homology, while its output consists of sets of patterns in the form of field correlations as generated by the field equations. These correlation functions fully encode information about patterns in data space, where the relevant information about the system which the data refer to is encoded. The field theory is self-consistent. It is topological because the data space features it resorts to are topological invariants, and because the gauge group embodies the most general transformations of data space, which leave such global topological features unchanged. Finally, the field evolution – due to the PL nature of the construct – has a natural implementation in terms of finite state automata, which maps both the emergence of patterns and the identification of correlations into well-defined formal language theoretical questions.

1.11 Acknowledgments

The financial support for this paper was provided by the Future and Emerging Technologies (FET) program within the Seventh Framework Programme (FP7) for Research of the European Commission, under the FET-Proactive grant agreement TOPDRIM, number FP7-ICT-318121.

References

[1] V. Cerf, *Where is the science in computer science?* Communications of the ACM. 5 (10), 5 (2012).
[2] Various Authors, *Special Section: Dealing with Data*, Science. 331 (2011).
[3] S. Barry Cooper, *Incomputability after Alan Turing*, Notices, AMS. 59(6), 776–784 (2012).
[4] J. Pearl, *Causality: models, reasoning, and inference.* Cambridge: Cambridge University Press, (2009).
[5] L. Wittgenstein, *Philosophical investigations*, transl. Anscombe, GEM. London: Blackwell Publishing, (1921).
[6] J. Hopcroft and R. Kannan, *Foundations of data science.* (2013). Available at http://blogs.siam.org/the-future-of-computer-science/.
[7] G. Petri, M. Scolamiero, I. Donato and F. Vaccarino, *Topological strata of weighted complex networks*, PLoS ONE. 8(6), (2013). e66506. DOI: 10.1371/journal.pone.0066506.
[8] G. Carlsson, *Topology and data*, Bulletin of the AMS. 46(2), 255–308 (2009).
[9] H. Edelsbrunner and J. Harer, *Computational topology: an introduction.* American Mathematical Society, (2010).
[10] A. J. Zomorodian, *Topology of computing.* Cambridge: Cambridge University Press, (2009).
[11] S. Basu, R. Pollack and M. F. Roy, *Algorithms in real algebraic geometry.* New York: Springer-Verlag, (2006).
[12] E. Witten, *Quantum field theory and the Jones polynomial*, Communications in Mathematical Physics, 121(3), 351–399 (1989).
[13] C. R. Shalizi and J. P. Crutchfield, *Computational mechanics: pattern and prediction, structure and simplicity*, Journal of Statistical Physics, 104(3–4), 816–879 (2001).
[14] S. P. Novikov, *On manifolds with free abelian fundamental group and applications*, Izv. Akad. Nauk SSSR ser. mat. 30(1), 208–246 (1966). English translation: A.M.S. Transl. 67(2) 1–42 (1967).
[15] J. Zinn-Justin, *Quantum field theory and critical phenomena.* Oxford: Clarendon Press, (2002).
[16] L. Vietoris, *Über den höeren Zusammenhang kompakter Räume und eine Klasse von zusammenhangstreuen Abbildungen*, Math. Ann. 97, 454–472 (1927).
[17] E. Čech, *Théorie générale de l'homologie dans un espace quelconque*, Fund. Math. 19, 149–183 (1932).
[18] V. de Silva, *A weak definition of Delaunay triangulation.* (2003). Preprint arXiv cs. CG/031003 v1.
[19] B. Auchmann and S. Kurz, *A geometrically defined discrete Hodge operator on simplicial cells*, IEEE Transactions on Magnetics. 42(4), 643–646 (2006).

[20] D. Battaglia and M. Rasetti, *Quantum-like diffusion over discrete sets*, Physics Letters, A. 313, 8–15 (2003).

[21] M. Harada and G. Wilkin, *Morse theory of the moment map for representations of quivers*, Geom. Dedicata. 150, 307–353 (2011). *Preprint* arXiv math.DG 0807.4734v3.)

[22] G. Harder and M. S. Narasimhan, *On the cohomology groups of moduli spaces of vector bundles on curves*, Math. Ann. 212, 215–248 (1974/5).

[23] M. Gromov, *Structures métriques pour les variétés Riemanniennes*. Paris: Conception Edition, Diffusion Information Communication, Nathan, (1981).

[24] K. Fukaya, *Hausdorff convergence of Riemannian manifolds and its applications*, Advanced Studies Pure Mathematics. 18(1), 143–238 (1990).

[25] J. Ambjørn, M. Carfora and A. Marzuoli, *The geometry of dynamical triangulations*, Lecture Notes in Physics. New York: Springer-Verlag, (1997).

[26] G. Wilkin, *Homotopy groups of moduli spaces of stable quiver representations*, Int. J. Math. (2009). *Preprint* arXiv 0901.4156.

[27] P. Diaconis, K. Khare and L. Saloff-Coste, *Gibbs sampling, exponential families and orthogonal polynomials*, Statistical Science. 23(2), 151–178 (2008).

[28] C. N. Yang and R. Mills, *Conservation of isotopic spin and isotopic gauge invariance*, Phys. Rev. 96(1), 191–195 (1954).

[29] T. Tao, *What is a gauge?* (2008). Available at http://terrytao.wordpress.com/2008/09/27/what-is-a-gauge/.

[30] M. F. Atiyah and R. Bott, *The Yang-Mills equations over Riemann surfaces*, Philos. Trans. Roy. Soc. London Ser. A. 308(1505), 523–615 (1983).

[31] C. P. Rourke and B. J. Sanderson, *Block bundles: I, II, III*. Annals Math. 87(1) 1–28, (2) 256–278, (3), 431–483 (1968).

[32] O. Knill, *The Dirac operator of a graph*. (2013). *Preprint* arXiv math.CO 1306.2166v1.

[33] W. V. D. Hodge, *The theory and applications of harmonic integrals*. Cambridge: Cambridge University Press, (1941).

[34] B. Farb and D. Margalit, *A primer on mapping class group*. Princeton: Princeton University Press, (2011).

[35] J. Baeten, *The history of process algebra*, Theoretical Computer Science. 335(2–3), 131–146 (2005).

[36] H. H. Lewis and C. H. Papadimitriou, *Elements of the theory of computation*. New Jersey: Prentice-Hall, (1998).

[37] J. D. McCarthy and A. Papadopoulos, *Simplicial actions of mapping class groups*, in *Handbook of Teichmüller Theory Vol. III* (A. Papadopoulos ed.). Zürich: European Mathematical Society Publishing House, 297–423 (2012).

[38] W. P. Thurston, *Three-dimensional geometry and topology*. Princeton: Princeton University Press, (1997).

[39] M. Rasetti, *Is quantum simulation of turbulence within reach?* International Journal Quantum Information, (2014). *In press* DOI: 10.1142/ S0219749915600084.

[40] A. Asok, B. Doran and F. Kirwan, *Yang-Mills theory and Tamagawa numbers: the fascination of unexpected links in mathematics*, Bull. London Mathematical Society. 40(4), 533–567 (2008).

[41] W. Crawley-Boevey, *Geometry of the moment map for representations of quivers*, Compositio Math. 126(3), 257–293 (2001).

[42] D. Zagier, *Elementary aspects of the Verlinde formula and of the Harder-Narasimhan-Atiyah-Bott formula*, Israel Mathematical Conference Proceedings. 445–462 (1996).

[43] E. M. Gold, *Complexity of automaton identification from given data*, Information and Control. 37, 302–320 (1978).

[44] J. Baeten, F. Corradini and C. A. Grabmayer, *A characterization of regular expression under bisimulation*, Journal of ACM. 54(2), 1–28 (2007).

[45] L. Mosher, *Mapping class groups are automatic*, Ann. of Math. 142(2), 303–384 (1995).

[46] V. Lapshin, *The topology of syntax relations of a formal language*. (2008). Preprint arXiv math.CT 0802.4181v1.

[47] M. Artin, A. Grothendieck and J. L. Verdier (eds.), *Théorie des topos et cohomologie étale des schémas Séminaire de Géométrie Algébrique du Bois Marie* 1963-64, (SGA 4) Vol. 1 Lecture notes in mathematics (in French). 269. Berlin: Springer-Verlag, xix, 525 (1972).

[48] E. Merelli, M. Pettini and M. Rasetti, *Topology driven modeling – the IS metaphor*, Natural Computing. (2014). *In press* DOI: 10.1007/s11047-014-9436-7.

[49] C. Barrett, H. B. Hunt, M. V. Marathe, S. S. Ravi, D. J. Rosenkrantz, R. E. Stearns and M. Thakur, *Predecessor existence problems for finite discrete dynamical systems*, Theor. Computer Sci. 386, 3–37 (2007).

[50] H. Haken, *Information and Self-Organization: a macroscopic approach to complex systems*, Series in Synergetics. New York: Springer-Verlag, (2010).

[51] J. A. S. Kelso, *Dynamic patterns: the self-organization of brain and behavior*. Cambridge: The MIT Press (1995).

[52] E. Frenkel, *Lectures on the langlands program and conformal field theory*. (2005) Preprint arXiv hep-th math.AG math.QA /0512172v1.

[53] I. Runkel, J. Fjelstad, J. Fuchs and C. Schweigert, *Topological and conformal field theory as Frobenius algebras*, Contemp. Math. 431, 225–248 (2007).

[54] S. Rees, *Hairdressing in groups: a survey of combings and formal languages*, in Geometry & Topology Monographs Vol. 1: The Epstein Birthday Schrift (I. Rivin, C. Rourke and C. Series, eds.). International Press, 493–509 (1998).

[55] D. B. A. Epstein, J. W. Cannon, D. F. Holt, S. V. F. Levy, M. S. Paterson and W. P. Thurston, *Word processing in groups*. Boston: Jones and Bartlett (1992).

[56] M. R. Bridson and R. H. Gilman, *Formal language theory and the geometry of 3-manifolds*, Comment. Math. Helv. 71, 525–555 (1996).

[57] G. Petri, P. Expert, F. Turkheimer, R. Carhart-Harris, D. Nutt, P. J. Hellyer and F. Vaccarino, *Homological scaffolds of brain functional networks*, J. Roy. Soc. Interface. 11 20140873 (2014). DOI: 10.1098/rsif.2014.0873.

2

A Random Walk in Diffusion Phenomena and Statistical Mechanics

ELENA AGLIARI

2.1 Introduction

In this chapter I provide a basic background on stochastic processes for diffusion problems and statistical mechanics techniques for cooperative systems. Although these fields may look quite distant, there exist many interesting, solid bridges among them. Thus, while much of the theory presented is well consolidated and already available in many wonderful reviews, what I wish to convey is just such an intrinsic connection, hoping that it may be useful for further advances and deepening. In writing this chapter I looked for a compromise between simplicity and rigor. This trade-off implies that, despite my efforts to keep the exposition as basic as possible, certain parts are still quite technical, and, while I tried to be as precise as possible, some parts are quite sloppy. Also, as I am spanning over a number of topics, the notation may change from one section to another seeking for functionality (but should be consistent within each single section).

The first part of this chapter is devoted to diffusion and, in particular, to random walks on graphs; the second part is more statistical-mechanics oriented and several models (e.g., the Gaussian model, the Curie-Weiss model) are treated and solved exploiting analogies with diffusion problems. More precisely, in Section 2.2 the phenomenology of diffusion is approached from several perspectives showing convergence to the same set of equations and behaviors. Focusing on the approach based on random walks, I discuss the emergence of anomalous diffusion and I review several analytical techniques for their investigation (e.g., generating functions and Tauberian theorems, algebraic analysis). Finally, I define and discuss the so-called Polya problem in the case of Euclidean lattices. In Section 2.3, I first provide basic definitions for graphs and their topological properties; in particular, I introduce the adjacency matrix and the Laplacian matrix, which algebraically describe the graph structure. Then, I formalize the problem of random walks on arbitrary graphs and show how the underlying topology can have dramatic effects on the properties of the walks (especially when considering in finite graphs, which are introduced to describe macroscopic systems in the thermodynamic limit). The asymptotic behavior of random walks on infinite graphs can be used to define the so called type problem and the spectral dimension. Non-trivial phenomenologies emerging in highly inhomogeneous structures (e.g., the splitting between local and average properties, the two-particles type problem) are also discussed. Finally, the quantal version of random walks is presented and the role of the underlying topology is as well discussed. In Section 2.4 I show that, for a series of fundamental statistical-mechanics models, the Hamiltonian is linear in the adjacency matrix related to the embedding structure. Thus, the main concepts and parameters characterizing random walks (e.g., recurrence and transience, spectral dimension) also affect the properties of these models, which may have very different physical origins. In this context I discuss the oscillating network, the free scalar field and the spherical model. I also mention at a generalization of the Mermin-Wagner theorem and of the Frlich-Simon-Spencer theorem on graphs to determine the magnetizability of a large number of structures with a minimal amount of computation. Finally, in Section 2.5 I deepen the strong analogies between diffusion theory and statistical mechanics from a methodological perspective. Using the Curie-Weiss model as a paradigmatic example, I show how to solve for its free energy by mapping this problem into a random-walk framework, so to use techniques originally meant for the latter. In particular, I will prove that the Cole-Hopf transform of the free energy obeys a Fourier equation, while the order parameter evolves according to a Burgers/Hopf streaming, whose shocks trigger phase transitions in the original framework. Further, this approach links symmetries (e.g., from continuity arguments) in the diffusion theory with constraints in statistical mechanics (e.g., self-averaging properties).

As mentioned above, there exists already many valuable reviews and books on diffusion and statistical mechanics by which I was inspired. In particular, in Section 2.2 many parts follow the perspective and the arguments given by Gilliespie and Seitaridou in [48] and by Weiss in [88] which provide incredibly clear presentations and explanations of the subject matter. In Sections 2.3 and 2.4 I mainly review the works by Cassi and collected in [17]. In the fourth lecture I especially refer to the works by Guerra and Barra, scattered in a number of scientific papers (see e.g., [21, 22, 26, 50, 92]).

2.2 Different Approaches to Diffusion

2.2.1 Macroscopic Perspective (Fick's Law)

A classical way to introduce diffusion is through Fick's law (derived by Adolf Fick in 1855), namely a differential equation describing the spread of solute molecules through random motion, typically from regions of higher concentration to regions of lower concentration. Fick's law is an empirical relationship, which, coupled with a continuity equation, leads to the diffusion equation. We now intend to briefly review the derivation of such a diffusion equation, referring to [48, 23] for more details.

First, we introduce the solute *molecular density* $\rho(\mathbf{r},t)$ and the *molecular flux* $\mathbf{J}(\mathbf{r},t)$. The former is defined as the average number of solute molecules per unit volume at position $\mathbf{r}$ at time t. The latter is defined as the vector whose component $\hat{\mathbf{n}} \cdot \mathbf{J}(\mathbf{r},t)$ in the direction of any unit vector $\hat{\mathbf{n}}$ gives the average net number of particles per unit time crossing a unit area normal to $\hat{\mathbf{n}}$, in the direction of $\hat{\mathbf{n}}$, at position $\mathbf{r}$ at time t (therefore $\mathbf{J}$ has unit mass $\times$ area^{-1} $\times$ time^{-1}). In particular, $\hat{\mathbf{x}} \cdot \mathbf{J}(\mathbf{r},t)$ will be referred to as J_x and similarly for J_y and J_z.

An exact mathematical relation between $\rho(\mathbf{r},t)$ and $\mathbf{J}(\mathbf{r},t)$ can be established since the average net increase in the number of particles inside the infinitesimal volume element $dx\,dy\,dz$ at position $\mathbf{r}$ during the infinitesimal time interval $[t, t+dt)$ must be equal to the average net influx of solute molecules into that volume element through its six sides during that time interval (see Figure 2.1), namely

$$\rho(\mathbf{r}, t+dt) \cdot dxdydz - \rho(\mathbf{r},t) \cdot dxdydz$$
$$= [J_x(\mathbf{r},t) \cdot dydz \cdot dt - J_x(\mathbf{r}+\hat{\mathbf{x}}dx, t) \cdot dydz \cdot dt]$$
$$+ [J_y(\mathbf{r},t) \cdot dxdz \cdot dt - J_y(\mathbf{r}+\hat{\mathbf{y}}dy, t) \cdot dxdz \cdot dt]$$
$$+ [J_z(\mathbf{r},t) \cdot dxdy \cdot dt - J_z(\mathbf{r}+\hat{\mathbf{z}}dz, t) \cdot dxdy \cdot dt]. \quad (2.1)$$

By dividing both sides by $dxdydz \cdot dt$ and then letting all infinitesimals approach zero, we obtain the well-known *continuity equation*

$$\frac{\partial \rho(\mathbf{r},t)}{\partial t} = -\left(\frac{\partial J_x(\mathbf{r},t)}{\partial x} + \frac{\partial J_y(\mathbf{r},t)}{\partial y} + \frac{\partial J_z(\mathbf{r},t)}{\partial z} \right) = -\nabla \cdot \mathbf{J}(\mathbf{r},t), \quad (2.2)$$

where ∇ is the divergence operator, i.e., $\nabla = (\partial/\partial x, \partial/\partial y, \partial/\partial z)$.

The classical diffusion equation is derived from the continuity equation (2.2) by assuming the validity of an empirical relation between ρ and the components (J_x, J_y, J_z) of $\mathbf{J}$, called *Fick's law*. The latter states that

$$J_u(\mathbf{r},t) = -D \frac{\partial \rho(\mathbf{r},t)}{\partial u} \quad (u = x, y, z), \quad (2.3)$$

where D is a positive constant called the *diffusion coefficient*. Now, before proceeding, we will try to justify this assertion.

Referring to Figure 2.1 and focusing for simplicity on the flow along the x direction, let us suppose that at time t, there are $n(x,t)$ particles at the left position

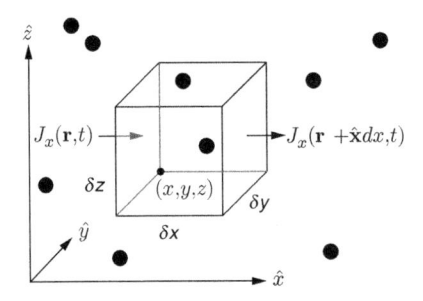

Figure 2.1. Sketch of the flow used for deriving the continuity equation (2.2). The average net increase during the next infinitesimal time interval dt in the number of particles within the infinitesimal volume element $dxdydz$ at space point (x,y,z) is equal to the average net number of particles entering the volume element through its left face in dt, namely $J_x(\mathbf{r},t)dydz \cdot dt$, minus the average net number leaving the volume element through its right face in dt, namely $J_x(\mathbf{r}+\hat{\mathbf{x}}dx,t)dydz \cdot dt$, plus analogous contributions from the other sides of the volume (back and front faces, bottom and top faces), not explicitly shown in figure.

x and $n(x+dx,t)$ particles at the right position $x+dx$. Since equal probabilities $(1/2)$ are assigned for the movement of the molecules (either to the right or to the left), half of the $n(x,t)$ and half of the $n(x+dx,t)$ molecules will cross the plane at the next instant of time $t+dt$, moving in opposing directions. The net number of molecules crossing the plane to the right is $-1/2[n(x+dx,t)-n(x,t)]$ and the corresponding net flux J_x is

$$J_x(x,t) = -\frac{1}{2Adt}[n(x+dx,t)-n(x,t)], \tag{2.4}$$

where $A \equiv dydz$ is the area of the infinitesimal plane that has been crossed and dt is the time interval. Multiplying and dividing the right part by $(dx)^2$ and rearranging, we get

$$J_x(x,t) = -\frac{(dx)^2}{2dt}\frac{1}{dx}\left[\frac{n(x+dx,t)}{Adx} - \frac{n(x,t)}{Adx}\right]. \tag{2.5}$$

The terms in the brackets express the concentration of particles per unit volume, i.e., $\rho(x,t) \equiv n(x,t)/(Adx)$, while the term $(dx)^2/(2dt)$ is the diffusion coefficient D[1]. We thus obtain

$$J_x(x,t) = -D\frac{\rho(x+dx,t)-\rho(x,t)}{dx}. \tag{2.6}$$

Since the fractional term in the limit $dx \to 0$ is the partial derivative of $\rho(x,t)$ with respect to x, one can write $J_x(x,t) = -D\partial\rho(x,t)/\partial x$, hence recovering equation (2.3). In vectorial notation one can write $\mathbf{J}(\mathbf{r},t) = -D\nabla \cdot \rho(\mathbf{r},t)$. Thus,

---

[1] The diffusion coefficient or diffusivity is, by definition, a proportionality constant between the mass flow due to diffusion and the driving force for diffusion (e.g., the gradient of concentration of the species).

Fick's law states that the net flux is proportional to the gradient of the concentration function and the minus sign indicates that the flow occurs from the concentrated to the diluted region of the solution. Therefore, the force acting to diffuse the material through the surface is the concentration gradient $\partial \rho / \partial x$.

We are now ready to derive the diffusion equation. In fact, by inserting equation (2.3) into the right side of equation (2.2), we straightforwardly get

$$\frac{\partial \rho(\mathbf{r},t)}{\partial t} = D\left(\frac{\partial^2 \rho(\mathbf{r},t)}{\partial x^2} + \frac{\partial^2 \rho(\mathbf{r},t)}{\partial y^2} + \frac{\partial^2 \rho(\mathbf{r},t)}{\partial z^2}\right) = D\nabla^2 \rho(\mathbf{r},t), \qquad (2.7)$$

where ∇^2 is the Laplacian operator $\nabla^2 = \sum_u \partial^2/\partial u^2$. The solution to this partial differential equation depends on the imposed initial condition and boundary conditions, which are determined by the specific features of the physical problem considered. A collection of problems and related solutions can be found in [48]. Here we just sketch the simplest case of unrestricted diffusion on a line.

Let us consider a system of N particles, initially set together at $x = 0$ and then allowed to diffuse unrestrictedly along the entire x-axis. For this problem the initial condition is therefore $\rho(x,0) = N\delta(x)$, where δ is the Dirac delta function, while the two boundary conditions are $\rho(\pm\infty,t) = 0$, which are required in order that $\int_{-\infty}^{\infty} \rho(x,t)dx = N < \infty$. The solution of equation (2.7) that satisfies these three conditions is just a Gaussian or "normal" function $\mathcal{N}(x;0,2Dt)$ of x with mean zero and variance $2Dt$, namely

$$\rho(x,t) = \frac{N}{\sqrt{4\pi Dt}} \exp\left(-\frac{x^2}{4Dt}\right) \quad (t \geq 0; -\infty < x < \infty). \qquad (2.8)$$

2.2.2 Physical Perspective (Einstein's Theory)

In 1855, Fick proposed an equation (2.3) that expresses the relation between the molecular flux and the gradient of the molecular density. Fifty years later, in 1905, Albert Einstein explained the physics of Fick's phenomenological diffusion theory in terms of Brownian motion, and three years later Jean Perrin experimentally confirmed that analysis. In a nutshell, Einstein derived the diffusion equation (2.7) from a random molecular model instead of from the continuity equation (2.2) and Fick's law (2.3). In this section Einstein's argument will be briefly presented, following the presentation of [48], while a more extensive treatment can be found in [41].

Let us still focus for simplicity on a one-dimensional system, and let us assume the existence of a time interval δt which can be considered infinitesimally small on a macroscopic scale, yet large enough that a solute molecule typically experiences many collisions with solvent molecules in that time. Then, let us denote with

$\psi(\xi;\delta t)$ a probability density function such that $\psi(\xi;\delta t)d\xi$ gives the probability that the solute molecule will move along the x-coordinate with a displacement length from ξ to $\xi + \delta\xi$ during the time interval δt. This stochastic change in the position of a solute molecule is assumed to stem from the combined effects of the many random collisions that the molecule has with the many smaller solvent molecules[2] in time δt. Notice that, in this derivation, time and space are homogeneous: the change in position of a solute molecule during any δt interval is independent of the current position and time (this assumption is another reason why δt cannot be taken arbitrarily small) and of the starting position.

Provided with the displacement probability distribution function ψ, we write the average number $\rho(x,t+\delta t)dx$ of solute molecules with an x-coordinate between x and $x+dx$ at time $t+\delta t$ in terms of the number at the earlier time t as follows:

$$\rho(x,t+\delta t)dx = \int_{-\infty}^{\infty} [\rho(x-\xi,t)dx] \times [\psi(\xi;\delta t)d\xi]. \tag{2.9}$$

In fact, in the right hand side of equation (2.9) the first factor in brackets is the average number of solute molecules in the dx-interval at $x-\xi$ at time t. The second factor is the average fraction of those that will, as a result of collisions with solvent molecules in the next δt, experience a change in the x-coordinate between ξ and $\xi+d\xi$, and will therefore end up in the dx interval at x at time $t+\delta t$. The product is then integrated over all possible values of ξ to get the average number of molecules with x-coordinate between x and $x+dx$ at time $t+\delta t$. Cancelling the dx's in equation (2.9), and then expanding $\rho(x-\xi)$ under the integral in a Taylor series around x, we obtain

$$\rho(x,t+\delta t) = \int_{-\infty}^{\infty} \psi(\xi;\delta t)\rho(x-\xi,t)d\xi$$

$$= \int_{-\infty}^{\infty} \psi(\xi;\delta t) \left[\rho(x,t) + \sum_{k=1}^{\infty} \frac{(-\xi)^k}{k!} \frac{\partial^k \rho(x,t)}{\partial x^k} \right] d\xi, \tag{2.10}$$

namely, highlighting the two contributions in the integral,

$$\rho(x,t+\delta t) = \rho(x,t) \int_{-\infty}^{\infty} \psi(\xi;\delta t)d\xi + \sum_{k=1}^{\infty} \frac{\partial^k \rho(x,t)}{\partial x^k} \left[\frac{1}{k!} \int_{-\infty}^{\infty} (-\xi)^k \psi(\xi;\delta t)d\xi \right].$$

$$\tag{2.11}$$

The integral in the first term on the right is unity due to the fact that ψ is a normalized probability distribution function in ξ. Moreover, $\psi(\xi;\delta t)$

---

[2] In treating diffusion one usually refers to a system where relatively large solute molecules move in a solvent made of relatively small molecules. On the other hand, the diffusion of small solute particles in a solvent of large particles is a much more complicated problem, still ill-understood (see e.g., [77]).

should be an even function of ξ (because diffusive moves in the positive and negative x-directions should be equally likely), so the integral under the summation vanishes for all odd integers k. Moving the first term on the right and dividing by δt, we obtain the following expression for the time derivative[3] of $\rho(x,t)$:

$$\frac{\partial \rho(x,t)}{\partial t} = \sum_{k=1}^{\infty} \left[\frac{1}{\delta t} \frac{1}{2k!} \int_{-\infty}^{\infty} \xi^{2k} \psi(\xi;\delta t) d\xi \right] \frac{\partial^{2k} \rho(x,t)}{\partial x^{2k}}. \qquad (2.12)$$

At this point Einstein assumed that, as the infinite series converges sufficiently rapidly, it can be approximated by its first ($k = 1$) term only[4], hence getting

$$\frac{\partial \rho(x,t)}{\partial t} = D \frac{\partial^2 \rho(x,t)}{\partial x^2}, \qquad (2.13)$$

where D is given by

$$D = \frac{1}{2\delta t} \int_{-\infty}^{\infty} \xi^2 \psi(\xi;\delta t) d\xi = \frac{1}{2\delta t} \langle \xi^2 \rangle; \qquad (2.14)$$

the last relation is often referred to as the *Einstein-Smoluchowski relation*. Notice that equation (2.13) is just the classical one-dimensional diffusion equation (2.7) and that Einstein recovered it just by effectively accounting for the collisional effects on solute molecules due to the surrounding solvent molecules.

Before concluding, a few remarks are in order (see also [48]).

According to the definition of diffusion coefficient, we expect that it depends on features such as particle size, solvent, temperature and pressure, and that it does not depend on the somewhat arbitrary value of δt. Thus, according to equation (2.14), the second moment of the distribution $\psi(\xi;\delta t)$ should be directly proportional to δt. Moreover, Einstein's probability distribution function ψ is actually a special case of the solute density function ρ. To see this, suppose there are N solute molecules in the system at the origin at time $t = 0$, i.e., $\rho(x,0) = N\delta(x)$. Then $N^{-1}\rho(x,\delta t)dx$ will give the average fraction of those N molecules whose x-coordinates will be in the interval $[x, x+dx)$ at time $t = \delta t$, and, given the fact that solute molecules move roughly independently of each other, one can write $\psi(x;\delta t) = N^{-1}\rho(x,\delta t)$. This implies a certain "circularity", since in assuming various properties for the probability distribution function ψ, Einstein was somehow assuming that those properties were possessed by the solute density function ρ, which was the target of his analysis.

---

[3] The resulting derivative is actually a macroscopic time derivative since δt cannot be taken arbitrarily small.
[4] This point actually constitutes one of the weaknesses of Einstein's derivation: although the factors $1/(2k)!$ in the infinite series in equation (2.12) might suggest that truncation is a reasonable approximation, the assumption that the sum of all terms in that series beyond the first is negligible is not rigorously justified (see also [48]).

2.2.3 Mathematical Perspective (Wiener Process)

In 1908, three years after Einstein's work on diffusion, a different analysis of diffusion was published by Paul Langevin. His approach includes Einstein's theory as a special limiting approximation and can be comprehensively framed within the mathematical machinery of *continuous Markov process theory*. The following description of the Langevin equation is a summary of a much wider theory which is presented in detail in e.g., [48].

Let us introduce a *process*, defined by a function $X(t)$, whose independent real variable represents time. A process is said to be *Markov* if and only if its future values depend only on its present value, namely, it is a "memoryless" process. Here we will focus on *stochastic* (i.e., future is only probabilistically determined by present), *real* (i.e., $X(t) \in \mathbb{R}$) processes, then $X(t)$ will be fully described by a singly-conditioned probability density function ϕ, which is defined so that

$$\phi(x,t|x_0,t_0)dx \equiv \text{probability that } X(t) \text{ will lie in the infinitesimal interval}$$
$$[x, x+dx), \text{ given that } X(t_0) = x_0 \text{ for } t_0 \leq t. \quad (2.15)$$

Of course, ϕ needs to satisfy a set of conditions for being a well-defined probability distribution function, namely

$$\phi(x_2,t_2|x_1,t_1) \geq 0, \quad (2.16)$$

$$\int_{-\infty}^{\infty} \phi(x_2,t_2|x_1,t_1)dx_2 = 1, \quad (2.17)$$

$$\phi(x_2,t_1|x_1,t_1) = \delta(x_2 - x_1), \quad (2.18)$$

as well as the condition

$$\phi(x_3,t_3|x_1,t_1) = \int_{-\infty}^{\infty} \phi(x_3,t_3|x_2,t_2)\phi(x_2,t_2|x_1,t_1)dx_2, \text{ for all } t_1 \leq t_2 \leq t_3, \quad (2.19)$$

also called the *Chapman-Kolmogorov equation*. The latter stems from the definition (2.15) as addition and multiplication laws of probability are applied (see [48] for more details).

By definition, a continuous Markov process $X(t)$ satisfies the *continuity condition*

$$X(t+dt) \to X(t) \text{ as } dt \to 0. \quad (2.20)$$

Posing $X(t) = x$ we define the process increment as

$$\Xi(dt;x,t) \equiv X(t+dt) - X(t), \quad (2.21)$$

in such a way that the condition (2.20) can be equivalently written as

$$\Xi(dt;x,t) \to 0 \text{ as } dt \to 0. \quad (2.22)$$

Of course, the increment Ξ is itself a random variable and the probability that $\Xi(dt;x,t)$ lies in $[\xi,\xi+d\xi]$ is just $\phi(x+\xi,t+dt|x,t)d\xi$. Now, if $\Xi(dt;x,t)$ simultaneously i) satisfies the continuity condition (2.22); ii) is a smooth function of its arguments dt, x and t; iii) has well defined first and second moments; and iv) satisfies the self-consistency condition[5], then $\Xi(dt;x,t)$ must have the mathematical form

$$\Xi(dt;x,t) = \mathcal{N}(A(x,t)dt, D(x,t)dt), \quad (2.23)$$

where A and $D \geq 0$ are smooth functions of x and t, while $\mathcal{N}(\mu,\sigma)$ is a Gaussian variable with mean μ and variance σ. We refer to [48] for a proof of this theorem. An immediate consequence of the theorem (combined with the definition (2.21) and normal random variable properties) is the following Langevin equation:

$$X(t+dt) - X(t) = \mathcal{N}(A(X(t),t)dt, D(X(t),t)dt)$$
$$= A(X(t),t)dt + \sqrt{D(X(t),t)dt}\mathcal{N}(0,1). \quad (2.24)$$

Let us now define $B_n(x,t)$ as the n-th moment of the increment $\Xi(dt;x,t)$, namely

$$B_n(x,t) \equiv \lim_{dt \to 0^+} \frac{1}{dt} \langle \Xi^n(dt;x,t) \rangle, \quad (2.25)$$

and, recalling (2.23), one has

$$B_1(x,t) = A(x,t), \; B_2(x,t) = D(x,t), \; B_{n \geq 3}(x,t) = 0. \quad (2.26)$$

Now, the Langevin process described by (2.24) can be restated in terms of a differential equation for its probability density $\phi(x,t|x_0,t_0)$, by replacing in equation (2.19) (x_1,t_1) with (x_0,t_0), (x_2,t_2) with $(x-\xi,t)$, and (x_3,t_3) with $(x,t+dt)$. Then, by Taylor-expanding the term under integration we get

$$\phi(x,t+dt|x_0,t_0) = \sum_{n=0}^{\infty} \frac{(-1)^n}{n!} \frac{\partial^n}{\partial x^n} \left[\int_{-\infty}^{+\infty} \xi^n \phi(x+\xi,t+dt|x,t)\phi(x,t|x_0,t_0) \right]. \quad (2.27)$$

With some algebraic manipulations the previous equation reduces to

$$\frac{\partial}{\partial t}\phi(x,t|x_0,t_0) = \sum_{n=1}^{\infty} \frac{(-1)^n}{n!} \frac{\partial^n}{\partial x^n} [B_n(x,t)\phi(x,t|x_0,t_0)] \quad (2.28)$$

$$= -\frac{\partial}{\partial x}[A(x,t)\phi(x,t|x_0,t_0)] + \frac{1}{2}\frac{\partial^2}{\partial x^2}[D(x,t)\phi(x,t|x_0,t_0)], \quad (2.29)$$

---

[5] This condition states that if the infinitesimal interval $[t,t+dt)$ is divided into two subintervals $[t,t+\alpha dt)$ and $[t+\alpha dt, t+dt)$, where $\alpha \in (0,1)$, then the increments incurred over $[t,t+dt)$ should be computable as the sum of the successive increments incurred over those subintervals, at least to lowest order in dt.

where, in the second line we exploited (2.26). This is the so called *forward Fokker-Planck equation* and it is exact for continuous Markov processes. This equation recovers Einstein's equation as long as $A(x,t) = 0$ and $D(x,t) = D$ is constant (but Einstein's equation was not exact due to a somehow arbitrary truncation).

The simplest of all stochastic continuous Markov processes is the *driftless Wiener process*. It is defined by

$$A(x,t) \equiv 0, \ D(x,t) \equiv c, \tag{2.30}$$

where c is a positive constant. The Langevin equation (2.24) for this process thus reads as

$$X(t+dt) = X(t) + \mathcal{N}(0,1)\sqrt{c\,dt}, \tag{2.31}$$

and the related Fokker-Planck equation (2.28) reads as

$$\frac{\partial}{\partial t}\phi(x,t|x_0,t_0) = \frac{c}{2}\frac{\partial^2 \phi(x,t|x_0,t_0)}{\partial x^2}, \tag{2.32}$$

which corresponds to Einstein's diffusion equation (2.13), with $c = 2D$, D being the solute coefficient of diffusion. By fixing as initial condition $\phi(x,t_0|x_0,t_0) = \delta(x-x_0)$, we get that the driftless Wiener process is nothing but the normal random variable with mean x_0 and variance $c(t-t_0)$.

2.2.4 Microscopic Perspective (Random Walks)

In the previous section we saw that the behavior of solute molecules or, more generally, agents, particles, etc., moving randomly, follows the so called diffusion equation (2.7) as long as the following set of simple assumptions concerning their microscopic motion is fulfilled:

- Each solute molecule steps randomly once every time interval δt, properly fixed. In other words, the displacement probability density function is homogeneous in time.
- The particles, by interacting with the molecules of solvent, forget what they did on the previous leg of their journey in such a way that successive steps are statistically independent. In particular, in the case of unbiased motion, the displacement probability density function is central and symmetric in such a way that, for a given direction, the probability of going to the right at each step is $1/2$, equal to the probability of going to the left.
- The solute molecules do not interact with one another and each particle moves independently of all other particles (in practice, this will be true provided that the suspension of particles is reasonably diluted).

Focusing on the unbiased case, these rules have two striking consequences: particles go nowhere on the average and their root-mean-square displacement is proportional not to the time, but rather to the square-root of the time.

It is possible to establish these propositions by using an iterative procedure, based on a microscopic perspective, where the position of the solute molecule is described by a discrete stochastic variable.

First, the volume where solute and solvent are confined has to be properly discretized into "cells"; in the simplest case of homogeneous, translation-invariant space this can be modeled in terms of a hypercubic d-dimensional lattice of fixed lattice spacing ℓ. The spatial coordinate of a cell is given by a d-dimensional vector $\mathbf{r} \in \mathbb{R}^d$, corresponding to the center of the cell itself. By taking ℓ sufficiently small we can make solute molecules be approximately uniformly distributed inside each cell (however, as we will see shortly, ℓ cannot be arbitrarily small).

Any solute molecule in any cell might, as a result of collisions with the solvent molecules, suddenly move into an adjacent cell and such transfers happen independently of each other according to the following rule: there exists a constant $p > 0$ such that $p\delta t$ is (neglecting orders > 1 in δt) the probability that a randomly chosen molecule in a given cell will move to a particular adjacent cell in the next macroscopically infinitesimal time δt. Notice that this is the coarse-grained analogy to Einstein's assumption in equation (2.9).

Before proceeding further, some remarks are in order.

First, the strict limit $\delta t \to 0$ is not allowed, because δt is a macroscopic infinitesimal, namely δt is infinitesimally small on a macroscopic scale, yet it must be large enough that a solute molecule will experience many collisions with solvent molecules in that time. Moreover, the hypothesis assumes that solute molecules inside a given cell are uniformly distributed over the cell volume in such a way that the probability of diffusing across the left boundary in the next δt is the same

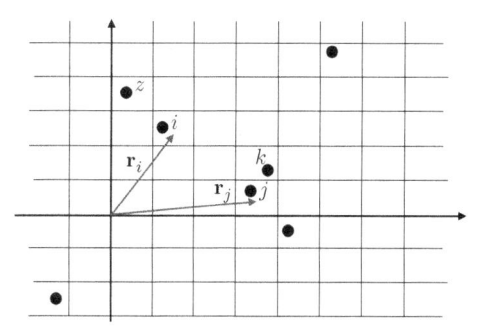

Figure 2.2. Example of space discretisation in $\mathbb{R}^2$: particles, labeled by lowercase letters, are associated to different cells and their position is therefore made to correspond to the center of the cell itself.

as for the right boundary. This also implies that ℓ should be much larger than the diameter of a solute molecule; in fact, this hypothesis implicitly requires that the probability that a solute molecule will jump to a particular adjacent cell is independent of the number of solute molecules that are already in that cell and this requires, in turn, that molecules already in that cell occlude a negligibly small fraction of the cell volume.

Finally, particles are not allowed to jump to non-adjacent cells in time δt, even though that is physically possible. Since the likelihood of jumps to non-adjacent cells increases with δt and decreases with ℓ, it follows that we are implicitly assuming that δt is not too large and ℓ is not too small.

Now, focusing on a single molecule in some cell belonging to the bulk (to avoid any boundary effects) the hypothesis above implies that the probability that in the next δt the molecule will jump to the cell to its right (respectively left), thereby augmenting its x-coordinate by $+\ell$ (respectively $-\ell$), is $p\delta t + o(\delta t)$ and the probability that it does not jump at all, therefore preserving its x-coordinate, is $(1 - 2p\delta t) + o(\delta t)$.

As a consequence, the average x-displacement of the molecule in the next δt is, to first order in δt,

$$\langle \delta x \rangle = (p\delta t)(+\ell) + (p\delta t)(-\ell) + (1 - 2p\delta t)(0) = 0, \qquad (2.33)$$

as expected. The average squared x-displacement reads as

$$\langle \delta x^2 \rangle = (p\delta t)(+\ell)^2 + (p\delta t)(-\ell)^2 + (1 - 2p\delta t)(0)^2 = 2(p\ell^2)\delta t, \qquad (2.34)$$

again to first order in δt. Now, Einstein's definition of the diffusion coefficient D in (2.14) implies that the mean-square x-displacement of a diffusing molecule in a sufficiently long time δt is $2D\delta t$. Agreement between that result and equation (2.34) can evidently be obtained if and only if we have $p\ell^2 = D$, or $p = D/\ell^2$ (see [48] for a more robust derivation).

From a more abstract viewpoint, the phenomenology just described is captured by a random walk model, which is probably the simplest and best known stochastic process. Actually, there exist several versions of random walk, according to whether one assumes either discrete or continuous space and time. Of course, under proper conditions, the related behaviors are (at least asymptotically) the same.

Let us now start from discrete space and discrete time. At each time unit the walker steps from its present position to one of the other sites of the underlying lattice according to a prescribed random rule. This rule is independent of the history of the walk, and so the process is Markovian. In the simplest version of a random walk, the walk is performed in a hypercubic d-dimensional lattice of unitary spacing. At each time step the walker hops to one of its nearest-neighbor sites, with equal probabilities. After n steps the net displacement is

$$\mathbf{r}(n) = \sum_{i=1}^{n} \mathbf{e}_i, \qquad (2.35)$$

where $\mathbf{e}_i$ is a unit vector giving the direction taken at the i-th step and it is drawn from an isotropic, homogeneous distribution. Because $\langle \mathbf{e}_i \rangle = 0$, the average displacement (averaged over many realizations of the walk) is

$$\langle \mathbf{r}(n) \rangle = 0. \qquad (2.36)$$

On the other hand, since $\langle \mathbf{e}_i \cdot \mathbf{e}_i \rangle = 1$, and $\langle \mathbf{e}_i \cdot \mathbf{e}_j \rangle = 0$ for $i \neq j$ (the steps are independent), the mean-squared displacement is

$$\langle r^2(n) \rangle = \left\langle \left(\sum_{i=1}^{n} \mathbf{e}_i \right)^2 \right\rangle = n + 2\sum_{i>j}^{n} \langle \mathbf{e}_i \cdot \mathbf{e}_j \rangle = n. \qquad (2.37)$$

More generally, let the lattice spacing be ℓ and let the step time unit be τ, in such a way that $t = n\tau$, then

$$\langle r^2(t) \rangle = \frac{\ell^2 t}{\tau} = (2d)Dt, \qquad (2.38)$$

where $D = \ell^2/[(2d)\tau]$ is the diffusion constant, in agreement with the previous expression $D = p\ell^2$ found for $\tau = 1$ and given that $p = 1/(2d)$ for isotropic diffusion. The expression in equation (2.38) is the hallmark of regular diffusion and it is worth noticing that the speed of the walker vanishes at long times, that is, $v \sim \sqrt{r^2}/t \sim t^{-1/2}$. Further, the time dependence highlighted is universal, regardless of the dimension of the substrate d. However, if the walk took place in fractals or disordered media, anomalous diffusion would emerge, leading to a qualitatively different diffusion law which depends functionally on the underlying topology (e.g., see [60]); otherwise stated, fractals or disordered media cannot be regarded as an extrapolation of regular Euclidean spaces (namely replacing d with the related fractal dimension).

The mean-square displacement $\langle r^2(n) \rangle$ can also be calculated from the probability density $P_n(\mathbf{r})$ that the walker has displaced to $\mathbf{r}$ after n time steps[6]: $\langle r^2(n) \rangle = \int r^2 P_n(\mathbf{r}) d^d \mathbf{r}$. In one dimension $P_n(x)$ can be calculated easily: assume that jumps to the right occur with probability p_r and jumps to the left with

---

[6] In the remainder of this section we will sometimes switch from discrete time to continuous time and, seeking for clarity, we highlight the dependence on *discrete* time by a subscript (e.g., $P_n(x)$), while the dependence on a *continuous* time is still highlighted in the parenthesis (e.g., $P(x,t)$).

probability $p_l = 1 - p_r$. If the walker moves m times to the right and $n - m$ times to the left, its displacement is $x = m - (n - m) = 2m - n$. The probability for this is given by the binomial distribution

$$P_n(x) = \binom{n}{\frac{n+x}{2}} p_r^{\frac{n+x}{2}} (1-p_r)^{\frac{n-x}{2}}. \tag{2.39}$$

Of course, due to parity effects $(n \pm x)/2$ must be integer. From this expression one gets

$$\langle r_n \rangle = n(2p_r - 1) \tag{2.40}$$

$$\langle r_n^2 \rangle = n^2(2p_r - 1)^2 + 4np_r(1 - p_r). \tag{2.41}$$

On making the substitution $p_r = 1/2$ (unbiased motion) and using the Stirling approximation $n \approx \sqrt{2\pi n}(n/e)^n$, being $x \ll n$, one obtains (posing $\tau = 1$)

$$P(x,t)dx \approx \frac{1}{\sqrt{2\pi t}} e^{-x^2/(2t)} dx. \tag{2.42}$$

In fact, when the number of steps is very large, one can properly move from the discrete random walk to a diffusion process as both the time and spatial coordinates are allowed to take a continuous range of values. The specific limiting process requires that the rate at which steps are made should tend to infinity, while, at the same time, the lattice spacing tends to zero. We will see in the next subsection that these limiting processes cannot be carried out independently of one another, but are constrained by the requirement that the probability density for the displacement at time t should satisfy a partial differential equation of the second-order. However, it should be stressed that a diffusion limit does not exist for all possible random walk models. A prime example of such an exception is the continuous-time random walk, where two subsequent steps are temporally separated by a so-called waiting time drawn from a heavy-tailed distribution (see Section 2.2.8).

Finally, we sketch another approach to the definition of random walks. In fact, the definition of random walk as a sum of random variables is probably one of the simplest ways to introduce diffusion from a mathematical perspective (see e.g., [63]), although the notion of random walks has to be found in the earliest investigations of gambling problems by the pioneers of probability theory [52, 66], hence in contexts not directly connected with diffusion.

If $X_1, X_2, \ldots$ are independent, identically distributed random variables in $\mathbb{R}$ with mean zero and variance σ^2, then the central limit theorem states that the distribution of

$$\frac{S_n}{\sqrt{n}} = \frac{X_1 + \ldots + X_n}{\sqrt{n}} \tag{2.43}$$

approaches that of a normal distribution with mean zero and variance σ^2. In other words, for $-\infty < r < s < \infty$,

$$\lim_{n\to\infty} P\left(r \leq \frac{X_1+\ldots+X_n}{\sqrt{n}} \leq s\right) = \int_r^s \frac{1}{\sqrt{2\pi\sigma^2}} e^{-\frac{y^2}{2\sigma^2}} dy. \tag{2.44}$$

More formally, one can show that

$$P(S_n = k) = P\left(\frac{k}{\sqrt{n}} \leq \frac{S_n}{\sqrt{n}} \leq \frac{k+1}{\sqrt{n}}\right)$$

$$\approx \int_{k/\sqrt{n}}^{(k+1)/\sqrt{n}} \frac{1}{\sqrt{2\pi\sigma^2}} e^{-\frac{y^2}{2\sigma^2}} dy \approx \frac{1}{\sqrt{2\pi\sigma^2 n}} e^{-\frac{k^2}{2\sigma^2 n}} \tag{2.45}$$

and extend the result to high-dimensional spaces (see e.g., [63] for a review of the joint normal distribution).

In conclusion, the general subject area of the asymptotic properties of sums of random variables is a subject widely studied in the literature of mathematics and statistics. Not only do these limit theorems attract the interest of pure mathematicians, but their significance in the application of probability in almost all sciences should not be underestimated [88]. In the absence of limit theorems, the theory of random walks would consist only of a large number of isolated examples. The existence of limit theorems means that there are universal properties which, at least at long times (i.e., after many steps of the random walk), apply to large classes of random walks and are independent of all but a few general properties of these walks. There is a close connection between limit theorems for random walks and the theory of diffusion processes. Thus, the diffusion process can also be regarded as a kind of generalized random walk defined by the passage to different kinds of limits in terms of space and time variables.

2.2.4.1 The Continuum Limit of Random Walks

We have previously shown that random walks constitute a suitable discrete model for diffusion. It is instructive to see this relation the other way round, namely to see how the continuum limit of random walks just recovers diffusion.

For simplicity, let us discuss this limit in a one-dimensional lattice of equal spacing. At each time step the walker hops to the nearest site to its right or left, with equal probability $p = 1/2$. The probability $P_n(m)$ of being at the site m at the n-th time step satisfies the equation

$$P_{n+1}(m) = \frac{1}{2}P_n(m-1) + \frac{1}{2}P_n(m+1), \tag{2.46}$$

or

$$P_{n+1}(m) - P_n(m) = \frac{1}{2}[P_n(m-1) - 2P_n(m) + P_n(m+1)], \tag{2.47}$$

On making the change of variables $x = m\ell$ and $t = n\tau$, and taking the limit $\ell \to 0$ and $\tau \to 0$, the probability $P_n(m)$ is replaced by the probability density $P(x,t)$. Provided that we keep $\ell^2/\tau = 2D$ constant in the limiting process, equation (2.47) becomes

$$\frac{\partial}{\partial t} P(x,t) = D \frac{\partial^2}{\partial x^2} P(x,t). \qquad (2.48)$$

Thus, we have recovered the diffusion equation in one dimension. Notice that equation (2.42) is a solution of the diffusion equation (2.48) with initial condition $P(x,0) = \delta(x)$. Indeed, in the long-time limit there is basically no difference between discrete random walks and diffusion.

The generalisation of diffusion to higher dimensions is straightforward, that is

$$\frac{\partial}{\partial t} P(\mathbf{r},t) = D \nabla^2 P(\mathbf{r},t), \qquad (2.49)$$

where ∇^2 is the d-dimensional Laplacian operator.

2.2.5 Continuous-time Random Walks

In 1965, Elliot Montroll and George Weiss [71] introduced the concept of continuous-time random walks (CTRWs) as a way to render time continuous, without appealing to the diffusion limit. However, the model achieves much more than this original goal: some forms of CTRW are fundamentally different than the classical diffusion model, and the theory has numerous important applications.

Imagine a random walk on a lattice, starting at the origin, but such that the steps are taken at random times. Let $\psi(t)$ be the probability density for the waiting time between successive steps. We assume that the waiting times for different jumps are statistically independent and are all characterized by $\psi(t)$. The probability that the waiting time between steps is greater than t is

$$\Psi(t) = \int_t^\infty \psi(t')dt'. \qquad (2.50)$$

Let us also define $\psi_n(t)$ as the probability density that the n-th jump occurs at time t. Clearly, $\psi_1(t) = \psi(t)$, and, since the waiting times between steps are independent,

$$\psi_{n+1}(t) = \int_0^t \psi_n(t')\psi(t-t')dt'. \qquad (2.51)$$

Using the Laplace transform $\tilde{\psi}(s) \equiv \int_0^\infty \psi(t)e^{-st}dt$ we find $\tilde{\psi}_n(s) = [\tilde{\psi}(s)]^n$ and

$$\tilde{\Psi}(s) = \int_0^\infty e^{-st}\Psi(t)dt = \int_0^\infty e^{-st}\int_t^\infty \psi(t')dt'dt = \frac{1-\tilde{\psi}(s)}{s}. \qquad (2.52)$$

Let us now move forward and introduce the probability $P_n(\mathbf{r})$ of being at $\mathbf{r}$ at the n-th time step. Then, we can express $P(\mathbf{r},t)$ as

$$P(\mathbf{r},t) = \sum_{n=0}^{\infty} P_n(\mathbf{r}) \int_0^t \psi_n(t')\psi(t-t')dt'. \tag{2.53}$$

The integral represents the probability that, once the walker has arrived at site $\mathbf{r}$ at time $t' < t$, it will remain there until time t. Taking the Laplace transform, we obtain

$$\tilde{P}(\mathbf{r},s) = \delta_{\mathbf{r},0}\tilde{\Psi}_0(s) + \tilde{\Psi}(s)\sum_{n=0}^{\infty} P_n(\mathbf{r})[\tilde{\psi}(s)]^n, \tag{2.54}$$

where $\tilde{\Psi}_0(s)$ is just the Laplace transform of the probability that the time of the first step is greater than t. Given the waiting time distribution $\psi(t)$ and the dynamics of the corresponding (discrete time) random walk $P_n(\mathbf{r})$, one can plug the related information in the previous equation, anti-transform and derive $P(\mathbf{r},t)$. The special case of $\psi(t) = \delta(t-\tau)$, with τ constant, reduces to the standard random walks discussed earlier, where τ plays the role of time unit. In fact, any $\psi(t)$ that falls off fast enough yields similar regular behavior. On the other hand, interesting anomalous behavior may be obtained with slow-decaying ψ, as will be shown in Section 2.2.8. See also [12, 69, 60] for further reviews on this topic.

2.2.6 Mathematical Tools: Characteristic Functions

In Section 2.2.4, we easily derived that, in one-dimensional systems, the probability for a random walker to reach a distance x in time t is given by a normal distribution with mean zero and variance t (see equation (2.42)). However, to establish results in more general contexts (e.g., in the presence of a stochastic waiting time between steps – as anticipated in Section 2.2.5, or when the underlying structure is topologically more complex – as we will see in Section 2.3), one needs more sophisticated machineries than those used before. For instance, Fourier and Laplace transform function is a heavily exploited technique which allows moving from one parameter space to another parameter space where a solution is more easily achievable; such a solution shall then be anti-transformed to return in the original space.

Here we will focus on Fourier transform, while in the next section we will mainly focus on Laplace transforms. Let us review this approach by means of a simple example. Let us consider a discrete time, continuous space random walk that takes place in $\mathbb{R}^d$ and whose step lengths are drawn from the probability density $p(\mathbf{r})$, and assume that the walk starts at the origin. The simple random walk is a special case, in which $p(\mathbf{r}) = [1/(2d)]\sum_i \delta(\mathbf{r} - \mathbf{e_i})$. We are interested in the probability density $P_n(\mathbf{r})$ that the walker is in $\mathbf{r}$ at the n-th step. The function

$$\hat{P}_n(\mathbf{k}) = \int P_n(\mathbf{r}) e^{i\mathbf{k}\cdot\mathbf{r}} d^d\mathbf{r} \tag{2.55}$$

is called the characteristic function of the probability density. In fact, it is the Fourier transform of $P_n(\mathbf{r})$, and hence it encompasses an equivalent amount of information. Similarly, the characteristic function of the step probability density is

$$\hat{p}(\mathbf{k}) = \int p(\mathbf{r}) e^{i\mathbf{k}\cdot\mathbf{r}} d^d\mathbf{r} \tag{2.56}$$

and is also known as the step structure function. Because the steps are independent, the process obeys the Markov property

$$P_{n+1}(\mathbf{r}) = \int P_n(\mathbf{r}') p(\mathbf{r}-\mathbf{r}') d^d\mathbf{r}'. \tag{2.57}$$

Taking the Fourier transform, namely applying the operator $\int d^d\mathbf{r} e^{i\mathbf{k}\cdot\mathbf{r}}$, results in

$$\hat{P}_{n+1}(\mathbf{k}) = \hat{P}_n(\mathbf{k}) \hat{p}(\mathbf{k}), \tag{2.58}$$

a recursive relation that yields

$$\hat{P}_n(\mathbf{k}) = \hat{p}(\mathbf{k})^n. \tag{2.59}$$

The probability density may now be obtained from the inverse transform

$$P_n(\mathbf{r}) = \frac{1}{(2\pi)^d} \int \hat{P}_n(\mathbf{k}) e^{-i\mathbf{k}\cdot\mathbf{r}} d^d\mathbf{k}. \tag{2.60}$$

Suppose that the step probability function has zero mean and finite variance, that is

$$\int \mathbf{r} p(\mathbf{r}) d^d\mathbf{r} = 0, \tag{2.61}$$

$$\int \mathbf{r}^2 p(\mathbf{r}) d^d\mathbf{r} = \ell^2 < \infty, \tag{2.62}$$

then, at long times, the distribution $P_n(\mathbf{r})$ tends to a Gaussian (this result stems from the central limit theorem). Let us see how this happens in the simple case of a walk on the line, $d=1$. We may then work with scalar variables instead of the d-dimensional vectors $\mathbf{r}$ and $\mathbf{k}$. The structure function is, according to equation (2.56) and (2.61–2.62),

$$\hat{p}(k) = 1 - \frac{1}{2} k^2 \ell^2 + O(k^2), \tag{2.63}$$

and so the characteristic function is

$$\hat{P}_n(k) = e^{-nk^2\ell^2/2 + n\mathcal{O}(k^2)}. \tag{2.64}$$

For very long times ($n \gg 1$) the main contribution to $\hat{P}_n(k)$ comes from $|k| < 1/\sqrt{n}$, and $n\mathcal{O}(k^2) \sim 1/\sqrt{n}$ may be neglected. We then have

$$P_n(\mathbf{r}) = \frac{1}{2\pi} \int_{-\infty}^{\infty} e^{-nk^2\ell^2/2 - ikr} dk = \frac{1}{\sqrt{2\pi\ell^2 n}} e^{-r^2/(2\ell^2 n)}. \tag{2.65}$$

This may be written in terms of the time variable $t = n\tau$ as

$$P(r,t) = \frac{1}{\sqrt{4\pi Dt}} e^{-r^2/(4Dt)}, \tag{2.66}$$

which recovers (equation 2.45).

2.2.7 Mathematical Tools: Generating Functions and Tauberian Theorems

The method of generating function is based on the discrete version of the Laplace transform introduced in Section 2.2.5: a sequence $\{f_n\}$, $n = 0, 1, 2, \ldots$, is replaced by the power series $f(\lambda) = \sum_{n=0}^{\infty} f_n \lambda^n$ [often recast in a slightly different notation as $f(s) = \sum_{n=0}^{\infty} f_n e^{-sn}$], with the magnitude of the complex variable λ sufficiently small to ensure convergence of the series. Recurrence relations or difference equations involving the sequence $\{f_n\}$ reduce to algebraic expressions for its generating function $f(\lambda)$, and once $f(\lambda)$ is found explicitly, the sequence $\{f_n\}$ can be recovered using the standard formula from complex analysis $f_n = \frac{1}{2\pi i} \int_\Gamma f(\lambda) d\lambda/\lambda^{n+1}$, where Γ is a simple closed contour encircling $\lambda = 0$ once, which lies within the circle of convergence of the power series and is traversed anti-clockwise.

Solving a problem for the generating functions rather than the original function itself can be much easier, yet in order to get the explicit solution in the original space/time one has to properly invert the expressions for the generating functions. Only for particularly simple models can one actually invert the Laplace transform explicitly and hence characterize the properties of the random walk. Nevertheless, it is possible to find asymptotic properties from these transforms through the use of a class of mathematical techniques summarized under the general heading of Abelian and Tauberian theorems [42].

In a nutshell, the content of Abelian and Tauberian theorems is that there is a close relation between the asymptotic dependence of f_n on n, and the singularity in its transform function $f(s)$ as the variable s approaches the radius of convergence. This class of relationships can be exploited in both directions. On the one hand, specific types of asymptotic behavior in the f_n lead to specific singularities in $f(s)$; these theorems, schematized in the form $\{f_n\} \to f(s)$, are known as Abelian theorems. On the other hand, under specific conditions, the singular behavior of $f(s)$ can be used to determine the asymptotic behavior of f_n: those, schematized in the form $f(s) \to \{f_n\}$, are addressed by Tauberian theorems and are, in general, harder. Both types of theorems have been exploited in the study of random walks, but Tauberian theorems play the more significant role because often one only has access to the characteristic function in a given problem.

A Tauberian theorem for power series that is very useful in treating random walks is the following: let f_n be a series of strictly positive elements and $f(s)$ its (discrete) Laplace transform, namely the power series in e^{-s}, s being a real variable, defined by

$$f(s) = \sum_{n=0}^{\infty} f_n e^{-ns}. \qquad (2.67)$$

Suppose that in the limit $s \to 0$ the behavior of $f(s)$ is singular in the sense that

$$f(s) \sim s^{-\alpha} L(s^{-1}), \qquad (2.68)$$

where $L(x)$ is a slowly varying function[7] and $\beta(x) \equiv x^{\alpha} L(x)$ is a positive, monotonically increasing function of x. Then, in the limit $n \to \infty$, the partial sum of the f_n is approximated by

$$\sum_{j=0}^{n} f_j \sim \frac{\beta(n)}{\Gamma(1+\alpha)} = \frac{n^{\alpha} L(n)}{\Gamma(1+\alpha)}, \qquad (2.69)$$

where $\Gamma(x)$ is the Gamma function[8].

It is not possible to infer the behavior of f_n themselves from this relation without imposing at least one further restriction on the f_n. The most useful of these requires that the f_n are monotonic functions of n, at least from some value $\tilde{n}$ onwards. This condition suffices to recover the asymptotic form of the f_n by differentiating (2.69). Under these circumstances we have the even more useful result

$$f_n \sim \frac{\beta'(n)}{\Gamma(1+\alpha)} = \frac{\alpha n^{\alpha-1} L(n) + n^{\alpha} L'(n)}{\Gamma(1+\alpha)}. \qquad (2.70)$$

In the following we present a number of examples in which the use of Tauberian theorems is crucial.

---

[7] A function $L(x)$ is said to be slowly varying at $x = \infty$ if, for every constant $c > 0$, it satisfies the condition

$$\lim_{x \to \infty} \frac{L(cx)}{L(x)} = 1.$$

For instance, the function $L(x) = \log(x)$ is slowly varying while $L(x) = x^{\beta}$ is not.
[8] $\Gamma(x) = (x-1)!$, if $x \in \mathbb{N}$; $\Gamma(x) = \int_0^{\infty} y^{x-1} e^{-y} dy$, if x is a complex number with a positive real part.

2.2.8 Anomalous Diffusion

As anticipated in Section 2.2.4, under proper conditions (e.g., presence of traps, topological disorder, etc.) the diffusion law $\langle r^2(t) \rangle \sim t$ may be replaced by a more general law $\langle r^2(t) \rangle \sim t^\alpha$, which, as long as $\alpha \neq 1$, is referred to as anomalous diffusion. More precisely, when $\alpha < 1$ we talk of subdiffusion, while when $\alpha > 1$ we talk of superdiffusion. Here we briefly sketch the emergence of anomalous diffusion for a one-dimensional CTRW with proper waiting time distribution $\psi(t)$ and proper jump length distribution $p(x)$. This also allows seeing Tauberian theorems at work.

In fact, in Section 2.2.5 we developed a formalism necessary for the analysis of properties of the CTRW, but found that results are most readily expressible in terms of Laplace transforms. We recall that $\psi_n(t)$ is the probability density function for the time at which the n-th step is made; this notation implies that $\psi_1(t) = \psi(t)$ and, for arbitrary n, the recursive relation (2.51) holds. The probability that exactly n steps have been made during the time interval $(0,t)$ will be denoted by $U_n(t)$, whose expression in terms of the $\psi_n(t)$ is

$$U_n(t) = \int_0^t \psi_n(\tau) \Psi(t-\tau) d\tau, \qquad (2.71)$$

where $\Psi(t) \equiv \int_t^\infty \psi(t') dt'$. Of course, the number of steps made by the random walker in time t is a random variable and its expectation value is

$$\langle n(t) \rangle = \sum_{n=0}^\infty n U_n(t) = \int_0^t \Psi(t-\tau) \sum_{n=0}^\infty n \psi_n(\tau) d\tau. \qquad (2.72)$$

At this point it is convenient to pass to the Laplace transform domain in which we have already shown in Section 2.2.5 that $\tilde{\psi}_n(s) = [\tilde{\psi}(s)]^n$. The Laplace transform of $\langle n(t) \rangle$ is found with some algebra from equation (2.72) as

$$\langle \tilde{n}(s) \rangle = \frac{\tilde{\psi}(s)}{s[1 - \tilde{\psi}(s)]}. \qquad (2.73)$$

When the expected value of the pausing time between successive steps is finite so that as $s \to 0$, $\tilde{\psi}(s) \sim 1 - s\langle t \rangle$, the function $\langle \tilde{n}(s) \rangle$ can be expanded to the lowest-order in $s\langle t \rangle$ as

$$\langle \tilde{n}(s) \rangle \sim \frac{1}{s^2 \langle t \rangle}, \qquad (2.74)$$

which, according to the Tauberian theorem discussed earlier, is equivalent to

$$\langle n(t) \rangle \sim \frac{t}{\langle t \rangle}, \quad t \to \infty. \qquad (2.75)$$

When $\psi(t)$ is asymptotically proportional to $\langle t \rangle^\alpha / t^{\alpha+1}$ as $t \to \infty$, where $0 < \alpha < 1$, the transform $\tilde{\psi}(s)$ can be expanded in the neighborhood in which s is small as $\tilde{\psi}(s) \sim 1 - (s\langle t \rangle)^\alpha$. Since $\langle \tilde{n}(s) \rangle \sim 1/(t^\alpha s^{\alpha+1})$ in the same limit, a Tauberian theorem allows us to conclude that

$$\langle n(t) \rangle \sim \frac{1}{\Gamma(\alpha)} \left(\frac{t}{\langle t \rangle} \right)^\alpha. \tag{2.76}$$

Having established these results for the expected number of steps made in time t, we can return to the problem of finding asymptotic properties of random walk observables. Recall that, provided that the origin in time coincides with the time of the first step of the CTRW, the exact expressions for the Laplace transforms of the mean displacement and the associated second moment of a CTRW in one dimension are

$$\langle \tilde{x}(s) \rangle = \frac{\langle x \rangle \tilde{\psi}(s)}{s[1 - \tilde{\psi}(s)]}, \tag{2.77}$$

$$\langle \tilde{x}^2(s) \rangle = \frac{\langle x \rangle^2 \tilde{\psi}(s)}{s[1 - \tilde{\psi}(s)]} + \frac{2 \langle x^2 \rangle \tilde{\psi}^2(s)}{s[1 - \tilde{\psi}(s)]^2}, \tag{2.78}$$

where we accounted for a distribution $p(x)$ of step length such that the first two moments of the displacement in a single step are $\langle x \rangle$ and $\langle x \rangle^2$ (the discrete, single-step case, i.e., the standard random walk, is trivially recovered by taking $p(x) = \delta(x - \ell)$).

Consider first the case in which the pausing time density has at least two finite moments, $\langle t \rangle$ and $\langle t^2 \rangle$ respectively. This is equivalent to stating that $\tilde{\psi}(s)$ can be expanded to the lowest-order around $s = 0$ as

$$\tilde{\psi}(s) \sim 1 - \langle t \rangle s + \frac{\langle t^2 \rangle}{2} s^2 - \ldots \tag{2.79}$$

Therefore, the most singular terms in an expansion of $\langle \tilde{x}(s) \rangle$ around $s = 0$ are

$$\langle \tilde{x}(s) \rangle \sim \langle x \rangle \left[\frac{1}{\langle t \rangle s^2} + \left(\frac{\langle t^2 \rangle}{2 \langle t \rangle^2} - 1 \right) \frac{1}{s} \right]. \tag{2.80}$$

The application of a Tauberian theorem yields

$$\langle x(t) \rangle \sim \langle x \rangle \left[\frac{t}{\langle t \rangle} + \left(\frac{\langle t^2 \rangle}{2 \langle t \rangle^2} - 1 \right) \right], \tag{2.81}$$

when $t/\langle t \rangle$ is large. The second term in the right hand side is negligible with respect to the first when $\langle x \rangle \neq 0$ and $t \gg \langle t \rangle$. Hence, equation (2.81) states that the average number of steps taken in a time t is equal to $t/\langle t \rangle$ and the length of each step is, on average, equal to $\langle x \rangle$ so that the asymptotic value of the average displacement is just the average number of steps multiplied by the average displacement per step.

A similar expansion can be made of the Laplace transform $\langle \tilde{x}^2(s) \rangle$ and we can then find that the lowest-order term in the asymptotic expression for the variance of the displacement is equal to

$$\langle x^2(t) \rangle - \langle x(t) \rangle^2 = \frac{[\sigma^2(x)\langle t \rangle^2 + \sigma^2(t)\langle x \rangle^2]}{\langle t \rangle^3} t, \qquad (2.82)$$

where both the variance $\sigma^2(x) = \langle x^2 \rangle - \langle x \rangle^2$ of the single-step displacement and the variance $\sigma^2(t) = \langle t^2 \rangle - \langle t \rangle^2$ of the pausing time density contribute. Of course, when the random walk is symmetric, the left hand side is simply $\sigma^2(x)t/\langle t \rangle \sim \sigma^2(x)\langle n(t) \rangle$.

Let us now consider the case where there are no finite integer moments of the pausing time; for instance, let us pose $\psi(t) \sim \langle t \rangle^\alpha / t^{\alpha+1}$, where $0 < \alpha < 1$. The lowest order term in the expansion of $\hat{\psi}(s)$ turns out to be $\hat{\psi}(s) \sim 1 - (s\langle t \rangle)^\alpha$. If we substitute this expansion into equation (2.77) and apply a Tauberian theorem we can show that, at long times,

$$\langle x(t) \rangle \sim \langle x \rangle \frac{t^\alpha}{\Gamma(1-\alpha)\Gamma(1+\alpha)\langle t \rangle^\alpha}. \qquad (2.83)$$

Because of the longer waiting times between successive steps the random walk slows, that is, the average displacement increases at a rate slower than t. This is equal to saying that the process subdiffuses. The asymptotic form for the variance depends on whether the walker is biased:

$$\langle x^2(t) \rangle - \langle x(t) \rangle^2 = \frac{\sigma^2(x)t^\alpha}{\Gamma(1-\alpha)\Gamma(1+\alpha)\langle t \rangle^\alpha}, \text{ if } \langle x \rangle = 0 \qquad (2.84)$$

$$\langle x^2(t) \rangle - \langle x(t) \rangle^2 = \frac{\langle x \rangle^2}{\Gamma^2(1-\alpha)} \left[\frac{2}{\Gamma(1+2\alpha)} - \frac{1}{\Gamma^2(1+\alpha)} \right] \left(\frac{t}{\langle t \rangle} \right)^{2\alpha}, \text{ if } \langle x \rangle \neq 0. \qquad (2.85)$$

More details on the derivation of the previous expressions can be found in [88, 13].

Beyond the presence of traps and/or of long jumps, a further model for anomalous diffusion relies on spatial inhomogeneities. In this case, the entire space is not accessible, namely the random walk evolves on a space where bottlenecks and dead-ends exist on all scales resulting in an effective subdiffusion. A discrete version is exemplified by diffusion in deterministic or random fractal networks. While the fractal dimension d_f characterizes the geometry of the fractal, the diffusive dynamics involves the random walk exponent d_w. The latter is related to the anomalous diffusion exponent through $\alpha = 2/d_w$ ($d_w \geq 2$ for subdiffusion). Fractals can be used to model complex networks and can mimic certain features of diffusion under conditions of molecular crowding [12] (see also the next section for a review on random walks in arbitrary topologies).

Finally, let us briefly discuss the case of finite characteristic waiting time and diverging variance for the jump length; for instance, let us take a Poissonian or delta waiting time and a heavy-tailed distribution for the jump length, that is

$$p(x) \sim \frac{1}{|x|^{1+\mu}} \quad \text{as } x \to \infty. \tag{2.86}$$

When $\mu \geq 2$ the motion displays (for asymptotic times) a Gaussian probability function with a variance scaling linearly with time, namely a normal diffusion; for $\mu < 2$ it can be shown (e.g., see [12]) that, for asymptotic times, the probability density $P(x,t)$ of the walk being at x at time t follows the Lévy distribution, which for $|x|^\mu \gg t$ (considering dimensionless units) scales as:

$$P(x,t) \sim \frac{t}{|x|^{1+\mu}}. \tag{2.87}$$

The resulting mean-square displacement $\langle x^2 \rangle \equiv \int_{-\infty}^{\infty} x^2 p(x,t)dx$ is therefore divergent while $\langle |x|^q \rangle \sim t^{q/\mu}$ ($0 < \mu < 2$) [37]. This is equal to super diffusion. More generally, when embedded in a d-dimensional lattice, a mean-square displacement $\langle r^2 \rangle$ scaling with time faster than linearly can be accomplished by taking a distribution for step lengths scaling like $p(\mathbf{r}) \sim 1/|r|^{\mu+d}$, with $0 < \mu < 2$ (e.g., see [91]). In this process, usually referred to as Lévy flight, due to the asymptotic property (2.86) of the jump length probability distribution function, very long jumps may occur with a significantly higher probability than for any finite variance jump length distribution. The scaling nature of $p(\mathbf{r})$ leads to the clustering nature of the Lévy flights, i.e., local motion is occasionally interrupted by long sojourns, on all length scales. That is, the Lévy flight is self-similar and, in fact, its trajectory can be assigned a fractal dimension $d_f = \mu$.

This kind of motion is commonly supposed to be an efficient search strategy of living organisms: the search process consists in straight ballistic phases (the jumps) that alternate with reorientation phases (between two jumps); contrarily, the trajectory drawn from a finite variance distribution fills the space completely, and features no distinguishable clusters, as all jumps are of about the same length [69].

Anomalous diffusion can elegantly be treated by means of fractional calculus which is a natural extension of the differential and integral classic calculus. Oversimplifying, let us consider a one-dimensional walker with jump distributions $p(x) \sim |x|^{-1-\mu}$; then, the probability distribution $P(x,t)$ can be described by

$$\frac{\partial P(x,t)}{\partial t} = \nabla^\mu \cdot \rho(x,t), \tag{2.88}$$

with $\mu \in \mathbb{R}$, and $\nabla^\mu = \partial^\mu \rho(x,t)/\partial x^\mu$ is a derivative of fractional order μ which encodes the information on the heavy-tailed distribution of jumps.

Unlike classical derivatives, there are several options for computing fractional derivatives (e.g., Riemann-Liouville, Caputo, Riesz, etc.). Diffusion equations with fractional derivatives on space are usually written in terms of the symmetric Riesz derivative on space. This operator has a relatively simple behavior under transformations, and, in fact, it is often defined in terms of its Fourier transform [61, 55] as

$$\frac{\partial^\mu f(x)}{\partial |x|^\mu} = \mathcal{F}^{-1}[-|k|^\mu \mathcal{F}[f(x)]], \tag{2.89}$$

where $\mathcal{F}$ and $\mathcal{F}^{-1}$ denote the Fourier transform and its inverse, respectively.

The interested reader is referred to [69, 60] for nice reviews on fractional calculus.

2.2.9 Mathematical Tools: an Algebraic Approach

When dealing with the discrete time random walker on a discrete structure such as a lattice or, more generally, a graph (see Section 2.3) one can look at it as a Markov chain and exploit the related tools to derive, for example, the asymptotic distribution, the absorption time, and the splitting probability. This approach is especially useful when the underlying structure is finite and not translationally invariant (as deepened in Section 2.3).

More precisely, each site (of a lattice or of any arbitrary graph), labeled as $i = 1, 2, \ldots$, corresponds to a different state and the probability to jump from one site, say j, to another site, say k, corresponds to the transition rate T_{jk}. We call $\mathcal{V}$ the set of possible states. Now, let us assume that the chain is *irreducible*, namely it is possible to get any state from any other state. One can then prove that there exists a unique *stationary distribution* $\pi = \{\pi_1, \pi_2, \ldots\}$, i.e., a unique probability distribution satisfying the *balance equations*

$$\pi_j = \sum_{i \in \mathcal{V}} \pi_i T_{ij} \ \forall j, \tag{2.90}$$

or, in vector notation, $\pi = \mathbf{T}\pi$, that is, π is the Perron-Frobenius eigenvector associated to $\mathbf{T}$. As is well known in probability theory, the stationary distribution plays the main role in asymptotic results. In particular, calling $N_i(t)$ the number of visits to state i during times $0, 1, \ldots, t$, then for any initial distribution,

$$\frac{N_i(t)}{t} \to \pi_i \ a.s., \text{ as } t \to \infty, \tag{2.91}$$

provided the chain is aperiodic; this theorem is the simplest illustration of the ergodic principle "time averages equal space averages." Moreover, the expected return time to i is $\tau_i = 1/\pi_i$.

We can see this problem from another perspective. Let us assume that π is a probability distribution satisfying $\pi_i T_{ij} = \pi_j T_{ji}$, for all i,j. Then, one can prove that π is the unique stationary distribution and the chain is *reversible*. In fact, the previous condition corresponds to the detailed balance equation as it implies $\sum_{i \in \mathcal{V}} \pi_i T_{ij} = \pi_j \sum_{i \in \mathcal{V}} T_{ji} = \pi_j$ for all j.

The combination of Markov theory and algebra leads to interesting properties of **T**. For instance, $(\mathbf{T}^n)_{ij}$ is just the probability that the walker starting from the site i is in site j after n steps. A clear treatment of the basic theory of Markov chains along with a wide selection of applications is available in several textbooks (e.g., see [76]). In the remainder of this subsection, we present a couple of problems which can be simply addressed via this algebraic approach.

Absorbing time. Let us assume that the space underlying diffusion contains a set $\mathcal{A} \subset \mathcal{V}$ of absorbing sites (i.e., traps) such that, for any $j \in \mathcal{A}$, $T_{ji} = 0, \forall i \in \mathcal{V}$. Also, let us denote with $V \equiv |\mathcal{V}|$ the cardinality of $\mathcal{V}$. A key timescale characterizing the trapping processes is the mean time to reach any of these sites. A powerful approach is based on the recurrent equations $\tau_i^{(j)} = \sum_k T_{i,k}^{(j)} \tau_k^{(j)} + 1$, $i \neq j, j$ being the trap location and $\tau_i^{(j)}$ the mean time to absorption from initial site $i \neq j$; moreover, $\mathbf{T}^{(j)}$ is obtained from **T** by deleting the j-th row and the j-th column (this equals to say that j is a trap). In matrix notation, $\tau^{(j)} = \mathbf{T}^{(j)} \tau^{(j)} + \mathbf{e}$, where $\tau^{(j)} = (\tau_1^{(j)}, \ldots, \tau_{j-1}^{(j)}, \tau_{j+1}^{(j)}, \ldots, \tau_V^{(j)})^\dagger$ is a $(V-1)$-dimensional vector and **e** is the $(V-1)$-dimensional "one" vector $\mathbf{e} = (1,1,\ldots,1)^\dagger$. Therefore,

$$\tau^{(j)} = (\mathbf{I} - \mathbf{T}^{(j)})^{-1} \mathbf{e} \equiv \mathbf{Ne}, \qquad (2.92)$$

where **N** is often referred to as a *fundamental matrix* (see e.g., [76]).

Splitting probability. Let us focus on a more complicated problem: let us divide the set of traps $\mathcal{A}$ into two subsets, i.e., $\mathcal{A}_1 \cup \mathcal{A}_2 = \mathcal{A}$, and ask for the probability $P_{1|2}$ of being trapped in any site in $\mathcal{A}_1$ without visiting any site in $\mathcal{A}_2$; *mutatis mutandis*, for $P_{2|1}$.

To calculate the splitting probabilities $P_{1|2}$ and $P_{2|1}$ we can exploit the properties of absorbing Markov chains (see e.g., [76]). In particular, for an arbitrary chain displaying a set $\mathcal{A} \in \mathcal{V}$, with $|\mathcal{A}| = k < V$, of absorbing states, and the remaining transient states, we properly reshuffle the columns and the rows of **T**, in such a way that the first $V - k$ states are transient and the last k states are absorbing. In this way we get the so-called canonical form

$$\mathbf{T} = \left(\begin{array}{c|c} \mathbf{Q} & \mathbf{R} \\ \hline \mathbf{0} & \mathbf{I} \end{array} \right), \qquad (2.93)$$

where $\mathbf{I}$ is a $k \times k$ identity matrix, $\mathbf{0}$ is a $k \times (V-k)$ zero matrix, $\mathbf{R}$ is a non-zero $(V-k) \times k$ matrix and $\mathbf{Q}$ is a $(V-k) \times (V-k)$ matrix.

Now, when building $\mathbf{T}^n$, the entries of $\mathbf{Q}^n$ give the probabilities of being in each of the transient states after n steps for each possible transient starting state. The matrix $\mathbf{N} \equiv (\mathbf{I} - \mathbf{Q})^{-1} = I + Q^2 + Q^3 + ...$, is the expected number of times that the chain is in state j, given that it starts in state i. We can therefore derive that $\tau^{(A)} = \mathbf{N}\mathbf{e}$ is the expected number of steps before the chain is absorbed. More precisely, $\tau_i^{(A)}$ is the expected time when the chain starts in i. We can also write $\mathbf{B} = \mathbf{N}\mathbf{R}$, where B_{ij} is the probability that an absorbing chain will be absorbed in the state j if it starts in the transient state i.

To conclude this subsection, we stress that the absorbing time is the first example of mean-first-passage problem we have seen so far, but we will devote much attention to this problem in the following section. Here we simply mention that first-passage properties are widely used in the context of diffusion-limited processes [78], in chemistry [80], in biology [82], in electrical black-out spreading [36], in epidemiology [64], for foraging animals [87], etc. In general, in these applications, there is a stochastic process, represented by a random walker, which triggers an event when reaching for the first time a given target, and it can be advantageous to minimize (e.g., in transport process) or to maximize (e.g., in viruses spreading) this first-passage time. When several targets are available, for instance when a given protein can react with several different targets in a living cell, it becomes important to know which one is first reached.

2.2.10 Resistance Networks

The connection between random walks and resistance networks allows analytical expressions to be derived for the absorbing probabilities [40]. Let us see how it works starting from the simplest example of one-dimensional lattices.

Consider a random walk moving on a chain of length N. Let $Q_{N|0}(x)$ be the probability, starting at x, of reaching N before 0, hereafter denoted as $q(x)$ to lighten the notation. We regard $q(x)$ as a function defined on the points $x = 0, 1, 2, ..., N$. The function $q(x)$ has the following properties:

a) $q(0) = 0$,
b) $q(N) = 1$,
c) $q(x) = \frac{1}{2}q(x-1) + \frac{1}{2}q(x+1)$, for $x = 1, 2, ..., N-1$.

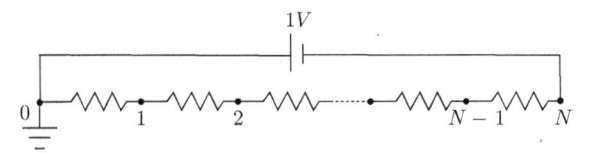

Figure 2.3. The problem of a random walk on the integers $0, 1, 2, \ldots, N$ can be mapped into an electric problem where we connect equal resistors in series and put a unit voltage $\Delta v = 1$ across the ends.

Properties (a) and (b) follow from our convention that 0 and N are traps; if the walker reaches one of these positions, it stops there. Property (c) states that, for an interior point, the probability $q(x)$ of reaching the trap set in 0 from x is the average of the probabilities $q(x-1)$ and $q(x+1)$ of reaching 0 from the points that the walker may go to from x.

Let us consider a second, apparently very different, problem. We connect equal resistors in series and put a unit voltage across the ends as in Figure 2.3. Voltages $v(x)$ will be established at the points $x = 0, 1, 2, \ldots, N$. We have grounded the point $x = 0$ so that $v(0) = 0$. We ask for the voltage $v(x)$ at the points x between the resistors. If we have N resistors, we make $v(0) = 0$ and $v(N) = 1$, so $v(x)$ satisfies properties (a) and (b) stated above. We now show that $v(x)$ also satisfies (c). By Kirchhoff's Laws, the current flowing into x must be equal to the current flowing out. By Ohm's Law, if points x and y are connected by a resistance of magnitude R, then the current i_{xy} that flows from x to y is equal to

$$i_{xy} = \frac{v(x) - v(y)}{R}. \qquad (2.94)$$

Thus, for $x = 1, 2, \ldots, N-1$, assuming that the resistances are equivalent,

$$\frac{v(x-1) - v(x)}{R} + \frac{v(x+1) - v(x)}{R} = 0. \qquad (2.95)$$

Multiplying by R and solving for $v(x)$ gives $v(x) = [v(x+1) + v(x-1)]/2$, for $x = 1, \ldots, N-1$. Therefore, $v(x)$ also satisfies property (c).

Now, since $q(x)$ and $v(x)$ both satisfy properties (a), (b) and (c), we naturally wonder whether $q(x)$ and $v(x)$ are equal. For this simple example, we can easily find $v(x)$ using Ohm's Law, find $q(x)$ using elementary probability, and see that they are the same. However, one can prove that for very general circuits there is only one function that satisfies these properties. The proof is based on the uniqueness principle of harmonic function as briefly sketched hereafter for the simple one-dimensional case.

Let $\mathcal{S}$ be the set of points $\mathcal{S} = \{0, 1, 2, \ldots, N\}$. We call the points of the set $\mathcal{D} = \{1, 2, \ldots, N-1\}$ the interior points of $\mathcal{S}$ and those of $\mathcal{B} = \{0, N\}$ the boundary points of $\mathcal{S}$. A function $f(x)$ defined on $\mathcal{S}$ is harmonic if, at points of $\mathcal{D}$, it satisfies the averaging property

$$f(x) = \frac{f(x-1) + f(x+1)}{2}. \tag{2.96}$$

As we have seen, $q(x)$ and $v(x)$ are harmonic functions on $\mathcal{S}$ having the same values on the boundary: $q(0) = v(0) = 0$; $q(N) = v(N) = 1$. Thus both $q(x)$ and $v(x)$ solve the problem of finding a harmonic function having these boundary values. The problem of finding a harmonic function given its boundary values is called the Dirichlet problem, and the Uniqueness Principle for the Dirichlet problem asserts that there cannot be two different harmonic functions having the same boundary values. In particular, it follows that $q(x)$ and $v(x)$ are really the same function. Of course, the problem can be extended to higher dimension lattices and even to arbitrary graphs (see [40] and Figure (2.4) for a sketch).

To summarize, in this mapping, being the two nodes a and b supplied with the boundary conditions $v(a) = 1$ and $v(b) = 0$, the voltage at a node x determines the probability that starting at $x \neq a, b$ the node a is reached before b. Also, the

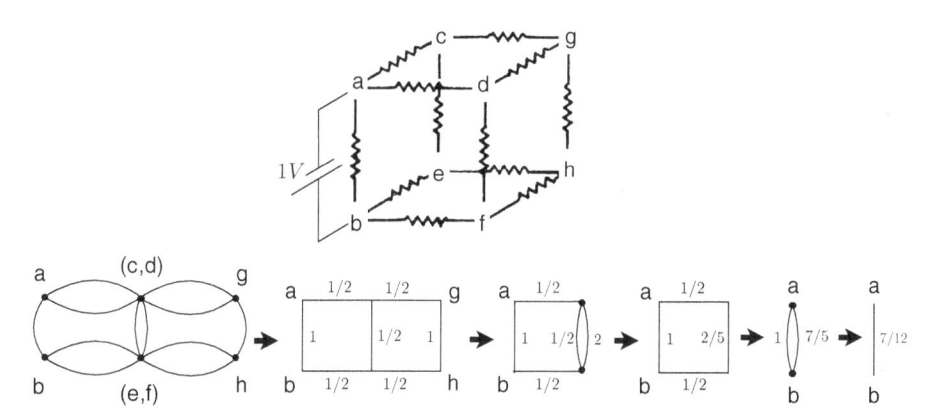

Figure 2.4. In a unit cube of unit resistors we put a unit battery between a and b, in such a way that $v(a) = 1$ and $v(b) = 0$. Then, a current $i_a = \sum_x i_{ax}$ will flow into the circuit from the outside source. The amount of current that flows depends upon the overall resistance in the circuit. We define the effective resistance R_{eff} between a and b by $R_{\text{eff}} = v(a)/i_a$. The reciprocal quantity $C_{\text{eff}} = 1/R_{\text{eff}}$ is the effective conductance. The effective resistance between the two adjacent points can be calculated as shown in the lower scheme: by symmetry the voltages at c and d are equivalent to those at e and f, then, using the laws for the effective resistance of resistors in series and parallel, the network can be successively reduced to a single resistor of resistance $R_{\text{eff}} = 7/12$. The effective conductance is $C_{\text{eff}} = 12/7$ and $C_a = \sum_x C_{ax} = 3$. Thus, the current flowing into a from the battery will be $i_a = R_{\text{eff}}^{-1} = 12/7$ and the probability that a walker starting at a will reach b before returning to a is $p = i_a/C_a = (12/7)/3 = 4/7$.

current i_{xy} is the expected number of times the walker passes along the edge (x,y). One therefore tries to calculate the effective resistances, possibly relying on useful tools such as the star-triangle equivalence or the Wheatstone-bridge-like tricks, and then map results in the original diffusive context.

2.2.11 The Pólya Problem and the Probability of Return to the Origin

Imagine a random walk in a d-dimensional infinite lattice; an important question is to see whether the walk[9] is *transient* or *recurrent*. Transience means that the walker has a finite probability of never returning to its starting point again, whereas recurrence means that it is certain to come back. One can prove that the random walk is recurrent in $d \leq 2$, but transient otherwise. This result is often referred to as Pólya's theorem.

The lattice on which the random walk takes place will be assumed to be both homogeneous and translationally invariant, but no further assumption about its properties is required in the following analysis (see [88] for more details). Two sets of probabilities will be required to make any statements about the return of a random walker to the origin. The first, denoted by $P_n(\mathbf{j})$, is the probability that a random walker is at lattice site $\mathbf{j}$ at step n (where $\mathbf{j}$ is a vector of d integers when the lattice is d-dimensional); the second, denoted by $F_n(\mathbf{j})$, is the probability that the random walker reaches $\mathbf{j}$ for the *first* time at the n-th step. The $F_n(\mathbf{j})$, by definition, differ from the $P_n(\mathbf{j})$ in excluding walks which have reached $\mathbf{j}$ before the n-th step.

A knowledge of the first-passage time probabilities leads directly to a solution to the original version of the Pólya problem. To see this we note that the probability of eventually reaching the origin is equal to

$$Q(\mathbf{0}) = \sum_{n=0}^{\infty} F_n(\mathbf{0}), \tag{2.97}$$

since this is the sum of probabilities of reaching it on step n ($n = 1, 2, 3, \ldots$). Equation (2.97) can be generalized to account for the probability that the random walker reaches an arbitrary site $\mathbf{j}$ at some time during the course of the walk, by simply replacing $\mathbf{0}$ with $\mathbf{j}$.

The $F_n(\mathbf{j})$ are directly related to the $P_n(\mathbf{j})$ by

$$P_n(\mathbf{j}) = \sum_{k=1}^{n} F_k(\mathbf{j}) P_{n-k}(\mathbf{0}), \tag{2.98}$$

---

[9] Actually, as will be made clear in the following section, it is the underlying graph that is either recurrent or transient, rather than the walk itself.

Diffusion Phenomena and Statistical Mechanics

which is derived for $\mathbf{j} \neq \mathbf{0}$ by noting that if the random walker is at $\mathbf{j}$ at the step n it must have been there for the first time at step k ($k \leq n$) and then possibly returned there in a time $n-k$ (of course, $P_0(\mathbf{0}) = 1$). When $\mathbf{j} = \mathbf{0}$, equation (2.98) is replaced by

$$P_n(\mathbf{0}) = \delta_{n,0} + \sum_{k=1}^{n} F_k(\mathbf{0}) P_{n-k}(\mathbf{0}), \tag{2.99}$$

where the Kronecker delta on the right-hand side accounts for the initial position of the random walker at the origin.

The fact that the sums in equations (2.98) and (2.99) have a discrete convolution structure allows us to write for the related generating functions $\tilde{P}(\mathbf{j}; z)$ and $\tilde{F}(\mathbf{j}; z)$ the following relations:

$$\tilde{F}(\mathbf{0}; z) \equiv \sum_{n=0}^{\infty} F_n(\mathbf{0}) \lambda^n = 1 - \frac{1}{\tilde{P}(\mathbf{0}; \lambda)}, \tag{2.100}$$

$$\tilde{F}(\mathbf{j}; z) \equiv \sum_{n=0}^{\infty} F_n(\mathbf{j}) \lambda^n = \frac{\tilde{P}(\mathbf{j}; \lambda)}{\tilde{P}(\mathbf{0}; \lambda)}, \quad \mathbf{j} \neq \mathbf{0}. \tag{2.101}$$

The connection between the generating functions in the last equation and the Pólya problem is that equation (2.97) for the probability of return to the origin can be rewritten as

$$Q(\mathbf{0}) = \tilde{F}(\mathbf{0}; 1). \tag{2.102}$$

We next find an integral representation for the generating function $\tilde{P}(\mathbf{j}; \lambda)$. Recall that for a random walk on a lattice the function $P_n(\mathbf{j})$ can be expressed in terms of the characteristic function as

$$P_n(\mathbf{j}) = \frac{1}{(2\pi)^d} \int_{-\pi}^{\pi} \cdots \int_{-\pi}^{\pi} \hat{p}^n(\theta) e^{-i \mathbf{j} \cdot \theta} d^d \theta, \tag{2.103}$$

where $\hat{p}(\theta)$ is the characteristic function of the single-step displacement probabilities. On multiplying both sides of this last equation by λ^n and summing over all n (remember that $|\hat{p}(\theta)| \leq 1$) we find that

$$\tilde{P}(\mathbf{j}; \lambda) = \frac{1}{(2\pi)^d} \int_{-\pi}^{\pi} \cdots \int_{-\pi}^{\pi} \frac{e^{-i \mathbf{j} \cdot \theta}}{1 - \lambda \hat{p}(\theta)} d^d \theta. \tag{2.104}$$

The generating functions for first-passage time probabilities can therefore be expressed in terms of ratios of these integrals, as a consequence of the identities in equation (2.100). In particular, the probability that the random walker eventually returns to the origin is

$$Q(\mathbf{0}) = 1 - \frac{1}{\tilde{P}(\mathbf{0}; 1)}, \tag{2.105}$$

and the probability that the random walker eventually reaches site **j** is

$$Q(\mathbf{j}) = \frac{\tilde{P}(\mathbf{j};1)}{\tilde{P}(\mathbf{0};1)}. \tag{2.106}$$

Equation (2.105) implies that the random walker will return to the origin with probability equal to 1 if and only if $\tilde{P}(\mathbf{0};1) = \infty$. In other words, when the return to the origin is certain the multiple integral in equation (2.104) must diverge when $\mathbf{j} = \mathbf{0}$ and $\lambda = 1$. We therefore consider the conditions under which such divergence is expected to occur. Since we assume that $|\hat{p}(\theta)| < 1$ for all $\theta \neq \mathbf{0}$ (while $\hat{p}(\mathbf{0}) = 1$ for the normalization condition), it is clear that a divergent integral representation of $\tilde{P}(\mathbf{0};1)$ can only be the result of the denominator of equation (2.104) vanishing at some point in the θ-space. This can only occur at $\theta = \mathbf{0}$. Hence we focus on the behavior of $\hat{p}(\theta)$ in the neighborhood of the origin.

First, we restrict our analysis to symmetric random walks (the Pólya problem is mainly interesting for symmetric random walks, since otherwise there is a net drift away from the origin, in which case it is intuitively reasonable that the return probability should be less than 1).

Then, we assume that all of the second moments of $p(\mathbf{j})$ are finite. This assumption allows us to expand $\hat{p}(\theta)$ in the neighborhood of the origin as

$$\hat{p}(\theta) \sim 1 - \frac{1}{2} \sum_{k=1}^{d} \sum_{m=1}^{d} \langle j_k j_m \rangle \theta_k \theta_m + o(\theta^2), \tag{2.107}$$

where the j_k is the k-th component of the displacement in a single step. Notice that the terms that are first-order in the θ_k are absent from the expansion because of the assumed symmetry of the random walk. When (2.107) is plugged into equation (2.104) with $\lambda = 1$ one finds

$$\tilde{P}(\mathbf{0};1) \sim \frac{2}{(2\pi)^d} \int_{-\pi}^{\pi} \cdots \int_{-\pi}^{\pi} \frac{d^d\theta}{\sum_{k=1}^{d}\sum_{m=1}^{d} \langle j_k j_m \rangle \theta_k \theta_m}. \tag{2.108}$$

Now, because the singularity of interest can only occur at the origin in θ-space, we can expand the limits of integration in each variable to $(-\infty, \infty)$ in order to simplify our analysis of the integral. The extension of the limits of integration is an artifice, since the resulting integrals will generally diverge, due to the behavior of the integrand at $\pm\infty$ (this is not the case in one dimension). However, the only source of the divergence that merits investigation is the one at $\theta = \mathbf{0}$, which does not depend on the limits set on the integrals. As the quadratic term in the denominator is nonnegative definite it can be diagonalized, with the eigenvalues ω_k

being positive. Hence the function $\tilde{P}(\mathbf{0};1)$ is proportional to an integral of the form

$$\tilde{P}(\mathbf{0};1) \propto \int_{-\infty}^{\infty} \cdots \int_{-\infty}^{\infty} \frac{d^d\alpha}{\sum_{k=1}^{d} \omega_k \alpha_k^2} = \frac{1}{\sqrt{\omega_1 \omega_2 \cdots \omega_k}} \int_{-\infty}^{\infty} \cdots \int_{-\infty}^{\infty} \frac{d^d\beta}{\sum_{k=1}^{d} \beta_k^2},$$

(2.109)

where we posed $\alpha_k = \beta_k/\sqrt{\omega_k}$. Since the denominator can be expressed as the square of the radius of a sphere in d dimensions, we can transform to spherical coordinates and integrate out all of the angle variables. The Jacobian of the transformation implies a term ρ^{d-1}, where ρ is the radial coordinate (this is a purely geometric factor). Hence we have, as a formal relation,

$$\tilde{P}(\mathbf{0};1) \propto \int_0^\infty \rho^{d-3} d\rho.$$

(2.110)

As mentioned earlier, the divergence at the upper limit is apparent and not a real one, since the only significant divergence is due to the behavior of the integrand at $\rho = 0$. We conclude from equation (2.110) that $\tilde{P}(\mathbf{0};1)$ diverges in $d=1$ and $d=2$ dimensions and converges in $d \geq 3$ dimensions. The probabilistic interpretation of this result is that the random walker is certain to return to the origin in one and two dimensions, and that there is a non-zero probability that the random walker does not return to the origin in three or more dimensions. Thus, the random walk is termed *recurrent* in one and two dimensions, and *transient* in three or more dimensions. The results just proved are valid for any random walk in which the second moments of single-step displacements are finite, and therefore constitute a slight generalization of Pólya's original theorem, which dealt with simple random walks (the generalization to the case of waiting time with first and second moments finite is also straightforward). The general case for an arbitrary site $\mathbf{j}$ can be resolved using essentially the same techniques as those used before. We refer to [87] for more general results.

2.3 Random Walks on Graphs

2.3.1 A Short Introduction to Graphs

Many systems can be naturally represented as networks, where the components of the system are represented by the vertices of the network and the interactions between the components are represented by edges connecting vertices in the network [75]. In the last decade the study of networks (or, in a more mathematical jargon, graphs) has become an important area of research in many disciplines, including physics, mathematics, biology, computer science and the social sciences [90, 85, 1, 74, 15].

Examples of technological networks are the Internet (where the vertices are computers and associated equipment while the edges are the data connections between them) and the World Wide Web (where the vertices are web pages and the edges are hyperlinks). Examples of biological networks include: the metabolic networks (in which the vertices represent metabolites and the edges connect any two metabolites that take part in the same reaction), the protein-protein interaction networks (where vertices represent proteins and two proteins that interact biologically are connected by an edge), the food webs (where species are represented by vertices and edges represent predator-prey relationships between the species), the immune networks (where vertices represents lymphocytes and edges represent the interactions via immunoglobulins or cytokines), and the neural networks (where the vertices represent neurons and the edges represent neural connections). Further examples are provided by social networks (where the vertices represent individuals or groups and the edges represent some type of connection between them, such as acquaintance or professional/familiar ties).

The function of graphs in physics, however, is not purely descriptive. Geometry and topology have a deep influence (often even more important than the interaction details) on the physical properties of the system. Here, we shall be mainly concerned with the properties of a graph which most affect the dynamical and thermodynamical behavior of the system it describes. In order to show these effects we especially focus on random walks, which is probably the simplest stochastic process affected by topology and, at the same time, the basic model of diffusion phenomena and non-deterministic motion.

2.3.2 Definitions

Let us define a graph $\mathcal{G}$ as a set $\mathcal{V}$ of sites connected pairwise by a set $\mathcal{E}$ of unoriented links. A couple of sites $i,j \in \mathcal{V}$ is said to be connected (or adjacent) whenever $(i,j) = (j,i) \in \mathcal{E}$. The overall number of sites making up the graph is called the volume N of the graph and it is given by the cardinality of the set $\mathcal{V}$, i.e. $N \equiv |\mathcal{V}|$. When N is finite (infinite), $\mathcal{G}$ is called a finite (infinite) graph.

A *subgraph* of $\mathcal{G}$ is constructed by taking a subset $\mathcal{S}$ of $\mathcal{E}$ together with all vertices incident in $\mathcal{G}$ with some edge belonging to $\mathcal{S}$.

A graph can be algebraically described through the *adjacency matrix* **A** of size $N \times N$, whose entry A_{ij} equals k if i and j are connected through k edges. It follows that $z_i = \sum_{j=1}^{N} A_{ij}$ is the number of links stemming from i, namely the *degree* (also called the coordination number) of i. A graph is said to be *locally finite* if $z_i < \infty, \forall i \in \mathcal{V}$. In the following we shall focus on graphs devoid of self-loops and of multiple edges in such a way that the entries of the adjacency matrix can be

either 0 or 1:
$$A_{ij} = \begin{cases} 1 & \text{if } i \text{ and } j \text{ are adjacent;} \\ 0 & \text{otherwise.} \end{cases} \quad (2.111)$$

Of course, by definition, **A** is a real symmetric matrix, and the trace of **A** is zero. Since the rows and the columns of **A** correspond to an arbitrary labelling of the vertices of $\mathcal{G}$, it is clear that the most interesting properties of the adjacency matrix are those which are invariant under permutations of the rows and columns. For instance, let h be a nonnegative integer. Then, $(\mathbf{A}^h)_{ij}$ is the number of walks of length h from i to j. In particular, $(\mathbf{A}^2)_{ii}$ is the degree of the vertex i, and $\text{Tr}(\mathbf{A}^2)$ equals twice the number of edges of $\mathcal{G}$; similarly, $\text{Tr}(\mathbf{A}^3)$ is six times the number of triangles in $\mathcal{G}$.

Moreover, a graph is said to be *connected* if each pair of vertices is joined by a walk and the number of edges traversed in the shortest walk connecting i and j is called the distance d_{ij} in $\mathcal{G}$ between i and j. The maximum value of the distance function in a connected graph $\mathcal{G}$ is called the diameter of $\mathcal{G}$, namely $d \equiv \max_{i,j \in V} d_{ij}$. In a connected graph with diameter d the spectrum of the adjacency matrix has at least $d+1$ distinct eigenvalues.

Indeed, among the above mentioned invariant properties are the spectral properties of **A**. For instance, the *characteristic polynomial* $\det(\lambda \mathbf{I} - \mathbf{A})$ can be written as

$$\chi(\mathbf{G};\lambda) = \lambda^N + c_1 \lambda^{N-1} + c_2 \lambda^{N-2} + c_3 \lambda^{N-3} + \ldots + c_N, \quad (2.112)$$

where the coefficients $c_1, c_2, \ldots, c_N$ can be related to the topological properties of $\mathcal{G}$ itself. In fact, it is proved in the theory of matrices that all the coefficients can be expressed in terms of the principal minors of **A**, where the principal minor is the determinant of a sub-matrix obtained by taking a subset of the rows and the same subset of the columns. Then, one can find that [25]

- $c_1 = 0$; as $-c_1$ is the sum of the zeros, that is, the sum of the eigenvalues; this is also the trace of **A** which, as we have already noted, is zero.
- $(-c_2)$ is the sum of edges in $\mathcal{G}$. In fact, for each $i \in \{1, 2, \ldots, N\}$, the number $(-1)^i c_i$ is the sum of those principal minors of **A** which have i rows and columns; in particular, for $i = 2$ any principal minor which has non-zero entry must be of the form $\left|\begin{smallmatrix} 0 & 1 \\ 1 & 0 \end{smallmatrix}\right|$ and there is one such minor for each pair of adjacent vertices of $\mathcal{G}$, and each has value -1.
- $(-c_3)$ is twice the number of triangles in $\mathcal{G}$; in fact, following the previous point we can figure out only three possibilities for nontrivial principal minors of size $i = 3$, namely $\left|\begin{smallmatrix} 0 & 1 & 0 \\ 1 & 0 & 0 \\ 0 & 0 & 0 \end{smallmatrix}\right|$; $\left|\begin{smallmatrix} 0 & 1 & 1 \\ 1 & 0 & 0 \\ 1 & 0 & 0 \end{smallmatrix}\right|$; $\left|\begin{smallmatrix} 0 & 1 & 1 \\ 1 & 0 & 1 \\ 1 & 1 & 0 \end{smallmatrix}\right|$ and, of these, the only non-zero one is the last, whose value is 2, corresponding to three mutually adjacent vertices in $\mathcal{G}$.

Moreover, since **A** is real and symmetric, its eigenvalues λ are real, and the multiplicity of λ as a root of the equation $\det(\lambda \mathbf{I} - \mathbf{A}) = 0$ is equal to the dimension of the space eigenvectors corresponding to λ.

In order to get more familiar with these concepts let us consider a few simple examples.

The *complete graph* K_N is the graph with N vertices where each distinct pair of nodes are adjacent, in such a way that the related adjacency matrix displays all entries equal to 1, except those on the diagonal which are zero. An easy calculation shows that the spectrum of K_N is $\lambda_1 = N - 1$ and $\lambda_2 = -1$, with multiplicity 1 and $N - 1$, respectively.

The spectrum of the *bipartite graph* $K_{M,N}$, made by two parties of size M and N, respectively, and $M \times N$ links connecting each couple of nodes belonging to different parties, includes $\pm\sqrt{M \times N}$ (each with multiplicity 1) and 0 (with multiplicity $M + N - 2$).

The *cycle* C_N has a spectrum consisting of the numbers $2\cos(2\pi j/N)$, with $j = 0, \ldots, N - 1$.

Finally, for the *regular graph* $\mathcal{G}_k$, where each node has degree k, one can see that k is an eigenvalue of $\mathcal{G}_k$ and its multiplicity is 1 if $\mathcal{G}_k$ is connected. Also, for any eigenvalue λ of $\mathcal{G}_k$ we have $|\lambda| \leq k$ (see [25] for a complete proof).

2.3.3 Elements of Spectral Graph Theory

Given a graph, beyond the adjacency matrix **A**, it is possible to define the corresponding Laplacian matrix, **L**, which takes values -1 for pairs of connected vertices and z_i (the degree of the corresponding node i) in diagonal sites. Obviously, $\mathbf{L} = \mathbf{Z} - \mathbf{A}$, where **Z** is the diagonal connectivity (or degree) matrix, namely $Z_{ij} = z_i \delta_{ij}$. If the graph is undirected both **A** and **L** are symmetric matrices.

The (ordinary) spectrum of a finite graph $\mathcal{G}$ is, by definition, the spectrum of the adjacency matrix **A**, that is, its set of eigenvalues together with their multiplicities. The *Laplace spectrum* of a finite undirected graph (without loops) is the spectrum of the Laplace matrix **L** and in the following we summarize a few examples of propositions proved in this context (see also [24] for a diffusive treatment).

Let $\mathcal{G}$ be an undirected graph with at least one vertex, and Laplace matrix **L** with eigenvalues $0 = \mu_1 \leq \mu_2 \leq \ldots \leq \mu_N$. It is easy to see that $\mu_1 = 0$ is a trivial eigenvalue of **L** with eigenvector $(1, 1, \ldots, 1)$ and that the eigenvalues μ_i satisfy

$$0 = \mu_1 \leq \mu_2 \leq \ldots \leq \mu_N \leq 2z_{\max}, \tag{2.113}$$

where $z_{\max}$ is the largest degree in the graph (see [35]). Also, the multiplicity of $\mu_1 = 0$ equals the number of connected components of $\mathcal{G}$.

For the complete graph K_N, the Laplace matrix is $(N-1)\mathbf{I} - \mathbf{A}$, which has spectrum 0 (with multiplicity 1) and N (with multiplicity $N-1$).

A graph $\mathcal{G}$ is bipartite if and only if the Laplace spectrum and the signless Laplace spectrum of $\mathcal{G}$ are equal. The Laplace spectrum of the complete bipartite graph $K_{M,N}$ is 0 (with multiplicity 1), M (with multiplicity $N-1$), N (with multiplicity $M-1$), and $M+N$ (with multiplicity 1).

The spectrum of the N-cycle C_N consists of the numbers $2 - 2\cos(2\pi j/N)$, $(j = 0,\ldots,N-1)$.

For the undirected, regular graph $\mathcal{G}_k$, k is the largest eigenvalue, and its multiplicity equals the number of connected components of $\mathcal{G}_k$.

Finally, given graphs Γ and Δ with vertex sets $\mathcal{V}$ and $\mathcal{W}$, respectively, their Cartesian product, denoted with $\mathcal{G} = \Gamma \times \Delta$, is the graph with vertex set $\mathcal{V} \times \mathcal{W}$, where $(v,w) \sim (v',w')$, when either $v = v'$ and $w \sim w'$ or $w = w'$ and $v \sim v'$. For the adjacency matrices we have $\mathbf{A}_{\mathcal{G}} = \mathbf{A}_\Gamma \otimes \mathbf{I} + \mathbf{I} \otimes \mathbf{A}_\Delta$. If u and v are eigenvectors for Γ and Δ with ordinary or Laplace eigenvalues θ and η, respectively, then the vector w defined by $w(x,y) = u_x v_y$ is an eigenvector of $\Gamma \times \Delta$ with ordinary or Laplace eigenvalue $\theta + \eta$. For example, the hypercube 2^n is the Cartesian product of n factors K_2. The spectrum of K_2 is $1, -1$, and hence the spectrum of 2^n consists of the numbers $n - 2i$ with multiplicity $\binom{n}{i}$, $(i = 0, 1, \ldots, n)$.

2.3.3.1 Spectral Gap

In order to better appreciate the power of the spectral analysis, in this subsection we focus on the so-called spectral gap μ_2, namely the smallest non-zero Laplace eigenvalue, and we discuss the information which can be accordingly inferred.

Let us start considering a network perfectly separated into a number of independent subsets or communities having only intra-subset links and not inter-subset connections. Obviously, its Laplacian matrix is block diagonal. Each subgraph has its own associated sub-matrix, and therefore zero is an eigenvalue of any of them. In this way, the degeneration of the lowest (trivial) eigenvalue of $\mathbf{L}$ coincides with the number of disconnected subgraphs. For each of the separated components, the corresponding eigenvector has constant components within each subgraph.

On the other hand, if the subgraphs are not perfectly separated, but a small number of inter-subset links exist, eigenvalues and eigenvectors will be slightly perturbed and the degeneracy will be shifted. In particular, relatively small eigenvalues will appear, and their corresponding eigenvectors will take "almost constant" values within each subgraph. Basically, a graph with a "small" first nontrivial Laplacian eigenvalue, customarily called the spectral gap (or also algebraic connectivity), has a relatively clean bisection. In other words, the smaller the spectral gap and the smaller the relative number of edges required to be cut away to generate a bipartition. Conversely, a large spectral gap characterizes non-structured networks, with poor modular structure. In fact, spectral methods provide an effective bipartitioning criterion (see e.g., [46, 39]).

The spectral gap also has remarkable effects on dynamical processes. For instance, let us consider the problem of synchronizability of the dynamical processes occurring at the nodes of a given network and described by

$$\dot{x}_i = F(x_i) + \beta \sum_{j|(i,j)\in \mathcal{E}} [H(x_j) - H(x_i)] = F(x_i) - \beta \sum_j L_{ij} H(x_j), \qquad (2.114)$$

where x_i with $i \in 1, 2, \ldots, N$ are dynamic variables, F and H are an evolution and a coupling function respectively, and β is a constant. A standard linear stability analysis can be performed by: i) expanding around a fully synchronized state $x_1 = x_2 = \cdots = x_N = x^s$, with x^s the solution of $\dot{x}^s = F(x^s)$; ii) diagonalizing $\mathbf{L}$ to find its N eigenvalues; and iii) writing equations for the normal modes y_i of perturbations $\dot{y}_i = [F'(x^s) - \beta \mu_i H'(x^s)] y_i$. As explained in [39], the interval in which the synchronized state is stable is larger for smaller eigenratios μ_N/μ_2, and therefore one concludes that a network has a more robust synchronized state if the ratio μ_N/μ_2 is as small as possible.

However, as said before, probably the easiest dynamical process affected by the underlying topology is the random walk. In this context, the spectral gap is associated with spreading efficiency: random walks move around quickly and disseminate fluently on graphs with a large spectral gap, in the sense that they are very unlikely to stay long within a given subset of vertices unless its complementary subgraph is very small [65].

2.3.4 The Random Walk Problem on Graphs

Let us now introduce the so-called simple random walk on an arbitrary graph $\mathcal{G}$. Assuming the time t to be discrete, we recall that at each time step t the jumping probability p_{ij} between nearest neighbor sites i and j is defined as:

$$p_{ij} = \frac{A_{ij}}{z_i} = (\mathbf{Z}^{-1} \mathbf{A})_{ij}. \qquad (2.115)$$

This is the simplest case we can consider: the jumping probabilities are isotropic at each point and they do not depend on time; in addition the walker is forced to jump at every time step. As we will see later, the last condition – i.e., the impossibility of staying on the site, although crucial for the short time behavior, has no significant influence on the long time regime.

Usually, the random walk problem is considered to be completely solvable if, for any $i, j \in \mathcal{G}$ and $t \in \mathbb{N}$, we are able to calculate the functions $P_{ij}(t)$, each representing the probability of being at site j at time t for a walker starting from site i at time 0. These probabilities are the elements of a matrix **P**, which is equal to the t-th power of the jumping probabilities matrix **p**, namely

$$P_{ij}(t) = (\mathbf{p}^t)_{ij}; \tag{2.116}$$

this relation can be easily proven by induction on t.

For finite systems, one can find a closed-form formula for $P_{ij}(t)$ by exploiting algebraic methods such as the following:

(1) compute the spectrum $\lambda_1, \lambda_2, \ldots, \lambda_N$ of **p**
(2) if any eigenvalue is repeated, then the general form includes polynomial factors in t; for instance, if an eigenvalue λ has multiplicity 2 then the general form includes the term $(at + b)\lambda^t$. The set of constants can be derived by imposing the initial conditions and the normalization.

In the case of large or infinite graphs, the exact calculation of all $P_{ij}(t)$ becomes practically impossible and, above all, of little significance. In fact, for large systems we are mainly interested in global and collective properties as typically happens in statistical physics. In particular, as we will see, the asymptotic behavior of the probability $P_{ii}(t)$ of returning to the starting point after t steps, gives the most direct characterization of the large scale topology for infinite graphs. A related quantity is the average number P_{ii} of returns to the starting point i, which can be generalized to the average number P_{ij} of passages through j starting from i:

$$P_{ij} \equiv \lim_{t \to \infty} \sum_{k=0}^{t} P_{ij}(k), \tag{2.117}$$

where the limit can be infinite.

Another related quantity is the mean displacement $r_i(t)$ from the starting site i after t steps, and it is defined as

$$r_i(t) \equiv \sum_{j \in \mathcal{V}} r_{ij} P_{ij}(t), \tag{2.118}$$

where r_{ij} is calculated according to the chemical distance and is defined as the number of links in the shortest path connecting vertices i and j. The mean-square displacement $r^2(t)$ is in turn deeply related to the diffusion properties (see Sections 2.2.4 and 2.2.8). Quantities such as $P_{ij}(t)$ and $r_i(t)$ are not sensible to the history, that is, they can be determined by considering the situation of the walker at an arbitrary time $t_0 < t$ regardless of its previous positions. This is not the case for a class of non-Markovian quantities which do keep track of what happened along the whole time span before t. Among these quantities, the first-passage probability $F_{ij}(t)$ denotes the probability for a walker starting from i of reaching for the *first* time the site $j \neq i$ in t steps. For $i = j$ the previous definition would not be interesting, the walker being at i at $t = 0$ by definition. Therefore, one defines $F_{ii}(t)$ to be the probability of returning to the starting point i for the first time after t steps and one sets $F_{ii}(0) = 0$. In spite of the deeply different nature of P and F, a fundamental relation can be established between them:

$$P_{ij}(t) = \sum_{k=0}^{t} F_{ij}(k) P_{jj}(t-k) + \delta_{ij} \delta_{t0}. \tag{2.119}$$

In fact, each walker which is at j at time t only has two possibilities: either it gets there for the first time, or it has reached j for the first time at a previous time k and then it has returned there after $t - k$ steps. The first-passage probability is in turn connected to other history dependent quantities. In particular, the probability F_{ij} of ever reaching the site j starting from i (or of ever returning to i, if $i = j$) is given by

$$F_{ij} = \sum_{t=0}^{\infty} F_{ij}(t). \tag{2.120}$$

Moreover, the average number $S_i(t)$ of different sites visited after t steps by a walker starting from i is related to $F_{ij}(t)$ by

$$S_i(t) = 1 + \sum_{k=1}^{t} \sum_{j \in \mathcal{V}} F_{ij}(k). \tag{2.121}$$

Finally, the first-passage time τ_{ij}, i.e., the average time at which a walker starting from i reaches j for the first time (or returns for the first time to i, if $i = j$) is

$$\tau_{ij} = \lim_{t \to \infty} \frac{\sum_{k=0}^{t} k F_{ij}(k)}{F_{ij}}. \tag{2.122}$$

Before proceeding we mention that the simple random walk can be modified to give a richer behavior and to describe more general physical problems. Indeed, one can introduce anisotropic jumping probabilities by substituting in (2.115) the adjacency matrix $\mathbf{A}$ with a nonnegative coupling matrix $\mathbf{J}$:

$$p_{ij} = \frac{J_{ij}}{I_i} = (\mathbf{I}^{-1} \mathbf{J})_{ij}, \tag{2.123}$$

where $I_{ij} = I_i \delta_{ij}$ and $I_i = \sum_{k \in \mathcal{V}} J_{ik}$. Depending on the specific properties of J_{ij}, this can produce only local effects or introduce a global bias which destroys the leading diffusive behavior giving rise to transport phenomena.

2.3.4.1 Generating Functions

It is worth recalling here a few remarks on generating functions (already introduced in Section 2.2.7) as they will be extremely useful to estimate the quantities presented in the last section (see equations (2.117–2.122)). The discrete Laplace transform maps a time function $f(t)$ onto its generating function $\tilde{f}(\lambda)$ defined by

$$\tilde{f}(\lambda) = \sum_{t=0}^{\infty} \lambda^t f(t), \qquad (2.124)$$

where λ is, in general, a complex number. The inverse equation giving $f(t)$ from $\tilde{f}(\lambda)$ is

$$f(t) = \frac{1}{t} \frac{\partial^t \tilde{f}(\lambda)}{\partial \lambda^t}\bigg|_{\lambda=0}. \qquad (2.125)$$

This equation is useful as far as we are interested in small t behavior, but it becomes ineffectual in the study of asymptotic regimes for $t \to \infty$. In this case a very powerful tool is provided by the Tauberian theorems (see Section 2.2.7), relating the singularities of $\tilde{f}(\lambda)$ to the leading large t behavior of $f(t)$.

A great advantage of this mathematical technique is that a series of relevant random walk parameters, which are non-local in time, can be obtained directly from a generating function, without calculating the corresponding time dependent quantities. For instance, P_{ij}, F_{ij} and τ_{ij} are related to $\tilde{P}_{ij}(\lambda)$ and $\tilde{F}_{ij}(\lambda)$ by

$$P_{ij} = \lim_{\lambda \to 1^-} \tilde{P}_{ij}(\lambda), \quad F_{ij} = \tilde{F}_{ij}(1), \quad \tau_{ij} = \lim_{\lambda \to 1^-} \frac{\partial \log \tilde{F}_{ij}(\lambda)}{\partial \lambda}. \qquad (2.126)$$

Remarkably, by transforming the convolution equation (2.119) we get:

$$\tilde{P}_{ij}(\lambda) = \tilde{F}_{ij}(\lambda)\tilde{P}_{jj}(\lambda) + \delta_{ij} \qquad (2.127)$$

and this relation is particularly useful in the analytical treatment of random walks and related applications. Notice that, under Laplace transform the relation (2.119) which was non-local in time becomes local in λ.

2.3.5 Random Walks on Finite Graphs

Finite graphs consist of a finite number of sites and links. The finite-size theory turns out to be appropriate when dealing with mesoscopic structures and finite-size effects. The random walk problem on finite graphs is simplified by the finiteness of the adjacency matrix. In fact, the analytical study is reducible to a spectral problem on a real finite-dimensional vector space, and numerical simulations are easily implemented by Monte Carlo techniques.

The arguments presented in Section 2.2.9 can be properly generalized to address an arbitrary underlying topology, where the transition matrix **p** is given by (2.115)[10]. In fact, a random walk on a graph is a Markov process and we can therefore apply the fundamental limit theorem which states, for regular, finite chains, the existence of a stationary distribution π corresponding to the unique (normalized such that the entries sum to 1) left eigenvector of **p** with eigenvalue 1. More explicitly, if $\mathcal{G}$ is not bipartite and is devoid of traps, one can easily show that the random walk is ergodic, i.e. that it admits limit probabilities for $t \to \infty$:

$$\pi_{ij} = \lim_{t \to \infty} P_{ij}(t) \quad \forall i,j \quad (2.128)$$

and that

$$\pi_{ij} = \frac{z_i}{2|\mathcal{E}|} \quad \forall i. \quad (2.129)$$

This uniform limit value is independent of the initial condition and it is reached exponentially, the exponential decay of each matrix element being no slower than λ_2^t, where λ_2 is the second largest eigenvalue of **p**.

Such stationary probabilities are directly related to the mean return times by the Kac formula [31]: τ_{ii} being the mean return time to i one has

$$\tau_{ii} = \frac{1}{\pi_i}. \quad (2.130)$$

Similarly, one can easily prove that

$$\lim_{t \to \infty} F_{ij}(t) = 0, \forall i,j, \quad \text{and} \quad \lim_{t \to \infty} S_i(t) = N, \forall i, \quad (2.131)$$

the limit values being reached exponentially. Moreover,

$$P_{ij} = \infty, \quad \text{and} \quad F_{ij} = 1, \quad \text{and} \quad t_{ij} < \infty, \quad \forall i,j. \quad (2.132)$$

2.3.6 Infinite Graphs

When dealing with macroscopic systems composed of a very large number N of sites, one usually takes the thermodynamic limit $N \to \infty$, hence allowing for infinite graphs, i.e. graphs composed by an infinite number of sites. This is particularly convenient for two main reasons: first of all, a single infinite structure effectively describes a very large (indeed, infinite) number of large structures

---

[10] For a random walk in the absence of traps the matrix elements p_{ij} satisfy the relations $p_{ij} \geq 0, \forall i,j \in \mathcal{V}$ and $\sum_{j=1}^{N} p_{ij} = 1, \forall i \in \mathcal{V}$, hence defining a stochastic matrix. The introduction of at least one trap on a site i (meaning that $\sum_{j=1}^{N} p_{ij} < 1$), dramatically changes the random walk behavior. The jumping probabilities matrix **p** is no longer stochastic, since the unitary condition is not satisfied, while the nonnegativity condition still holds.

having different sizes, but similar geometrical features; moreover, the singularities in thermodynamic potentials that are typical of critical phenomena, as well as a series of universal asymptotic behaviors, only occur on infinite structures [17].

As for random walks on large real structures, the time dependence of physical quantities exhibits different features according to the time scale considered. For very long times, the walker can explore every site and its behavior is described by the finite graph laws introduced in the previous section. However, if the time is long enough to explore large portions of the system, but still too short to experience the finite size effects, many significant quantities are quite insensitive to local details and exhibit power law time dependence with universal exponents. Often, this is the most interesting regime in physical applications.

In order to deal with infinite graphs, it is useful to define the generalized Van Hove sphere: a generalized Van Hove sphere $S_{o,r}$ of centre o and radius[11] r is the subgraph of $\mathcal{G}$, given by the set of vertices $V_{o,r} = \{i \in \mathcal{V} | r_{i,o} \leq r\}$ and by the set of links $E_{o,r} = \{(i,j) \in \mathcal{E} | i \in V_{o,r}, j \in V_{o,r}\}$. Then, the growth rate of the graph at large scales is described by the number of elements in $V_{o,r}$, denoted as $|V_{o,r}|$, as r increases [73]. A polynomial growth such as $|V_{o,r}| \sim r^{d_g}$ allows introducing the connectivity dimension d_g.[12] In general, we can think of it as the analogue of the fractal dimension, when the chemical distance metric is considered instead of the usual Euclidean metric. However, d_g does not constitute a good generalization of d, as the physical properties of models are typically independent of d_g.

Before proceeding with the analysis of random walks and statistical-mechanics models on infinite, connected graphs it is worth recalling that a *physical graph* [17] is expected to: i) display bounded coordination numbers (i.e., $\exists z_{max} | z_i \leq z_{max} \forall i \in \mathcal{V}$) since physical interactions are always bounded; and ii) be embedded in finite-dimensional spaces, which requires (a) that $\mathcal{G}$ has a polynomial growth (i.e. $\forall o \in \mathcal{V} \exists c, k$, such that $|V_{o,r}| < c r^k$) *and* (b) that boundary conditions are negligible in the thermodynamic limit (i.e, $\lim_{r \to \infty} |\partial V_{o,r}|/|V_{o,r}| = 0$, where $\partial V_{o,r}$ denotes the border of $V_{o,r}$, namely the set of points of $V_{o,r}$ not belonging to $V_{o,r-1}$).

Note that some graphs studied in the physical literature, such as the Bethe lattice, do not satisfy (a) and (b), while many random graphs do not fulfill i.

Let us now focus on the random walk problem on infinite graphs. Despite the mathematical complexity, many general results about infinite graphs have been rigorously proven. Some have correspondents in the finite graph case, but most

---

[11] A connected graph is endowed with an intrinsic metric generated by the chemical distance r_{ij}, which is defined as the number of links in the shortest path connecting vertices i and j.
[12] Of course, on lattices, the connectivity dimension coincides with the usual Euclidean dimension d, and for many fractals it has been exactly evaluated [45].

of them concern quantities and properties which cannot even be defined on finite structures. We now summarize the main differences with respect to the finite case.

For random walks on infinite, connected graphs devoid of traps satisfying *i* and *ii*, one can show that

$$P_{ij}^{\infty} = \lim_{t\to\infty} P_{ij}(t) = 0, \quad \forall i,j. \tag{2.133}$$

Unlike the finite case, the limit in (2.133) in general is not reached exponentially. Indeed, if (b) also holds, i.e., for physical graphs, the asymptotic behavior is typically a power law, whose exponent only depends on topology, as we will discuss in detail in the next subsections[13]. Similarly, one can prove that

$$\lim_{t\to\infty} F_{ij}(t) = 0, \quad \forall i,j, \quad \text{and} \quad \lim_{t\to\infty} S_i(t) = \infty, \quad \forall i. \tag{2.134}$$

As for the quantities concerning the number of visits and the first visit probabilities, the situation is far more complex: P_{ij} and τ_{ij} can be either finite or infinite, and F_{ij} can be either 1 or smaller than 1.

2.3.7 Again on the Type Problem

In 1921 Pólya showed that on low-dimensional lattices (i.e., $d \leq 2$) a random walk surely visits each site, while on high-dimensional lattices (i.e., $d \geq 3$) a random walk has a finite probability of never returning to the starting site and, more generally, of never visiting a given site, no matter how long the walk is run. These two cases are referred to as recurrent and transient, respectively. This distinction can be straightforwardly generalized to infinite arbitrary graphs where the walker can either be sure to return to the starting point (i.e., $F_{ii} = 1$) and any other site (i.e., $F_{ij} = 1$), or have the chance to escape forever from its starting point (i.e., $F_{ii} < 1$), and never reach a given site (i.e., $F_{ij} < 1$), even in the absence of traps. Of course, transience is an exclusive property of infinite graphs and it is due fundamentally to large scale topology. In other words, in the transient case, it happens that the number of paths leading the walker away from its starting point is large enough, with respect to the number of returning paths, to act as an asymptotic trap (still conserving the total probability).

For physical graphs transience and recurrence depend only on the graph topology, that is, they are intrinsic properties of a discrete structure and the classification of infinite graphs according to them is also known as the type problem. We therefore generalize the discussion of Section 2.2.11 to the case of arbitrary, infinite, physical graphs.

---

[13] Note that the widely studied case of Bethe lattices, not satisfying *ii*, is still characterized by an exponential decay.

Let us define the problem mathematically. First of all, a very general theorem on Markov chains states the following:

$$\exists i,j \in \mathcal{G} | F_{ij} = 1 \Rightarrow F_{hk} = 1 \forall h,k \in \mathcal{G} \qquad (2.135)$$

(where the sites h and k can coincide). This means that recurrence is point independent, or, in other words, that if a walker surely reaches a point j starting from a given point i, then it surely reaches any point k starting from any point h. It is straightforward to see that an analogous result follows for the case $F_{ij} < 1$. Another important result relates F_{ij} and P_{ij}. Indeed, from (2.126) and (2.127), it follows that

$$F_{ij} = 1 \Leftrightarrow P_{ij} = \infty \qquad (2.136)$$

and

$$F_{ij} < 1 \Leftrightarrow P_{ij} < \infty \qquad (2.137)$$

i.e., a walk is recurrent (transient) if and only if any site is visited an infinite (finite) number of times. The latter can be taken as an alternative definition of recurrence and transience. On the other hand, a similar relation does not hold for the mean first-passage time, as τ_{ij} can diverge even in recurrent graphs (e.g., in the infinite one-dimensional lattice $\tau_{ij} \to \infty$, $\forall i,j$). A consequent property concerns the way the walker explores the sites of $\mathcal{G}$:

$$F_{ij} = 1 \Leftrightarrow \lim_{t \to \infty} \frac{S_i(t)}{t} = 0, \qquad (2.138)$$

while

$$F_{ij} < 1 \Leftrightarrow 0 < \lim_{t \to \infty} \frac{S_i(t)}{t} < 1. \qquad (2.139)$$

Therefore, in recurrent graphs the number of distinct visited sites increases slower than the number of steps (compact exploration). Recurrent graphs exhibit a further relevant property:

$$\lim_{\lambda \to 1^-} \frac{\tilde{P}_{ij}(\lambda)}{\tilde{P}_{hk}(\lambda)} = \lim_{t \to \infty} \frac{P_{ij}(t)}{P_{hk}(t)} = \frac{z_j}{z_k} \quad \forall i,j,h,k. \qquad (2.140)$$

Remarkably, the type problem is invariant with respect to a wide class of dynamical and topological transformations; that is to say, it is independent of the graph "details". For example, any local bounded rescaling of nonnegative couplings and waiting probabilities leave the random walk type unchanged. Moreover, even the local topological details are irrelevant to determining the type of a graph: recurrence and transience are left invariant by adding and cutting of links satisfying the quasi-isometry conditions (for more details see [17] and references therein).

Finally, it is worth recalling that all results presented so far refer to random walks without traps. The introduction of at least one trap, leaves transience unchanged, while recurrent random walks always become transient.

2.3.8 The (Local) Spectral Dimension

As underlined above, recurrence and transience are large-scale properties of a graph. On regular (translation invariant) lattices, the asymptotic behavior of random walks (such as the probability of eventually returning to the starting point) is ultimately controlled by the lattice (Euclidean) dimension d. For example,

$$P_{ii}(t) \sim t^{-d/2} \text{ for } t \to \infty, \forall i, \tag{2.141}$$

$$F_{ii}(t) \sim \begin{cases} t^{\min(d/2-2,-d/2)} & \text{for } t \to \infty, \forall i, \text{ for } d \neq 2, \\ 1/(t \log^2 t) & \text{for } t \to \infty, \forall i, \text{ for } d = 2. \end{cases} \tag{2.142}$$

$$S_i(t) \sim \begin{cases} t^{\min(1,d/2)} & \text{for } t \to \infty, \forall i, \text{ for } d \neq 2, \\ t/\log t & \text{for } t \to \infty, \forall i, \text{ for } d = 2. \end{cases} \tag{2.143}$$

From these expressions, given the definition of transience and recurrence (see also equations (2.136) and (2.139)), one can derive that structures with $d \leq 2$ are recurrent, while structures with $d > 2$ are transient[14].

As we mentioned before, these scalings typically present power laws behavior even on general physical graphs, and the exponents of such powers can be used to define a generalized dimension. In analogy with Equation (2.141), let us suppose that for a random walk on $\mathcal{G}$ (devoid of traps) the return probability to the starting site i is

$$P_{ii}(t) \sim t^{-\tilde{d}/2} \quad \text{for } t \to \infty, \tag{2.144}$$

then, it can be shown that

$$P_{hk}(t) \sim t^{-\tilde{d}/2} \quad \text{for } t \to \infty, \forall h, k. \tag{2.145}$$

This means that the exponent of the power law is site independent and, therefore, it is a parameter characterizing the whole random walk and it can be considered as a dimension associated with the random walk on $\mathcal{G}$. More precisely, we shall call

---

[14] The fact that the leading terms of $F_{ii}(t)$ in Equation (2.142) for $d = 1$ and $d = 3$ are the same should not be misleading. In fact, short time behaviors, as well as multiplicative factors are different, ultimately leading to $F_{ii} = 1$ in $d = 1$ and $F_{ii} < 1$ in $d = 3$.

local spectral dimension[15] the limit

$$\tilde{d} = -2 \lim_{t \to \infty} \frac{\log P_{ii}(t)}{\log t}, \qquad (2.146)$$

when it exists. Note that the existence of this limit for a given site i implies it exists and has the same value for any $j \in \mathcal{G}$. Moreover, the definition given in (2.146) is more general than (2.144), since it includes the case of possible multiplicative corrections to the asymptotic behaviour, provided they are slower than any power law (e.g., logarithmic corrections).

Another way to define the local spectral dimension is by studying the behaviour of $\tilde{P}_{ii}(\lambda)$ as $\lambda \to 1$ (of course, this limit is strictly related to the asymptotic limit of $P_{ii}(t)$ as $t \to \infty$ as stressed in Section 2.2.7). In particular, one finds [51]

$$\tilde{P}_{ii}(\lambda) \sim (1-\lambda)^{\tilde{d}/2-1} [\log(1-\lambda)]^{I(\tilde{d}/2)} \quad \lambda \to 1, \qquad (2.147)$$

where $\sim$ denotes the behavior of the singular part around $\lambda = 1$ and $I(x)$ equals 1 if $x \in \mathbb{N}$, otherwise it is zero.

From now on, we shall consider random walks on graphs where $\tilde{d}$ is defined[16]. Then, one can easily derive a number of results, summarized hereafter [17].

- A graph[17] is recurrent if $\tilde{d} < 2$ and transient if $\tilde{d} > 2$; for $\tilde{d} = 2$, if (2.144) holds, the graph is recurrent. However, sub-leading corrections to the power law can change the type to transient.
- When (2.144) holds,

$$S_i(t) \sim t^{\min(1,\tilde{d}/2)} \text{ for } t \to \infty, \forall i, \text{ for } \tilde{d} \neq 2, \qquad (2.148)$$

otherwise, in general,

$$\lim_{t \to \infty} \frac{\log S_i(t)}{\log t} = \min(1,\tilde{d}/2) \forall i, \text{ for } \tilde{d} \neq 2, \qquad (2.149)$$

- When (2.144) holds,

$$F_{ij}(t) \sim t^{\min(\tilde{d}/2-2,-\tilde{d}/2)} \text{ for } t \to \infty, \forall i,j \text{ for } \tilde{d} \neq 2, \qquad (2.150)$$

---

[15] The adjective "local" is used to specify that the definition is given considering the local probability to reach a site, in contrast with the average probability (obtained by averaging over all sites) which provides the average spectral dimension (see Section 2.3.10). The adjective "spectral" has historical reasons since $\tilde{d}$ was thought to be connected to the vibrational dynamics on fractals; actually, it was later shown that the anomalous dimension involved in vibrational dynamics is the average spectral dimension (see Section 2.3.10).
[16] In any case, on all known cases of random walks on physical graphs, the local spectral dimension has been shown to exist. As for the existence of the limit (2.146), a general theorem is still lacking.
[17] The graph is not meant to display any strong inhomogeneities in such a way that local and average properties are analogous, see also Section 2.3.10.2.

otherwise, in general,

$$\lim_{t \to \infty} \frac{\log F_{ij}(t)}{\log t} = \min(\tilde{d}/2 - 2, -\tilde{d}/2), \forall i,j \text{ for } \tilde{d} \neq 2. \quad (2.151)$$

The expressions above generalize equations (2.142) and (2.143).

Remarkably, the asymptotic behaviour of $S_i(t)$ and $F_{ij}(t)$ have a different dependence on $\tilde{d}$ for $\tilde{d} < 2$ and $\tilde{d} > 2$; the case $\tilde{d} = 2$ is critical and discriminate recurrence from transience. Also, the case $\tilde{d} < 2$ is often referred to as compact exploration. We can understand this notion as follows: for a compact exploration, once a site is visited, the near vicinity will almost certainly be visited. For a non-compact exploration, one site can be visited while the neighboring one has a probability strictly smaller than one to be visited.

2.3.8.1 An Alternative Approach to Local Spectral Dimension

Many of the properties of random walkers can be described directly in terms of properties of the matrices $\mathbf{A}$ and $\mathbf{L}$; this is also the case for the local spectral dimension.

For instance, one can see that

$$\tilde{P}_{ii}(\lambda) = \sum_{t=0}^{\infty} \lambda^t P_{ii}(t) = \sum_{t=0}^{\infty} \lambda^t (\mathbf{Z}^{-1}\mathbf{A})_{ii}^t = (1 - \lambda \mathbf{Z}^{-1}\mathbf{A})_{ii}^{-1}$$
$$= [\mathbf{Z}\lambda^{-1}(\mathbf{L} + (1-\lambda)\lambda^{-1}\mathbf{Z})^{-1}]_{ii}, \quad (2.152)$$

and, posing $\mu = 1 - \lambda$, in the limit $\lambda \to 1^-$, we get

$$\lim_{\lambda \to 1^-} \tilde{P}_{ii}(\lambda) = z_i \lim_{\mu \to 0^+} (\mathbf{L} + \mu \mathbf{Z})_{ii}^{-1}. \quad (2.153)$$

Actually, one can show [53] that $\lim_{\mu \to 0^+}(\mathbf{L} + \mu \mathbf{Z})_{ii}^{-1}$ is finite if and only if $\lim_{\mu \to 0^+}(\mathbf{L} + \mu)_{ii}^{-1}$ is finite as well. Therefore, $\lim_{\mu \to 0^+}(\mathbf{L} + \mu)_{ii}^{-1}$ diverges for graph (locally) recurrent, while it is finite for graph (locally) transient[18]. Still exploiting the results in [53] and equation (2.152) we get that the definition of local spectral dimension (2.147) is equivalent to

$$(\mathbf{L} + \mu)_{ii}^{-1} \sim \mu^{\tilde{d}/2 - 1}[\log(\mu)]^{I(\tilde{d}/2)}, \mu \to 0^+, I(x) = 1 \text{ if } i \in \mathbb{N}, 0 \text{ otherwise}, \quad (2.154)$$

where, again, one can prove that the asymptotic behavior is independent of the site i. Equation (2.154) provides a definition of $\tilde{d}$ directly in terms of the Laplacian matrix.

---

[18] We specify *local* transience and *local* recurrence to distinguish from transience *on average* and recurrence *on average*, which shall be treated in Section 2.3.10.

Before concluding this section it is worth discussing briefly the main properties of the spectral dimension, which can be highlighted by exploiting either of the two alternative methods outlined above to determine $\tilde{d}$:

1. The spectral dimension does not change if we deform the graph over which it was defined.

    This property derives from the fact that $\tilde{d}$ is related to the adjacency matrix of the graph and it is therefore not affected by the whole set of deformations (twisting, folding), which leave **A** invariant, that is, which do not cut links or delete nodes.

2. The spectral dimension is affected only by the large-scale topology of the graph, and it is not influenced by details of short scale.

3. The spectral dimension of a graph does not depend on the particular point chosen to apply the definitions (2.147) and (2.154).

    In fact, the choice of the point i influences the asymptotic behaviors only via a multiplicative factor, while it does not alter the exponent $\tilde{d}$.

4. The spectral dimension of a lattice coincides with its Euclidean dimension.

5. The spectral dimension evaluated through the analysis of random walks is not altered by the introduction of any probability distribution of waiting times. The spectral dimension evaluated through the matrix analysis of the model does not change if we introduce multiplicative factors (i.e., if we change the mass distribution $\{m_i^2\}$, see the Gaussian model in Section 2.4.1.2). More precisely, the hypotheses to fulfill are that the constant (i.e. the masses) must be upper and lower bounded by positive constants and that the stopping probabilities must be smaller than 1.

6. For random walkers, the spectral dimension $\tilde{d} < 2$ denotes that the walker will eventually return to its starting point. Conversely, in the case $\tilde{d} > 2$ there is no certainty of return. The case $\tilde{d} = 2$ is critical, in fact, in that case recurrence can be destroyed by proper logarithmic corrections.

2.3.9 Averages in Infinite Graphs

In Section 2.4 we will deal with statistical-mechanics models (e.g., spin models, harmonic oscillators) embedded in infinite graphs; the investigation of such models crucially relies on averaged extensive quantities whose definition and behavior may be rather subtle. In fact, while in models defined on lattices, due to translation invariance, the local quantities are the same in any point and therefore equal to their average, on an arbitrary graph, in general, these quantities are site dependent and the average behavior cannot be reduced to the local behavior.

Usually, infinite graphs describing real systems are inhomogeneous, i.e., in mathematical terms, they are not invariant with respect to a transitive symmetry group. This means that the topology is seen in a different way from every site. The main effect of inhomogeneity is that the numerical values of physical quantities are site dependent. Therefore, one is typically interested in taking averages over all sites. This requires the introduction of suitable mathematical tools and definitions. First of all, the average in the thermodynamic limit $\bar{\phi}$ of a function ϕ_i defined on each site i of the infinite graph $\mathcal{G}$ is defined by

$$\bar{\phi} \equiv \lim_{r \to \infty} \frac{\sum_{i \in S_{o,r}} \phi_i}{N_{o,r}}, \qquad (2.155)$$

where we posed $N_{o,r} = |V_{o,r}|$. The measure $|S|$ of a subset $S \subseteq \mathcal{V}$ is the average $\overline{\chi(S)}$ of its characteristic function $\chi_i(S)$, being $\chi_i(S) = 1$ if $i \in S$ and $\chi_i(S) = 0$ is $i \notin S$. The normalized trace $\overline{\text{Tr}\mathbf{B}}$ of a matrix $\mathbf{B}$ is $\overline{\text{Tr}\mathbf{B}} \equiv \bar{b}_i$, with $b_i \equiv B_{ii}$.

An interesting property of the definition (2.155) is that, given a function ϕ which is lower-bounded, if its average $\bar{\phi}$ exists, then $\bar{\phi}$ is independent of the choice of the center o of the sequence of spheres. In fact, exploiting the fact that the coordination number is limited and that boundary conditions are negligible in the thermodynamic limit (see Section 2.3.6), one can get

$$\lim_{r \to \infty} \frac{\sum_{i \in S_{o',r-r_{o,o'}}} \phi_i}{N_{o',r-r_{o,o'}}} \leq \lim_{r \to \infty} \frac{\sum_{i \in S_{o,r}} \phi_i}{N_{o,r}} \leq \lim_{r \to \infty} \frac{\sum_{i \in S_{o',r+r_{o,o'}}} \phi_i}{N_{o',r+r_{o,o'}}}. \qquad (2.156)$$

Therefore, if there exists the limit for the spheres centered in point o', then the same result is expected by using any other center o.

It is worth stressing that the condition of negligible boundary conditions (i.e, $|\partial S_{o,r}|/N_{o,r} \to 0$) and the existence of the lower bound for ϕ_i are both necessary to prove the independence of the thermodynamic limit with respect to o. As a counterexample, let us consider a Bethe lattice with $z_i = 3, \forall i$, in such a way that $|\partial S_{o,r}| \sim N_{o,r}$. If we pose $\phi_i = 1$ for all points of one branch and $\phi_i = -1$ for all points of the other branch, we have that the average is null if we choose as center the root of the lattice, while it is positive (negative) if we choose a point in the former (latter) branch. On the other hand, let us consider a function ϕ non-limited from below; for instance, on a chain we choose $\phi = \exp(r_{i,p})$ for points on the right of p and $\phi = -\exp(r_{i,p})$ for points on the left of p. If we choose as origin the point p the thermodynamic limit is null, but if we take a point in the right (left) of p the limit is (minus) infinite.

2.3.10 Average Properties of Random Walkers

When studying statistical models on graphs we are interested in the thermodynamic behavior of average quantities and these can be related to the average behavior of random walks. For instance, as we will see in the next section, the spontaneous breaking of continuous symmetries, the critical exponents of the spherical model and the harmonic vibrational spectra are related to the large-scale geometry of the graph through the average behavior of random walkers at long and very long time scales. In particular, we define

$$\bar{P} = \lim_{\lambda \to 1^-} \overline{\tilde{P}(\lambda)} \equiv \lim_{\lambda \to 1^-} \overline{\mathrm{Tr}\tilde{P}(\lambda)} \qquad (2.157)$$

$$\bar{F} = \lim_{\lambda \to 1} \overline{\tilde{F}(\lambda)} \equiv \lim_{\lambda \to 1^-} \overline{\mathrm{Tr}\tilde{F}(\lambda)}. \qquad (2.158)$$

From these definitions we can properly generalize the concept of transience and recurrence: a graph $\mathcal{G}$ is said to be *recurrent on average* (ROA) if $\bar{F} = 1$ and *transient on average* (TOA) if $\bar{F} < 1$. Recurrence and transience on average are generally independent of the related local properties. For instance, the so-called NT_D graphs (see Figure 2.5) are locally transient but recurrent on average.

This "promiscuity" can be understood by noticing that, while for local properties the convolution relation $\tilde{P}_i(\lambda) = \tilde{F}_i(\lambda)\tilde{P}_i(\lambda) + 1$ holds, an analogous relation for (2.157) and (2.158) cannot be proved because the average of $\tilde{F}_i(\lambda)\tilde{P}_i(\lambda)$ over all sites is, in general, different from the product of the averages, due to correlations. Therefore, the double implication $\tilde{F}_i(1) = 1 \iff \lim_{\lambda \to 1} \tilde{P}_i(\lambda) = \infty$ cannot be extended to the average properties. Indeed, there exist graphs where $\bar{F} < 1$ but $\bar{P} = \infty$ (see [17]) and, in general, investigating the relation between $\bar{P}$ and $\bar{F}$ is by far a nontrivial problem.

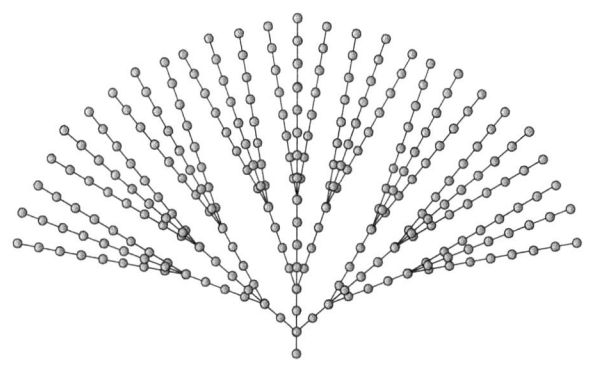

Figure 2.5. NT_D fractal trees can be defined recursively as follows: an origin point is connected to a point by a link of unitary length; to the latter point $z-1$ chains of length 2 are attached; the ends of these branches split into $z-1$ branches of length 4 and so on; each endpoint of a branch of length 2^g splits into $z-1$ branches of length 2^{g+1}. In this figure we show an example of NT_D of generation $g = 4$ and $z = 4$.

Now, in analogy with equation (2.147), the behaviour of $\overline{\tilde{P}(\lambda)}$ as $\lambda \to 1$ allows the defining of the *average spectral dimension* $\bar{d}$; more precisely

$$\overline{\tilde{P}(\lambda)} \sim (1-\lambda)^{\bar{d}/2-1}[\log(1-\lambda)]^{I(\bar{d}/2)} \quad \lambda \to 1. \qquad (2.159)$$

Of course, if $\bar{d} \leq 2$ then $\bar{P} = \infty$, while if $\bar{d} > 2$ then $\bar{P} < \infty$. However, in this case a dimension $\bar{d}$ smaller than 2 does not necessarily imply that the graph is recurrent on average.

The average spectral dimension, recurrence and transience on average also fulfill important universality criteria. In fact, these properties are invariant under a wide class of local topological transformations. For instance, one can show that the invariance holds under rescaling the jump probabilities, adding or cutting a set of zero measure of sites or links, the introduction of links among next-nearest-neighbors, and the introduction of long-range jump probabilities which decrease with the distance according to a power law with sufficiently large exponent (see [17] for more details). Therefore, the dimension $\bar{d}$ describes the global properties which depend exclusively on the large-scale geometry of the graph. As a consequence, if one proves that some characteristics of a model depend only on the dimension or on the recurrence on the average, then one immediately obtains that these properties are invariant for models where the interactions are locally rescaled or where impurities (i.e., perturbations of null measure) of next-nearest neighbors interaction are included.

Of course, on lattices $\bar{d} = \tilde{d} = d_g = d$; that is, all definitions of dimension given so far coincide with the Euclidean dimension. However, in general, due to inhomogeneity, $\bar{d}, \tilde{d}$ and d_g can be different from each other. For instance, on comb lattices one has $\bar{d} = 1, \tilde{d} = 3/2$ and $d_g = 2$.

2.3.10.1 Alternative Approaches to Average Spectral Dimension

Similarly to the local spectral dimension $\tilde{d}$, the average spectral dimension $\bar{d}$ allows for alternative definitions. In particular, the definition (2.159) is equivalent to

$$\overline{P(t)} \sim t^{\bar{d}-2}, \quad t \to \infty. \qquad (2.160)$$

In order to prove the previous equation we first need to prove that

$$\overline{\tilde{P}(\lambda)} = \sum_{t=0}^{\infty} \lambda^t \overline{P(t)}, \qquad (2.161)$$

which means that, for $\lambda < 1$, the thermodynamic average commutes with the sum over discrete times. In fact, one can write

$$\overline{\tilde{P}(\lambda)} = \lim_{r \to \infty} \sum_{i \in S_{o,r}} N_{o,r}^{-1} \left(\sum_{t=0}^{t'} \lambda^t P_{ii}(t) + \sum_{t=t'}^{\infty} \lambda^t P_{ii}(t) \right)$$

$$= \sum_{t=0}^{t'} \lambda^t \overline{P(t)} + \lim_{r \to \infty} \sum_{i \in S_{o,r}} N_{o,r}^{-1} \sum_{t=t'}^{\infty} \lambda^t P_{ii}(t). \quad (2.162)$$

Now, $\sum_{i \in S_{o,r}} N_{o,r}^{-1} \sum_{t=t'}^{\infty} \lambda^t P_{ii}(t) \leq \lambda^{t'}/(1-\lambda)$ and, by taking in (2.162) the limit $t' \to \infty$ we get (2.161). Of course, an analogous equation can be proven for $\overline{\tilde{F}(\lambda)}$.

An alternative definition based on $\mathbf{A}$ and $\mathbf{L}$ can be introduced also for $\bar{d}$. By averaging the expression in equation (2.152) over all sites we get

$$\lim_{\lambda \to 1^-} \overline{\tilde{P}(\lambda)} = \lim_{\mu \to 0^+} \overline{\text{Tr}[\mathbf{Z}(\mathbf{L}+\mu\mathbf{Z})^{-1}]} \quad (2.163)$$

and by exploiting the finite connectivity of $\mathcal{G}$ we get

$$\lim_{\mu \to 0^+} \overline{\text{Tr}[(\mathbf{L}+\mu\mathbf{Z})^{-1}]} \leq \lim_{\lambda \to 1^-} \overline{\tilde{P}(\lambda)} \leq z_{\max} \lim_{\mu \to 0^+} \overline{\text{Tr}[(\mathbf{L}+\mu\mathbf{Z})^{-1}]}. \quad (2.164)$$

Exploiting the properties of invariance under local rescaling [16], we get that the asymptotic behavior of $\overline{\text{Tr}}(\mathbf{L}+\mu\mathbf{Z})^{-1}$ is given by

$$\overline{\text{Tr}}(\mathbf{L}+\mu)^{-1} \sim \mu^{\bar{d}/2-1}[\log(\mu)]^{l(\bar{d}/2)}, \mu \to 0^+, \quad (2.165)$$

and, clearly, $\lim_{\mu \to 0^+} \overline{\text{Tr}}(\mathbf{L}+\mu)^{-1}$ is finite for graphs where $\bar{P}$ is finite, while it diverges for graphs where $\bar{P}$ diverges. Now, due to the invariance of $\bar{d}$ under local rescaling of the jump probabilities (2.123), equation (2.165) can be straightforwardly generalized to the case of $\mathbf{L}$ replaced by the generalized Laplacian.

The trace in (2.165) can be calculated in the base where the matrix $\mathbf{L}$ is diagonal and, calling $\rho(l)$ the normalized spectral density of $\mathbf{L}$ we get

$$\int_0^{l_{\max}} \frac{1}{l+\mu} \rho(l) dl \sim \mu^{\bar{d}/2-1}[\log(\mu)]^{l(\bar{d}/2)}, \mu \to 0^+. \quad (2.166)$$

Thus, it must be

$$\rho(l) \sim l^{\bar{d}/2-1}[\log(l)]^{l(\bar{d}/2)}, l \to 0^+. \quad (2.167)$$

In this way, we got a definition for $\bar{d}$ directly in terms of the spectral properties of the matrix $\mathbf{L}$. The average spectral dimension is therefore closely related to the density of infrared modes in the Laplacian matrix and these do not depend on the local details of the graph, but only on the topological properties at large scale.

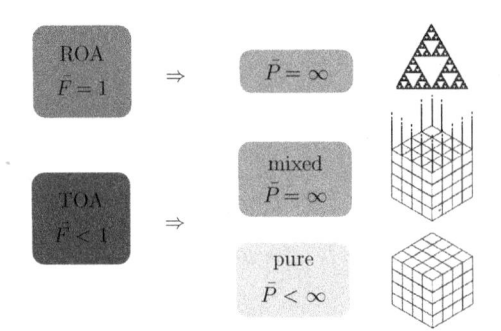

Figure 2.6. Schematic representation of average recurrence/transience properties.

2.3.10.2 Pure and Mixed Transience on Average

As already stressed, for average quantities there exists no simple equation connecting $\bar{P}$ and $\bar{F}$, similar to (2.127) for local quantities. Therefore, in order to provide a complete description able to account for the fact that $\bar{F} = 1 \not\Leftrightarrow \bar{P} = \infty$, it is necessary to classify graphs into *pure* transient (i.e., $\bar{F} < 1$ and $\bar{P} < \infty$) and *mixed* transient (i.e., $\bar{F} = 1$ and $\bar{P} < \infty$) on average (see Figure 2.6). Hereafter, just the key points will be sketched to lead to such a characterization, while we refer to [17] for more details.

First, one can prove that the relation between the local generating functions can be extended to the average case at least in one direction, namely $\bar{F} = 1 \Rightarrow \bar{P} = \infty$. In fact, if $\bar{F} = 1$ we have that for any $\delta > 0$ there exists an $\epsilon > 0$ such that if $1 - \epsilon \leq \lambda < 1$, then $1 - \delta \leq \overline{\tilde{F}(\lambda)} < 1$. Let us consider the subset $S \subseteq \mathcal{V}$ of sites i such that $\tilde{F}_i(1-\epsilon) < 1 - \sqrt{\delta}$ and let us call $\bar{S}$ its complementary set. One has [17]

$$\overline{\tilde{P}(\lambda)} \geq \overline{\tilde{P}(1-\epsilon)} \geq \overline{\chi(\bar{S})[1-\tilde{F}(1-\epsilon)]^{-1}} \geq |\bar{S}|\delta^{-1/2} \geq (1-\sqrt{\delta})\delta^{-1/2}. \quad (2.168)$$

Therefore, for any value, arbitrary large, of $(1-\sqrt{\delta})\delta^{-1/2}(\delta \to \infty)$, there exists a value of ϵ such that for any $\lambda \geq 1 - \epsilon$ we have $\overline{\tilde{P}(\lambda)} \geq (1-\sqrt{\delta})\delta^{-1/2}$ and, as a consequence, $\bar{P} = \lim_{\lambda \to 1} \overline{\tilde{P}(\lambda)} = \infty$. In this way, we have proven that for ROA graphs $\bar{P} = \infty$ or, equivalently, $\lim_{\mu \to 0^+} \overline{\text{Tr}}(\mathbf{L} + \mu)^{-1} = \infty$.

Then, one can define mixed TOA graphs as those graphs where there exists a subset $S \subseteq \mathcal{V}, |S| > 0$ such that $\lim_{\lambda \to 1} \chi(S)\bar{F}(\lambda) = |S|$. On the other hand, pure TOA graphs are those graphs where $\lim_{\lambda \to 1} \chi(S)\bar{F}(\lambda) < |S|$ for any subset $S \subseteq \mathcal{V}, |S| > 0$; in this case we have $\bar{P} < \infty$.

Now, the proof reported above can be extended to any structure where there exists a subset S of non-null measure such that $\lim_{\lambda \to 1} \chi(S)\tilde{F}(\lambda) = |S|$, even if $\bar{F} < 1$. In this case, analogously to what is done before, we can show that

$$\bar{P} \geq \lim_{\lambda \to 1} \overline{\chi(S')\tilde{P}(\lambda)} = \infty \ \forall S' \subseteq S, |S'| > 0. \quad (2.169)$$

Therefore, for mixed TOA we have $\bar{F} < 1$, but $\bar{P} = \infty$ and $\lim_{\mu \to 0^+} \overline{\text{Tr}}$ $(\mathbf{L} + \mu)^{-1} = \infty$. As for pure TOA graphs, analogous proofs show that $\bar{P} < \infty$ and therefore $\lim_{\mu \to 0^+} \overline{\text{Tr}}(\mathbf{L} + \mu)^{-1} < \infty$ (see [17] for an extensive treatment).

Remarkably, from the definition of mixed TOA there follows an important property which allows simplifying the study of statistical models on such particularly inhomogeneous structures. More precisely, by cutting a null-measure set of links it is always possible to divide the graph $\mathcal{G}$ into two subgraphs, one being pure TOA and the other being ROA; moreover, the jump probabilities are defined independently in the two subgraphs. The separability (or more precisely, the fact that it is sufficient to cut a null-measure set) implies that on the two subgraphs the thermodynamic properties of the statistical models are independent and can therefore be studied separately. In fact, in this case the partition functions can be factorized [17].

Before concluding, some remarks are in order.

As we have previously seen, both the spectral dimension and the average spectral dimension can be derived from the analysis of random walks, yet they are in principle different. In order to understand such a difference, let us discuss a simple example. Let us consider a comb lattice (see Figure 2.7) and let us set on each of its points a random walker which we allow to move diffusively. We fix a time t in which each walker must explore its neighborhood and return to its starting point. Now, walkers that started from points belonging to the backbone or points of teeth at a distance from the backbone smaller than $t/2$ have had the possibility to meet one or more junctions and therefore have realized that they are on a branched structure. The remaining class of walkers are those that started on points of teeth at a distance from the backbone larger than $t/2$. These will not have had any chance to meet a junction and therefore they are sure to be on a chain. It is easy to see that, whatever t, most of the walkers will belong to the second class as teeth have an infinite length. Therefore, on average, the behavior of walkers is the one we would expect if walkers were on a linear chain. We stress that the

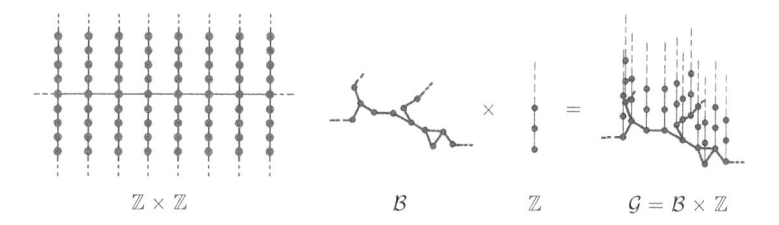

Figure 2.7. Left panel: simple 2-dimensional comb obtained by taking as base a linear chain $\mathbb{Z}$ and attaching a linear chain $\mathbb{Z}$ to any site of the base. Right panel: generic comb obtained by attaching to any site of the arbitrary base $\mathcal{B}$ a linear chain $\mathbb{Z}$.

concept of average includes a limit as the number of points of the graph is meant as infinite.

Once seen that walkers, after an exploration of t steps, think, in the average, to be on a linear chain, we can take the limit $t \to \infty$. This limit provides the asymptotic behavior of random walkers on a linear chain which, we know, is governed by a spectral dimension equal to 1. In this context this value is just the average spectral dimension of the comb lattice and it is different than the local spectral dimension.

We stress that the limits $N \to \infty$ and $t \to \infty$ do not commute: if we invert the order we would assign to each walker an infinite time to return to the origin and, as we saw, under this condition each walker realizes they are on a comb and therefore the spectral dimension resulting for each walker would be 3/2. At this point, averaging over all N points ($N \to \infty$) would be irrelevant, as each walker would see the same things in its trip.

Therefore, on comb lattices the average spectral dimension is $\bar{d} = 1$, while the local spectral dimension is $\tilde{d} = 3/2$.

In general, for exactly decimable fractals one can prove that $\tilde{d} = \bar{d} < 2$.

2.3.11 Non-uniqueness of the Thermodynamic Limit in Inhomogeneous Graphs

As explained above, inhomogeneous structures may display a nontrivial splitting between local and average properties. This crucial point lies in the non-commutability between the limits $t \to \infty$ and $N \to \infty$. In fact, on (statistically) homogeneous networks, i.e., networks where all points are statistically equivalent as far as large scale properties are concerned, the results obtained in the thermodynamic limit are unambiguous, and differ from the exact results for finite structures only for small quantities that vanish with increasing number of sites. On the other hand, when considering strongly inhomogeneous structures, it can happen that two (or more) points, with a topologically different neighborhood, become more and more distant from each other when increasing the network size, and their neighborhoods evolve to two deeply different infinite graphs in the thermodynamic limit. In such cases, the thermodynamic limit is not unique and it depends on the specific point chosen as the origin of the intrinsic coordinate system. Of course, when this happens, the choice of the starting point gives rise to a completely different time evolution of random walks [89].

In the following we review some basic examples of structures where the thermodynamic limit is non-unique. Of course, starting from these examples one may figure out several variations on a theme exhibiting analogous ambiguities.

Probably, the easiest example of structure where the thermodynamic limit can not be uniquely defined is the semi-infinite chain: here, nodes belonging to the "bulk" of the structure will see a long (eventually infinite) chain on both sides, while nodes finitely close to the origin will see a long (eventually infinite) chain on one side but the other side will remain finite and unchanged.

When considering, for instance, the mean time τ to first return to the starting point (when starting from a bulk node) we get τ_{bulk}, which, in the thermodynamic limit, recovers the result expected for an infinite chain. On the other hand, when starting from a peripheral node, say the one at distance 1 from the origin, we get $\tau_{\text{periph}} \sim \tau_{\text{bulk}}/2$. This example can be easily generalized to the case of semi-infinite Euclidean lattices. The T-fractal (compare the central node with any node farthest from the center) and the Sierpinski gasket (compare the apex with the node in the middle of any of the three main sides) also work like that.

In general, such structures display a "soft" inhomogeneity, which, once the thermodynamic limit is taken, may affect quantitatively the main physical quantities, although their asymptotic laws are preserved. When inhomogeneity is strong, the difference between the asymptotic behaviors cannot be absorbed into a simple factor. For instance, this is the case for the hierarchical comb, namely a comb displaying side branches with a hierarchical distribution of lengths (see Figure 2.8), that is used as a model to mimic the organization of percolation clusters and ultrametricity in spin glasses and related systems [79]. Interestingly, the moments of the first-passage time of random walks on such structures scale with the size of the system with an infinite hierarchy of exponents, characterizing a multifractal behavior [54]. Moreover, it was found that a transition occurs from

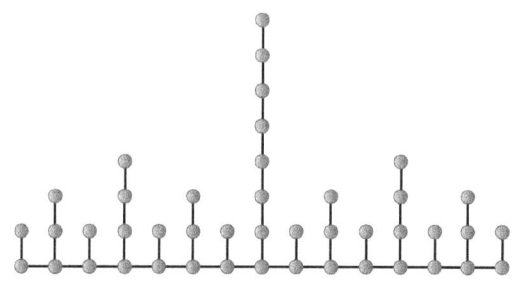

Figure 2.8. Hierarchical combs can be built by starting from a chain of length $2^{g+1} - 1$, which plays as a backbone. Then, we attach to its middle node a chain of length $L_g = R^g$; in both half-sides of the backbone, we can outline a middle point, and we attach a chain of length R^{g-1} to both. Now, the backbone is divided into four "bare" segments, we detect the related middle points and we attach a chain of length R^{g-2} to each of them. We proceed recursively, up to the exhaustion of all bare nodes of the backbone. In the last step there will be 2^g nodes and these will be connected to a chain of length R^0, i.e., to a single node. This figure shows an example of hierarchical comb with $R = 2$ and $g = 3$.

ordinary to anomalous diffusion at $R = 2$ [59]. The thermodynamic limit of this structure is non-unique: according to the node where we sit, we may see a semi-infinite chain, or an infinite chain, or a comb with teeth of different (possibly infinite) length.

Other examples are provided by the NT_D and by the Cayley tree (and its extended versions) (see Figures 2.5 and 2.9 respectively, and [70]).

2.3.12 The Two-particle Problem

In the previous sections we have shown that topological inhomogeneity can qualitatively affect the behavior of dynamic processes. Here, we present another nontrivial effect induced by the underlying inhomogeneous structure and concerning the problem of the encounter between two simple random walkers.

We recall that, in homogeneous structures, the two-particle problem (i.e., the problem of finding out how likely it is that two particles eventually meet) can be mapped into a one-particle problem (i.e., the problem of finding out how likely it is that one particle eventually reaches a given fixed target). In general, this map does not hold in inhomogeneous structures, where the two problems are not only intrinsically distinct but, also, their solutions can be strikingly different [58, 32].

It is easy to see that on a line the probability of encounter and the probability of first encounter have the same asymptotic behavior as the probability of return to the origin and of first return for a single random walker on a line. The same result is valid for all Euclidean lattices. Indeed in a d-dimensional Euclidean lattice we can consider only the relative distances along the axes: $D_1, D_2, \ldots, D_d$. The mapping is therefore a d-dimensional Euclidean lattice where we are interested in the return probability to the origin: $D_1 = D_2 = \ldots = D_d = 0$. Thus in these lattices the probability of encounter for two particles has the same asymptotic behavior as the

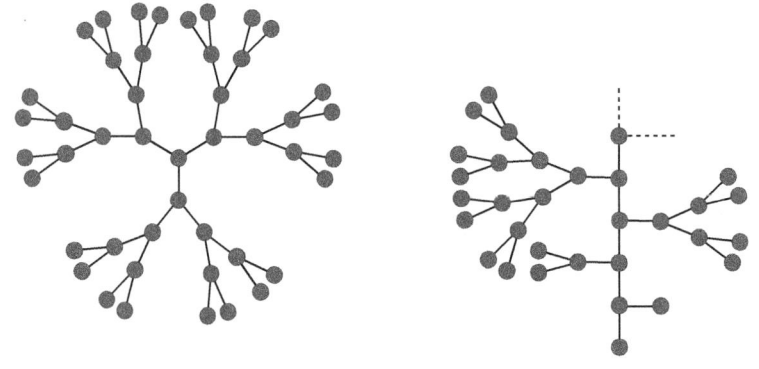

Figure 2.9. Cayley tree of generation $g = 4$ and $z = 3$. The topological perspective from the central node (left panel) and from a rim (right panel) are compared.

probability of return to the origin for a single walker and, similarly, if the mean number of returns to the origin is infinite (i.e. the graph is recurrent), then even the mean number of encounters is infinite. Interestingly, Peres and Krishnapur [58] presented a class of graphs where the simple random walk is recurrent, yet two walkers meet only finitely many times almost surely; this result has been rigorously proven for infinite combs (where the base is an arbitrary recurrent graph, see Figure 2.7) [58, 32]. If two independent simple random walks on a graph starting from the same vertex meet only finitely many times almost surely, the graph is said to exhibit the finite collision property.

The finite collision property was proven to be equivalent to the fact that two random walkers have a finite probability of never meeting [33]. The latter property is referred to as *two-particle transience* and, similarly, one can denote structures where two random walkers certainly meet as *two-particle recurrent*. As underlined above, in general, the (one-particle) recurrence can be compatible with the finite collision property; therefore, the (one-particle) recurrence does not imply the two-particle recurrence. Let us now formalize these concepts.

When considering two independent walkers, starting at time 0 from the vertices v and w, one can compute the joint probability of motion to, respectively, v' and w' in t steps as

$$\mathcal{P}_{(vw) \to (v'w')}(t) = P_{(vv')}(t) P_{(ww')}(t). \quad (2.170)$$

If we let the final position be the same (i.e., $v' = w'$), we obtain the probability that the two walkers meet at time t. In turn, it is possible to link the latter quantity to the probability that the first encounter between the walkers happens at time t at vertex v', which we denote as $\mathcal{F}_{(vw) \to v'}(t)$. If the agents meet at t in v', either it is their first encounter, or they have already clashed somewhere else at an earlier time, so that the corresponding probability can be written as

$$\mathcal{P}_{(vw) \to (v'w')}(t) = \sum_{t'=1}^{t} \sum_{l \in \mathcal{V}} \mathcal{F}_{(vw) \to v'}(t') \mathcal{P}_{(ll) \to (v'v')}(t - t') + \delta(t) \delta_{vv'} \delta_{wv'}. \quad (2.171)$$

Passing to generating functions:

$$\tilde{\mathcal{P}}_{(vw) \to (v'w')}(\lambda) = \sum_{l \in \mathcal{V}} \tilde{\mathcal{F}}_{(vw) \to l}(\lambda) \tilde{\mathcal{P}}_{(ll) \to (v'v')}(\lambda) + \delta_{vv'} \delta_{wv'}. \quad (2.172)$$

Now, we can give the following mathematical definition: a graph is two-particle recurrent if the probability that two particles will ever meet is 1, i.e.,

$$\sum_{t=0}^{\infty} \sum_{v' \in \mathcal{V}} \mathcal{F}_{(vw) \to v'}(t) = \sum_{v' \in \mathcal{V}} \tilde{\mathcal{F}}_{(vw) \to v'}(1) = 1, \quad (2.173)$$

for all (vw). Should the graph not satisfy this condition, we term it two-particle transient.

By definition, transience and recurrence (either one-particle or two-particle) are concepts living in the thermodynamic limit. However, real phenomena necessarily occur in *finite* structures, and one may wonder whether and, if so, to what extent, similar effects may take place in finite structures.

Indeed, in finite structures, at intermediate times (i.e., times long enough to see the emergence of asymptotic behaviors, but not so long that the random walk realizes the finiteness of the substrate), the two-particle problem does exhibit nontrivial features [4, 30]. In particular, the encounter experiences a qualitative slowing down with respect to the mean time to first reach a fixed point. This is expected to have many interesting applications. For instance, in pharmacokinetics, drugs will affect mobile and static targets differently; chemical reactions are favoured when either of the reagents is immobilized; in prey-predator models, the prey is more likely to survive if it keeps moving.

2.3.13 Other First-passage Quantities

We will now introduce some supplementary first-passage quantities to offer a wider overview of what a first-passage observable could be. It is important to note that all these quantities are intrinsically non-Markovian.

Up to now we have always talked about first-passage, but for a random walker performing jumps in a continuous space, two quantities can be defined: the *first-passage time* and the *first-arrival time*. The former is the first time the random walk trajectory crosses the target, and this could occur during a jump, the latter is the first time the random walker hits the target. By definition, the first arrival time is always larger than (or equal to) the first-passage time.

The *first-exit time* is defined when the random walker starts in a given subset of space. As shown in Figure 2.10, the first time the random walker moves outside this subset is the first exit time. It is equivalent to a first-passage time if we define the subset boundary as the target.

The *maximal excursion* is a related concept: $r_{\max}(t)$ is the maximal distance from the origin the random walker has ever reached at time t. If we denote with $\mathbf{r}(t)$ the random walker position at time t, we can define this maximal excursion as $r_{\max}(t) = \max_{0 \leq t' \leq t} ||\mathbf{r}(t') - \mathbf{r}(t=0)||$. This maximal excursion is a growing function of time, and is closely related to the first-exit time. If the first-exit time of a sphere centered on $\mathbf{r}(t=0)$ and of radius r_0 is t, then $r_{\max}(t) = r_0$. The first-exit time focuses on the time needed to exit a given subset while the maximal excursion focuses on the maximal distance reached at a given time.

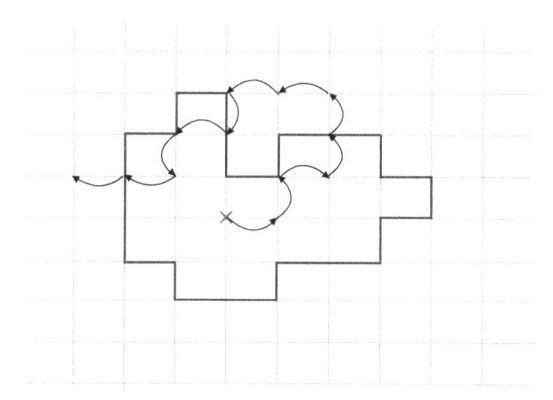

Figure 2.10. Example of a random walker embedded in a two-dimensional lattice in the presence of a boundary denoted with a bold line. Here the walker is started at a node denoted with a cross and it takes 2 time units before encountering for the first time the domain boundary. For this example the maximal excursion $r_{max}(t)$ grows linearly with time up to 5 at time $t = 5$ and then remains constant.

Finally, we define the *occupation time*, as the number of times N_i a random walker visits a given site *i* before reaching a given target site. This quantity is useful in the context of reactions occurring with a finite probability per unit of time.

An example of application of these quantities to data analysis in the context of social sciences can be found in [3].

2.3.14 Quantum Walks

In this final subsection devoted to random walks on graphs, we address the quantum analogue of classical random walks, i.e., quantum walks. These constitute an advanced tool for building quantum algorithms and for modeling the coherent transport of mass, charge or energy and have been diffusively treated in e.g., [56, 67, 86]. Whatever the context considered, the quantum "agent" is moving in a potential which specifies the Hamiltonian of the system, which, in turn, determines the time evolution. For instance, the dynamics of electrons in a simple crystal is described by the Bloch ansatz, which mirrors their behavior in metals quite accurately; more generally, the tight-binding approximation used in solid state physics as well as in Hückel's theory is equivalent to quantum walks. In quantum-information theory the most prominent examples of algorithms which can be efficiently implemented in terms of quantum walks are the Shor's algorithm (devised to find the prime factors of an integer number) [83] and Grover's algorithm (devised to find an item in an (unsorted) database of qubits) [47, 34, 7].

Now, the extension of classical random walks to the quantum domain is not unique and allows variants, of which two types capture most of the interest:

discrete-time quantum walks (introduced by Aharonov et al., using an additional internal "coin" degree of freedom [8]), and continuous-time quantum walks (CTQW) (introduced by Farhi and Gutmann, where the connection to CTRW uses the analogy between the quantum mechanical Hamiltonian and the classical transfer matrix [43]).

For both quantum walk variants, experimental implementations have been proposed, based on microwave cavities, on Rydberg atoms or on ground state atoms in optical lattices or in optical cavities, or using the orbital angular momentum of photons, just to cite a few (see [67] and references therein).

2.3.14.1 Definitions

Given a graph $\mathcal{G}$, algebraically described by the adjacency matrix **A** and by the Laplacian matrix **L**, the classical continuous-time random walk (CTRW) moving in $\mathcal{G}$ is described by the following Master equation [88]:

$$\frac{d}{dt}P_{k,j}(t) = \sum_{l=1}^{N} T_{kl} P_{l,j}(t), \qquad (2.174)$$

where $P_{k,j}(t)$ is the conditional probability that the walker is on node k when it started from node j at time 0. If the walk is symmetric with a site-independent transmission rate γ, then the transfer matrix **T** is simply related to the Laplacian operator through $\mathbf{T} = -\gamma \mathbf{L}$.

The CTQW, the quantum-mechanical counterpart of the CTRW, is introduced by identifying the Hamiltonian of the system with the classical transfer matrix, $\mathbf{H} = -\mathbf{T}$ [67] (in the following we will set $\hbar \equiv 1$). The set of states $|j\rangle$, representing the walker localized at the node j, spans the whole accessible Hilbert space and also provides an orthonormal basis set. Therefore, the behavior of the walker can be described by the transition amplitude $\alpha_{k,j}(t)$ from state $|j\rangle$ to state $|k\rangle$, which obeys the following Schrödinger equation:

$$\frac{d}{dt}\alpha_{k,j}(t) = -i \sum_{l=1}^{N} H_{kl} \alpha_{l,j}(t). \qquad (2.175)$$

If at the initial time $t_0 = 0$ only the state $|j\rangle$ is populated, then the formal solution to equation (2.175) can be written as

$$\alpha_{k,j}(t) = \langle k | \exp(-i\mathbf{H}t) | j \rangle, \qquad (2.176)$$

whose squared magnitude provides the quantum mechanical transition probability $\pi_{k,j}(t) \equiv |\alpha_{k,j}(t)|^2$. In general, it is convenient to introduce the orthonormal basis $|\psi_n\rangle, n \in [1, N]$ which diagonalizes **T** (and, clearly, also **H**); the correspondent set of eigenvalues is denoted by $\{\lambda_n\}_{n=1,\ldots,N}$. Thus, we can write

$$\pi_{k,j}(t) = \left| \sum_{n=1}^{N} \langle k|e^{-i\lambda_n t}|\psi_n\rangle\langle\psi_n|j\rangle \right|^2. \tag{2.177}$$

Despite the apparent similarity between equation (2.174) and (2.175), some important differences are worth being recalled. First of all, the imaginary unit makes the time evolution operator $\mathbf{U}(t) = \exp(-i\mathbf{H}t)$ unitary, which prevents the quantum mechanical transition probability from having a definite limit[19] as $t \to \infty$. On the other hand, a particle performing a CTRW is asymptotically equally likely to be found on any site of the structure: the classical $P_{k,j}(t)$ admits a stationary distribution which is independent of initial and final sites, $\lim_{t\to\infty} P_{k,j}(t) = 1/N$. Moreover, the normalization conditions for $P_{k,j}(t)$ and $\alpha_{k,j}(t)$ read respectively $\sum_{k=1}^{N} P_{k,j}(t) = 1$ and $\sum_{k=1}^{N} |\alpha_{k,j}(t)|^2 = 1$.

2.3.14.2 Average Displacement

The average displacement performed by a quantum walker until time t allows a straightforward comparison with the classical case; it is also more directly related to transport properties than the transfer probability $\pi_{k,j}(t)$. It constitutes the expectation value of the distance reached by the particle after a time t and its time dependence provides information on how fast the particle propagates over the substrate.

For CTQW (subscript q) starting at node j, we define the *average (chemical) displacement* $\langle r_j(t)\rangle_q$ performed until time t as

$$\langle r_j(t)\rangle_q = \sum_{k=1}^{N} r_{kj}\pi_{k,j}(t), \tag{2.178}$$

where r_{kj} is the chemical distance between the sites j and k, i.e., the length of the shortest path connecting j and k. We can average over all starting points to obtain

$$\overline{\langle r(t)\rangle}_q = \frac{1}{N}\sum_{j=1}^{N}\langle r_j(t)\rangle_q. \tag{2.179}$$

For classical (subscript c) regular diffusion (on infinite lattices) the average displacement $\langle r(t)\rangle_c$ depends on time t according to

$$\langle r(t)\rangle_c \sim t^{2/d_w}, \tag{2.180}$$

---

[19] In order to compare classical long time probabilities with quantum mechanical ones, one often relies on the so-called long time averages [8].

where d_w is the walk dimension of the underlying structure resulting as $d_w = 2d_f/\tilde{d}$; for regular diffusion $d_w = 2$, but in fractals $d_w \neq 2$ and one talks of anomalous diffusion.

The behavior of $\langle r(t) \rangle_q$ and $\langle r(t) \rangle_c$ has been shown to differ qualitatively (e.g., see [6]). In particular, in homogeneous structures, at short times, quantum walks appear to be more effective (i.e. they spread faster), while this is no longer true for inhomogeneous structures and/or at long time scales.

2.3.14.3 Return Probability

As it is well known, for a diffusive particle the probability of returning to the starting point is topology sensitive, and it can indeed be used to extract information about the underlying structure [88]. It is therefore interesting to compare the classical return probability $P_{k,k}(t)$ with the quantum mechanical $\pi_{k,k}(t)$ (see also [67]). One has

$$P_{k,k}(t) = \langle k | \exp(\mathbf{T}t) | k \rangle = \sum_{n=1}^{N} |\langle k | \psi_n \rangle|^2 \exp(-\gamma t \lambda_n) \qquad (2.181)$$

and

$$\pi_{k,k}(t) = |\alpha_{k,k}(t)|^2 = \left| \sum_{n=1}^{N} |\langle k | \psi_n \rangle|^2 \exp(-i\gamma t \lambda_n) \right|^2. \qquad (2.182)$$

In order to get global information about the likelihood to be (return or stay) at the origin, independent of the starting site, we average over all sites of the graph, obtaining

$$\bar{p}(t) = \frac{1}{N} \sum_{k=1}^{N} P_{k,k}(t) = \frac{1}{N} \sum_{n=1}^{N} e^{-\gamma \lambda_n t} \qquad (2.183)$$

and

$$\bar{\pi}(t) = \frac{1}{N} \sum_{k=1}^{N} \pi_{k,k}(t) = \frac{1}{N} \sum_{n,m=1}^{N} e^{-i\gamma(\lambda_n - \lambda_m)t} \sum_{k=1}^{N} |\langle k | \psi_n \rangle|^2 |\langle k | \psi_m \rangle|^2. \qquad (2.184)$$

For finite substrates, the classical $\bar{p}(t)$ decays monotonically to the equipartition limit, and it only depends on the eigenvalues of $\mathbf{T}$. On the other hand, $\bar{\pi}(t)$ depends explicitly on the eigenvectors of $\mathbf{H}$ [67]. By means of the Cauchy-Schwarz inequality we can obtain a lower bound for $\bar{\pi}(t)$ which does not depend on the eigenvectors [67]:

$$\bar{\pi}(t) \geq \left| \frac{1}{N} \sum_{k=1}^{N} \alpha_{k,k}(t) \right| \equiv |\bar{\alpha}(t)|^2 = \frac{1}{N^2} \sum_{m,n=1}^{N} e^{-i\gamma(\lambda_n - \lambda_m)t}. \qquad (2.185)$$

Notice that equations (2.183) and (2.184) can serve as measures of the efficiency of the transport process performed by CTRW and CTQW, respectively. In fact, the faster $\bar{p}(t)$ decreases towards its asymptotic value, the more efficient the transport. Analogously, a more rapid decay of the envelope of $\bar{\pi}(t)$ (or of $|\bar{\alpha}(t)|^2$) implies a faster delocalization of the quantum walker over the graph. Also, for quantum transport processes the degeneracy of the eigenvalues plays an important role, as the differences between eigenvalues determine the temporal behavior, while for classical transport the long time behavior is dominated by the smallest eigenvalue. Situations in which only a few, highly degenerate eigenvalues are present are related to slow CTQW dynamics, while when all eigenvalues are non-degenerate the transport turns out to be efficient [67].

For extensions to systems with long-range interactions and systems with disorder and localization we refer again to [67].

2.3.14.4 Systems with Absorption

In general, an excitation does not stay forever in the system in which it was created; the excitation either decays (radiatively or by exciton recombination) or, e.g., in the case of biological light-harvesting systems, it gets absorbed at the reaction center, where it is transformed into chemical energy. In such cases, the total probability of finding the excitation within the network is not conserved. Such loss processes can be modeled phenomenologically by changing the transfer matrix or the Hamiltonian. To fix the ideas, we consider networks in which the excitation can only vanish at certain nodes. These nodes will be called trap-nodes or traps. In the absence of traps, let the transfer-matrix and the Hamiltonian of the corresponding network be T_0 and H_0, respectively. Take now M out of the N total nodes to be traps and denote them by m, so that $m \in \mathcal{M}$, with $\mathcal{M} \subset \mathcal{V}$. The trapping process is now modeled by introducing a trapping matrix Γ which is given by a sum over all trap nodes; Γ has only diagonal elements, i.e.,

$$\Gamma \equiv \sum_m \Gamma_m |m\rangle\langle m|, \qquad (2.186)$$

(in the following we assume that $\Gamma_m = \Gamma > 0, \forall m$). CTRWs with decreasing exciton probabilities due to trapping are well described through the following transfer matrix:

$$\mathbf{T} \equiv \mathbf{T_0} - \mathbf{\Gamma}. \qquad (2.187)$$

The total Hamiltonian **H** corresponding to trapping is then:

$$\mathbf{H} \equiv \mathbf{H_0} - i\mathbf{\Gamma}. \qquad (2.188)$$

Note that the connection between CTRW and CTQW is now less direct than before. For CTRW the term corresponding to trapping has only real elements and the total transfer matrix stays real. For CTQW, however, the trapping term has purely imaginary elements. As a result, **H** is non-Hermitian and has N complex eigenvalues, $E_l = \epsilon_l - i\gamma_l (l = 1,\ldots,N)$. In general, **H** has N left and N right eigenstates $|\Phi_l\rangle$ and $\langle\tilde{\Phi}_l|$ respectively. It turns out that in most cases the eigenstates of **H** form a complete and biorthonormal set

$$\sum_{l=1}^{N} |\Phi_l\rangle\langle\tilde{\Phi}_l| = 1 \text{ and } |\tilde{\Phi}_l\rangle\langle\Phi_{l'}| = \delta_{ll'}. \quad (2.189)$$

Both eigenvalues and eigenstates will be different for CTRW and CTQW because the incorporation of the trapping process is different. If the trapping strength Γ is small compared to the couplings between neighboring nodes, perturbation theory allows relating the real part of the eigenvalues to the eigenvalues of the unperturbed Hamiltonian H_0. Let $|\Phi_l^{(0)}\rangle$ be the l-th eigenstate and $E_l^{(0)} \in \mathbb{R}$ be the l-th eigenvalue of the unperturbed system with Hamiltonian H_0. Up to first-order, the eigenvalues of the perturbed system are given by [67]

$$E_l = E_l^{(0)} + E_l^{(1)} = E_l^{(0)} - i\Gamma \sum_{m \in \mathcal{M}} \left|\langle m|\Psi_l^{(0)}\rangle\right|^2. \quad (2.190)$$

Therefore, the correction term determines the imaginary parts γ_l, while the unperturbed eigenvalues are the real parts $\epsilon_l = E_l^{(0)}$. Moreover, the imaginary parts are solely determined by the contribution of the eigenstates of the network without traps at the trap nodes m.

In an ideal experiment one would excite exactly one node, say $j \notin \mathcal{M}$, and read out the outcome $\pi_{kj}(t)$, i.e., the probability of being at node $k \notin \mathcal{M}$ at time t. However, it is easier to keep track of the total outcome at all nodes $k \notin \mathcal{M}$, namely of $\sum_{k \notin \mathcal{M}} \pi_{kj}(t)$. Since the states $|k\rangle$ form a complete, orthonormal basis set, one has $\sum_{k \notin \mathcal{M}} |k\rangle\langle k| = \mathbf{1} - \sum_{k \in \mathcal{M}} |k\rangle\langle k|$. By averaging over all $j \notin \mathcal{M}$, the mean survival probability is given by

$$\Pi_M \equiv \frac{1}{N-M} \sum_{j \notin \mathcal{M}} \sum_{k \notin \mathcal{M}} \pi_{kj}(t)$$

$$= \frac{1}{N-M} \sum_{l=1}^{N} e^{-2\gamma_l t}\left(1 - 2\sum_{m \in \mathcal{M}} \langle\tilde{\Phi}_l|m\rangle\langle m|\Phi_l\rangle\right)$$

$$+ \frac{1}{N-M} \sum_{l,l'=1}^{N} e^{-i(E_l - E_{l'}^*)}\left(\sum_{m \in \mathcal{M}} \langle\tilde{\Phi}_{l'}|m\rangle\langle m|\Phi_l\rangle\right)^2. \quad (2.191)$$

At long times the oscillating term on the right hand side of the previous expression drops out and for small $M \ll N$ one has $2\sum_{m \in \mathcal{M}} \langle\tilde{\Phi}_l|m\rangle\langle m|\Phi_l\rangle \ll 1$. Thus, $\Pi_M(t)$

can be approximated by a sum of exponentially decaying terms:

$$\Pi_M \approx \frac{1}{N-M} \sum_{l=1}^{N} e^{-2\gamma_l t}, \qquad (2.192)$$

and is dominated asymptotically by the smallest γ_l values. If the smallest one, $\gamma_{\min}$, is well separated from the other values, for $t \gg 1/\gamma_{\min}$ one has $\Pi(t) \approx \exp(-2\gamma_{\min})$. Such long times are not of much experimental relevance (see also below), since most measurements highlight shorter times, at which many γ_l contribute. In the corresponding energy range the γ_l often scale, so that in a large l range one finds $\gamma_l \sim a l^\mu$. The prefactor a depends only on Γ and N. For densely distributed γ_l and at intermediate times one has, from equation (2.192),

$$\Pi_M(t) \approx \int dx e^{-2atx^\mu} = \int dy \frac{e^{-y^\mu}}{(2at)^{-1/\mu}} \sim t^{-1/\mu}. \qquad (2.193)$$

Analogously, the mean survival probability for CTRW is given by

$$P_M(t) \equiv \sum_{j \notin \mathcal{M}} \sum_{k \notin \mathcal{M}} p_{kj}(t) = \frac{1}{N-M} \sum_{l=1}^{N} e^{-\lambda_l t} \left| \sum_{k \notin \mathcal{M}} \langle k | \phi_l \rangle \right|^2. \qquad (2.194)$$

If the smallest eigenvalue, λ_1, is well separated from the rest, $P_M(t)$ turns very quickly into a simple exponential decay. Then, for not too small times, it can be shown that [67]

$$P_M(t) \approx \frac{1}{N-M} e^{-\lambda_1 t} \left| \sum_{k \notin \mathcal{M}} \langle k | \phi_1 \rangle \right|^2. \qquad (2.195)$$

2.4 A Brief Introduction to Statistical Mechanics Models on Graphs (and Their Relation to Random Walk)

2.4.1 A Survey of Models and Results

As we have shown in the previous sections, the random walk problem is strictly related to the topology of the underlying graph. Indeed, the main physical quantities are simple functions of the adjacency matrix **A**, which algebraically describes the graph structure. Now, the Hamiltonians of a series of fundamental statistical models are linear in **A**, and therefore even their behavior is deeply influenced by topology and it can be expressed in terms of random walk functions. For this reason, the main concepts and parameters characterizing random walks, such as recurrence and transience, as well as the spectral dimension, also determine the properties of these models, which may have very different physical origins.

This provides a very powerful tool to investigate and classify geometrically disordered and inhomogeneous systems, where the usual techniques and ideas developed for lattices do not apply.

In the following, we briefly review some models and their related results. For now, the underlying concepts of statistical mechanics will be taken for granted, while in Section 2.5 they will be explained in more details.

2.4.1.1 The Oscillating Network

Let us consider a generic network of masses m linked by springs of elastic constant K. The harmonic oscillations of such a system can be studied by writing the equations of motion for the displacements x of each mass from its equilibrium position:

$$m\ddot{x}_i = -K\sum_{j\in\mathcal{V}} A_{ij}(x_i - x_j) = -K\sum_{j\in\mathcal{V}} L_{ij}x_j, \quad (2.196)$$

which, after Fourier transforming with respect to the time, reads

$$\frac{\omega^2}{\omega_0^2}\tilde{x}_i = \sum_{j\in\mathcal{V}} L_{ij}\tilde{x}_j, \quad (2.197)$$

with $\omega_0^2 \equiv K/m$. Thus, the determination of the normal modes and of the normal frequencies of the oscillating network reduces to the diagonalization of the Laplacian operator $\mathbf{L}$. Noting that $\mathbf{L} = \mathbf{Z}(\mathbf{I} - \mathbf{P})$, where $\mathbf{I}$ is the identity matrix and $\mathbf{P}$ is given by (2.115), one can establish mathematical correspondences with random walks. In particular, the density $\rho(\omega)$ of normal modes at low frequencies scales as

$$\rho(\omega) \sim \omega^{\bar{d}-1} \quad \text{for } \omega \to 0. \quad (2.198)$$

In fact, posing $\omega^2 = l$, we denote with $\rho_l(l)$ the density of eigenstates with eigenvalue equal to l; if $\rho(\omega) \sim \omega^{\bar{d}-1}$, then $\rho_l(l) \sim l^{\bar{d}/2-1}$, as $l \to 0$. Under this hypothesis one has

$$\int \frac{\rho_l(l)}{l+\epsilon}dl \sim \epsilon^{\bar{d}/2-1} \quad \epsilon \to 0. \quad (2.199)$$

Comparing this expression with equations (2.166) and (2.167), it follows that the parameter $\bar{d}$ appearing in equation (2.198) is exactly the average spectral dimension.

Due to the already mentioned universality properties, the above result holds for the very general case where oscillating masses and elastic constants may have different values on different sites and links, provided they are bounded by positive numbers; otherwise stated, this analysis can be extended to the case of weighted graphs with ferromagnetic coupling $\mathbf{J}$ (see [17] for more details).

We close with a historical note: this basic connection between random walks and harmonic oscillations was first introduced by Alexander and Orbach in 1982 for the case of fractals and, at that time the splitting between local and average spectral dimensions on inhomogeneous structures was not yet known, in such a way that the exponent describing the scaling of the density of states at low frequencies was simply called spectral dimension, since it was related to the vibrational spectrum.

2.4.1.2 The Gaussian Model

The Gaussian model (also known as "free scalar field" in field theory) is the simplest statistical-mechanics model used to study magnetic systems on lattices; even if it is not realistic, its properties are fundamental to understand more complex and phenomenologically significant models. Moreover, as mentioned previously in Section 2.3.10, the rigorous mathematical definition of spectral dimension of a graph is based on the analysis of infrared singularities of the Gaussian model defined on the graph itself. Let us now review the main passages.

First, we introduce the Hamiltonian for the Gaussian model on $\mathcal{G}$, as

$$\mathcal{H}(\phi|J,\{m_i\}) = \frac{1}{4}\sum_{(i,j)\in\mathcal{E}}(\phi_i - \phi_j)^2 + \sum_{i\in\mathcal{V}}m_i^2\phi_i^2 = \frac{1}{2}\sum_{(i,j)\in\mathcal{E}}\phi_i(JL_{ij} + m_i^2\delta_{ij})\phi_j$$
$$= \frac{1}{2}\phi \cdot (\mathbf{L} + m^2) \cdot \phi, \qquad (2.200)$$

where ϕ_i is a real field, $J > 0$ is a homogeneous ferromagnetic coupling, and the squared masses m_i^2 are upper and lower bounded by some positive numbers.

Any field configuration $\{\phi_i\}$ is associated to the Gibbs measure[20] given by exp $(-\mathcal{H})/Z$. For the investigation of the model in the thermodynamic limit it is convenient to introduce the Hamiltonian $\mathcal{H}_r$ that is restricted to the Van Hove sphere $S_{o,r}$ of radius r, in such a way that the intensive free energy for the infinite graph can be written as

$$f \equiv -\lim_{r\to\infty}\frac{1}{N_{o,r}}\log\int\prod_{i\in S_{o,r}}d\phi e^{-\mathcal{H}_r}. \qquad (2.201)$$

Also, the thermodynamic average value for an arbitrary function $g(\phi)$ of the fields is given by

$$\langle g(\phi)\rangle \equiv \lim_{r\to\infty}\frac{\int\prod_{i\in S_{o,r}}d\phi_i g(\phi)e^{-\mathcal{H}_r}}{\int\prod_{i\in S_{o,r}}d\phi_i e^{-\mathcal{H}_r}}. \qquad (2.202)$$

---

[20] Here, we are not inserting the temperature β^{-1} as, exploiting the fact that the Hamiltonian is quadratic, the parameter β can be easily reabsorbed in the definition of the fields as $\phi' = \beta^{-1/2}\phi$. The symbol Z in the denominator represents the partition function.

In particular, we are interested in the expectation value for $\overline{\phi^2}$ that is

$$\langle \overline{\phi^2} \rangle = \lim_{r \to \infty} \frac{1}{N_{o,r}} \frac{\int \prod_{i \in S_{o,r}} d\phi_i \left(\sum_{i \in S_{o,r}} \phi_i^2 \right) e^{-\mathcal{H}_r}}{\int \prod_{i \in S_{o,r}} d\phi_i e^{-\mathcal{H}_r}}. \quad (2.203)$$

Now, the intensive free energy can be easily calculated by exploiting the well-known property of multivariable Gaussian integrals, that is $\int dx \exp(-1/2 \mathbf{x} \cdot \mathbf{B} \cdot \mathbf{x}) = (2\pi)^{N/2} (\det \mathbf{B})^{-1/2}$, where $\mathbf{x}$ is an N-component vector and $\mathbf{B}$ is a symmetric, positive-definite $N \times N$ matrix. Therefore, we get

$$f = -\lim_{r \to \infty} \frac{1}{N_{o,r}} \log\{(2\pi)^{N_{o,r}/2} [\det(\mathbf{L}+m)]^{-1/2}\}$$

$$= -\frac{1}{2} \log(2\pi) + \lim_{r \to \infty} \frac{1}{2N_{o,r}} \mathrm{Tr}[\log(\mathbf{L}+m^2)]$$

$$= -\frac{1}{2} \log(2\pi) + \frac{1}{2} \overline{\mathrm{Tr}}[\log(\mathbf{L}+m^2)], \quad (2.204)$$

where we also exploited the property $\log(\det \mathbf{B}) = \mathrm{Tr}[\log(\mathbf{B})]$.

On the other hand, $\langle \overline{\phi^2} \rangle$ can be calculated starting from equation (2.203), by simply performing an integration by parts with respect to the whole set of fields ϕ_i appearing in the integral at the numerator:

$$\langle \overline{\phi^2} \rangle = \lim_{r \to \infty} \frac{1}{N_{o,r}} \sum_{i \in S_{o,r}} (\mathbf{L}+m^2)^{-1}_{ii} = \overline{\mathrm{Tr}}(\mathbf{L}+m^2)^{-1}. \quad (2.205)$$

Now, exploiting equation (2.163), and replacing μ with m^2, we get an expression for $\langle \overline{\phi^2} \rangle$ in terms of random walks and the singular behavior for $m \to 0$ can be described in terms of the average spectral dimension:

$$\langle \overline{\phi^2} \rangle \sim m^{\bar{d}-2} (\log m)^{I(\bar{d}/2)}, \quad m \to 0^+. \quad (2.206)$$

Moreover, $\langle \overline{\phi^2} \rangle$ and f are related by

$$\langle \overline{\phi^2} \rangle = \frac{1}{m} \frac{\partial f}{\partial m}, \quad (2.207)$$

in such a way that, from (2.206) and (2.207), we finally get

$$f \sim m^{\bar{d}} (\log m)^{I(\bar{d}/2)}, \quad m \to 0^+. \quad (2.208)$$

In this way we obtained the singular behavior of the free energy as $m \to 0$ directly as a function of the average spectral dimension. Therefore, $\bar{d}$ properly describes the critical behavior of the model considered.

Notice that equation (2.208) provides a possible further definition for $\bar{d}$ and such a definition can be useful to prove the universality properties of $\bar{d}$ [17].

Finally, we mention another correspondence between the Gaussian model and random walks: the correlation function $\langle \phi_i \phi_j \rangle$ results as

$$\langle \phi_i \phi_j \rangle = (\mathbf{L} + m^2)^{-1}_{ij}, \qquad (2.209)$$

and can be written as a sum of walks connecting i and j (see [53]).

2.4.1.3 The Spherical Model and O(n) Models

The spherical model is again a magnetic model and it can be defined on a generic graph through the Hamiltonian (2.200) with the generalized spherical constraint $\sum_i z_i \phi_i^2 = n$. We assume the coordination numbers to be bound as $1 \leq z_i \leq z_{\max}$.

As reviewed in [17], the free energy and the correlation functions of the spherical model can be expressed in terms of the Gaussian ones. Then, the critical behavior is obtained from the infrared singularities of the latter, i.e., in terms of the long time behavior of random walks. Also, the spherical model exhibits phase transitions at a finite temperature for $\bar{d} > 2$; its critical exponents can be exactly determined and they turn out to be simple functions of $\bar{d}$, pointing out the crucial role of the average spectral dimension in phase transitions and critical phenomena.

The so-called $O(n)$ model (or n-vector model) [84] is defined, for positive integer n, by the Hamiltonian $\mathcal{H}_n$ given by

$$\mathcal{H}_n(\{\mathbf{S}\}|\mathbf{J}) = \frac{1}{2} \sum_{(i,j) \in \mathcal{E}} J_{ij} (\mathbf{S}_i - \mathbf{S}_j)^2, \qquad (2.210)$$

where $J_{ij} > 0$ are ferromagnetic interactions, which may vary from link to link, and $\mathbf{S}_i$ is an n-dimensional vector of magnitude $\sqrt{n}$ (that is, $\mathbf{S}_i \mathbf{S}_i = n$). Again, the Gibbs measure $\exp(-\mathcal{H}_n/T)$ is adopted. This is a more realistic magnetic model with respect to the Gaussian and the spherical one, but its exact solution is in general not known. However, by means of a series of inequalities, relating the spin correlation functions to the random walk generating functions, one can derive some very general results concerning phase transitions on graphs. In particular it has been proven that:

- they cannot have phase transitions at finite temperature if $\mathcal{G}$ is recurrent on the average [29];
- they exhibit phase transitions at finite temperature if $\mathcal{G}$ is transient on the average [19];
- for $n \to \infty$ their critical exponents tend to the spherical ones [18].

Finally, notice that special cases of the *n*-vector model are:

- $n = 0$, the self-avoiding walk
- $n = 1$, the Ising model
- $n = 2$, the XY model
- $n = 3$, the Heisenberg model
- $n = 4$, the toy model for the Higgs sector of the Standard Model.

2.4.2 Generalization of the Mermin-Wagner Theorem

Few exact results are known in the field of phase transitions and critical phenomena, and almost all of them concern mean field models or translationally invariant systems. However, in recent years the study of non-translationally-invariant lattices has become of primary importance in many fields of physics, ranging from fractal structures in condensed matter to random graphs exhibiting scale-free and/or small-world features in social and biological applications [75, 15, 2, 5]. These geometrical structures are much more difficult to study from a theoretical point of view since conventional mathematical techniques (e.g., Fourier transform) cannot be applied to them. Thus, any rigorous and general result about universal properties of models defined on arbitrary structure is of great importance.

In 1966 Mermin and Wagner [72] obtained a fundamental result concerning translationally invariant lattices, showing that on these structures a quantum Heisenberg model with limited range couplings cannot have spontaneous magnetization in one and two dimensions (in the thermodynamic limit). An analogous theorem was proven a year later by Mermin [68] about classical Heisenberg and more generally $\mathcal{O}(n)$ models and, by further generalizations, it became clear that a continuous symmetry cannot be spontaneously broken in $d = 1$ and $d = 2$. A fundamental mathematical tool in proving such theorems is the Fourier transform, which does not apply when translational invariance is lost.

In 1992 Cassi [28] showed that these theorems can be generalized to non-translationally-invariant structures, proving that classical $\mathcal{O}(n)$ and quantum Heisenberg ferromagnetic models on a graph cannot have a spontaneous symmetry breaking if random walks on the same network are recursive (on average).

In order to deepen this result, let us consider a finite connected graph $\mathcal{G}$ consisting of N sites $i = 1,\ldots,N$ and of bonds (i,j) joining them, whose topology is described by its adjacency matrix $\mathbf{A}$ (see equation (2.111)). The $\mathcal{O}(n)$ model ($n \geq 2$) on $\mathcal{G} = (\mathcal{V},\mathcal{E})$ is defined by the Hamiltonian

$$\mathcal{H}(\{\mathbf{S}\}|\mathbf{J},\mathbf{h}) = -\sum_{i,j}^{N}\sum_{\mu=1}^{n} S_i^{\mu} J_{ij} S_j^{\mu} - \sum_{i=1}^{N}\sum_{\mu=1}^{n} h^{\mu} S_i^{\mu}, \tag{2.211}$$

with the spin components satisfying the constraint

$$\sum_{\mu=1}^{n} S_i^{\mu} S_i^{\mu} = 1, \qquad (2.212)$$

and with ferromagnetic couplings

$$J_{ij} = J_{ji} \begin{cases} = 0 \text{ if } A_{ij} = 0 \\ \neq 0 \text{ if } A_{ij} = 1 \end{cases} \qquad (2.213)$$

satisfying the further constraints

$$J_{ij} \geq d > 0, \qquad (2.214)$$

$$d_i \equiv \sum_{j=1}^{N} J_{ij} \leq J < \infty, \qquad (2.215)$$

for each bond $(i,j) \in \mathcal{E}$ and each site $i \in \mathcal{V}$. Notice that this condition also implies that the degree of any node is bounded.

Without loss of generality, let us choose the direction of the magnetic field $\mathbf{h}$ as the n-th coordinate axis in the spin space and define $h \equiv |\mathbf{h}|$. The magnetization per site m, which is the order parameter of the model, is defined as

$$m(h) \equiv \frac{1}{N}\sum_{i=1}^{N} \langle S_i^n \rangle \equiv \frac{1}{N}\sum_{i=1}^{N} m_i, \qquad (2.216)$$

where the average is in respect to the usual Gibbs measure. Now, even after taking the thermodynamic limit $N \to \infty$, we have $\lim_{h \to 0} m(h) = 0$ if random walks on $\mathcal{G}$ are recursive.

In fact, one can show that [28, 30]

$$\beta^{-1}(n-1)\frac{d}{J^2}\frac{\left[\overline{m_i \tilde{P}_{ii}(1;h)}\right]^2}{\overline{\tilde{P}_{ii}(1;h)}} \leq 1, \qquad (2.217)$$

where $\tilde{P}_{ii}(1;h)$ is the generating function evaluated in $\lambda = 1$ of the probability $P_{ii}(t;h)$ for a random walk to return at its initial site, being $S_{ij}(h) = J_{ij}/(h+d_i)$ its hopping probability from i to j. Then, starting from the inequality (2.217) one can prove that, for ROA graphs, given some $\epsilon > 0$, there exists no subgraph $\mathcal{G}_\epsilon \subseteq \mathcal{G}$ such that for $h \to 0$, $m_i > \epsilon$ for $i \in \mathcal{G}_\epsilon$. Otherwise stated, ROA and the existence of a positive magnetization in an infinite subgraph are incompatible (see [29] for the extensive proof including also the Heisenberg model). Notice that the local recurrence conditions $\tilde{F}_{ii}(1;h \to 0) = 1$ and $\tilde{P}_{ii}(1;h \to 0) = \infty$ are equivalent by

(2.136), but this is not the case for the average ones. In fact, even if $\bar{F}(1, h \to 0) = 1$ implies $\bar{P}(1, h \to 0) = \infty$, the inverse is not necessarily true. The definition of ROA, recalled here

$$\bar{F} = \overline{\tilde{F}_{ii}}(\lambda = 1, h = 0) = 1, \tag{2.218}$$

is necessary to prove the theorem, since only this definition implies the divergence on the average of $\tilde{P}_{ii}(1, h \to 0)$ on every subgraph with non-zero measure. On the other hand, structures where $\overline{\tilde{P}_{ii}}(\lambda = 1; h \to 0) = \infty$ but (2.218) does not hold, show spontaneous symmetry breaking at finite temperature.

To summarize, continuous symmetry models, such as the $\mathcal{O}(n)$ ($n > 1$) and the Heisenberg models on a general network with nearest-neighbor bounded ferromagnetic couplings, cannot have a phase transition with spontaneous symmetry breaking if random walks (without traps and with bounded hopping probabilities) on that network are recursive on average, i.e., with average return probability $\bar{F} = 1$. The proof holds not only for constant h, but also for an arbitrary distribution of positive h_i, i.e., for a site dependent magnetic field.

Of course, this theorem includes the Mermin-Wagner theorem for ferromagnetic models with nearest-neighbor interaction on usual ordered lattices, since, as shown by Pólya, random walks are recursive in one and two dimensions and transient in all the other dimensions. Moreover a simple corollary allows us to deal with fractal structures: indeed it is well known that random walks on fractals are recursive if and only if the spectral dimension of the structure is less or equal to 2 [53, 9]; so we can conclude that a continuous symmetry on fractals can be spontaneously broken only when spectral dimension is greater than 2. The same statement holds for all structures that are not geometrically fractals, but have a spectral dimension different from their space dimension, such as comb lattices. Actually, the theorem is even more general since, while there are infinite discrete structures for which spectral dimension cannot be defined, such as Bethe lattices [27], it is always possible to check whether random walks are either recursive or transient; for example, Bethe lattices are transient and phase transitions on them indeed occur. Notice that recursivity can also be experimentally proven by studying whether diffusing objects make a compact exploration of the structure.

As mentioned above, structures where (2.218) does not hold (even if $\overline{\tilde{P}_{ii}}(\lambda = 1; h \to 0)$ is infinite), show spontaneous symmetry breaking at finite temperature. This result has been formalized in [20], where $O(n)$ classical ferromagnetic spin models are proven to always display a broken symmetry phase at finite temperature on transient on average graphs, i.e., on graphs where (2.218) is not true. The proof, inspired by the classical result of Fröhlich, Simon and Spencer on lattices, holds for $n \geq 1$, hence including as a particular case the Ising model. As a consequence, for continuous symmetry modes (i.e., $n > 1$) we have

Models O(n), n≥2

ROA
$\bar{F}=1$ ⇒ ∄ SSB   Generalization of the Mermin-Wagner theorem

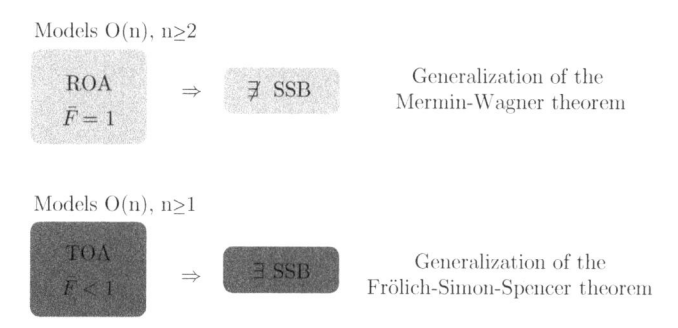

Models O(n), n≥1

TOA
$\bar{F}<1$ ⇒ ∃ SSB   Generalization of the Frölich-Simon-Spencer theorem

Figure 2.11. Schematic representation for the emergence of spontaneous symmetry breaking (SSB) in continuous and discrete symmetry models.

both a necessary and a sufficient condition for the existence of a spontaneous symmetry breaking, while for discrete symmetry models (i.e., $n = 1$) only a sufficient condition is available (see Figure 2.11).

Thus, from this point of view, all infinite networks can be divided into two classes: the *recursive on average* one and the *transient on average* one. On the networks belonging to the recursive class spontaneous breaking of a continuous symmetry is impossible at any finite temperature and the reason for this fact is the excessive topological correlation, meaning that the mean number of walks returning to the starting site at large times is too great with respect to the other ones.

We conclude with a remark. The definition of spontaneous symmetry breaking is based on the average magnetization m. However, on inhomogeneous structures it is possible to have $m = 0$ but all local magnetization $m_i > 0$ (if 0 is the only accumulation point for the m_i). This is another consequence of the splitting between local and average behavior on inhomogeneous structures. In general, critical phenomena on inhomogeneous structures should be classified according to both local and average behavior, and two different classes of critical exponents should be given to describe them completely. Even from an experimental point of view, one should consider the possibility of having different results studying localized or bulk physical quantities. This is particularly important in disordered systems and polymer physics, where inhomogeneous structures often occur.

2.5 Statistical Mechanics of Simple Mean Field Models through Analogies with Mechanical and Diffusive Dynamical Systems

2.5.1 Statistical Mechanics in a Nutshell

Statistical mechanics emerged in the last decades of the nineteenth century thanks to its founding fathers Ludwig Boltzmann, James Clerk Maxwell and Josiah Willard Gibbs [57]. Its "sole" scope (at that time) was to act as a theoretical ground for the already existing empirical thermodynamics, so to reconcile its noisy and irreversible macroscopic behavior to a deterministic and time reversal microscopic dynamics. Assuming the reader to be already familiar with the subject, we sketch hereafter a quick and informal summary.

Let us consider a very simple system, for example, a perfect gas: its molecules obey a Newton-like microscopic dynamics[21] and, instead of focusing on each particular trajectory for characterizing the state of the system, we define order parameters (e.g., the density) in terms of microscopic variables (e.g., the instantaneous position of particles). By averaging their evolution over suitable probability measures, and imposing on these averages energy minimization and entropy maximization, it is possible to infer the macroscopic behavior in agreement with thermodynamics, hence bringing together the microscopic deterministic and time reversal mechanics with the macroscopic strong dictates stemming from the second principle (i.e., the arrow of time coded in the entropy growth).

To infer which probability measure to use, let us assume that the system can be described by an ensemble of N possible values of the energy E, which accounts for kinetic terms only[22] as perfect gases do not interact by definition (collisions apart clearly), labeled as E_i, $i \in (1,\ldots,N)$ and let us associate a probability p_i to the configurations corresponding to these states, hence $E_i \to p_i$. Now, we require that the value of the entropy $S = -\sum_{i=1}^{N} p_i \log p_i$ of this system is the maximum allowed, and is compatible with the fact that the average energy must be $E = \sum_{i=1}^{N} p_i E_i$ and that p is effectively a probability, hence $\sum_{i=1}^{N} p_i = 1$. We can do this by introducing two Lagrange multipliers λ and β in the following functional $\phi_{\lambda,\beta}(p)$ and then extremizing the latter over the probability, in formulae

$$\phi_{\lambda,\beta}(p) = -\sum_{k=1}^{N} p_k \log p_k - \lambda \left(\sum_{k=1}^{N} p_k - 1 \right) - \beta \left(\sum_{k=1}^{N} p_k E_k - E \right), \quad (2.219)$$

$$\partial_{p_j} \phi_{\lambda,\beta}(p) = 0 \Rightarrow -\log p_j - 1 - \lambda - \beta E_j = 0, \quad (2.220)$$

that implies

$$p_j = \frac{e^{-\beta E_j}}{\sum_{i=1}^{N} e^{-\beta E_i}}, \quad (2.221)$$

which is called *Maxwell-Boltzmann distribution* and where $\beta = 1/k_B T$; T is the temperature of the thermal bath in equilibrium with the system and k_B

---

[21] Friction is neglected, as we are at the molecular level and thus time-reversal as dissipative terms in differential equations capturing system's evolution are coupled to odd derivatives.
[22] We recall that in a perfect (or ideal) gas all collisions between atoms or molecules are perfectly elastic and there are no intermolecular attractive forces. In such a gas, all the internal energy is in the form of kinetic energy and any change in internal energy is accompanied by a change in temperature.

is known as Boltzmann constant. Note further that for linear forces,[23] the Maxwell-Boltzmann distribution coincides from a statistical perspective with the Gaussian distribution where the role of standard deviation is played by the temperature. As a consequence, a small temperature corresponds to a narrow distribution, in agreement with the Second Principle of Thermodynamics (as expected because (2.221) stems from Entropy maximization).

A step forward beyond the perfect gas was achieved by Van der Waals and Maxwell in their pioneering works focused on *real gases* (see e.g., [81]), where particle interactions were finally considered by introducing a non-zero potential in the microscopic Hamiltonian describing the system. This extension implied fifty years of deep changes in the theoretical-physics perspective in order to be able to face new classes of questions. The remarkable reward lies in a theory of phase transitions where the focus is no longer on details regarding the system constituents, but rather on the characteristics of their interactions. Indeed, phase transitions, namely (abrupt) changes in the macroscopic state of the whole system, are not due to the particular system considered, but are primarily due to the ability of its constituents to perceive interactions over the thermal noise. For instance, when considering a system made of a large number of water molecules, whatever the level of resolution to describe the single molecule (ranging from classical to quantum), by properly varying the external tunable parameters (e.g., the temperature), this *system* eventually changes its state (e.g., from liquid to vapour).

Physicists were aware that the purely kinetic Hamiltonian, that was introduced for perfect gases (or Hamiltonian with mild potentials allowing for real gases), was no longer suitable for solids, where atoms do not move freely and the main energy contributions stem from potentials. An ensemble of harmonic oscillators (mimicking atomic oscillations of the nuclei around their rest positions) was the first scenario for understanding condensed matter: nuclei are arranged according to regular lattices, but merging statistical mechanics with lattice theories soon resulted in practically intractable models.[24]

As a paradigmatic example, let us consider the one-dimensional Ising model, originally introduced to investigate magnetic properties of matter: the generic, out of N, nucleus labeled as i is schematically represented by a spin σ_i, which can assume only two values ($\sigma_i = -1$, spin down and $\sigma_i = +1$, spin up); nearest-neighbor spins interact reciprocally through positive (i.e., ferromagnetic) interactions $J_{i,i+1} > 0$, hence the Hamiltonian of this system can be written as

---

[23] Linear forces correspond to potentials which are quadratic forms and these potentials constitute the hard core of classical physics.
[24] For instance the famous Ising model, dated 1920 (and actually invented by Lenz) and whose properties are known in dimensions one and two, is still waiting for a solution in three dimensions [14].

$\mathcal{H}_N(\sigma|\mathbf{J},h) \propto -\sum_i^N J_{i,i+1}\sigma_i\sigma_{i+1} - h\sum_i^N \sigma_i$, where h tunes the external magnetic field and the minus sign in front of each term of the Hamiltonian ensures that spins try to align with the external field and to get parallel with each other in order to fulfill the minimum energy principle.

The formal extension of this model to higher dimensions looks trivial, yet prohibitive difficulties emerge in facing the topological constraint due to short-range interactions and shortcuts, such as the mean field approximation, were properly implemented to turn this hurdle around.

The "mean field approximation" consists in extending the sum on nearest-neighbor couples (which are $\mathcal{O}(N)$) to include all possible couples in the system (which are $\mathcal{O}(N^2)$), properly rescaling the coupling ($J \to J/N$) in order to keep thermodynamical observables linearly extensive. If we consider a ferromagnet built of N Ising spins $\sigma_i = \pm 1$ with $i \in (1,\ldots,N)$, we can then write

$$H_N(\sigma|J) = -\frac{1}{N}\sum_{i<j}^{N,N} J_{ij}\sigma_i\sigma_j \sim -\frac{1}{2N}\sum_{i,j}^{N,N} \sigma_i\sigma_j, \qquad (2.222)$$

where in the last expression we included also the diagonal terms ($i=j$) as, providing a contribution $\sim \mathcal{O}(N^0)$, they are irrelevant for large N. From a topological perspective the mean field approximation is equal to abandoning the lattice structure in favour of a complete graph (see Figure 2.12). When the coupling matrix has only positive entries, e.g. $P(J_{ij}) = \delta(J_{ij} - J)$, this model is named the Curie-Weiss model and acts as the simplest microscopic Hamiltonian that is able to describe the paramagnetic-ferromagnetic transitions experienced by materials when temperature is properly lowered. An external (magnetic) field h can be accounted for by adding in the Hamiltonian an extra term $\propto -h\sum_i^N \sigma_i$.

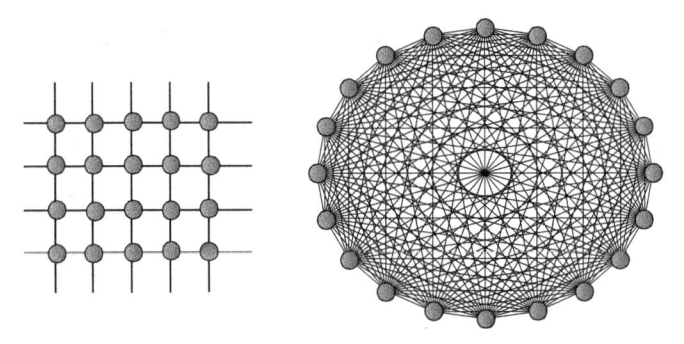

Figure 2.12. Example of regular lattice (left) and complete graph (right) with $N = 20$ nodes. In the former only nearest-neighbors are connected in such a way that the number of links scales linearly with N, while in the latter each node is connected with all the remaining $N-1$ in such a way that the number of links scales quadratically with N.

According to the principle of minimum energy, the two-body interaction appearing in the Hamiltonian in equation (2.222) tends to make spins parallel with each other and aligned with the external field, if it is present. However, in the presence of noise (i.e., if temperature T is strictly positive), maximization of entropy must also be taken into account. When the noise level is much higher than the characteristic energy (roughly, if $T \gg J$), noise and entropy-driven disorder prevail and spins are not able to "feel" reciprocally; as a result, they flip randomly and the system behaves as a *paramagnet*. Conversely, if noise is not too loud, spins start to interact possibly giving rise to a phase transition; as a result the system globally rearranges and all the spins get oriented in the same direction, which is the one selected by the external field if present, thus we have a *ferromagnet*.

Overall, the statistical mechanical procedure to investigate the thermodynamic behavior of the Curie-Weiss model (CW) can be summarized as follows: starting from a microscopic formulation of the system, i.e., N spins labeled as $i,j,\ldots$, their pairwise couplings $J_{ij} \equiv J$, and possibly an external field h, we derive an explicit expression for its (macroscopic) free energy[25] $f(\beta,h)$. Note that, for merely practical reasons, usually we will deal with the following modified expression for the free energy: $\alpha(\beta,h) = -\beta f(\beta,h)$. This quantity (in both its versions) is the effective energy, namely the difference between the internal energy E, divided by the temperature $T = 1/\beta$, and the entropy S, namely $\alpha(\beta,h) = S(\beta,h) - \beta U(\beta,h)$; in fact, S is the "penalty" to be paid to the Second Principle for being in a configuration that amounts to an energy E at noise level β. We can therefore link macroscopic free energy with microscopic dynamics via the fundamental relation

$$\alpha(\beta,h) = \lim_{N \to \infty} \frac{1}{N} \log Z_N(\beta,h) = \lim_{N \to \infty} \frac{1}{N} \log \sum_{\{\sigma\}}^{2^N} \exp[-\beta H_N(\sigma|J,h)], \quad (2.223)$$

where the sum is performed over the set $\{\sigma\}$ of all 2^N possible spin configurations, each weighted by the Boltzmann factor $\exp[-\beta H_N(\sigma|J,h)]$ that tests the likelihood of the related configuration. Notice that we use the subscript to denote that the observable is evaluated at finite size N, while whenever absent it means that the observable is evaluated at the thermodynamic limit $N \to \infty$. From expression (2.223), we can derive the whole thermodynamics and in particular phase-diagrams; that is, we are able to discern regions in the space of tunable parameters (e.g., temperature/noise level) where the system behaves as a paramagnet or as a ferromagnet.

Thermodynamical averages, denoted with the symbol $\langle . \rangle$, provide a given observable with an expected value, namely the value to be compared with measures

---

[25] The dependence on J is not made explicit since, being always coupled with β, having in mind a homogeneous system, one can always rescale the temperature to account for J.

in an experiment. For instance, for the magnetization $m(\sigma) \equiv \sum_{i=1}^{N} \sigma_i/N$ we have

$$\langle m(\beta,J,h) \rangle = \frac{\sum_\sigma m(\sigma) e^{-\beta H_N(\sigma|J,h)}}{\sum_\sigma e^{-\beta H_N(\sigma|J,h)}}. \qquad (2.224)$$

In the absence of a field ($h = 0$), when $\beta \to \infty$ the system is noiseless (zero temperature), hence, spins can arrange according to an imitative pattern without errors and the system behaves ferromagnetically ($|\langle m \rangle| \to 1$), while when $\beta \to 0$ the system behaves completely randomly (infinite temperature), thus interactions cannot be felt and the system is a paramagnet ($\langle m \rangle \to 0$). In between these regimes a phase transition happens.

In the Curie-Weiss model the magnetization $\langle m \rangle$ works as an *order parameter*: its thermodynamical average is zero when the system is in a paramagnetic (disordered) state, while it is different from zero in a ferromagnetic state (where it can be either positive or negative, depending on the sign of the external field). Dealing with order parameters allows us to avoid managing an extensive number of variables σ_i, which is practically impossible and, even more important, is not strictly necessary.

An explicit expression for $\alpha(\beta,J,h)$ can be obtained by carrying out summations in equation (2.223) and taking the *thermodynamic limit* $N \to \infty$, finding

$$\alpha(\beta,J,h) = \log 2 + \log \cosh[\beta(J\langle m \rangle + h)] - \frac{\beta J}{2} \langle m \rangle^2. \qquad (2.225)$$

(These calculations shall be expanded upon in the next section.) In order to impose thermodynamical principles, i.e., energy minimization and entropy maximization, we need to find the extrema of this expression with respect to $\langle m \rangle$ requiring $\partial_{\langle m(\beta) \rangle} \alpha(\beta,h) = 0$. This results in a *self-consistency* relation reading as

$$\partial_{\langle m \rangle} \alpha(\beta,J,h) = 0 \Rightarrow \langle m \rangle = \tanh[\beta(J\langle m \rangle + h)]. \qquad (2.226)$$

This expression returns the average behavior of a spin in a magnetic field. In order to see that a phase transition between paramagnetic and ferromagnetic states actually exists, we can fix $h = 0$ (and pose $J = 1$ for simplicity) and expand the right hand side of equation (2.226) to get

$$\langle m \rangle \propto \pm \sqrt{\beta J - 1}. \qquad (2.227)$$

Thus, while the noise level is higher than one ($\beta < \beta_c \equiv 1$ or, equivalently, $T > T_c \equiv 1$) the only solution is $\langle m \rangle = 0$, while, as the noise is lowered below its critical threshold β_c, two different-from-zero branches of solutions appear for the magnetization and the system becomes a ferromagnet (see Figure 2.13). The

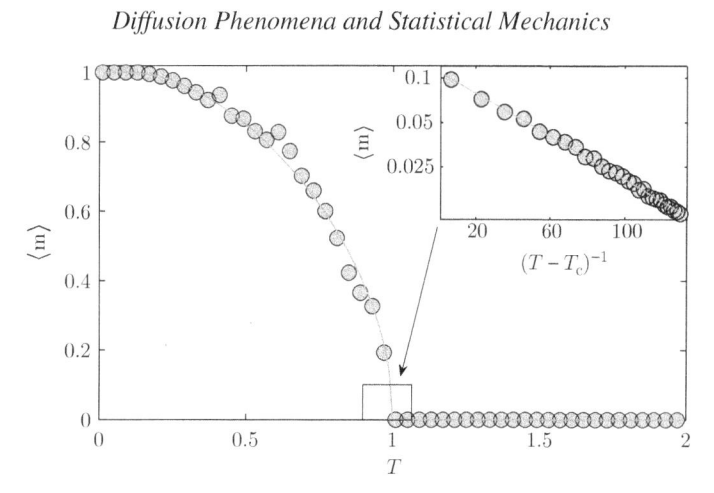

Figure 2.13. Average magnetization $\langle m \rangle$ versus temperature T for a Curie-Weiss model in the absence of a field ($h = 0$). The critical temperature $T_c = 1$ separates a magnetized region ($|\langle m \rangle| > 0$, only one branch shown) from a non-magnetized region ($\langle m \rangle = 0$). The box zooms over the critical region (notice the logarithmic scale) and highlights the power-law behavior $m \sim (T - T_c)^\beta$, where $\beta = 1/2$ is also referred to as a critical exponent (see also equation (2.227)). The data shown here (●) are obtained via Monte Carlo simulations for a system of $N = 10^5$ spins and is compared with the theoretical curve (solid line).

branch effectively chosen by the system usually depends on the sign of the external field or boundary fluctuations: $\langle m \rangle > 0$ for $h > 0$ and vice versa for $h < 0$.

Clearly, the lowest energy minima correspond to the two configurations with all spins aligned, either upwards ($\sigma_i = +1, \forall i$) or downwards ($\sigma_i = -1, \forall i$), these configurations being symmetric under spin-flip $\sigma_i \to -\sigma_i$. Therefore, the thermodynamics of the Curie-Weiss model is solved: energy minimization tends to align the spins (as the lowest energy states are the two ordered ones); however, entropy maximization tends to randomize the spins (as the highest entropy corresponds to the most disordered states, with half spins up and half spins down). The interplay between the two principles is driven by the level of noise introduced in the system and this is in turn ruled by the tuneable parameter $\beta \equiv 1/T$.

2.5.2 Getting Familiar with the Curie-Weiss Model

Before starting to explore the various methodologies driven by stochastic processes that allow solving for the Curie-Weiss free energy, we collect in this section a few general remarks which could help in the sections that follow.

The statistical mechanics package first requires us to define the relevant observables and to check that the model is well defined, namely, that it admits a good but nontrivial thermodynamic limit. An intuitive meaning behind the last condition can be streamlined as follows:

- We are working within an equilibrium picture. The equilibrium condition has been tacitly required in the Curie-Weiss analysis, when we have extremized the free energy over the order parameter to impose minimum energy and maximum entropy, and, at a more fundamental level, when we obtained the Maxwell-Boltzmann distribution via a variational procedure.
- For mean field systems at equilibrium, the probability of a state $P(\sigma_1,\ldots,\sigma_N)$ is the product of the probability of each single variable because of the causal independency among these degrees of freedom, thus $P(\sigma_1,\ldots,\sigma_N) = \prod_{i=1}^{N} P_i(\sigma_i)$, hence (as $\log \prod = \sum \log$) the entropy scales linearly with the system size: $S \propto N^1$.
- This in turn implies that we require systems whose averaged value of the Hamiltonian must scale linearly with N too, otherwise if $E \propto N^{1+\epsilon}$, in the thermodynamic limit, the entropic contribution is lost (as the free energy is the difference between the entropy – divided by the temperature – and the energy), or if $E \propto N^{1-\epsilon}$, in the thermodynamic limit, the energetic contribution is lost. Indeed, if we use the magnetization $m = N^{-1} \sum_{i=1}^{N} \sigma_i$ to write the Curie-Weiss Hamiltonian $H = -\frac{1}{2N} \sum_{i,j}^{N,N} \sigma_i \sigma_j$, we see that the latter can be written as $H = -\frac{N}{2} m^2$; remembering that $-1 \leq m \leq +1$ we get that the energy is a linear function of the system size as it should be.

It is important to become acquainted with upper and lower bounds (in the system size) for the Curie-Weiss free energy, hence we calculate them explicitly hereafter.

In order to get the upper bound, let us consider the trivial estimate of the magnetization m, valid for all trial fixed magnetization M,

$$(m-M)^2 \geq 0 \Rightarrow m^2 \geq 2mM - M^2, \qquad (2.228)$$

and plug it into the partition function to get (neglecting terms vanishing in the thermodynamic limit)

$$Z_N(\beta) = \sum_\sigma e^{\frac{\beta}{N} \sum_{1 \leq i < j \leq N} J \sigma_i \sigma_j} = \sum_\sigma e^{\frac{\beta J N m^2}{2}} \geq \sum_\sigma e^{\beta J m M N} e^{-\frac{1}{2} \beta J M^2 N}.$$

Now, the last expression can be calculated easily since the magnetization appears linearly and therefore the sum factorizes in each spin. Physically speaking, in the last passage we replaced the two-body interaction (encoded by m^2), which is difficult to treat, with a one-body interaction (encoded by m). This required the introduction of a trial magnetization M which mimics the field acting on each spin: a term M is coupled to m and a correction term quadratic in this trial magnetization M is inserted.[26]

---

[26] This idea is reminiscent of a recent powerful method introduced by Aizenman and co-workers [11] for the spin-glass theory, in which the key idea is letting the system interact with an external structure in such a

The result is the following bound

$$\frac{1}{N}\log Z_N(\beta) \geq \sup_M \left\{ \log 2 + \log\cosh(\beta JM) - \frac{1}{2}\beta JM^2 \right\}, \qquad (2.229)$$

which holds for any size N of the system. The result is quite typical: the term $\log 2 + \log\cosh(\beta JM)$ stems from the sum over spin configurations of a Boltzmann factor including only one-body interactions, and essentially gives the entropy, while the third term is the internal energy (multiplied by $-\beta$ as we are not dealing with the free energy f but with $\alpha = -\beta f$).

In order to get the opposite bound to (2.229), i.e., the lower bound on the free energy, let us notice that the magnetization m can take only $2N+1$ distinct values. We can therefore split the partition function into sums over configurations with constant magnetization in the following way

$$Z_N(\beta) = \sum_{\{\sigma\}} \sum_M \delta_{mM} e^{\frac{1}{2}\beta J N m^2}, \qquad (2.230)$$

where we used the trivial identity $\sum_M \delta_{mM} = 1$. Now, within the sum $m = M$ holds, and this means that we can replace m^2 with $2mM - M^2$. Using also the trivial inequality $\delta_{mM} \leq 1$ we can rewrite equation (2.230) as

$$Z_N(\beta) \leq \sum_M \sum_{\{\sigma\}} e^{\beta J N m M} e^{-\frac{1}{2}\beta J N M^2}. \qquad (2.231)$$

Now, one can carry out the sum over the state configurations $\{\sigma\}$ bounding the remaining sum over M by $2N+1$ times its largest term, hence getting

$$Z_N(\beta) \leq \sup_M \exp\left\{ \log 2 + \log\cosh(\beta JM) - \frac{1}{2}\beta JM^2 \right\} \qquad (2.232)$$

from which

$$\frac{1}{N}\log Z_N(\beta) \leq \frac{\log(2N+1)}{N} + \sup_M \left\{ \log 2 + \log\cosh(\beta JM) - \frac{1}{2}\beta JM^2 \right\}. \qquad (2.233)$$

This gives, together with (2.229), the exact value of the free energy per site, at least in the thermodynamic limit.

Before moving to more sophisticated techniques, it is important to re-obtain the free energy of the model with a sum rule technique, as this will introduce the concept of self-averaging from a non-statistical perspective. For this task let us

---

way that, sending the size of this structure to infinity, thanks to the mean-field nature of the interaction, the couplings within our system get negligible with respect to the couplings with the external world, thus (under the assumption that the external world shares with the model the same statistical properties) making the mathematical control simpler.

assume $h = 0$ (as it is not essential to understand the following concept) and let us introduce an interpolating free energy structure as

$$\alpha(\beta, t) = \frac{1}{N} \log \sum_{\sigma} e^{t\left(\frac{\beta J}{2N} \sum_{ij} \sigma_i \sigma_j\right) + (1-t)(\beta J \psi \sum_i \sigma_i)}, \quad (2.234)$$

where ψ is a generic field felt by any spin. This free energy interpolates between the original Curie-Weiss model (recovered for $t = 1$) and an ensemble of spins living under the field ψ modulated by J (for mathematical simplicity). We are interested in the solution of the Curie-Weiss free energy, namely $\alpha(\beta, t = 1)$, which we can conveniently write as a sum of two contributions, that is

$$\alpha(\beta, t = 1) = \alpha(\beta, t = 0) + \int_0^1 \frac{d\alpha}{dt}\bigg|_{t=t'} dt'. \quad (2.235)$$

In fact, evaluating $\alpha(\beta, t = 0)$ is straightforward and returns

$$\alpha(\beta, t = 0) = \log 2 + \log \cosh(\beta J \psi), \quad (2.236)$$

while the calculation of the derivative of the free energy is

$$\frac{d\alpha(\beta, t)}{dt} = \frac{\beta J}{2} \left(\langle m^2 \rangle - \langle m\psi \rangle\right) = \frac{\beta J}{2} \left(\langle (m - \psi)^2 \rangle\right) - \frac{\beta J}{2} \psi^2. \quad (2.237)$$

If we now choose $\psi \equiv m$, we get $\langle (m - \psi)^2 \rangle = 0$, therefore the extra term $-\frac{\beta J}{2} \psi^2$ introduced to build up the perfect square $\langle (m - \psi)^2 \rangle$, correctly recovers the Curie-Weiss free energy. Therefore, we can conclude that the magnetization is self-averaging, namely that at finite N the magnetization is a random variable, with its probability distribution function $P(m)$, but in the thermodynamic limit $P(m) \to \delta(m - \langle m \rangle)$, or, in other words, its variance goes to zero as a proper power of the volume (i.e., as $1/\sqrt{N}$).

2.5.3 Statistical Mechanics via Mechanical Analogy

Now we start to discuss *non-canonical* approaches to the equilibrium statistical mechanics of mean-field spin systems. First, we will show that it is possible to map the problem of evaluating explicitly the free energy for the Curie-Weiss model into the problem of solving the temporal evolution of a mechanical system in classical mechanics. Once this bridge is built, one can take advantage of techniques and methodologies in one field and translate results in the other field with little effort.

To complete this task, we need to map the space of the parameters (β, h) as a real $1 + 1$ space-time (t, x), where t stands for a temporal coordinate and x for a spatial one. Then, we can investigate the partial derivative equations that are fulfilled by the model observables in such a space and try to solve them. In order to exploit our idea, let us introduce the following action $S(t,x)$ as

$$S_N(t,x) = -\frac{1}{N} \log \sum_\sigma \exp\left(\frac{t}{2N} \sum_{ij} \sigma_i \sigma_j + x \sum_i \sigma_i\right), \qquad (2.238)$$

where the variables t, x can be thought of as fictitious time and space, and such that $\lim_{N\to\infty} S_N(t,x) = S(t,x)$ and $S(t,x) = \beta f(\beta) = -\alpha(\beta)$ whenever evaluated at $t = \beta$ and $x = 0$ (here $J = 1$ has been fixed).

In order to highlight our approach we need to work out the derivatives of $S_N(t,x)$ which read as

$$\frac{\partial S_N(t,x)}{\partial t} = -\frac{1}{2} \langle m_N^2 \rangle, \qquad (2.239)$$

$$\frac{\partial S_N(t,x)}{\partial x} = -\langle m_N \rangle, \qquad (2.240)$$

$$\frac{1}{2N} \frac{\partial^2 S_N(t,x)}{\partial x^2} = \frac{1}{2}\left(\langle m_N^2 \rangle - \langle m_N \rangle^2\right). \qquad (2.241)$$

Following Guerra prescriptions [50, 21] and noticing the form of the derivatives (2.239–2.241), it is possible to build (by direct construction) a Hamilton-Jacobi equation for $S_N(t,x)$ as

$$\partial_t S_N(t,x) + \frac{1}{2}(\partial_x S_N(t,x))^2 + V(t,x) = 0, \qquad (2.242)$$

where we remark that

$$V(t,x) = -\frac{1}{2N} \partial_{xx}^2 S_N(t,x) = \frac{1}{2}(\langle m_N^2 \rangle - \langle m_N \rangle^2); \qquad (2.243)$$

namely the potential where the motion evolves is nothing but the variance of the magnetization, which goes to zero in the thermodynamic limit (see the last paragraph of Section 2.5.2).

Before solving the Hamilton-Jacobi equation (2.242), we derive it with respect to x and, calling $u_N(t,x) = \partial_x S_N(t,x) = -\langle m_N \rangle$, we get the following Burger equation for the velocity (i.e., the magnetization in the statistical mechanics framework, minus sign apart)

$$\partial_t u_N(t,x) + u_N(t,x) \partial_x u(t,x) - \frac{1}{2N} \partial_{xx}^2 u_N(t,x) = 0. \qquad (2.244)$$

Let us point out that such an equation becomes naturally inviscid in the thermodynamic limit. In fact, $S(t,x)$ (that morally is a free energy beyond the principal

Hamilton function) admits the thermodynamic limit, hence $\frac{1}{2N}\partial_{xx}^2 u_N(t,x) \to 0$ almost everywhere. Actually, the only point where it does not converge to zero, not even in the $N \to \infty$ limit, is on the phase transition, exactly where the (rescaled) fluctuations of the order parameter diverge because the system is starting to "decide" on which of the two branches of the free energy (positive or negative magnetizations) it will live once ergodicity is definitively broken (we will deepen this point in the next section).

It is now time to obtain an explicit expression for the Principal Hamilton Function $S(t,x)$. The solution of the Hamilton-Jacobi equation can be achieved in many ways (e.g., via the characteristics as deepened in the next subsection); here we use the canonical transformation, and we write

$$S(t,x) = S(t_0,x_0) + \int_{t_0}^{t} \mathcal{L}(t',x) dt', \qquad (2.245)$$

where $\mathcal{L}$ is the Lagrangian associated with the system, that is $\mathcal{L}(t,x) = \frac{\langle m \rangle^2}{2} + V(t,x)$: a kinetic term (remember that the magnetization plays as the velocity in this analogy) and a potential term. The initial point (t_0,x_0) can be chosen arbitrarily.

As we are interested in the thermodynamic limit $N \to \infty$, where $V(t,x) \propto (\langle m^2 \rangle - \langle m \rangle^2) \to 0$, the motion is Galilean (time translational invariant) and occurs on straight lines $x(t) = x_0 + \langle m \rangle t$. As a consequence, the integral of the Lagrangian becomes trivial and returns (choosing $t_0 = 0$)

$$\int_0^t \mathcal{L}(t',x) dt' = \frac{t}{2} \langle m^2 \rangle. \qquad (2.246)$$

It is important to stress that, as anticipated, the Cauchy condition can be fixed arbitrarily and, of course, a clever choice is $t = 0$ because it allows the two-body interactions to be neglected and the calculations to become straightforward, reducing to one-body expressions that always factorize. The evaluation of the Cauchy condition thus gives

$$S(t=0, x=x_0) = -\log 2 - \log\cosh(x_0) = -\log 2 - \log\cosh(x - \langle m \rangle t). \quad (2.247)$$

Therefore, using equations (2.245–2.247), we have the complete solution

$$S(t,x) = -\log 2 - \log\cosh(x - \langle m \rangle t) + \frac{t}{2}\langle m^2 \rangle, \qquad (2.248)$$

and, fixing $t = \beta$, $x = \beta h$, we finally recover the solution of the Curie-Weiss free energy (2.225), without requiring any computation of statistical mechanical flavour!

In the two next subsections the properties of this mechanical analogy will be studied in more depth. In particular, in Section 2.5.3.1, we will push forward the equivalence at the level of the order parameter, showing that the velocity field in mechanics (that evolves according to a nonlinear equation) eventually develops a shock wave and this happens exactly when, in the statistical mechanics counterpart, the system undergoes a phase transition. Further, in Section 2.5.3.2 we analyze the existence of conserved quantities: the idea is that a symmetry (a conserved quantity) can be expressed by equating to zero its derivative, but, as quantities in mechanics involve the magnetization (i.e., the velocity) and its momenta, this will result in a class of polynomial equations for the magnetization. Basically we show that conserved quantities in mechanics mirror self-averaging properties in statistical mechanics.

2.5.3.1 Shock Waves and Spontaneous Symmetry Breaking

While it is well-known from classical arguments of statistical mechanics that the Curie-Weiss model undergoes a phase transition from an ergodic (paramagnetic) phase to a ferromagnetic one at $\beta_c = 1/J$, it is very instructive to tackle this phenomenon within a mechanical framework, where such a phase transition is obtained as a shock wave for the Burger equation (2.244) that holds for the magnetization.

In order to see this, the simplest approach is to investigate the mass conservation coupled to the "fictitious particle" of the mechanical analogy: let us fix $J = 1$ for simplicity (such that the critical noise becomes $\beta_c = 1$) and consider the evolution of the system (hence the propagation of the mass) from a generic point x_0 to a generic point x, as

$$\rho(x)dx = \rho(x_0)dx_0. \tag{2.249}$$

By the equation of motion $x = x_0 + u(0, y)t = x_0 - \tanh(x_0)t$ we get

$$\frac{dx}{dx_0} = 1 + \partial_{x_0} u(0, x_0)t = 1 - [1 - \tanh^2(x_0)]t, \tag{2.250}$$

thus, for the mass density in a generic point x, we get

$$\rho(x) = \rho(x_0)\frac{1}{1 - [1 - \tanh^2(x_0)]t}. \tag{2.251}$$

At $x = 0$ (which is the point of interest in the statistical mechanical counterpart), $\rho(x_0) = \rho(0)/(1-t)$ which diverges for $t_c = 1$, thus, as $t_c = \beta_c$, exactly where the phase transition happens in statistical mechanics.

To see the shock wave from another perspective, we can consider in more detail the Burger equation (2.244) that we rewrite for practical convenience (but whose solution is left for the next subsection as it will bring us to the connection with diffusion):

$$\frac{\partial u_N(t,x)}{\partial t} + u_N(t,x)\frac{\partial u_N(t,x)}{\partial x} - \frac{1}{2N}\frac{\partial^2 u_N(t,x)}{\partial x^2} = 0. \quad (2.252)$$

First, it is important to see that, as in the Hamilton-Jacobi equation (2.242) the potential $V_N(t,x)$ is vanishing for $N \to \infty$ (because it is the self-averaging of the magnetization, the factor $1/2$ apart) almost everywhere (the critical point apart), and as $V_N(t,x) = \partial_{xx}^2 S(t,x)/2N$, the corresponding Burger equation (2.252) becomes inviscid in such a limit, or in other words, the infinite volume limit of the magnetization evolves according to the following Riemann-Hopf equation

$$\frac{\partial u_N(t,x)}{\partial t} + u_N(t,x)\frac{\partial u_N(t,x)}{\partial x} = 0, \quad (2.253)$$

that represents the two branches of the magnetization once ergodicity is broken as Hopf bifurcations. Indeed, we can have a representation of its solutions as characteristics. We get

$$u(x,t) = -\tanh[x - u(x,t)t], \quad (2.254)$$

i.e., the well known self-consistence relation for the Curie-Weiss model (2.226), with trajectories (parameterized by $s \in \mathbb{R}$)

$$\begin{cases} t = s \\ x = x_0 - s\tanh x_0. \end{cases} \quad (2.255)$$

Then, we can immediately state that in the region of the plane (x,t), defined by

$$x \geq -\sqrt{t(t-1)} + \operatorname{arctanh}\left(\sqrt{\frac{t-1}{t}}\right), \quad \text{for } x_0 \geq 0$$

and

$$x \leq -\sqrt{t(t-1)} + \operatorname{arctanh}\left(\sqrt{\frac{t-1}{t}}\right), \quad \text{for } x_0 \leq 0$$

trajectories (2.255) have no intersection points.

This last statement defines in another way the onset of ergodicity breaking, that is the phase transition, in the statistical mechanical counterpart, and is presented in the following section.

Let us set, for instance, that $x_0 \geq 0$. Once fixed $s = \bar{s}$, let us investigate the position at time $\bar{s}$ as a function of the starting point x_0. We have

$$x(x_0) = x_0 - \bar{s}\tanh x_0.$$

If $x(x_0)$ is monotone with respect to x_0, then for every starting point there is a unique position at time t. In other words, two trajectories born in different points of the boundary cannot, at the same time, assume the same position (i.e., they cannot intersect). Hence we have

$$x'(x_0) = 1 - \bar{s}(1 - \tanh^2 x_0) \geq 0 \,\forall x_0,$$

only if

$$\bar{s} \leq \frac{1}{1 - \tanh^2 x_0},$$

as $1 - \tanh^2 x_0$ always belongs to $[0,1]$. The last formula implies

$$x_0 \geq \operatorname{arctanh}\left(\sqrt{\frac{t-1}{t}}\right),$$

and bearing in mind the form of trajectories (2.255) we get

$$x \geq \operatorname{arctanh}\left(\sqrt{\frac{t-1}{t}}\right) - \sqrt{t(t-1)}. \tag{2.256}$$

Of course, the proof is analogous for $x_0 \leq 0$.

We notice that the previous argument gives the region of the (x,t) plane in which the invertibility of the motion fails. On the other hand, every trajectory has its end point at the intersection with the x-axes, or they are all merged in a unique line, that is $(x = 0, t > 1)$.

More rigorously, the curve $(x = 0, t > 1)$ is a discontinuity line for our solution, since it is easily seen that every point of such a line is an intersection point of the trajectories (2.255). Also, by direct calculation, we can get by (2.254)

$$\partial_x u(x,t) = -\frac{1 - u^2}{1 + t(1 - u^2)} < 0, \tag{2.257}$$

i.e., the velocity field is strictly decreasing along the x direction.[27]

To characterize formally the phase transition, if we now name u_+ the limiting value from positive x, and u_- the one from negative x, we obtain that $0 < u_- = -u_+ < 0$ almost everywhere for $t > 1$. Indeed, the curve of discontinuity can be parameterized as

$$\begin{cases} t > 1 \\ x = 0, \end{cases}$$

and we can use the Rankine-Hugoniot condition to state

$$u_+^2 = u_-^2.$$

---

[27] This is a particular case of a more general property of the Lax-Oleink solution [62], named *entropy condition*, that ensures that $u(x,t)$ never increases along x. The general form will not be supplied here as it is redundant in this context, but it is simply mentioned because it can be very useful when studying generalized ferromagnets.

Since $u_+ < u_-$ for (2.257), the phase transition must exist and $(x = 0, t > 1)$ is a shock line for the Burger equation (2.253).

We stress that the relation $u_+^2 = u_-^2$, in this context, mirrors the spin-flip symmetry shared by the two minima of the Curie-Weiss model in the broken ergodicity phase, i.e. $|+\langle m \rangle| = |-\langle m \rangle|$.

2.5.3.2 Conservation Laws: Noether Invariants in Mechanics and Self-averaging in Statistical Mechanics

The Hamilton-Jacobi equation can be rewritten from a mechanical point of view as

$$\partial_t S_N(x,t) + H_N(\partial_x S_N(x,t), x, t) = 0,$$

where the Hamiltonian function reads off as

$$H_N(\partial_x S_N(x,t), x, t) = \frac{p^2(x,t)}{2} + V_N(x,t). \tag{2.258}$$

In other words, we fixed the mass unitary in such a way that the velocity field u coincides with the generalized time dependent momentum p. The Hamilton equations are nothing but characteristics[28] of equation (2.242):

$$\begin{cases} \dot{x} &= u_N(x,t) \\ \dot{t} &= 1 \\ \dot{p} &= -u_N(x,t)\partial_x u_N(x,t) - \partial_x V_N(x,t) \\ \dot{E} &= -u_N(x,t)\partial_x (\partial_t S_N(x,t)) - \partial_t V_N(x,t). \end{cases} \tag{2.259}$$

The latter two equations express the conservation laws for momentum p and energy E for the system evolving in the mechanical analogy, and can be written in the form of streaming equations as

$$\begin{cases} D_N u_N(x,t) &= -\partial_x V_N(x,t) \\ D_N(\partial_t S_N(x,t)) &= -\partial_t V_N(x,t), \end{cases}$$

where D_N is the hydrodynamic transport operator. Since in the thermodynamic limit the potential is vanishing, the system approaches a free motion and, recalling that $u_N(x,t) = -\langle m_N \rangle$ and $\partial_t S_N(x,t) = -\frac{1}{2}\langle m_N^2 \rangle$, so $D_N = \partial_t - \langle m_N \rangle \partial_x$, for $N \to \infty$, we conclude

$$\begin{cases} D_N \langle m_N \rangle &= 0 \\ D_N \langle m_N^2 \rangle &= 0, \end{cases} \tag{2.260}$$

---

[28] See e.g., [10] for a textbook on differential equations, including the method of characteristics.

namely,

$$\begin{cases} \langle m_N^3 \rangle - 3 \langle m_N \rangle \langle m_N^2 \rangle + 2 \langle m_N \rangle^3 &= O(\frac{1}{N}) \\ (\langle m_N^4 \rangle - \langle m_N^2 \rangle^2) - 2 \langle m_N \rangle \langle m_N^3 \rangle + 2 \langle m_N \rangle^2 \langle m_N^2 \rangle &= O(\frac{1}{N}). \end{cases}$$

(2.261)

These equations do not provide evidence of any ergodicity breakdown in their range of validity; in fact, we recall that in this analysis we excluded the line ($x = 0, t > 1$) because in this segment shocks emerge, while the characteristics method only holds in a continuity regime. Therefore, we have that $\langle m_N^2 \rangle = \langle m_N \rangle^2 + O(\frac{1}{N})$ everywhere but on the line ($x = 0, t > 1$) and the phase transition is the only point where self-averaging breaks down.

2.5.4 Statistical Mechanics via Analogy with a Diffusion Problem

So far we have examined the properties of the free energy using a mechanical analogy. In particular, we have shown that the free energy of the Curie-Weiss model plays as the principal Hamilton function in an equivalent dynamics problem, and that solving the mechanical problem implies, in turn, the solving of the statistical-mechanics problem itself. Now, we can focus on the properties of the magnetization. We already know that the magnetization obeys a Burger equation (2.244), which collapses on the Riemann-Hopf equation (2.253) in the thermodynamic limit, but, so far, we have not tried to solve it. The plan for this section is to show that there exists a transformation, namely the Cole-Hopf transform, that maps the Burger equation into the Fourier equation (the prototype of simple diffusion): the Fourier equation will be solved and we will re-obtain the thermodynamics of the Curie-Weiss via this alternative route.

In order to start the plan, let us recall that the Burger equation for the velocity field $u_N(t,x) = -\langle m_N \rangle$ is

$$\partial_t u_N(t,x) + u_N(t,x) \partial_x u(t,x) - \frac{1}{2N} \partial_{xx}^2 u_N(t,x) = 0, \qquad (2.262)$$

and it is important to point out explicitly that such an equation becomes naturally inviscid (i.e., it approaches the Riemann-Hopf equation) in the thermodynamic limit, as $S(t,x)$ admits the thermodynamic limit as we have already seen in Section 2.5.3.

If we now perform the Cole-Hopf transform, defined as

$$\Psi_N(t,x) = \exp\left(-N \int dx\, u_N(t,x)\right), \qquad (2.263)$$

it is immediate to check that $\Psi_N(t,x)$ satisfies the following diffusion equation

$$\frac{\partial \Psi_N(t,x)}{\partial t} - \frac{1}{2N}\frac{\partial \Psi_N(t,x)}{\partial x^2} = 0, \qquad (2.264)$$

which will now be solved in the Fourier space, through the Green propagator and the convolution theorem.

Before approaching the formal solution, the reader is invited to realize that the Cole-Hopf auxiliary function Ψ (defined in equation (2.263)) is nothing but the partition function in statistical mechanics and that, as a consequence of this observation, we now know that the Curie-Weiss partition function *evolves by diffusion* in the $(t = \beta, x = \beta h)$ space.

In the Fourier space we deal with $\hat{\Psi}_N(t,k)$ defined as

$$\hat{\Psi}_N(t,k) = \int dk\, e^{-ikx} \Psi_N(t,x), \qquad (2.265)$$

and equation (2.264) in the impulse space reads off as

$$\partial_t \hat{\Psi}_N(t,k) + \frac{k^2}{2N}\hat{\Psi}_N(t,k) = 0, \qquad (2.266)$$

whose solution is

$$\hat{\Psi}_N(t,k) = \hat{\Psi}_N(0,k)\exp\left(-\frac{k^2}{2N}t\right), \qquad (2.267)$$

and translates in the original space as

$$\Psi_N(t,x) = \int dy\, G_t(x-y)\Psi_0(y), \qquad (2.268)$$

where the Green propagator is given by

$$G_t(x-y) = \sqrt{\frac{N}{2\pi t}}\exp\left[-N\frac{(x-y)^2}{2t}\right]. \qquad (2.269)$$

Overall we get

$$S_N(t,x) = -\frac{1}{N}\log\sqrt{\frac{N}{2\pi t}}\int dy\, e^{-N\left[\frac{(x-y)^2}{2t} - \log 2 - \log\cosh(y)\right]}, \qquad (2.270)$$

where we used $S_N(0,x) = -\log 2 - \log\cosh(x)$ and the definition of the Cole-Hopf transform (2.263).

As the exponent in equation (2.270) is proportional to the volume, for large N, we can apply the saddle point argument to get

$$\alpha(t,x) = \sup_y \left\{ -\frac{(x-y)^2}{2t} + \log 2 + \log \cosh(y) \right\}$$

$$= -\frac{(x-\hat{y})^2}{2t} + \log 2 + \log \cosh(\hat{y}), \qquad (2.271)$$

with $\hat{y}$ maximizer.

2.5.5 Generalized Models and Techniques for Mean Field Many-body Problems

At this point one may wonder whether the equivalence between classical mechanics and statistical mechanics can be pushed even further, beyond the Curie-Weiss scheme. Indeed this is the case, although this field has not been completely investigated so far. Hereafter, for the sake of completeness, the simplest generalizations to the problem will be presented.

Given N Ising spins $\sigma_i = \pm 1$, $i \in (1,\ldots,N)$, let us consider a general ferromagnetic model with interactions among spins described by the Hamiltonian

$$\frac{H_N}{N} = -F(m_N) - hG(m_N), \qquad (2.272)$$

where, as before,

$$m_N = \frac{1}{N} \sum_{i=1}^{N} \sigma_i$$

is the magnetization, $F(m_N)$ models the generic p–spin mean field interaction[29], and $G(m_N)$ accounts for the interaction with an external magnetic field h (and it is usually one-body, i.e., $G(m_N) = m_N$).

Let us consider an even, real-valued and strictly convex Hamiltonian $-N[F(m) + hG(m)] \equiv H : (0,1) \ni m \to H(m)$, such that $F(+m) = F(-m)$, $\partial_{xx}^2 F(m) > 0$ and $F(0) = 0$ (notice that the subscript N to m has been dropped, to lighten the notation); let us also assume $G(m) \equiv m$ for the sake of simplicity. Now two interpolating (scalar) parameters t, x can be introduced, which can be thought of

---

[29] In a nutshell, a p-spin ferromagnet is a system where the interactions among spins happen in p-plets, instead of more classical couples. The model is described by the Hamiltonian $H_N(\sigma|J) = -N^{-(p-1)} \sum_{i_1 < i_2 < \ldots < i_p}^{N} J_{i_1 i_2 \ldots i_N} \sigma_{i_1} \sigma_{i_2} \ldots \sigma_{i_N}$. As standard ferromagnets, the model is shown to exhibit two phases, a paramagnetic one and a (replica symmetric) ferromagnetic one. On the other hand, different to the standard ferromagnet, the phase transition does not display criticality for $p > 2$ [44].

as fictitious time and space respectively. The interpolating procedure is fixed as

$$Z_N(t,x) = \sum_{\{\sigma\}} \exp[N(tF(m)+xm)], \qquad (2.273)$$

such that we can show that the thermodynamic limit for the free energy, defined by $H(m)/N = F(m)+hG(m)$, exists and reads as

$$\lim_{N\to\infty} \frac{1}{N} \ln Z_N(t,x) = \inf \frac{1}{N} \ln Z_N(t,x) = \alpha(t,x). \qquad (2.274)$$

The proof of this statement works within the classical Guerra-Toninelli scheme [49]: it is enough to prove the model subadditivity, namely that

$$\ln Z_N(t,x) \leq \ln Z_{N_1}(t,x) + \ln Z_{N_2}(t,x). \qquad (2.275)$$

Once this last relation is proved, the thermodynamic limit is well posed by definition. To prove subadditivity, it is enough to split the system into two subsystems built of N_1 and N_2 spins respectively, such that $N = N_1 + N_2$ and, calling m_1 the magnetization of the first subsystem (built of spins $i = 1,\ldots,N_1$) and m_2 the magnetization related to the second subsystem (built of spins $i = N_1 + 1,\ldots,N$), note that

$$Nm = N_1 m_1 + N_2 m_2 \Rightarrow m = \frac{N_1}{N} m_1 + \frac{N_2}{N} m_2,$$

and that, thanks to convexity, we can write

$$F(m) = F\left(\frac{N_1}{N} m_1 + \frac{N_2}{N} m_2\right) \leq \frac{N_1}{N} F(m_1) + \frac{N_2}{N} F(m_2), \qquad (2.276)$$

hence $NF(m) \leq N_1 F(m_1) + N_2 F(m_2)$.

We can paste this expression in the definition of $Z_N(t,x)$ and note further that, as far as $G(m) \equiv m$, the term depending on x factorizes so that, overall, we have

$$Z_N(t,x) \leq Z_{N_1}(t,x) \cdot Z_{N_2}(t,x), \qquad (2.277)$$

and, consequently, this class of models admits a proper thermodynamic limit.

However, as we will see, we cannot explore the whole space of functions that fulfill this framework: using the dispersion curve to catalogue models in statistical mechanics, one can see that the Curie-Weiss is only a particular case (one out of three possible cases) of the second-order dispersion curves. The kind of emergent dispersion curve is determined by the functional form of the potential $W(u)$, namely

$$W(u) = -\frac{u^2}{2} \rightarrow \text{F-type} \qquad (2.278)$$

$$W(u) = \sqrt{1+u^2} \rightarrow \text{K-type} \qquad (2.279)$$

$$W(u) = -\sqrt{1-u^2} \rightarrow \text{P-type} \qquad (2.280)$$

The simplest statistical mechanical system example of the F-type corresponds to the Curie-Weiss model studied previously, while we refer to [22] for an extensive treatment of K-type and of the P-type cases, as outlined above.

2.5.6 Further Problems Solved through the Mechanical Analogy

Beyond the "simple" Curie-Weiss model described above, variational principles of classical mechanics have been used to obtain the full Parisi solution of the Sherrington-Kirkpatrick model without the use of statistical mechanics [21]. The general scheme is that the free energy obeys a Hamilton-Jacobi partial differential equation and the order parameter (this time encoded by the overlaps) evolves under shocks within a Riemann-Hopf partial differential equation. In this framework it is possible to prove that each step K of replica symmetry breaking mirrors the motion of a system in a $K+1$ Euclidean space, while the full Parisi solution has its convergence on a Hilbert space. As a result, the corresponding order parameter becomes a functional as in the original framework.

Moreover, the analogy between solving for the free energy of a mean field system and obtaining the explicit motion of a suitable dynamical system has been extended to relativistic and quantum mechanics [22]. Interestingly, it was shown that for mean field systems that involve two-body interactions (e.g., the Sherrington-Kirkpatrick and the Curie-Weiss models) the related dynamical system lies in a Euclidean, classic space, while the quantum relativistic extension (i.e., the Klein-Gordon field) accounts for all the general p-spin models.

Finally, the approach based on the combination of statistical mechanics and non-linear partial-differential equations theory provides a novel and powerful tool to tackle phase transitions. In [26], this method was tested on the van der Waals model, which constitutes a paradigmatic example of first-order phase transition (semi-heuristically described). In particular, the first global mean field partition function for a system of finite number of particles was obtained. The partition function is a solution to the Klein-Gordon equation, which reproduces the van der Waals isotherms away from the critical region and, in the thermodynamic limit $N \rightarrow \infty$, automatically encodes the Maxwell equal areas rule.

Acknowledgments

This work has greatly benefited from discussions and interactions with several colleagues and friends whom I am proud to thank: Adriano Barra, Olivier Benichou, Alexander Blumen, Davide Cassi, Pierluigi Contucci, Ton Coolen, Francesco Guerra, Oliver Mulken, Sydney Redner, Cecilia Vernia, Raphael Voiturez. I am particularly indebted with Adriano Barra who introduced me to "serious" statistical mechanics, and taught me so much about its depth and modelling power; beyond the scientific aspects, I wish to thank him also for his constant, generous, energetic support.

References

[1] R. Albert and A. Barabási, *Statistical mechanics of complex networks*, Reviews of Modern Physics. 74(47), (2002).
[2] E. Agliari and A. Barra, *A Hebbian approach to complex-network generation*, Europhys. Lett. 94(1), 10002 (2011).
[3] E. Agliari, A. Barra, P. Contucci, R. Sandell, and C. Vernia, *A stochastic approach for quantifying immigrant integration: the Spanish test case*, New Journal of Physics. 16(10), 103034 (2014).
[4] E. Agliari, A. Blumen, and D. Cassi, *Slow encounters of particle pairs in branched structures*, Phys. Rev. E. 89(5), 052147 (2014).
[5] E. Agliari, A. Barra, A. Galluzzi, F. Guerra, D. Tantari, and F. Tavani, *Retrieval capabilities of hierarchical networks: from Dyson to Hopfield*, Phys. Rev. Lett. 114(2), 028103 (2015).
[6] E. Agliari, A. Blumen, and O. Muelken, Journal of Physics A. 39(48), (2008).
[7] E. Agliari, A. Blumen, and O. Muelken, *Dynamics of continuous-time quantum walks in restricted geometries*, Physical Review A. 82(1), 012305 (2010).
[8] Y. Aharonov, L. Davidovich, and N. Zagury, *Quantum random walks*, Physical Review A. 48(2), 1687 (1993).
[9] A. Alexander and R. Orbach, *Density of states on fractals: "fractons"* J. Phys. (Paris), Lett. 43, L625 (1982).
[10] V.I. Arnold, *Lectures on partial differential equations*. New York: Springer, (2004).
[11] M. Aizenman, R. Sims, and S. L. Starr, *Extended variational principle for the Sherrington-Kirkpatrick spin-glass model*, Phys. Rev. B. 68, 214403 (2003).
[12] D. ben Avraham and S. Havlin, *Diffusion and reactions in fractals and disordered systems*. Cambridge: Cambridge University Press, 17 (2000).
[13] E. Barkai, *Stochastic processes in physics*. Lecture Notes for the course "Stochastic processes in physics" held at Bar Ilan University, Israel, (2014).
[14] R. J. Baxter, *Exactly solved models in statistical mechanics*. United States: Courier Dover Publications, (2007).
[15] A. Barrat, M. Barthélemy, and A. Vespignani, *Dynamical processes on complex networks*. New York: Cambridge University Press, 36 (2008).
[16] R. Burioni and D. Cassi, *Universal properties of spectral dimension*, Phys. Rev. Lett. 76, 1091 (1996).
[17] R. Burioni and D. Cassi, *Random walks on graphs: ideas, techniques and results*, Journal of Physics A: Mathematical and General. 38(8), R45 (2005).

[18] R. Burioni, D. Cassi, and C. Destri, Phys. Rev. Lett. 85, 1496 (2000).
[19] R. Burioni, D. Cassi, and A. Vezzani, *Transience on the average and spontaneous symmetry breaking on graphs*, J. Phys. A. 32, 5539 (1999).
[20] R. Burioni, D. Cassi, and A. Vezzani, *Transience on the average and spontaneous symmetry breaking on graphs*, J. Phys. A: Math. Gen. 32(30), 5539 (1999).
[21] A. Barra, A. Di Biasio, and F. Guerra, *Replica symmetry breaking in mean-field spin glasses through the Hamilton-Jacobi technique*, Journal of Statistical Mechanics. P09006 (2010).
[22] A. Barra, A. Di Lorenzo, F. Guerra, and A. Moro, *On quantum and relativistic mechanical analogues in mean field spin models*, Proceeding of the Mathematical Royal Society. 470(2172), (2014).
[23] H. C. Berg, *Random walks in biology*. Princeton: Princeton University Press, (1993).
[24] A. E. Brouwer and W. H. Haemers, *Spectra of graphs*, New York: Springer, (2011).
[25] N. Biggs, *Algebraic graph theory*, New York: Cambridge University Press, (1993).
[26] A. Barra and A. Moro, *Exact solution of the van der Waals model in the critical region*, Ann. Phys. 359, 290–299 (2015).
[27] D. Cassi, *Existence of infinite networks without spectral dimension*, Int. J. Mod. Phys. B. 6 (1992).
[28] D. Cassi, *Phase transitions and random walks on graphs: a generalization of the mermin-wagner theorem to disordered lattices, fractals and other discrete structures*, Phys. Rev. Lett. 68, 3631 (1992).
[29] D. Cassi, *Local vs average behavior on inhomogeneous structures: recurrence on the average and a further extension of Mermin-Wagner theorem on graps*, Phys. Rev. Lett. 76(16), 2941 (1996).
[30] L. Cattivelli, E. Agliari, F. Sartori, and D. Cassi, *Lévy flights with power-law absorption*, Phys. Rev. E 92, 042156 (2015).
[31] S. Condamin, O. Bénichou, and M. Moreau, *Random walks and brownian motion: a method of computation for first-passage times and related quantities in confined geometries*, Physical Review E 75(2), 021111 (2007).
[32] X. Chen and D. Chen, *Some sufficient conditions for infinite collisions of simple random walks on a wedge comb*, Elect. J. Prob. 16, 1341–1355 (2011).
[33] R. Campari and D. Cassi, *Random collisions on branched networks: how simultaneous diffusion prevents encounters in inhomogeneous structures*, Physical Review E. 86, 021110 (2012).
[34] A. M. Childs and J. Goldstone, *Spatial search by quantum walk*, Physical Review A. 70(2), 022314 (2004).
[35] F. Chung, *Spectral Graph Theory*, CBMS Regional Conference Series in Mathematics, (vol. 92). Providence: American Mathematical Society, (1997).
[36] B. A. Carreras, V. E. Lynch, I. Dobson, and D. E. Newman, *Critical points and transitions in electrical power transmission model for cascading failure blackouts*, Chaos 12, 987–994 (2002).
[37] A. V. Chechkin, R. Metzler, J. Klafter, and V. Yu Gonchar, *Introduction to the theory of Lévy flights*, in *Anomalous Transport: Foundations and Applications* (R. Klages, G. Radons and I. M. Sokolov (eds.). Weinheim: Wiley-VCH, (2008).
[38] K. A. Dill and S. Bromberg, *Molecular driving forces*. New York: Garland Science, (2003).
[39] L. Donetti and M. Munoz, *Detecting network communities: a new systematic and efficient algorithm*, J. Stat. Mech. 10012, (2004).
[40] P. G. Doyle and L. Snell, *Random walks and electric networks*, The Carus Mathematical Monographs. 28 (1984).

[41] R. Fürth and A. Cowper (eds.), *Albert Einstein, investigation on the theory of the brownian movement*, United States: Dover, 1956.
[42] W. Feller, *An introduction to probability theory and its applications*, (vol. 2). United States: John Wiley, (1971).
[43] E. Farhi and S. Gutmann, *Quantum computation and decision trees*, Physical Review A. 58, 915 (1998).
[44] S. Franz, M. Mézard, F. Ricci-Tersenghi, M. Weigt, and R. Zecchina, *A ferromagnet with a glass transition*, Europhys. Lett. 55, 465 (2001).
[45] Y. Gefen, A. Aharonov, and B. B. Mandelbrot, *Critical phenomena on fractal lattices*, Phys. Rev. Lett. 45, 855 (1980).
[46] M. Girvan and M. E. J. Newman, *Community structure in social and biological networks*, Proceedings of the National Academy of Sciences. 99, 7821 (2002).
[47] L. K. Grover, *Quantum computer can search rapidly by using almost any transformation*, Phys. Rev. Lett. 79, 4329 (1997).
[48] D. T. Gillespie and E. Seitaridou, *Simple brownian diffusion*, New York: Oxford University Press, (2013).
[49] F. Guerra and F. L. Toninelli, *The thermodynamic limit in mean field spin glass models*, Commun. Math. Phys. 230(1), 71–79 (2002).
[50] F. Guerra, *Sum rules for the free energy in the mean field spin glass model*, Mathematical Physics in Mathematics and Physics, Field Inst. Comm. 30, 161 (2001).
[51] P.J. Grabner and W. Woess, *Functional iterations and periodic oscillations for simple random walk on the sierpinski graph*, Stochastic Process. Appl. 69, 127–138 (1997).
[52] A. Hald, *A history of probability and statistics and their applications before 1750.* New York: John Wiley and Sons, (1990).
[53] K. Hattori, T. Hattori, and H. Watanabe, *Gaussian field theories on general networks and the spectral dimensions*, Prog. Theor. Phys. Suppl. 92, 108–143 (1987).
[54] S. Havlin and O. Matan, *Multifractal nature of diffusion on hierarchical structures*, J. Phys. A: Math. Gen. 21, L307 (1988).
[55] S. Jespersen, R. Metzler, and H. C. Fogedby, *Lévy flights in external force fields: Langevin and fractional Fokker-Planck equations and their solutions*, Phys. Rev. E. 59(3), 2736 (1999).
[56] J. Kempe, *Quantum random walks: an introductory overview*, Contemporary Physics. 44(4), 307 (2003).
[57] C. Kittel, *Elementary statistical physics.* Courier Dover Publications, (2004).
[58] M. Krishnapur and Y. Peres, *Recurrent graphs where two independent random walks collide finitely often*, Elect. Commun. Probab. 9(8), 72–81 (2004).
[59] B. Kahng and S. Redner, *Scaling of the first-passage time and the survival probability on exact and quasi-exact self-similar structures*, J. Phys. A. 22(7), 887 (1989).
[60] R. Klages, G. Radons, and I. M. Sokolov, *Anomalous transport: foundations and applications.* Weiheim: Wiley-VCH, (2008).
[61] A.A. Kilbas, H.M. Srivastava, and J.J. Trujillo, *Theory and applications of fractional differential equations*, (vol. 204). San Diego: Elsevier Science Limited, (2006).
[62] P. Lax, *Hyperbolic systems of conservation laws and the mathematical theory of shock waves.* SIAM, (1973).
[63] G. Lawler and V. Limic, *Random walk: a modern introduction*, New York: Cambridge University Press, 123 (2010).
[64] A. L. Lloyd and R. M. May, *Epidemiology: how viruses spread among computers and people*, Science. 292, 1316–1317 (2001).
[65] L. Lovász, *Combinatorics, Paul Erdös is eighty*, (vol. 2). Hungary: Janos Bolyai Mathematical Society, (1993).
[66] L. E. Maistrov, *Probability theory, a historical sketch.* New York: Academic Press, (1974).

[67] O. Mülken and A. Blumen, *Continuous-time quantum walks: models for coherent transport on complex networks*, Physics Reports. 502, 37–87 (2011).

[68] N. D. Mermin, *Absence of ordering in certain classical systems*, J. Math. Phys. 8, 1061 (1967).

[69] R. Metzler and J. Klafter, *The random walk's guide to anomalous diffusion: a fractional dynamics approach*, Physics Reports. 339, 1–77 (2000).

[70] R. Metzler, G. Oshanin, and S. Redner (eds.), *First-passage phenomena and their applications*. Singapore: World Scientific, (2014).

[71] E. W. Montroll and G. H. Weiss, *Random walks on lattices (II)*, Journal of Mathematical Physics. 6, 167 (1965).

[72] N. D. Mermin and H. Wagner, *Absence of ferromagnetism in one- or two-dimensional isotropic Heisenberg models*, Phys. Rev. Lett. 17, 1133 (1966).

[73] B. Mohar and W. Woess, *A survey on spectra of infinite graphs*, Bull. Lond. Math. Soc. 21, 209 (1989).

[74] M. E. J. Newman, *The structure and function of complex networks*, SIAM review 45, 167–256 (2003).

[75] M. E. J. Newman, *Networks: an introduction*. Oxford: Oxford University Press, (2010).

[76] J. R. Norris, *Markov Chains*. Cambridge: Cambridge University Press, (1997).

[77] I. Prigogine and S. A. Rice (eds.), *Advances in chemical physics*, (vol. 116). United States: John Wiley and Sons, (2001).

[78] S. Redner, *A guide to first-passage processes*. Cambridge: Cambridge University Press, (2001).

[79] S. Redner, *A guide to first-passage processes*. Cambridge: Cambridge University Press, (2001).

[80] S. A. Rice, *Diffusion-limited reactions*. Amsterdam: Elsevier, (1985).

[81] L. E. Reichl and I. Prigogine, *A modern course in statistical physics*, (vol. 71). Austin: University of Texas Press, (1980).

[82] M. J. Saxton, *A biological interpretation of transient anomalous subdiffusion. II. Reaction kinetics*, Biophysical Journal. 94, 760–771 (2008).

[83] P. W. Shor, 35th IEEE Symposium on Foundations of Computer Science (Los Alamos), IEEE. 124 (1994).

[84] H. E. Stanley, *Dependence of Critical properties upon dimensionality of spins*, Phys. Rev. Lett. 20, 589–592 (1968).

[85] S. Strogatz, *Emergence of scaling in random networks*, Science. 286, 509–512 (1999).

[86] S. E. Venegas-Andraca, *Quantum walks: a comprehensive review*, Quantum Information Processing. 11(5), 1015–1106 (2012).

[87] G. M. Viswanathan, E. P. Raposo, and M. G. E. da Luz, *Lévy flights and superdiffusion in the context of biological encounters and random searches*, Physics of Life Reviews. 5, 133–150 (2008).

[88] G. H. Weiss, *Aspects and applications of the random walk*. Amsterdam: North Holland Press, (1994).

[89] W. Woess, *Random walks on infinite graphs and groups*, Cambridge Tracts in Mathematics 138. New York: Cambridge University Press, (2000).

[90] D. J. Watts and S. H. Strogatz, *Collective dynamics of small world networks*, Nature 393, 440–442 (1998).

[91] D. H. Zanette, *Statistical-thermodynamical foundations of anomalous diffusion*, Brazilian Journal of Physics. 29(1), 108–124 (1999).

[92] A. Barra, *The mean field Ising trough interpolating techniques*, J. Stat. Phys. 132, 787–809 (2008).

3

Legendre Structures in Statistical Mechanics for Ordered and Disordered Systems

FRANCESCO GUERRA

3.1 Introduction

The aim of this paper is to report on some aspects of the Lagrange variational principles that arise in ordered and disordered statistical mechanics models.

Firstly, in Section 3.2, by working in the frame of the ensemble theory, we consider the general case of a system whose microscopic configurations are cells in phase space. Then the free energy, as a convex functional of the interaction, is conjugated to the entropy as a concave functional of the state. A dual system of variational principles connects entropy and free energy. These considerations are very instructive, because they show that the variational principle for the free energy gives the complete physical content of the theory at thermodynamic equilibrium, as the dual functional is the whole state. Of course we must allow a sufficiently detailed expression of the interaction.

In Section 3.3, we consider the particular case of mean field models, even of multispecies type. It will show how the entropic principle gives a full characterization of the free energy for very generic interactions in the infinite volume limit. The order parameters are given by interspecies magnetizations.

It is very well known (see for example [9]) that in some particular cases it is possible to exploit the convexity of the interaction, with respect to the magnetization, in order to prove firstly that the free energy density does exist in the infinite volume limit, by exploiting a kind of subadditivity in the volume, and then have the expression of the free energy through an appropriate variational principle. Some simple cases will be considered in Section 3.4. However, the variational principles that come from convexity are not in the Legendre form. We will show how to reduce them to the Legendre form through a simple procedure, which will be the prototype for a similar procedure to be exploited in the much more difficult case of the mean field disordered models, as will be explained in Section 3.9.

Section 3.5 will be devoted to a simple outline of the entropic principle for the random energy model (REM), that is introduced by Bernard Derrida in [5]. The full dual Legendre structure will be reconstructed.

In Section 3.6 we will recall that the disordered mean field models require a novel variational structure, where the trial order parameters acquire a functional form, and the variational principle for the free energy is given in terms of an optimization procedure inverted with respect to the usual entropic one.

The novel variational principle is explained in the particular case of the REM in Section 3.7. Its very basic origin comes from the fact that the free energy, which is convex in terms of the interaction, is also concave in terms of the covariance of the random interaction. The inverted variational principle finds a very simple and natural interpretation through the introduction of some auxiliary models where the random interaction is squared, as shown in Section 3.8.

In Section 3.9, we treat mean field disordered models in general. We show how the inverted dual Lagrange structure arises and it will provide a very natural description of the order parameters. Moreover, we show how the broken replica symmetry bounds can be connected with the general Legendre structure, in full analogy with the simple case already met for ordered models in Section 3.4. The role of the models with squared random interaction will also be discussed.

Finally, Section 3.10 is dedicated to an outline of the perspectives for future developments.

3.2 The General Legendre Structure for Statistical Mechanics Systems: the Entropic Principle

The general Legendre structure for statistical mechanics systems is at the basis of everything. It encodes the essence of the second principle of thermodynamics.

I would like to take the opportunity to describe the thermodynamic system in the frame of the *ensemble* theory, as developed by Boltzmann, Gibbs, Einstein, and others.

In an appropriate *coarse graining* the microscopic configurations of the systems are described by a large number of cells $(1, 2, \ldots, K) \ni i$.

A **thermodynamic state** (not necessarily at equilibrium) is described by a probability distribution on cells

$$\mathcal{S} \ni p : (1, 2, \ldots, K) \ni i \to p_i, 0 \leq p_i \leq 1, \sum_i p_i = 1. \tag{3.1}$$

We call $\mathcal{S}$ the simplex of all states.

The entropy of the state p is defined as

$$p \to S(p) = -\sum_i p_i \log p_i, \qquad (3.2)$$

and turns out to be strictly concave on $\mathcal{S}$.

In fact, let the state p be the mixture of two states $p^{(1)}$ and $p^{(2)}$, in the sense that

$$p_i = \alpha p_i^{(1)} + (1-\alpha) p_i^{(2)}, \qquad (3.3)$$

for all i, with the mixing parameter $0 \le \alpha \le 1$. Then

$$S(p) \ge \alpha S(p^{(1)}) + (1-\alpha) S(p^{(2)}), \qquad (3.4)$$

where $>$ holds in the generic case, and $=$ is true only for degenerate mixtures, as for example if $p^{(1)} = p^{(2)}$, or $\alpha = 0$, or $\alpha = 1$.

Let us now attribute an energy to each cell

$$E : (1, 2, \ldots, K) \ni i \to E_i, \qquad (3.5)$$

and define the partition function

$$Z(E) = \sum_i e^{-E_i}, \qquad (3.6)$$

where the temperature has been incorporated into E.

The free energy is $F(E) = -\log Z(E)$, while the internal energy in a given state p is

$$U(E, p) = \sum_i E_i p_i. \qquad (3.7)$$

Obviously $\log Z(E)$ is convex in E, while $U(E, p)$ is linear in E and affine in p.

The basic Legendre structure (the second principle of thermodymics) is based on the recognition that the state p and the energy E can be assumed as dual variables, while the entropy $S(p)$ and $\log Z(E)$ are dual functionals, involved in the dual Legendre variational principles

$$\log Z(E) = \sup_p \left(S(p) - \sum_i E_i p_i \right), \qquad (3.8)$$

and

$$S(p) = \inf_E \left(\log Z(E) + \sum_i E_i p_i \right). \qquad (3.9)$$

Legendre Structures in Statistical Mechanics 145

Therefore, $\log Z$ is the Legendre transform of S, and vice versa.

Notice that writing $F(E) = -\log Z(E)$ the variational principle for the free energy involves a *min* over all states, as is well known from elementary thermodynamics.

At equilibrium, where the optimal values are enforced, the state has the Boltzmann-Gibbs expression in terms of E

$$p_i = -\frac{\partial}{\partial E_i} \log Z(E) = e^{-E_i}/Z, \tag{3.10}$$

while the E_i are determined up to a constant. This is due to the constraint $\sum_i p_i = 1$.

Notice that the term $\sum_i E_i p_i$ can be also interpreted as a Lagrangian multiplier for the entropy principle. As a matter of fact the optimal state realizes the maximum for the entropy under the constraint of fixed internal energy, in pure Boltzmann spirit.

This general Legendre structure assumes various shapes in different models in the infinite volume limit.

The case of ordered mean field models, and of REM and the generalized random energy model (GREM), will be considered in the following section.

3.3 Ordered Mean Field Models

Now we will take Ising configurations on N sites

$$\sigma : (1, 2, \ldots, N) \ni i \to \sigma_i = \pm 1. \tag{3.11}$$

Note that each σ denotes a *cell* of the system.

The interaction is specified by a function $u : [-1, 1] \ni M \to u(M)$, where u is assumed bounded and Lipschitz, i.e.,

$$u(M) \leq \bar{u}, \ |u(M) - u(M')| \leq C|M - M'|, \tag{3.12}$$

for given constants $\bar{u}$ and C.

Now the partition function is defined as

$$Z_N(u) = \sum_\sigma e^{Nu(m)}, \tag{3.13}$$

where $m = \sum_i \sigma_i / N$.

Notice the high degeneracy of the energy function. Many cells σ have the same energy if they have the same value of the magnetization m. There are only $N+1$ possible values for m, while the number of cells is 2^N.

The term N in the *Boltzmannfaktor* has been introduced in order to have good thermodynamic behavior in the infinite volume limit. In fact, now $\log Z_N(u)$ behaves proportionally to N for large N, and we will be particularly interested in the normalized value $N^{-1} \log Z_N(u)$. Notice that $\log Z_N(u)$ is convex in u.

For finite N the Legendre structure is a particular case of that which was introduced before. But in the infinite volume limit there is a strong degeneracy, due to the mean field character of the interaction.

First of all let us introduce the entropy for a single spin configuration with magnetization M

$$s(M) = -\frac{1+M}{2} \log \frac{1+M}{2} - \frac{1-M}{2} \log \frac{1-M}{2}. \qquad (3.14)$$

Notice that s is symmetric under $M \to -M$, and strictly concave in M.

We can easily establish the following: let u be bounded and Lipschitz, then the infinite volume limit of $N^{-1} \log Z_N(u)$ does exist and has the following value

$$A(u) = \lim_{N \to \infty} \frac{1}{N} \log Z_N(u) = \sup_M \left(s(M) + u(M) \right). \qquad (3.15)$$

In general many different values for M realize the *sup*. There are multiple phases. In each phase the state is factorized as a product state on the corresponding value of M.

Due to the high degeneracy in the infinite volume limit, M must now be considered as the dual variable to the interaction u. Moreover, the convex function $A(u)$ of u and the concave function $s(M)$ of M participate to the dual variational principles

$$A(u) = \sup_M \left(s(M) + u(M) \right), \qquad (3.16)$$

and

$$s(M) = \inf_u \left(A(u) - u(M) \right). \qquad (3.17)$$

We see that duality is fully preserved, while the Legendre structure is defaced by degeneracy. In fact, the Lagrange multiplier assumes the form $u(M)$ (the energy read at the value M) which has the right linear structure in u, but not in M.

Recall that in the finite volume case the Lagrange multiplier has the form $\sum_i E_i p_i$, which is linear in the energy, and affine over the states. Here the states are product states, for which affinity does not hold.

These considerations are easily expanded to multispecies mean field models. For example, we can consider the following simple model: let us split the sites $i = 1, 2, \ldots, N$ into K non-overlapping blocks $A_1, A_2, \ldots, A_K$, each made by

$N_1, N_2, \ldots, N_K$ sites, respectively, with $N_1 + N_2 + \cdots + N_K = N$. Now the interaction will involve K different magnetizations, as a function

$$u : [-1,1]^K \ni M_1, M_2, \ldots, M_K \to u(M_1, M_2, \ldots, M_K). \tag{3.18}$$

For any Ising configuration on N sites

$$\sigma : (1, 2, \ldots, N) \ni i \to \sigma_i = \pm 1, \tag{3.19}$$

define the partial block magnetizations as

$$m_a = N_a^{-1} \sum_{i \in A_a} \sigma_i, \quad a = 1, 2, \ldots, K. \tag{3.20}$$

Now the partition function is

$$Z_{N_1, N_2, \ldots, N_K}(u) = \sum_\sigma e^{Nu(m_1, m_2, \ldots, m_K)}. \tag{3.21}$$

Consider the $\lim_{N \to \infty}$, in such a way that $N_a/N \to \alpha_a$, $a = 1, 2, \ldots, K$, with $\sum_a \alpha_a = 1$.

Then the following can be easily established

$$\lim_{N \to \infty} \frac{1}{N} \log Z_{N_1, N_2, \ldots, N_K}(u)$$
$$= \sup_{M_1, M_2, \ldots, M_K} (u(M_1, M_2, \ldots, M_K) + \alpha_1 s(M_1) + \alpha_2 s(M_2) + \cdots + \alpha_K s(M_K)).$$
$$\tag{3.22}$$

Therefore, the entropic principle holds in a simple form also in the case for the multispecies model. Each block gives its own contribution to entropy, proportional to its size.

3.4 Direct Methods versus Convexity and Interpolation Methods

In the previous section, we have seen how it is possible to follow direct methods to prove the existence of the infinite volume limit for the free energy in some ordered mean field models.

In some particular cases, it is also possible to exploit, for example, the convexity of the interaction u with respect to the magnetization m, in order to prove the a priori existence of the infinite volume limit for the free energy, and then proceed to its calculation by using interpolation arguments, as shown for example in [9].

Let us recall the following simple case of the Curie-Weiss model, where the interaction is defined by $u(M) = \frac{1}{2}\beta M^2$, so that the partition function is

$$Z_N(\beta) = \sum_\sigma e^{N\frac{1}{2}\beta m^2}. \qquad (3.23)$$

Here the parameter β plays the role of inverse temperature. The factor $\frac{1}{2}$ has been inserted for esthetic reasons, so that the critical point for β will be $\beta_c = 1$. Notice that u is convex in the magnetization.

By following the methods expounded in [9], a simple interpolation argument gives subadditivity of $\log Z_N$, in the form $\log Z_N \leq \log Z_{N_1} + \log Z_{N_2}$. Therefore the existence of the infinite volume limit is assured in the form

$$A(\beta) = \lim_{N\to\infty} \frac{1}{N}\log Z_N(\beta) = \inf_N \frac{1}{N}\log Z_N(\beta). \qquad (3.24)$$

The value of this limit can be found through a variational principle which exploits the convexity of the interaction as a function of the magnetization. We recall here the argument and its simple consequences, since we will meet a similar situation in the extremely more complicated case of the disordered models.

For any value of the order parameter M, clearly we have $e^{-\frac{1}{2}N\beta(m-M)^2} \leq 1$. We insert this expression in the definition of the partition function, and derive the following estimate

$$Z_N(\beta) = \sum_\sigma e^{N\frac{1}{2}\beta m^2} \geq \sum_\sigma e^{N\frac{1}{2}\beta m^2} e^{-\frac{1}{2}N\beta(m-M)^2} = \sum_\sigma e^{\beta mM - \frac{1}{2}\beta M^2}. \qquad (3.25)$$

The interaction has been linearized. The exponent is factorized with respect to the sites, and the sum can be explicitly calculated. We end up with the estimate

$$\frac{1}{N}\log Z_N(\beta) \geq A(\beta, M), \qquad (3.26)$$

where A is the trial function

$$A(\beta, M) = \log 2 + \log\cosh(\beta M) - \frac{1}{2}\beta M^2. \qquad (3.27)$$

Optimization with respect to the free parameter M gives

$$\frac{1}{N}\log Z_N(\beta) \geq \bar{A}(\beta) = \sup_M A(\beta, M), \qquad (3.28)$$

uniformly in N.

It is easy to show that $\bar{A}(\beta)$ gives the exact value of the infinite volume limit $A(\beta)$. In fact, consider any of the $N+1$ possible values M' taken by m in the definition of the partition function. Clearly we have

$$\sum_{M'} e^{-\frac{1}{2}\beta(m-M')^2} \geq 1, \qquad (3.29)$$

since all terms are positive, and the term, where the running M' is exactly equal to m, is equal to one. By inserting this inequality into the definition of the partition function we have

$$Z_N(\beta) = \sum_{\sigma} e^{N\frac{1}{2}\beta m^2} \leq \sum_{\sigma} e^{N\frac{1}{2}\beta m^2} \sum_{M'} e^{-\frac{1}{2}\beta(m-M')^2} \qquad (3.30)$$

$$= \sum_{M'} e^{N(\log 2 + \log\cosh(\beta M') - \frac{1}{2}\beta M'^2)} \leq e^{N\bar{A}(\beta)}(N+1), \qquad (3.31)$$

where we have interchanged the sums over M' and σ, developed the square, performed the sum over σ, and exploited the definition of $\bar{A}(\beta)$ as a supremum.

At the end we have

$$\frac{1}{N}\log Z_N \leq \bar{A}(\beta) + \frac{1}{N}\log(N+1), \qquad (3.32)$$

where the correction term disappears in the infinite volume limit.

With this method the variational principle for the free energy per site appears as

$$A(\beta) = \sup_M \left(\log 2 + \log\cosh(\beta M) - \frac{1}{2}\beta M^2 \right), \qquad (3.33)$$

in a form which is not of Legendre type, and quite different from what is expected from the entropic principle

$$A(\beta) = \sup_M \left(\frac{1}{2}\beta M^2 + s(M) \right). \qquad (3.34)$$

Here there is a nice Legendre form. The parameters β and M^2 are conjugated. The functions $A(\beta)$, convex in β, and $\tilde{s}(M^2) = s(M)$, concave in M^2, are connected by the dual Legendre variational principles

$$A(\beta) = \sup_{M^2} \left(\frac{1}{2}\beta M^2 + \tilde{s}(M^2) \right), \qquad (3.35)$$

and

$$\tilde{s}(M^2) = \inf_{\beta} \left(-\frac{1}{2}\beta M^2 + A(\beta) \right). \qquad (3.36)$$

However, the two variational principles for $A(\beta)$ are in fact equivalent. This is easily shown by taking the variational expression coming from the convexity strategy, adding and subtracting a term so as to have the right Lagrange multiplier, and "worsening" the resulting expression through an additional *inf* over β, according to the chain

$$\frac{1}{N}\log Z_N \geq \log 2 + \log\cosh(\beta M) - \frac{1}{2}\beta M^2$$

$$= \frac{1}{2}\beta M^2 + \log 2 + \log\cosh(\beta M) - \beta M^2 \qquad (3.37)$$

$$\geq \frac{1}{2}\beta M^2 + \inf_{\beta}(\log 2 + \log\cosh(\beta M) - \beta M^2). \qquad (3.38)$$

One can immediately find that

$$\inf_{\beta}(\log 2 + \log\cosh(\beta M) - \beta M^2) = s(M). \qquad (3.39)$$

Therefore, the equivalence between the two variational principles is established.

Similar considerations will be developed in Section 3.9, in the case of disordered models, in the frame of an inverted variational principle.

3.5 The Entropic Principle in the Random Energy Model

As is very well known [5], this is a disordered model characterized by the random partition function

$$Z_N(\beta, J) = \sum_{\sigma} e^{\beta\sqrt{\frac{N}{2}}J_{\sigma}}, \qquad (3.40)$$

where β is the inverse temperature, and the J's are a family of centered normalized independent Gaussian random variables, with averages $\mathbb{E}J_\sigma = 0$ and $\mathbb{E}(J_\sigma J_{\sigma'}) = \delta_{\sigma\sigma'}$. Obviously, $\log Z_N(\beta, J)$ is convex in β for each sample of the J's.

The infinite volume limit can be controlled by standard interpolation arguments [7, 4]. Therefore, we have superadditivity for the quenched averages ($N = N_1 + N_2$)

$$\mathbb{E}\log Z_N(\beta, J) \geq \mathbb{E}\log Z_{N_1}(\beta, J) + \mathbb{E}\log Z_{N_2}(\beta, J), \qquad (3.41)$$

which implies convergence for the quenched densities

$$\lim_{N\to\infty}\frac{1}{N}\mathbb{E}\log Z_N(\beta, J) = \sup_N \frac{1}{N}\mathbb{E}\log Z_N(\beta, J) = A(\beta). \qquad (3.42)$$

Moreover, by a concentration of measure argument and Borel-Cantelli lemma, we also have the convergence of the densities

$$\lim_{N\to\infty} \frac{1}{N} \log Z_N(\beta, J) = A(\beta), \tag{3.43}$$

with probability one.

The entropy can even be explicitly calculated in the infinite volume limit, as Derrida has shown in his original paper [5].

For a parameter ϵ (with the meaning of minus the internal energy density), let us define the random microcanonical entropy density

$$\frac{1}{N} \log \sum_\sigma \chi(J_\sigma \geq \epsilon\sqrt{2N}), \tag{3.44}$$

where $\chi(\mathcal{A})$ is the truth function for the event $\mathcal{A}$ appearing.

Of course here we cannot take the quenched average, which is $-\infty$ because the χ's turn out to be all zero on a set of finite measure.

On the other hand, the annealed average is well defined. An easy calculation shows

$$\lim_{N\to\infty} \frac{1}{N} \log \mathbb{E} \sum_\sigma \chi(J_\sigma \geq \epsilon\sqrt{2N}) = \log 2 - \epsilon^2, \tag{3.45}$$

for any $\epsilon \geq 0$, while the limit equals $\log 2$ for negative values of ϵ.

With the help of the Borel-Cantelli lemma, we can easily establish the convergence of the random entropy density with probability one in the form

$$\lim_{N\to\infty} \frac{1}{N} \log \sum_\sigma \chi(J_\sigma \geq \epsilon\sqrt{2N}) = s(\epsilon), \tag{3.46}$$

where $s(\epsilon) = \log 2$ for $\epsilon \leq 0$, $s(\epsilon) = \log 2 - \epsilon^2$ for $0 \leq \epsilon \leq \sqrt{\log 2}$ (as in the annealed case), $s(\epsilon) = -\infty$, for $\epsilon > \sqrt{\log 2}$. Notice that $s(\epsilon)$ is decreasing and concave in the parameter ϵ.

The usual chain of arguments connecting the microcanonical ensemble with the canonical one, as described in [11], holds also in the disordered case, almost surely with respect to the quenched noise. Therefore we have the dual variational principles

$$A(\beta) = \sup_\epsilon (s(\epsilon) + \beta\epsilon), \tag{3.47}$$

where the *sup* is reached in the interval $0 \leq \epsilon \leq \sqrt{\log 2}$, and

$$s(\epsilon) = \inf_\beta (A(\beta) - \beta\epsilon). \tag{3.48}$$

Since $s(\epsilon)$ is completely known, the first variational principle gives us the explicit form of $A(\beta)$ as

$$A(\beta) = \log 2 + \frac{\beta^2}{4}, \tag{3.49}$$

for $0 \leq \beta \leq \beta_c = 2\sqrt{\log 2}$, and

$$A(\beta) = \beta \sqrt{\log 2}, \tag{3.50}$$

for $\beta \geq \beta_c$.

In conclusion we see that there is a full Legendre structure, where β and ϵ are conjugated parameters, while the convex $A(\beta)$ and the concave $s(\epsilon)$ are conjugated functions related by the dual variational principles. By the way, the recognition of the Legendre structure is useful because, for example, we derive the free energy from the known entropy.

These considerations can be easily extended to the generalized random energy model.

3.6 Functional Order Parameter and the Inverted Variational Principle

It is very well known that a functional order parameter, connected to the overlap distribution, is necessary for the description of disordered mean field models [10], as for example, the Sherrington-Kirkpatrick model for a spin glass [12].

Moreover, in this scheme, the free energy is given by a *sup* on the functional order parameter for an appropriately chosen trial functional. Therefore, the variational principle for the free energy appears in a form inverted with respect to the usual entropic principle. It is our purpose to interpret these features in the frame of a generalized Legendre structure.

First of all we give a simple explanation of the inversion, by exploiting the simple case of the REM, where everything can be explicitly calculated. It turns out that the inversion is due to the fact that the free energy (with the usual minus sign) $\log Z$, convex in the interaction, is also concave in the **covariance** of the random interaction.

The functional order parameter turns out to be dual, in the Legendre sense, to the covariance of the random interaction. The two variational principles, for the dual functionals – one convex in the functional order parameter, the other concave in the covariance – can be also established in general.

Legendre Structures in Statistical Mechanics 153

3.7 The Inverted Principle in the Random Energy Model

First of all let us establish a general fact concerning the (-)free energy density in the infinite volume limit

$$A(\beta) = \lim_{N \to \infty} \frac{1}{N} \mathbb{E} \log \sum_\sigma e^{\beta \sqrt{\frac{N}{2}} J_\sigma}. \qquad (3.51)$$

The covariance of the interaction is $\beta^2 \frac{N}{2} \delta_{\sigma\sigma'}$. Therefore it is linear in β^2. For our purpose, it is convenient to make the change of variables $\beta^2 = t$, $\beta = \sqrt{t}$, and interpret $A(\beta)$ as a function of t, by introducing $\tilde{A}(t) = A(\beta)$.

Even without knowing the explicit form of $A(\beta)$, it is easy to establish that $\tilde{A}(t)$ must be **concave** in t, through a simple interpolation argument.

Here is the simple proof: for any rational number $0 \leq \alpha \leq 1$ let us split a system made by an increasing number N of sites in two systems of N_1 and N_2 sites respectively, such that $N = N_1 + N_2$, and $N_1/N = \alpha, N_2/N = 1 - \alpha$. For example, if $\alpha = \bar{N}_1/\bar{N}, \bar{N}_1 \leq \bar{N}$, we can take $N = k\bar{N}$, $N_1 = k\bar{N}_1$ $N_2 = k(\bar{N} - \bar{N}_1)$, with $k \to \infty$.

Let t, t_1, t_2 be related by $t = \alpha t_1 + (1 - \alpha)t_2$. The configuration space of the $\sigma = (\sigma_1, \sigma_2, \ldots, \sigma_N)$ can be seen as a Cartesian product of the two block configurations $\sigma^{(1)} = (\sigma_1, \sigma_2, \ldots, \sigma_{N_1})$ and $\sigma^{(2)} = (\sigma_{N_1+1}, \sigma_{N_1+2}, \ldots, \sigma_N)$.

Introduce three systems of independent Gaussian random variables, $J_\sigma, J_\sigma^{(1)}, J_\sigma^{(2)}$, indexed by σ, with mean zero and covariances given by

$$\mathbb{E}(J_\sigma J_{\sigma'}) = \delta_{\sigma\sigma'}, \ \mathbb{E}(J_\sigma^{(1)} J_{\sigma'}^{(1)}) = \delta_{\sigma\sigma'}^{(1)}, \ \mathbb{E}(J_\sigma^{(2)} J_{\sigma'}^{(2)}) = \delta_{\sigma\sigma'}^{(2)}, \qquad (3.52)$$

where $\delta_{\sigma\sigma'} = 1$ if the two configurations σ, σ' are equal, and zero otherwise, $\delta_{\sigma\sigma'}^{(1)} = 1$ only if the two configurations restricted to the first block are equal, and analogously for $\delta^{(2)}$. As a matter of fact, $J^{(1)}$ and $J^{(2)}$ do depend only on the configurations of the first and the second block, respectively. Moreover $\delta_{\sigma\sigma'} = \delta_{\sigma\sigma'}^{(1)} \delta_{\sigma\sigma'}^{(2)}$.

Let $0 \leq s \leq 1$ be an interpolating parameter. Introduce the auxiliary function as

$$\phi_N(s) = \frac{1}{N} \mathbb{E} \log \sum_\sigma e^{\sqrt{s}\sqrt{t}\sqrt{\frac{N}{2}} J_\sigma} e^{\sqrt{1-s}\sqrt{t_1}\sqrt{\frac{N_1}{2}} J_\sigma^{(1)}} e^{\sqrt{1-s}\sqrt{t_2}\sqrt{\frac{N_2}{2}} J_\sigma^{(2)}}. \qquad (3.53)$$

Through a well known direct calculation [9], involving integration by parts with respect to the external noises, we have

$$\frac{d}{ds}\phi_N(s) = \frac{t}{4}(1 - \langle \delta_{\sigma\sigma'} \rangle_s) - \frac{N_1}{N}\frac{t_1}{4}(1 - \langle \delta_{\sigma\sigma'}^{(1)} \rangle_s) - \frac{N_2}{N}\frac{t_2}{4}(1 - \langle \delta_{\sigma\sigma'}^{(2)} \rangle_s), \qquad (3.54)$$

where $\langle \ \rangle_s$ denotes the quenched average on the replicated state with variables σ, σ'. By collecting all terms, and recalling the definitions introduced above, we have

$$\frac{d}{ds}\phi_N(s) = \frac{1}{4}\alpha t_1 \langle \delta^{(1)}_{\sigma\sigma'} - \delta_{\sigma\sigma'} \rangle_s + \frac{1}{4}(1-\alpha)t_2 \langle \delta^{(2)}_{\sigma\sigma'} - \delta_{\sigma\sigma'} \rangle_s, \qquad (3.55)$$

which is clearly nonnegative, since for example $\delta^{(1)}_{\sigma\sigma'} - \delta_{\sigma\sigma'} = \delta^{(1)}_{\sigma\sigma'}(1 - \delta^{(2)}_{\sigma\sigma'}) \geq 0$. Therefore $\phi_N(s)$ is non-decreasing in s. By taking into account that the sum over σ is factorized in two blocks at $s=0$, we immediately get

$$\phi_N(1) - \phi_N(0) = \frac{1}{N}\mathbb{E}\log\sum_\sigma e^{\sqrt{t}\sqrt{\frac{N}{2}}J_\sigma}$$
$$- \frac{N_1}{N}\frac{1}{N_1}\mathbb{E}\log\sum_{\sigma^{(1)}} e^{\sqrt{t_1}\sqrt{\frac{N_1}{2}}J^{(1)}_{\sigma^{(1)}}} - \frac{N_2}{N}\frac{1}{N_2}\mathbb{E}\log\sum_{\sigma^{(2)}} e^{\sqrt{t_2}\sqrt{\frac{N_2}{2}}J^{(2)}_{\sigma^{(2)}}}. \qquad (3.56)$$

By taking the infinite volume limit, we reach the announced concavity

$$\tilde{A}(t) \geq \alpha\tilde{A}(t_1) + (1-\alpha)\tilde{A}(t_2), \qquad (3.57)$$

for $t = \alpha t_1 + (1-\alpha)t_2$.

The concavity of $\tilde{A}(t)$ allows us to introduce the following clean Legendre structure. Let m be the parameter conjugated to t. Only the values $0 \leq m \leq 1$ will matter. Introduce the convex funcion $\psi(m)$ as the Legendre transform of $\tilde{A}(t)$

$$\psi(m) = \sup_t \left(\tilde{A}(t) - \frac{1}{4}tm \right). \qquad (3.58)$$

The conjugate inverse transform will give $\tilde{A}$ in terms of ψ in the form

$$\tilde{A}(t) = \inf_m \left(\psi(m) + \frac{1}{4}tm \right). \qquad (3.59)$$

For each t at the optimal value for $m(t)$ we have

$$\tilde{A}(t) = \psi(m(t)) + \frac{1}{4}tm(t). \qquad (3.60)$$

If we take the t derivative, only the term where t appears explicitly must be taken into account

$$\frac{d}{dt}\tilde{A}(t) = \frac{1}{4}m(t). \qquad (3.61)$$

It is easy to give the physical interpretation of the optimal $m(t)$. In fact, by a direct calculation we have

$$\frac{d}{dt}\tilde{A}(t) = \frac{1}{4}(1 - \langle \delta_{\sigma\sigma'} \rangle_t), \qquad (3.62)$$

Legendre Structures in Statistical Mechanics   155

where $\langle \ \rangle_t$ denotes the quenched average on the replicated state with variables σ, σ'. By a comparison with the previous one we get

$$m(t) = 1 - \langle \delta_{\sigma\sigma'} \rangle_t. \tag{3.63}$$

Since the replica "overlap" δ can take only the values $0, 1$, the meaning of the order parameter $m(t)$ is obvious: it is the probability that the two replicas are different, i.e., $\delta = 0$.

Therefore, in this Legendre structure, with the inverted variational principle, the dual parameter m is connected with the overlap distribution at the optimal value.

Let us notice that this inverted variational principle is deeply different from the entropic principle that we have introduced above. From a historical point of view, the entropy principle was firstly exploited for the solution of the REM [5], but it is the inverted variational principle which gives the solution to more complicated disordered models [10, 8, 14], as for example the Sherrington-Kirkpatrick mean field spin glass [12].

From a pedagogical point of view, it is important to pursue the inverted variational principle strategy also in the case of the REM, independently from the entropy principle.

It is easy to obtain the following *a priori bound* on $\tilde{A}(t)$ in the right direction for an inverted variational principle:

$$\tilde{A}(t) \leq \frac{\log 2}{m} + \frac{1}{4} tm, \tag{3.64}$$

holding for any $0 < m \leq 1$.

The proof is simple: for any $0 < m \leq 1$, we have for the quenched free energy per site the inequality

$$N^{-1} \mathbb{E} \log \sum_{\sigma} \exp\left(\sqrt{t}\sqrt{\frac{N}{2}} J_\sigma\right)$$
$$\leq m^{-1} N^{-1} \mathbb{E} \log \sum_{\sigma} \exp(m\sqrt{t}\sqrt{\frac{N}{2}} J_\sigma), \tag{3.65}$$

which holds for any spin system for purely thermodynamic reasons (positivity of the entropy).

Now we apply the annealed inequality, coming from convexity, $\mathbb{E} \log .. \leq \log \mathbb{E} ...$ The $\mathbb{E}$ of the *Boltzmannfaktor* is immediately calculated as

$$\mathbb{E} \exp\left(\sqrt{t}\sqrt{\frac{N}{2}} J_\sigma\right) = \exp\left(\frac{1}{4} tN\right). \tag{3.66}$$

By taking into account that $\sum_\sigma = 2^N$, we end with the inequality, uniform in N,

$$N^{-1}\mathbb{E}\log\sum_\sigma \exp\left(\sqrt{t}\sqrt{\frac{N}{2}}J_\sigma\right) \leq \frac{\log 2}{m} + \frac{1}{4}tm. \tag{3.67}$$

By taking the limit $N \to \infty$ we find the stated inequality for $\tilde{A}(t)$.

Of course, the optimal value for m is found if we take the *inf* (inverted principle). Now

$$\inf_m \left(\frac{\log 2}{m} + \frac{1}{4}tm\right) \tag{3.68}$$

is precisely in the form of a good Lagrange variational principle. In fact, $\frac{\log 2}{m}$ is convex in m, and the second term is linear in both t and m separately.

It is easy to find the optimal m. It turns out that $m = 1$ for $0 \leq t \leq t_c = 4\log 2$, while $m = \sqrt{t_c/t}$ for $t \geq t_c$. Moreover, by an explicit calculation, based on the results found from the entropy principle, the optimal value of the trial function turns out to be exactly $\tilde{A}(t)$.

It is also possible to reach this result working always in the frame of the inverted variational principle. Details are elementary and will not be developed here. Therefore the explicit form of the dual function is

$$\psi(m) = \frac{\log 2}{m}, \tag{3.69}$$

for $0 < m \leq 1$. In the Legendre frame, we can extend the definition of the dual function to $\psi(m) = \log 2$ for $m \geq 1$.

It is also possible to connect the entropy s and the function ψ in the frame of a Legendre scheme by an appropriate change of the conjugated variables.

The inverted variational principle has a very simple, and quite unexpected interpretation, expounded in next section. This interpretation is easily extended to the more general cases of disordered systems.

3.8 The Squared Interaction in the Random Energy Model

The dual function $\psi(m)$ has been introduced as a consequence of the previously stated a priori bound

$$\tilde{A}(t) \leq \frac{\log 2}{m} + \frac{1}{4}tm. \tag{3.70}$$

However, we can easily see that $\psi(m)$ can be recognized to be connected to the infinite volume limit of the free energy density for a modified model, where the

Legendre Structures in Statistical Mechanics

random energy interaction is squared. This quite unexpected and surprising fact is at the very basic origin of the inverted variational principle, and can be easily generalized to more general cases, as will now be shown.

Let us start from the expression for the partition function

$$Z_N(\beta, J) = \sum_\sigma e^{\beta \sqrt{\frac{N}{2}} J_\sigma}. \tag{3.71}$$

We recall that general principles of statistical mechanics suggest that any term appearing at the exponent in the *Boltzmannfaktor*, if properly normalized, behaves as a constant in the infinite volume limit with respect to the Boltzmann-Gibbs distribution. These terms in fact act as Lagrange multipliers. For example, the whole energy in the exponent will enforce the entropic principle for the minimum of the entropy at given internal energy. This fact holds even if the terms in the exponent are random, and form the basis of the so called Ghirlanda-Guerra identities in disordered systems [6].

With this motivation in mind, we evaluate the exponent in the following way by using Schwarz inequality

$$\beta \sqrt{\frac{N}{2}} J_\sigma = \beta \sqrt{m} \sqrt{\frac{N}{2}} \cdot \frac{J_\sigma}{\sqrt{m}}$$
$$\leq \frac{1}{2} \beta^2 m \frac{N}{2} + \frac{1}{2} \frac{J_\sigma^2}{m}, \tag{3.72}$$

where the parameter m has been introduced in order to reproduce the best value of J_σ. In fact, the error term going from the first to the second line is proportional to

$$\left(\frac{J_\sigma}{\sqrt{2N}} - \frac{1}{2} \beta m \right)^2, \tag{3.73}$$

and m will be chosen to make minimum this error term under the Boltzmann-Gibbs average.

By using the stated inequality, we can establish

$$\frac{1}{N} \log \sum_\sigma e^{\beta \sqrt{\frac{N}{2}} J_\sigma} \leq \frac{1}{4} \beta^2 m + \frac{1}{N} \log \sum_\sigma e^{\frac{1}{2} \frac{J_\sigma^2}{m}}. \tag{3.74}$$

We see that a model with interaction J_σ^2 appears in a very natural way. Now it is easy to prove that

$$\lim_{N\to\infty} \frac{1}{N} \log \sum_\sigma e^{\frac{1}{2}\frac{J_\sigma^2}{m}} = \psi(m) \tag{3.75}$$

$$= \frac{\log 2}{m}, \quad m \leq 1, \tag{3.76}$$

$$= \log 2, \quad m \geq 1, \tag{3.77}$$

which is exactly the function conjugated to the free energy in the inverted variational principle. Therefore, we find that the whole Legendre structure of the inverted Legendre variational principle rests on models with the squared interaction.

We will see in the following section how this generalizes to more general models.

3.9 The Legendre Structure in Mean Field Spin Glass Models

We consider models where the (-)free energy per site is defined as

$$A_N(g) = \frac{1}{N} \mathbb{E} \log \sum_\sigma e^{\sqrt{\frac{N}{2}}\mathcal{K}(\sigma)}, \tag{3.78}$$

where $\sigma \to \mathcal{K}(\sigma)$ is a family of centered Gaussian random variables with variances given by

$$\mathbb{E}(\mathcal{K}(\sigma)\mathcal{K}(\sigma')) = g(q(\sigma,\sigma')). \tag{3.79}$$

Here q is the overlap between two configurations,

$$q(\sigma,\sigma') = \frac{1}{N} \sum_i \sigma_i \sigma_i'. \tag{3.80}$$

Of course g must be a positive definite kernel as a function of the two configurations. This can be easily obtained by taking $\mathcal{K}$ as a sum over p-spin glass interactions ($p \geq 2$). For the sake of notational simplification we neglect any external field.

There is no problem with the infinite volume limit of the (-)free energy per site, $\lim_{N\to\infty} A_N(g) = A(g)$, by following the strategy introduced in [7].

To proceed toward the Legendre structure we must recognize the order parameter and the associated Lagrange multiplier.

Legendre Structures in Statistical Mechanics

In order to evaluate how $A(g)$ changes with g, let us rescale $\mathcal{K}$ to the form $\sqrt{t}\mathcal{K}$. Then a standard calculation gives

$$\frac{d}{dt}A_N(g) = \frac{1}{4}\langle (g(1) - g(q(\sigma,\sigma'))) \rangle, \qquad (3.81)$$

where $\langle \, \rangle$ as usual is the quenched average on the two replica Boltzmann-Gibbs states.

According to a very general principle (Talagrand positivity, Section 6.6 in [13], arising from the Ghirlanda-Guerra identities [6]), to the effect of the free energy evaluation, only the region where the overlaps are nonnegative give a contribution. Therefore, by calling $\rho(q)$ the positive overlap distribution, we have

$$\frac{1}{4}\langle (g(1) - g(q(\sigma,\sigma')))\rangle = \frac{1}{4}\int_0^1 (g(1) - g(q))\rho(q)\,dq \qquad (3.82)$$

$$= \frac{1}{4}\int_0^1 g'(q)x(q)\,dq, \qquad (3.83)$$

where we have written $(g(1) - g(q)) = \int_q^1 g'(q')\,dq'$, ($g'$ is the derivative of $g(q)$), have defined $x(q) = \int_0^q \rho(q')\,dq'$, and have exchanged the integrations on q and q'.

From the obtained expression, we can immediately recognize the two conjugated variables which will be involved in the dual variational principles: the first is $g' : q \to g'(q)$, which is connected with the interaction; the second is a general functional order parameter $x : q \to x(q)$, which only at the optimal value is connected with the overlap distribution.

Of course g' is sufficient to characterize the interaction, because in any case we can take $g(0) = 0$, without loss of generality. In conclusion, we see that the Lagrange multiplier must have the form

$$\frac{1}{4}\int_0^1 g'(q)x(q)\,dq, \qquad (3.84)$$

for a generic x.

The Legendre transform of $A(g)$ is defined as

$$\psi(x) = \sup_g (A(g) - \frac{1}{4}\int_0^1 g'(q)x(q)\,dq), \qquad (3.85)$$

and turns out to be automatically convex in x.

The inverse Legendre transform is

$$\underline{A}(g) = \inf_x (\psi(x) + \frac{1}{4}\int_0^1 g'(q)x(q)\,dq). \qquad (3.86)$$

Of course, $\underline{A}(g)$ turns out to be concave in g, and the full Legendre structure is established only if $A(g)$ is also concave in g, in which case $A(g) = \underline{A}(g)$, and effectively

$$A(g) = \inf_x (\psi(x) + \frac{1}{4}\int_0^1 g'(q)x(q)\,dq). \tag{3.87}$$

Therefore, the concavity of $A(g)$ in g plays an important role here.

Up to this point we have seen that the whole structure is a simple generalization of that found for the REM. In this simple case however, it was possible to prove directly the a priori concavity, and an upper bound in the Legendre form was easily found. The present scheme reproduces that in the REM, by taking $g'(q) = \delta(1-q)$.

Let us see what can be said about the upper bound for the general model. First of all let us recall that by using a simple interpolation argument [8] it was possible to obtain the following uniform upper bound:

$$A_N(g) \le \phi(g,x), \tag{3.88}$$

where the trial functional has the form

$$\phi(g,x) = \log 2 + f(0,0;x,g) - \frac{1}{4}\int_0^1 qg''(q)x(q)\,dq. \tag{3.89}$$

Here g'' is the second derivative of the covariance g with respect to q, and $f(0,0;x,g)$ is the value at $q=0, y=0$ of the function $[0,1] \times \mathbb{R} \ni (q,y) \to f(q,y;x,g)$ and is defined as a solution to the differential equation

$$\partial_q f + \frac{g''}{4}(\partial_{yy}^2 + x(q)(\partial_y f)^2) = 0, \tag{3.90}$$

with final condition at $q=1$

$$f(1,y;x,g) = \log\cosh y. \tag{3.91}$$

The trial functional, which comes for free from the interpolation method, is identical to that found in the frame of the replica trick [10]. The optimization is reached by taking the $\inf_x$, in complete coherence with the nature of the *upper* bound. Of course the bound holds also for the infinite volume limit

$$A(g) \le \phi(g,x). \tag{3.92}$$

We define

$$A_P(g) = \inf_x \phi(g,x). \tag{3.93}$$

Then Michel Talagrand [14] was able to establish that $A(g) = A_P(g)$.

Recently Antonio Auffinger and Wei-Kuo Chen [1] have established that $f(0,0;x,g)$ as a functional of x is strictly convex. Since the other term in $\phi(g,x)$ is affine in x, we immediately have that the functional order parameter x coming from optimization is uniquely defined. A truly remarkable result!

In particular we notice that a phase transition is **not** characterized by multiple values of the order parameter, as in the ordered models.

Of course, the form of the trial functional $\phi(g,x)$ is not consistent with a Legendre form. The term

$$-\frac{1}{4}\int_0^1 qg''(q)x(q)\,dq \qquad (3.94)$$

has the right properties of being affine in x and linear in g, but it is not in the expected form

$$+\frac{1}{4}\int_0^1 g'(q)x(q)\,dq. \qquad (3.95)$$

Moreover, $f(0,0;x,g)$ depends heavily on both x and g. Therefore, the variational principle does not give $A(g)$ as the Legendre transform of some $\psi(x)$ in the expected form

$$A(g) = \inf_x(\psi(x) + \frac{1}{4}\int_0^1 g'(q)x(q)\,dq). \qquad (3.96)$$

However, there is a way out. Let us start from the bound

$$A(g) \leq \phi(g,x), \qquad (3.97)$$

written in the identical form

$$A(g) \leq \phi(g,x) - \frac{1}{4}\int_0^1 g'(q)x(q)\,dq + \frac{1}{4}\int_0^1 g'(q)x(q)\,dq, \qquad (3.98)$$

where we have subtracted and added the expected Lagrange multiplier. Now let us *define*

$$\psi(x) = \sup_g(\phi(g,x) - \frac{1}{4}\int_0^1 g'(q)x(q)\,dq). \qquad (3.99)$$

Then the bound, slightly worsened, becomes

$$A(g) \leq \psi(x) + \frac{1}{4}\int_0^1 g'(q)x(q)\,dq, \qquad (3.100)$$

which is in the Legendre form. It can be easily seen that the optimization gives the same value as in the original bound. Therefore, the Legendre structure can be enforced also in the general case

$$A(g) = \inf_x (\psi(x) + \frac{1}{4} \int_0^1 g'(q) x(q) \, dq). \tag{3.101}$$

In a recent preprint [2] Antonio Auffinger and Wei-Kuo Chen have shown that the effectiveness of the procedure based on taking the $\sup_g$ rests on solid ground, since the strict concavity of the involved functional can be rigorously shown.

Up to this point we have seen that the whole structure is a simple generalization of that found for the REM. In the simple case of the REM, however, it was possible to prove directly the a priori concavity, and the upper bound in the Legendre form was directly found. The present scheme reproduces that found in the REM, by taking $g'(q) = \delta(1-q)$.

If we adopt the view based on squared interactions, as explained for the REM, then we can write in the exponent of the *Boltzmannfaktor*

$$\sqrt{\frac{N}{2}} \mathcal{K}(\sigma) = \sqrt{\frac{N}{2}} \sum_{p=2} \beta_p \mathcal{K}_p(\sigma) \tag{3.102}$$

$$= \sum_{p=2} \sqrt{\frac{N}{2}} \sqrt{m_p} \beta_p \cdot \frac{\mathcal{K}_p(\sigma)}{\sqrt{m_p}} \tag{3.103}$$

$$\leq \frac{1}{2} \sum_{p=2} \frac{N}{2} m_p \beta_p^2 + \frac{1}{2} \sum_{p=2} \frac{\mathcal{K}_p^2(\sigma)}{m_p}, \tag{3.104}$$

where m_p play the role of trial parameters. We now denote by β, t, m, x, respectively, the sets of all $\beta_p, t_p = \beta_p^2, m_p, x_p = 1/m_p$.

By defining $\tilde{A}(t) = A(g)$, therefore we have the bound

$$\tilde{A}(t) \leq \frac{1}{4} \sum_p t_p m_p + \psi_N(m), \tag{3.105}$$

where

$$\psi_N(m) = \frac{1}{N} \mathbb{E} \log \sum_\sigma \exp\left(\frac{1}{2} \sum_{p=2} \frac{\mathcal{K}_p^2(\sigma)}{m_p}\right), \tag{3.106}$$

and the analogous bound in the infinite volume limit.

The resulting dual Legendre structure recognizes the sets t_p, m_p as dual parameters. The functionals $\psi(m)$, convex in m, and $\tilde{A}(t)$, concave in t, are connected through the dual variational principles

$$\tilde{A}(t) = \inf_m \left(\frac{1}{4} \sum_p t_p m_p + \psi(m) \right), \tag{3.107}$$

and

$$\psi(m) = \sup_t \left(\tilde{A}(t) - \frac{1}{4} \sum_p t_p m_p \right). \tag{3.108}$$

By comparison with the previously established inverted variational principle, we can identify the Lagrangian multipliers

$$\frac{1}{4} \int_0^1 g'(q) x(q) \, dq = \frac{1}{4} \sum_p t_p m_p, \tag{3.109}$$

so that the trial functional order parameter $q \to x(q)$ and the trial parameters m_p are connected by

$$m_p = p \int_0^1 p^{p-1} x(q) \, dq, \tag{3.110}$$

since

$$g'(q) = \sum_p \beta_p^2 p q^{p-1}. \tag{3.111}$$

However, the variational picture based on the squared interactions, where the set of m_p plays the role of order parameter, appears to be more general than the picture based on the functional order parameter $x(q)$. For example, if $t_{\bar{p}} = 0$, for some particular $\bar{p}$, then the optimization on the corresponding $m_{\bar{p}}$ gives $m_{\bar{p}} = \infty$, a value which cannot be understood in the frame where there is the functional order parameter $x(q)$. In general, the exploitation of the order parameter $x(q)$ is based on the tacit understanding that the case $t_{\bar{p}} = 0$ is realized by taking the infinite volume limit while $t_{\bar{p}} > 0$, and then letting $t_{\bar{p}} \to 0$ after.

3.10 Outlook and Perspectives

We start this section with a simple observation. Let us consider the interpolating bound,

$$A_N(g) \leq \phi(g, x), \tag{3.112}$$

as explained in Section 3.9, and apply it to a sequence of p-spin models with $p \to \infty$. It is very well known that the REM will be obtained in the limit. Consider an order parameter of the constant shape $x(q) = m$. Through a long and painful calculation it can be proven that if we take $g(q) = tq^p$ and a constant order parameter $x(q) = m$, then we have

$$\lim_{p \to \infty} \phi(g,x) = \log 2 + \frac{1}{4}mt, \qquad (3.113)$$

which is the bound for the REM and is easily obtained through simple thermodynamic arguments, as given in Section 3.6. Therefore, the broken replica symmetry bounds also give the simple Legendre bound in REM through a cumbersome limiting procedure. Research is under way on how to modify the broken replica symmetry bound procedure of [8] in order to get bounds directly in the Legendre form.

Another perspective for future developments is to establish the Legendre structure in the case of multispecies spin glass models, as in the bipartite ones. This would have interesting applications for neural networks (see for example [3]).

This paper has considered the case of disordered mean field models only for the sake of simplicity. In the ordered case, the dual Legendre structure for entropy and free energy density can be established also for short range models, as shown for example in [11]. Similarly, inverted variational principles can be constructed in the disordered short range interaction case by introducing models with the random interactions squared. Of course, in this case, the functional order parameter will be much more complicated. But the general structure is the same.

References

[1] A. Auffinger and Wei-Kuo Chen, *The Parisi formula has a unique minimizer*, Communications in Mathematical Physics. 335, 1429–1444 (2014).

[2] A. Auffinger and Wei-Kuo Chen, *The Legendre structure of the Parisi formula*. (2015). arXiv:1510.03414.

[3] A. Barra, G. Genovese and F. Guerra, *The replica symmetric approximation of the analogical neural network*, Journal of Statistical Physics. 140, 784–796 (2010).

[4] P. Contucci, M. Degli Esposti, C. Giardinà and S. Graffi, *Thermodynamical limit for correlated Gaussian random energy models*, Commun. Math. Phys. 236, 55–63 (2003).

[5] B. Derrida, *Random-energy model: an exactly solvable model of disordered systems*, Phys. Rev. B24, 2613 (1981).

[6] S. Ghirlanda and F. Guerra, *General properties of overlap probability distributions in disordered spin systems. Towards Parisi ultrametricity*, J. Phys. A: Math. Gen. 31, 9149 (1998).

[7] F. Guerra and F. L. Toninelli, *The thermodynamic limit in mean field spin glass models*, Comm. Math. Phys. 230, 71–79 (2002).

[8] F. Guerra, *Broken replica symmetry bounds in the mean field spin glass model*, Comm. Math. Phys. 233, 1–12 (2003).

[9] F. Guerra, *An introduction to mean field spin glass theory: methods and results*, in *Mathematical Statistical Physics* (A. Bovier et al. eds.). Oxford: Elsevier, 243–271 (2006).

[10] M. Mézard, G. Parisi and M. A. Virasoro, *Spin glass theory and beyond*. Singapore: World Scientific, (1987).

[11] D. Ruelle, *Statistical mechanics: rigorous results*. New York: W.A. Benjamin Inc., (1969).

[12] D. Sherrington and S. Kirkpatrick, *Solvable model of a spin-glass*. Phys. Rev. Lett. 35, 1792–1796 (1975).

[13] M. Talagrand, *Spin glasses: a challenge for mathematicians*. Berlin: Springer-Verlag, (2003).

[14] M. Talagrand, *The Parisi formula*, Annals of Mathematics. 163, 221–263 (2006).

4
Extrema of Log-correlated Random Variables
Principles and Examples

LOUIS-PIERRE ARGUIN

4.1 Introduction

There has been tremendous progress recently in the understanding of the extreme value statistics of stochastic processes whose variables exhibit strong correlations. The purpose of this paper is to survey the recent progress in this field for processes with logarithmically decaying correlations. The material has been developed for the mini-course *Extrema of log-correlated random variables: principles and examples* at the Introductory School held in January 2015 at the Centre International de Rencontres Mathématiques in Marseille for the trimester *Disordered systems, Random spatial processes and Applications* of the Institut Henri Poincaré.

The study of extreme values of stochastic processes goes back to the early twentieth century. The theory was developed in the context of independent or weakly correlated random variables. The first book regrouping the early advances in the field is the still relevant *Statistics of Extremes* by E. J. Gumbel [43] published in 1958. Gumbel credits the first work on extreme values to Bortkiewicz in 1898, which is one of the first papers emphasizing the importance of Poisson statistics. After progress by Fisher and Tippett in understanding the limit law of the maximum of a collection of independent and identically distributed (IID) random variables, the theory culminated in 1943 with the classification theorem of Gnedenko [42] which showed that the distribution of the maximum of N IID random variables when properly recentered and rescaled can only be of three types: Fréchet, Weibull or Gumbel. Prior to this, von Mises had shown sufficient conditions for convergence to the three types. All in all, it took the theory sixty years between the first rigorous results and an essentially complete theory as it appears in Gumbel's review. There are now many excellent textbooks on extreme value statistics, see e.g., [19, 28, 49, 54].

The classical theory, however, does not apply to collections of random variables that exhibit strong correlations, i.e., correlations of the order of the variance. As Gumbel points out at the very beginning of *Statistics of Extremes*:

Another limitation of the theory is the condition that the observations from which the extremes are taken should be independent. This assumption, made in most statistical work, is hardly ever realized.

In the last thirty years, there have been important breakthroughs in mathematics and in physics to identify *universality classes* for the distributions of extreme values of strongly correlated stochastic processes. One such class that has attracted much interest in recent years is the class of *log-correlated fields*. Essentially, these are stochastic processes for which the correlations decay logarithmically with the distance. Even though the theory has not reached a complete status as in the IID case, there has been tremendous progress in identifying the distribution of extremes in this case and to develop rich heuristics that describe accurately the statistics of large values. There are already excellent survey papers on the subject by Bovier [20] and Zeitouni [57], all offering a different perspective on log-correlated fields. The present paper modestly aims at complementing the growing literature on the subject by focusing on the techniques to obtain precise estimates on the order of the maximum of log-correlated fields and their potential applications to seemingly unrelated problems: the maxima of the characteristic polynomials of random matrices and the maxima of the Riemann zeta function on an interval of the critical line. The approach we take is based on Kistler's multiscale refinement of the second moment method introduced in [46].

The presentation of this paper is organized as follows. In Section 4.1, we review the basic theory in the case of IID random variables and will introduce log-correlated fields in generality. Section 4.2 focuses on two important examples of log-correlated fields: *branching random walks* and *the two-dimensional Gaussian free field*. In particular, three properties of these fields (and of log-correlated fields in general) are singled out as they play a crucial role in the analysis of extremes. In Section 4.3, we describe the general method of Kistler to prove fine asymptotics for the maximum of log-correlated fields [46]. Finally, in Section 4.4, we will explain how the method is expected to be applicable to study the high values of the characteristic polynomials of random matrix ensembles as well as the large values of the Riemann zeta function on an interval of the critical line.

4.1.1 Statistics of Extremes

We are interested in the problem of describing the maxima of a stochastic process in the limit where there are a large number of random variables. For example, consider the process consisting of IID centered Gaussian random variables. One

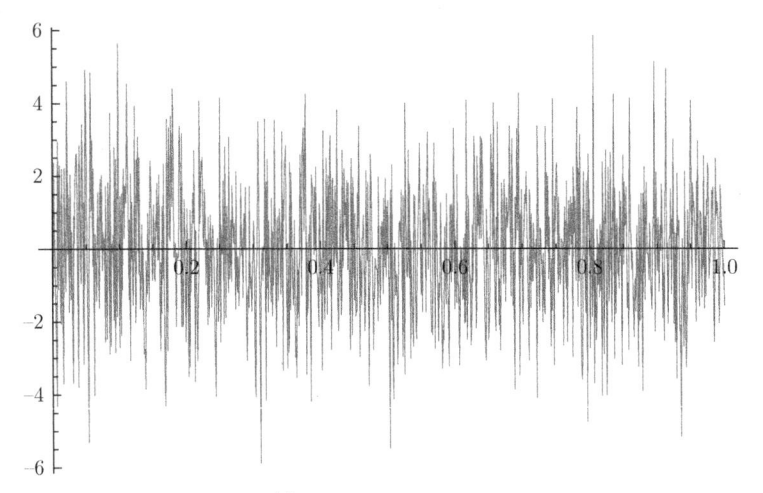

Figure 4.1. A realization of 2^{10} centered Gaussian random variables of variance $\frac{1}{2}\log 2^{10}$.

realization of this process for 2^{10} variables is depicted in Figure 4.1 with a particular choice of the variance. The maximum of the rugged landscape of this realization lies around 6. If we repeat this exercise for several realizations, we would observe that this is the case for every realization: the maximum has fluctuations roughly of order one around a deterministic value which is close to 6.

Now consider a drastically different stochastic process: let U be a random $2^{10} \times 2^{10}$ unitary matrix sampled uniformly from the unitary group and consider the modulus of the characteristic polynomial on the unit circle

$$P_U(\theta) = \left|\det(e^{i\theta} - U)\right|.$$

A realization of the logarithm of this quantity on the interval $[0, 2\pi]$ is given in Figure 4.2. This random process is of course qualitatively very different from the IID process above. It is not Gaussian yet it is a continuous process with singularities at the eigenvalues of U. However, perhaps by coincidence, the maximum of the landscape lies also around 6. In fact, if we repeat the experiment, we would again observe that the maximum fluctuates around a deterministic value close to 6. If we were precise enough, we would see that the deterministic value is close to the one for IID variables but there is a slight discrepancy. The conjecture of Fyodorov-Hiary-Keating [39, 40] makes a detailed prediction for the maximum of $P_U(\theta)$. The distribution of the maximum is different in the limit of a large number of variables from the one of IID, though it is close. However, it is a perfect match with the distribution of a log-correlated Gaussian field. This conjecture is still open (even to first order). However, the mechanism that would be responsible for this conjecture to hold is now quite clear. This is the goal of this review: to illustrate the

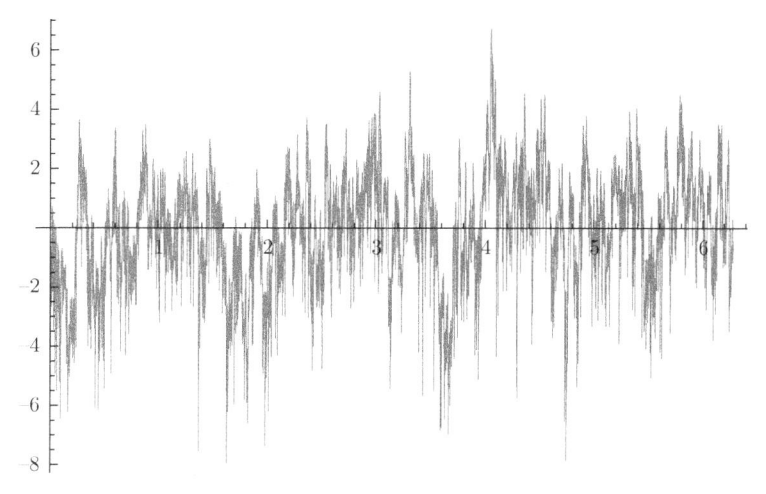

Figure 4.2. A realization of $P_U(\theta)$ on $[0, 2\pi]$ of a random unitary matrix U of size $2^{10} \times 2^{10}$.

techniques for log-correlated fields in the forthcoming sections and their relations to $P_U(\theta)$ in Section 4.4.

Throughout the paper, the framework will be as follows: the random field will be indexed by a discrete space V_n with 2^n points. One important example is $V_n = \{-1, 1\}^n$ representing the configurations of n spins. In most examples, V_n will be equipped with a metric that we denote $d(v, v')$, $v, v' \in V_n$. We consider the random field

$$X(n) = \{X_v(n), v \in V_n\}.$$

defined on a probability space (Ω, P). We will suppose without loss of generality that $E[X_v(n)] = 0$ for all $v \in V_n$. In most examples, the variance will be the same for every $v \in V_n$. This is not true for the two-dimensional Gaussian free field because of the boundary effect. However, the variance is almost uniform for most points which ultimately does not affect the results, but instead complicates the analysis. With this in mind, we choose to parametrize the variance as

$$E[X_v(n)] = \sigma^2 n,$$

for some σ^2. This turns out to be the correct physical parametrization: it gives a nontrivial limit for the free energy per spin in the limit $n \to \infty$, cf. Section 4.1.3.

Even though the analysis presented here is not limited to Gaussian fields, it is convenient to present the ideas in this setting. There are many advantages to doing so. First, the distribution of the field $X(n)$ is then entirely determined by the covariances

$$E[X_v(n)X_{v'}(n)].$$

Second, estimates for the probability of large values for Gaussians are very precise: it is not hard prove the Gaussian estimate for $X_v(n) \sim \mathcal{N}(0, \sigma^2 n)$,

$$P(X_v(n) > a) = \frac{(1+o(1))}{\sqrt{2\pi}} \frac{\sqrt{\sigma^2 n}}{a} \exp\left(-\frac{a^2}{2\sigma^2 n}\right) \quad \text{for } a > \sqrt{\sigma^2 n}. \quad (4.1)$$

Finally, comparison results for Gaussian processes can be used in some instances to bound functionals of the processes above and below with functionals of processes with simpler correlation structures.

The central question of extreme value theory is to describe the functional

$$\max_{v \in V_n} X_v(n) \text{ in the limit } n \to \infty.$$

As for other global functionals of a stochastic process such as the sum, it is reasonable to expect universality results for the limit, i.e., results that depend mildly on the fine details of the distributions of $X(n)$. From a physics perspective, this question corresponds to describing the ground state energy of a system with state space V_n and energy $-X_v(n)$ for the state v. The system is said to be *disordered* because the energies depend on the realization of the sample space Ω. A realization of the process as depicted in Figures 4.1 and 4.2 from this point of view represents the energy landscape of the system for a given disorder.

With a perhaps naive analogy with the limit theorems for sum of IID random variables, one might expect a corresponding *law of large numbers* and *central limit theorems* for the maximum. More precisely, the problem is to:

| Find a recentering a_n and rescaling b_n such that $\max_{v \in V_n} \frac{X_v(n) - a_n}{b_n}$ converges in law. | (4.2) |

If these sequences exist, it suggests that there is a nontrivial random variable $\mathcal{M}$ independent of n such that the approximate equality holds:

$$\max_{v \in V_n} X_v(n) \approx a_n + b_n \mathcal{M}.$$

In this regard, the recentering a_n is the term analogous to the law of large numbers, whereas b_n sizes the magnitude of the fluctuations with distribution $\mathcal{M}$, and thus it belongs to the central limit theorem regime. Of course, a_n should be unique up to an additive constant and b_n up to a multiplicative constant. We refer to a_n as the *order of the maximum*. We shall focus on general techniques to rigorously obtain the order of the maximum a_n for log-correlated fields in Section 4.4. The proofs of fluctuations are technically more involved, but the heuristics developed for the order of the maximum are crucial to the analysis at a more refined level.

The starting point of our analysis is the result when $X_v(n)$ are IID Gaussians with variance $\sigma^2 n$. In this case, the problem (4.2) is a simple exercise and has a complete answer.

Proposition 4.1.1 *Let $X(n) = (X_v(n), v = 1, \ldots, 2^n)$ be IID centered Gaussians with variance $\sigma^2 n$. Then for $c = \sqrt{2 \log 2}\, \sigma$ and*

$$a_n = cn - \frac{1}{2}\frac{\sigma^2}{c}\log n, \qquad b_n = 1, \tag{4.3}$$

we have

$$\lim_{n \to \infty} P\left(\max_{v \in V_n} X_v(n) \leq a_n + b_n x\right) = \exp(-Ce^{-cx}), \quad \text{with } C = \frac{\sigma/c}{\sqrt{2\pi}}.$$

In other words, the maximum converges in distribution to a *Gumbel random variable* also known as *double-exponential*. The order of the maximum is linear in n at first order with velocity $c = \sqrt{2 \log 2}\, \sigma$. The subleading order is logarithmic in n. The proof of this is simple but is instructive for the case of log-correlated fields. The same type of arguments can be used for the *order statistics*, i.e., the joint distribution of first, second, third, etc. maxima. It converges to a Poisson point process with exponential density.

Proof of Proposition 4.1.1. Clearly, since the random variables are IID, we have

$$P\left(\max_{v \in V_n} X_v(n) \leq a_n + b_n x\right) = \left(1 - \frac{2^n P(X_v(n) > a_n + b_n x)}{2^n}\right)^{2^n}.$$

This implies that the convergence of the distribution reduces to show that

$2^n P(X_v(n) > a_n + b_n x)$ converges for an appropriate choice of a_n and b_n.

The right side is exactly the expected *number of exceedances*, that is

$$E[\mathcal{N}(y)], \quad \text{where } \mathcal{N}(y) = \#\{v \in V_n : X_v(n) > y\}$$

for $y = a_n + b_n x$. It is not hard to prove using the Gaussian estimate (4.1) that the choice $a_n = cn - \frac{1}{2}\frac{\sigma^2}{c}\log n$ and $b_n = 1$ ensures the convergence. In fact, the leading order cn lies in the large deviation regime for the distribution $X_v(n)$. It compensates exactly the *entropy* term 2^n. As for the subleading order $-\frac{\sigma^2}{2c}$, it balances out the fine asymptotics

$$\frac{\sigma^2 n}{a_n} = \frac{\sqrt{\sigma^2 n}}{cn(1 + o(1))} = O(n^{-1/2})$$

in the Gaussian estimate. $\square$

It is a real challenge to achieve precise results of the kind of Proposition 4.1.1 for a stochastic $X(n)$ with correlations of the order of the variance. Correlations should affect the order of the maximum and its fluctuations, but to which extent? It is reasonable to expect the statistics of extremes of such processes to retain features of the IID statistics if we look at a coarse-enough scale. When correlations are present the expected number of exceedances is no longer a good proxy for the correct level of the maximum, since, by linearity of the expectations, it is blind to correlations and is the same as IID. The expected number of exceedances turns out to be inflated by the rare events of exceedances far beyond the right level of the maximum. As we will see in Section 4.3, the number of exceedances needs to be modified to reflect the correct behavior of the maxima.

4.1.2 Log-correlated Fields

In the framework where $X(n) = (X_v(n), v \in V_n)$ is a stochastic process on a metric space V_n with distance d, a *log-correlated* field has correlations

$$E[X_v(n)X_{v'}(n)] \approx -\log d(v, v'). \tag{4.4}$$

In particular, the correlation decays very slowly with the distance! Slower than any power. We emphasize that the covariance might not be exactly logarithmic for the field to exhibit a log-correlated behavior. For one, the covariance matrix must be positive definite, hence the covariance might be exactly the logarithm of the distance. Moreover, the logarithmic behavior might only be exhibited in the bulk due to some boundary effect. As we will see in Section 4.2, this is the case for the two-dimensional Gaussian free field. Other important examples of processes in this class are branching random walks and models of Gaussian multiplicative chaos. It is good to re-express (4.4) in terms of the size of neighborhoods with strong correlations with a fixed point v. More precisely, for $r > 0$, consider the number of points v' whose covariance is a fraction r of the variance. The logarithmic nature implies that

$$\frac{1}{2^n}\#\{v' \in V_n : \frac{E[X_v(n)X_{v'}(n)]}{E[X_v(n)^2]} > r\} \approx 2^{-rn}. \tag{4.5}$$

In other words, it takes approximately 2^{rn} balls to cover the space V_n with neighborhoods where correlations between points are greater than r times the variance. In particular, the size of these neighborhoods lies at mesoscopic scales compared to the size of the systems. This will play an important role in the analysis of Sections 4.2 and 4.3.

The motivations for the study of log-correlated fields are now plenty. Firstly, in the physics literature, the work of Carpentier and Le Doussal [26] spurred a lot of interest in the study of such processes as energy landscapes of disordered systems. We mention in particular the works of Fyodorov and Bouchaud [38] and Fyodorov et al. [41]. These papers develop a statistical mechanics approach to the problem of describing the extreme value statistics of the systems that are applicable to a wide range of systems. Secondly, these processes play an essential role in Liouville quantum gravity as well as models of three-dimensional turbulence. We refer to the excellent review on Gaussian multiplicative chaos of Rhodes and Vargas [55] and references therein for details in these directions. Thirdly, many models of volatility in finance are now built on the assumption that the time-correlations of the returns are log-correlated, see also [55] for more details on this. Finally, as we shall see in Section 4.4, log-correlated fields seem to provide the right structure to study the local maxima of the Riemann zeta function on the critical line as suggested in [39, 40].

It is argued in the physics literature that the distribution of the maximum of log-correlated fields is a borderline case where the features of IID statistics should still be apparent. In fact, it is expected that the following should hold for log-correlated fields that are close to Gaussians in some suitable sense.

Conjecture 4.1.2 (Informal) *Let $X(n) = \{X_v(n), v \in V_n\}$ be a log-correlated field in the sense of* (4.4) *with $E[X_v(n)] = 0$ and $E[X_v(n)^2] \approx \sigma^2 n$. Then for $c = \sqrt{2\log 2}\sigma$ and*

$$a_n = cn - \frac{3}{2}\frac{\sigma^2}{c}\log n, \qquad b_n = 1, \tag{4.6}$$

we have

$$\lim_{n \to \infty} P\left(\max_{v \in V_n} X_v(n) \le a_n + b_n x\right) = E[e^{-CZe^{-cx}}],$$

for some constant C and random variable Z called the derivative martingale.

This is to be compared with Proposition 4.1.1 for IID random variables. Perhaps surprisingly, the first order of the maximum is the same as for IID with the same velocity. The correlations start having an effect in the subleading where $\frac{1}{2}$ for IID is changed to $\frac{3}{2}$. This seemingly small change hides an important mechanism of the extreme value statistics that will be explained in Section 4.3.2. For now, let us just observe that under the recentering $a_n = cn - \frac{3}{2}\frac{\sigma^2}{c}\log n$, the expected number of exceedances diverges like n. To get the correct order, it will be necessary to modify the exceedances with a description of the values at each scale of the field. Finally, the fluctuations are not exactly Gumbel as for IID Gaussians. The distribution is *a mixture of Gumbels*, the mixture being on the different realizations of the random variables. This in fact changes qualitatively the nature of the distribution as it is expected that the right tail behaves like xe^{-cx} for x large whereas as it is e^{-cx} for double exponential distribution.

Conjecture 4.1.2 was proved in many instances of log-correlated fields. The most important contribution is without a doubt the seminal work of Bramson [25] who proved the result for branching Brownian motion (see Lalley and Sellke [47] for the expression in terms of the derivative martingale). For branching random walks, it was proved in great generality by Aïdekon [3] (see also [1, 14, 24]); for the two-dimensional Gaussian free field, by Bramson et al. [23] (see Biskup and Louidor [17] for the expression in terms of the derivative martingale); for a type of Gaussian multiplicative chaos by Madaule [50]; and finally in great generality for log-correlated Gaussian fields by Ding et al. [35].

We will not have time in this review paper to touch the subject of the order statistics or *extremal process* of log-correlated fields. Again, it turns out that the statistics retain features of the IID case. In fact, it is expected that the extremal process is a Poisson cluster process or Poisson decorated point process. This was proved for branching Brownian motion in [4, 8–10], for branching random walks in [51], and partially for the two-dimensional Gaussian free field in [17].

4.1.3 Relations to Statistical Physics

Questions of extreme value statistics such as Conjecture 4.1.2 can be addressed from a statistical physics point of view. In that case the object of interest is the *partition function*

$$Z_n(\beta) = \sum_{v \in V_n} \exp(\beta X_v(n)), \quad \beta > 0. \tag{4.7}$$

(Observe that we use β as opposed to the customary $-\beta$ since we are interested in the maximum and not the minimum of $X(n)$.) By design, the points v with a high value of the field have a greater contribution to the partition function, and the continuous parameter β (the inverse temperature) adjusts the magnitude of the contribution. The *free energy per particle* or *normalized log-partition function* is also of importance:

$$f_n(\beta) = \frac{1}{\beta n} \log Z_n(\beta). \tag{4.8}$$

In particular, in the limit $n \to \infty$, the free energy contains the information on the first order of the maximum. Indeed, we have the elementary inequalities

$$\frac{\max_{v \in V_n} X_v(n)}{n} \leq f_n(\beta) \leq \frac{\log 2}{\beta} + \frac{\max_{v \in V_n} X_v(n)}{n},$$

therefore

$$\lim_{n \to \infty} \frac{\max_{v \in V_n} X_v(n)}{n} = \lim_{\beta \to \infty} \lim_{n \to \infty} f_n(\beta), \quad \text{whenever the limits exist.}$$

To obtain finer information on the maximum, such as subleading orders or fluctuations, one could study directly the partition functions $Z_n(\beta)$. For example, one could compute the moments of $Z_n(\beta)$ and try to infer the distribution of $Z_n(\beta)$ from them. This is the approach taken in [38, 41] for example, but there are major obstacles to overcome to make this type of argument fully rigorous.

For log-correlated fields, it is possible to compute rigorously the free energy by the Laplace method. We have by rearranging the sum over v that

$$f_n(\beta) \approx \frac{1}{n} \log \left(\sum_{E \in [0,c]} \exp(\beta n E + n S_n(E)) \right), \tag{4.9}$$

where

$$S_n(E) = \frac{1}{n} \log \mathcal{N}_n(En) = \frac{1}{n} \log \#\{v \in V_n : X_v(n) > En\} \tag{4.10}$$

is the *entropy* or *log-number of high points*, i.e., the log-number of exceedances at level En. In the limit, it is possible to prove in some cases (see e.g., [27] and [12, 13]) that for log-correlated fields

$$S(E) = \lim_{n \to \infty} S_n(E) = \begin{cases} 0 & \text{if } E \geq c \\ \log 2 - \frac{E^2}{2\sigma^2} & \text{if } E \in [0, c] \end{cases} \quad \text{in probability,} \tag{4.11}$$

where $c = \sqrt{2 \log 2} \sigma$. We will see a general method to prove this in Section 4.3.1. Note that this is exactly the result one would obtain for IID random variables by applying (4.1)! The free energy is then easily calculated like a Gibbs variational principle

$$\lim_{n \to \infty} f_n(\beta) = \max_{E \in [0,c]} \left\{ \beta E + \log 2 - \frac{E^2}{2\sigma^2} \right\} = \begin{cases} \log 2 + \frac{\beta^2 \sigma^2}{2} & \text{if } \beta < \frac{c}{\sigma^2} \\ c\beta & \text{if } \beta \geq \frac{c}{\sigma^2}. \end{cases} \tag{4.12}$$

Again, not suprisingly, the free energy of log-correlated fields corresponds to the one for IID variables, also known as the *random energy model* (REM) as introduced by Derrida [30]. In particular, the free energy exhibits a *freezing* phase transition, i.e., for $\beta > \frac{c}{\sigma^2}$, the free energy divided by β is constantly reflecting the fact the partition function is supported on the maximal values of the field. Equation (4.12) was first proved for branching random walks by Derrida and Spohn [32] who pioneered the study of log-correlated random fields as disordered systems. Equation (4.12) was also proved for other log-correlated Gaussian fields in [12,13].

Another important object in statistical physics is the *Gibbs measure*, which in our context is a random probability measure on V_n, defined as

$$G_{n,\beta}(v) = \frac{\exp \beta X_v(n)}{Z_n(\beta)}, \quad v \in V_n. \tag{4.13}$$

This contains refined information on the order statistics of the field $X(n)$. We only mention here that the limiting distribution of the Gibbs measure depends on the parameter β. In particular, the *annealed Gibbs measure*

$$G_{n,\beta}^{ann}(v) = \frac{\exp \beta X_v(n)}{E[Z_n(\beta)]}$$

should have a nontrivial continuous limit for high temperature $\beta < \frac{c}{\sigma^2}$ [55], whereas for low temperature $\beta \geq \beta_c$, the quenched Gibbs measure as in (4.13) is the relevant limiting object and should be singular [12, 13].

4.2 Examples and General Properties

Branching random walk serves as a guiding example to prove rigorous results for log-correlated fields. In particular, we will use it to illustrate three important properties of log-correlated fields. We will refer to these as the *multiscale decomposition*, *the dichotomy of scales* and *the self-similarity of scales*. As we will see, these properties do not hold exactly for general log-correlated fields, like the 2D Gaussian free field. However, they hold approximately enough to reproduce the extremal behavior.

4.2.1 Branching Random Walk

We will focus on a simple example of branching random walk (BRW), where the branching is binary and the increments are Gaussian. This process is sometimes called *hierarchical Gaussian field*. More specifically, let V_n be the leaves of a binary rooted tree with n generations. The field is indexed by V_n and thus has 2^n random variables. As before, we denote the field by

$$X(n) = \{X_v(n) : v \in V_n\}.$$

The random variables $X_v(n)$ are constructed as follows. We consider $\{Y_e\}$ IID centered Gaussian random variables of variance σ^2, which is indexed by the edges of the binary tree. The parameter σ^2 will be adjusted when we compare log-correlated fields to approximate branching random walks. For a given leaf $v \in V_n$, the random variable $X_v(n)$ is given by the sum of the Y_e's for the edges e on the path from the root $\emptyset$ to the leaf v, i.e.,

$$X_v(n) = \sum_{e:\, \emptyset \to v} Y_e.$$

We will often abuse notation and write $Y_v(l)$ for the variable Y_e for the leaf v at level l in the tree. In other words,

$$X_v(n) = \sum_{l=1}^{n} Y_v(l). \tag{4.14}$$

We refer to $Y_v(l)$ as the *increment at scale l* of $X_v(n)$. With this definition, it is easy to see that the variance of each variable is

$$E[X_v(n)^2] = \sum_{l=1}^{n} E[Y_v(l)^2] = \sigma^2 n.$$

As for the covariance for $v, v' \in V_n$, we need to define

$v \wedge v'$, *the level of the tree where the paths of v and v' to the root split.*

We call $v \wedge v'$ the *branching time* or *branching scale* of v and v'. Note that $0 \leq v \wedge v' \leq n$ (see Figure 4.3 for an illustration). It then follows directly from the definition (4.14) that

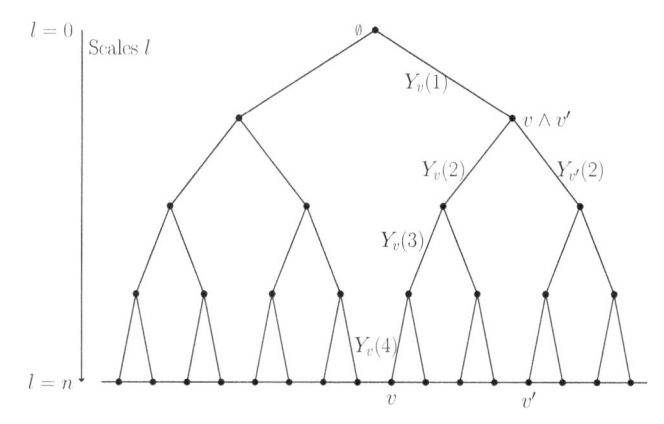

$$E[X_v(n)X_{v'}(n)] = \sum_{l=1}^{v \wedge v'} E[Y_v^2(l)] = \sigma^2 (v \wedge v').$$

Figure 4.3. An illustration of a binary branching random walk (BRW) with increments $Y_v(l)$ at each scale $0 \leq l \leq n$.

The covariance matrix completely determines the distribution of the centered Gaussian field $X(n)$.

Why is the field $X(n)$ log-correlated? The best way to see this is to think in terms of a neighborhood with a certain covariance as in equation (4.5). Indeed, let $0 \leq r \leq 1$, and suppose for simplicity that rn is an integer. Then the size of the neighborhood of a given v with covariance $r\sigma^2 n$ with v is exactly

$$\#\{v' \in V_n : \frac{E[X_v(n)X_{v'}(n)]}{\sigma^2 n} \geq r\} = 2^{n-rn}.$$

It corresponds to those v' which branched out from v at level rn! It is a ball of size $2^{n(1-r)}$ and is of mesoscopic scale compared to the size of the systems.

There are three elementary properties of BRW which will have their counterpart for any log-correlated field. These properties will be key to guide our analysis in more complicated models.

- *Multiscale decomposition:* The first property simply refers to the decomposition (4.14) into independent increments.

$$X_v(n) = \sum_{l=1}^{n} Y_v(l).$$

- *Dichotomy of scales:* Another simple consequence of the definition of BRW is the fact that for scales before the branching points of v and v', the increments are perfectly correlated whereas after the branching point, they are exactly independent. In other words:

$$E[Y_v(l)Y_l(v')] = \begin{cases} \sigma^2 & \text{if } l \leq v \wedge v', \\ 0 & \text{if } l > v \wedge v'. \end{cases} \quad (4.15)$$

We refer to this abrupt change in correlations for the increments as the *dichotomy of scales*.

- *Self-similarity of scales:* In the case of BRW, the increments at each scale are identically distributed. In particular, together with the property of dichotomy, this property implies that for a given $0 < l < n$, the variables $X_v(l) = \sum_{k=1}^{l} Y_v(k)$ define a process $X(l)$ with 2^l distinct values. Moreover, this process is exactly a BRW with l levels. Similarly, for a given v, we can define the variables

$$\{X_{v'}(n) - X_{v'}(l) : v' \wedge v \geq l\}.$$

This defines a BRW on 2^{n-l} points corresponding to the subtree with the common ancestor at level l acting as the root.

We point out that the above construction leads to a natural generalization of fields by modifying the variance of the increments in the multiscale decomposition at each scale. The self-similarity is lost in these models. The study of such types of models goes back to Derrida who introduced the *generalized random energy models* [31]. In the case of BRW, these are called time-inhomogeneous BRW. They were studied in [21, 22, 36, 37, 52]. A similar *scale-inhomogenous* field for the Gaussian free field was introduced and studied in [11, 13].

4.2.2 2D Gaussian Free Field

The discrete two-dimensional Gaussian free field (2DGFF) is an important model of random surfaces. Let V_n be a finite square box of $\mathbb{Z}^2$. To emphasize the similarities with BRW, we suppose that $\#V_n = 2^n$. The 2DGFF is a centered Gaussian field $X(n) = \{X_v(n), v \in V_n\}$ defined as follows.

Let $\mathbb{P}_v$ be the law of a simple random walk $(S_k, k \geq 0)$ starting at $v \in V_n$ and $\mathbb{E}_v$, the corresponding expectation. Denote by τ_n the first-exit time of the random walk from V_n. We consider the covariance matrix given by

$$E[X_v(n)X_{v'}(n)] = \mathbb{E}_v\left[\sum_{k=0}^{\tau_n} 1_{S_k = v'}\right] = G_n(v, v'). \tag{4.16}$$

The right-hand side is the expected number of visits to v' of a random walk starting at v before exiting V_n. It is not hard to show, using the Markov property of the random walk, that the right-hand side is the Green's function of the discrete Laplacian on V_n with Dirichlet boundary conditions. More precisely, for f a function on the vertices of V_n, the discrete Laplacian is

$$-\Delta f(v) = \frac{1}{4}\sum_{\omega \sim v} f(\omega) - f(v) = \mathbb{E}_v[f(S_1) - f(v)],$$

where $\omega \sim v$ means that ω and v share an edge in $\mathbb{Z}^2$. The Green's function can be seen as the inverse of the discrete Laplacian in the sense that it satisfies

$$-\Delta G_n(v, v') = \delta_{v'}(v) = \begin{cases} 1 & \text{if } v = v', \\ 0 & \text{if } v \neq v'. \end{cases}$$

In particular, the Green's function is symmetric and positive definite, since the Laplacian is. The density of the 2DGFF is easily obtained from the covariance (4.16). Indeed, by rearranging the sum, we have

$$\sum_{v \sim v'}(x_v - x_{v'})^2 = \frac{1}{4}\sum_v x_v(-\Delta)x_v.$$

Again, since the Green's function is the inverse of $-\Delta$, we have that

$$P(X_v(n) \in dx_v, v \in V_n) = \frac{1}{Z} \exp\left(-\frac{1}{8} \sum_{v \sim v'} (x_v - x_{v'})^2\right), \quad (4.17)$$

for the appropriate normalization Z.

Why is this field log-correlated? It is reasonable to expect this since the Green's function of the Laplacian in the continuous setting decays logarithmically with the distance. This analogy can be made rigorous using random walk estimates. In fact, it is possible to show, (see e.g., [48]) that

$$G_n(v, v') = E_v\left[\sum_{k=0}^{\tau_n} 1_{S_k = v'}\right] = E_v[a(v', S_{\tau_n})] - a(v, v'), \quad (4.18)$$

where $a(v, v) = 0$, and if $v \neq v'$

$$a(v, v') = \frac{1}{\pi} \log d(v, v')^2 + O(1) + O(d(v, v')^{-2}),$$

where $d(v, v')$ is the Euclidean distance in $\mathbb{Z}^2$. In particular, this implies that for points not too close to the boundary, the field is log-correlated: $G_n(v, v') = \frac{1}{\pi} \log \frac{2^n}{d(v,v')^2} + O(d(v, v')^{-2})$. Moreover, we get the following upper bound on the variance:

$$G_n(v, v) \leq \frac{1}{\pi} \log 2^n + O(1).$$

Again, a matching lower bound follows from (4.18), but only for points far from the boundary. As a first step towards a connection to BRW, these estimates already suggest that the BRW parameter σ^2 should be taken to be

$$\sigma^2 \longrightarrow \frac{\log 2}{\pi}.$$

To make the connection with an approximate BRW more precise, we shall need three fundamental facts. First, it is not hard to prove from the density (4.17) that the field satisfies the *Markov property*, that is for any finite box B in V_n,

$\{X_v(n), v \in B\}$ is independent of the field in B^c given the field on the boundary ∂B.
(4.19)

Here the boundary ∂B refers to the vertices in B^c that share an edge with a vertex in B. This property essentially follows from (4.17) since the correlations are expressed in terms of nearest neighbors. Second, it follows from the Markov property that the conditional expectation of the field inside v given the boundary is simply a linear combination of the field on the boundary. More precisely, we have

$$E[X_v(n)|\{X_{v'}(n), v' \in B^c\}] = \sum_{u \in \partial B} p_u(v) X_u(n), \qquad (4.20)$$

where $p_u(v)$ is the probability of a simple random walk starting at v to exit B at u. The specific form of the coefficient of the linear combination is proved using the strong Markov property of the random walk. A similar argument shows that the process is self-similar. Namely, if we write $X_v(B) = E[X_v(n)|\{X_u(n), u \in \partial B\}]$ for the harmonic average of $X_v(n)$ on the boundary of B, we have

$$\{X_v(n) - X_v(B), v \in B\} \quad \text{is a 2DGFF on } B, \qquad (4.21)$$

in the sense that it is a Gaussian free field with covariance (4.16) restricted to B.

The third property shows that if v and v' are in B, then $X_v(B)$ and $X_{v'}(B)$ should be close, if they are not too close to the boundary of B. More precisely, say B is a square box or a disc containing 2^l points for some l. Estimates of the Green's function as in (4.18) can be used to show that (see e.g., Lemma 12 in [18])

$$E[(X_v(B) - X_{v'}(B))^2] \leq O(d(v, v') 2^{-l/2}) \qquad (4.22)$$

whenever the distance of v and v' to the center of B is of the order of $2^{(l-1)/2}$ or less, to avoid boundary effects.

The connection between the 2D Gaussian free field and an approximate BRW is made through equations (4.19), (4.21), and (4.22). For each $v \in V_n$ and $0 \leq l \leq n$, we define $[v]_l$, a neighborhood of v containing 2^{n-l} points. For simplicity, suppose that $[v]_l$ is a square box. If such a neighborhood is not entirely contained in V_n, we define $[v]_l$ to be the intersection of the neighborhood with V_n. Note that these neighborhoods shrink as l increases. By convention, we take $[v]_0 = V_n$ and $[v]_n = \{v\}$. The boundary is denoted by $\partial [v]_l$ and the union of $[v]_l$ with its boundary will be denoted by $\overline{[v]}_l$. The connection of 2DGFF with BRW is illustrated in Figure 4.4 and is as follows. For simplicity, we assume that all vertices v are not too close to the boundary of V_n.

Multiscale decomposition: The decomposition here is simply a *martingale decomposition*. Namely, we define the *field of $v \in V_n$ at scale l* as

$$X_v(l) = E[X_v(n)|\{X_{v'}(n) : v' \in [v]_l^c\}] = E[X_v(n)|\{X_u(n) : u \in \partial [v]_l\}],$$

where the second equality is by the Markov property. Note that for each $v \in V_n$, $(X_v(l), 0 \leq l \leq n)$ is a martingale by construction. Define the *increments at scale l* as the martingale difference

$$Y_v(l) = X_v(l) - X_v(l-1), \quad 1 \leq l \leq n.$$

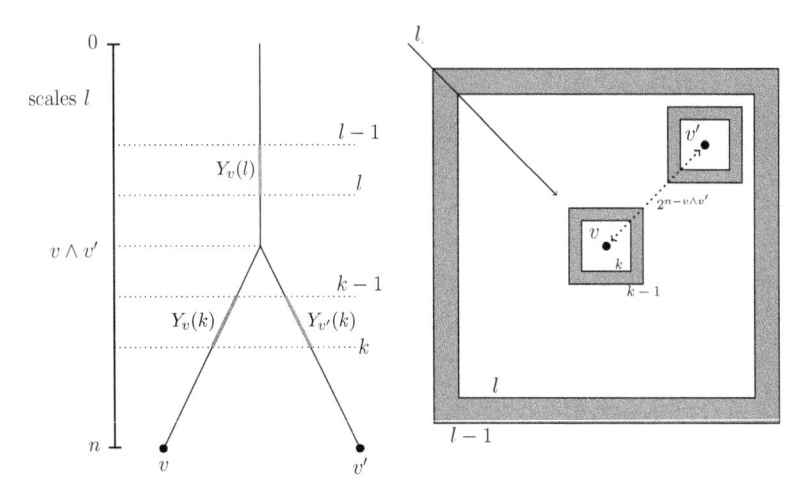

Figure 4.4. An illustration of the approximate branching structure defined by the 2DGFF. The gray area represents the increments at each scale.

These increments are orthogonal by construction. Since the field is Gaussian, so are the conditional expectations. In particular, the increments are independent of each other for a given v. Thus, the decomposition

$$X_v(n) = \sum_{l=1}^{n} Y_v(l), \quad v \in V_n, \tag{4.23}$$

seems so far to be the right analogue of the multiscale decomposition of BRW. It remains to check the other two properties.

- *Self-similarity of scales:* First, note that the increments are Gaussian by definition. The fact that the increments have approximately the same variance follows from (4.21). Indeed, for $0 \leq l \leq n$, the variance of $X_v(n) - X_v(l)$ is by (4.18)

$$E[(X_v(n) - X_v(l))^2] = (n-l)\frac{\log 2}{\pi} + O(1).$$

Because $X_l(v)$ is orthogonal to $X_v(n) - X_v(l)$, we deduce that

$$E[X_v(l)^2] = l\frac{\log 2}{\pi} + O(1).$$

In particular, since $X_v(l+1)$ admits the orthogonal decomposition $X_v(l) + Y_v(l+1)$, this implies

$$E[Y_v(l)^2] = \frac{\log 2}{\pi} + o(1),$$

where $o(1)$ is summable, so the contribution of the sum over l of the error is $O(1)$. This justifies the choice $\sigma^2 = \log 2/\pi$ for the comparison with BRW. Of

course, here, the increments are not exactly identically distributed, but since the error is summable, it will not affect the behavior of the extremes.
- *Dichotomy of scales:* It remains to establish the correlations of the increments between two distinct v, v'. For this purpose, it is useful to define the *branching scale* $v \wedge v'$ as follows

$$v \wedge v' = \min\{0 \leq l \leq n : \overline{[v]}_l \cap \overline{[v']}_l = \emptyset\}, \qquad (4.24)$$

that is the smallest scale for which the neighborhoods of v and v' (including their boundaries) are disjoint. Note that because the neighborhoods have size 2^{n-l}, the branching scale $v \wedge v'$ is related to the distance by

$$d(v, v')^2 \approx 2^{n - v \wedge v'}.$$

For small scales, a direct application of the Markov property (4.19) yields

$$E[Y_v(l) Y_{v'}(l)] = 0 \quad \text{if } l > v \wedge v'.$$

This is because if $l > v \wedge v'$, then the neighborhoods of v and v' do not intersect and are contained in each other's complement. For scales larger than the branching scale, we expect the increments to be almost perfectly correlated. To see this, note that by definition, for $l = v \wedge v' - 1$, then the neighborhoods intersect but for larger neighborhoods, that is $l < v \wedge v' - 1$, v and v' must be contained in each other's neighborhood. In particular, an application of (4.22) gives

$$E[(X_v(l) - X_l(v'))^2] = O(1).$$

In other words, the vectors $X_v(l)$ and $X_l(v')$ are close in L^2. Since these vectors are sums of orthogonal increments, these increments must also be close. In particular, we get

$$E[Y_v(l) Y_{v'}(l)] = \frac{\log 2}{\pi} + o(1) \quad \text{if } l < v \wedge v' - 1,$$

where $o(1)$ is summable in l.

This completes the construction of an approximate BRW embedded in 2DGFF. We remark that the dichotomy of scales is not as clean cut as the one of BRW in (4.15). In particular, nothing precise can be said for the correlation exactly at the branching scale $l = v \wedge v'$. It turns out that the coupling and decoupling of the increments do not have to be perfect as long as they occur fast enough. The same will be true for non-Gaussian models in Section 4.4.

4.3 Order of the Maximum

In this section, we explain a general method to show the leading and subleading orders of the maximum for log-correlated fields in Conjecture 4.1.2. We will prove them in the case of Gaussian branching random walk on the binary tree, following the treatment in [46]. We will also prove the convergence of the log-number of high points or entropy, cf. (4.11). The heuristic is (or is expected to be) the same for more complicated log-correlated models, such as the two-dimensional Gaussian free field. The multiscale decomposition of the field plays a fundamental role in the proof. This is also true when proving more refined results such as the convergence of the maximum and of the extremal process. In a nutshell, to prove such results on the maximum, one needs to understand the contribution of the increments at each scale to a large value of the field.

4.3.1 Leading Order of the Maximum

Recall the definition of the Gaussian BRW $X(n)$ on a binary tree defined in Section 4.2.1. The following result yields the leading order of the maximum.

Theorem 4.3.1 *Let $X(n) = \{X_v(n) : v \in V_n\}$ be a branching random walk on a binary tree with n levels as in (4.23) with Gaussian increments of variance σ^2. Then*

$$\lim_{n \to \infty} \frac{\max_{v \in V_n} X_v(n)}{n} = \sqrt{2 \log 2}\, \sigma \quad \text{in probability.}$$

The theorem goes back to Biggins for general BRW [16]. The analogue for the 2DGFF was proved by Bolthausen, Deuschel and Giacomin [18]. We point out that Kistler greatly simplified the proof in both cases [46] using a *multiscale refinement of the second moment method*. This approach also makes the analysis amenable to non-Gaussian models. We refer the reader directly to [46] for a proof of the theorem. We shall instead use Kistler's method to prove a very similar result on the *log-number of high points or exceedances*.

Theorem 4.3.2 *Let $X(n) = \{X_v(n) : v \in V_n\}$ be a branching random walk on a binary tree with n levels as in (4.23) with Gaussian increments of variance σ^2. Then for $0 \leq E < c = \sqrt{2 \log 2}\,\sigma$,*

$$\lim_{n \to \infty} \frac{1}{n} \log \#\{v \in V_n : X_v(n) > En\} = \log 2 \left(1 - \frac{E^2}{c^2}\right) \quad \text{in probability.}$$

In other words, the number of points whose field value is larger than En is approximately $2^{n(1-E^2/c^2)}$ with large probability for E smaller than the maximal level c. We stress that this is the same result for 2^n IID Gaussian random variables of variance σ^2. The analogue of this result for 2DGFF was shown by Daviaud in [27].

Proof. Upper bound: To bound the number of points beyond a given level, it suffices to use Markov's inequality. Let

$$\mathcal{N}_n(En) = \#\{v \in V_n : X_v(n) > En\}.$$

Then for a given $\varepsilon > 0$, we have

$$P\left(\mathcal{N}_n(En) > 2^{n(1+\varepsilon)\left(1-\frac{E^2}{c^2}\right)}\right) \leq 2^{-n(1+\varepsilon)\left(1-\frac{E^2}{c^2}\right)} E[\mathcal{N}_n(En)]. \qquad (4.25)$$

By linearity, the expectation is simply

$$E[\mathcal{N}_n(En)] = 2^n P(X_v(n) > En) \leq 2^n \exp\left(-\frac{E^2 n}{2\sigma^2}\right) = 2^{n\left(1-\frac{E^2}{2c^2}\right)},$$

where we use the Gaussian estimate (4.1) with $X_v(n) \sim \mathcal{N}(0, \sigma^2 n)$. It follows that the probability in (4.25) goes to 0. This proves the upper bound.

Lower bound: It remains to show that

$$P\left(\mathcal{N}_n(En) > 2^{n(1-\varepsilon)\left(1-\frac{E^2}{c^2}\right)}\right) \to 1.$$

The lower bound is an application of the *Paley-Zygmund inequality*, which states that for any random variable $\mathcal{N} > 0$ and $0 \leq \delta \leq 1$,

$$P(\mathcal{N} > \delta E[\mathcal{N}]) \geq (1-\delta)^2 \frac{E[\mathcal{N}]^2}{E[\mathcal{N}^2]}. \qquad (4.26)$$

So, ultimately, one needs to prove that in the limit the second moment matches the first moment squared. This would certainly be the case if the variables were independent. In the case of BRW, this is not so. However, we can create enough independence and not lose too much precision by dropping the increments at lower scales. In addition, as observed in [46], the matching of the moments is greatly simplified if one considers increments rather than the sum of increments. More precisely, since the leading of the maximum is linear in the scales, it is reasonable to expect that each scale contributes equally to the maximal value to leading order (since they have the same variance). It turns out this heuristic is correct and extends to the high points.

With this in mind, we introduce a parameter K which divides the n scales coarser K levels. We define accordingly, for $m = 1, \ldots, K$, the increments

$$W_m(v) = \sum_{\frac{m-1}{K} < l \leq \frac{m}{K}} Y_l(v).$$

The idea is that for a high point at level En, each variable $W_m(v)$ should contribute $E\frac{n}{K}$. Without loss of generality, we can suppose ε is small enough that $(1+\varepsilon)E < c$. We consider the event

$$\mathcal{E}_m(v) = \{W_m(v) > (1+\varepsilon)\frac{n}{K} E\}, \quad m = 1, \ldots, K,$$

and the *modified number of exceedances*,

$$\widetilde{\mathcal{N}}_n = \sum_{v \in V_n} \prod_{m=2}^{K} 1_{\mathcal{E}_m(v)}.$$

Note that if v is counted in $\widetilde{\mathcal{N}}_n$, then each $W_m(v)$ for $m \geq 2$ exceeds $(1+\varepsilon)K^{-1}nE$. Moreover, it is easy to check using the Gaussian estimate (4.1) that $|W_1(v)|$ is less than $K^{-1}n$ with large probability. In particular, $X_v(n) > (1+\varepsilon)(1-K^{-1})En$ on the events, so that we can take K large enough depending on ε so that $X_v(n) > En$. We conclude that

$$\widetilde{\mathcal{N}}_n \leq \mathcal{N}_n(En),$$

hence it suffices to show

$$P\left(\widetilde{\mathcal{N}}_n > 2^{n(1-\varepsilon)\left(1 - \frac{E^2}{c^2}\right)}\right) \to 1.$$

We apply the inequality (4.26) with $\delta = 2^{-n\varepsilon\left(1 - \frac{E^2}{c^2}\right)}$. We have by the Gaussian estimate (4.1)

$$E[\widetilde{\mathcal{N}}_n] = 2^n \frac{(1+o(1))}{\sqrt{2\pi}} \exp\left(-\frac{(1-K^{-1})(1+\varepsilon)^2 E^2 n}{2\sigma^2}\right) \geq 2^{n\left(1 - \frac{E^2}{c^2}\right)},$$

for n large enough. Thus

$$P\left(\widetilde{\mathcal{N}}_n > 2^{n(1-\varepsilon)\left(1 - \frac{E^2}{c^2}\right)}\right) \geq P\left(\widetilde{\mathcal{N}}_n > \delta E[\widetilde{\mathcal{N}}_n]\right).$$

It remains to show

$$\frac{(E[\widetilde{\mathcal{N}}_n])^2}{E[(\widetilde{\mathcal{N}}_n)^2]} \to 1 \quad \text{as } n \to \infty. \tag{4.27}$$

The idea is to split the second moment according to the branching scale. More precisely, we have

$$E[(\widetilde{\mathcal{N}}_n)^2] = \sum_{\substack{v,v' \in V_n \\ v \wedge v' \leq \frac{n}{K}}} P\left(\bigcap_{m=2}^{K} \mathcal{E}_m(v) \cap \mathcal{E}_m(v')\right)$$

$$+ \sum_{r=2}^{K} \sum_{\substack{v,v' \in V_n \\ \frac{(r-1)n}{K} < v \wedge v' \leq \frac{rn}{K}}} P\left(\bigcap_{m=2}^{K} \mathcal{E}_m(v) \cap \mathcal{E}_m(v')\right)$$

$$= \sum_{\substack{v,v' \in V_n \\ v \wedge v' \leq \frac{n}{K}}} \prod_{m=2}^{K} P(\mathcal{E}_m(v))^2 + \sum_{r=2}^{K} \sum_{\substack{v,v' \in V_n \\ \frac{(r-1)n}{K} < v \wedge v' \leq \frac{rn}{K}}} \prod_{m=2}^{K} P(\mathcal{E}_m(v) \cap \mathcal{E}_m(v')),$$

(4.28)

where the last line holds by the independence of the increments between the scale and the fact that $W_m(v) = W_m(v')$ for all $m \geq 2$ if $v \wedge v' \leq n/K$ by the dichotomy of scales. Since there are at least $2^n(2^n - 2^{n-n/K}) = 2^{2n}(1 - o(1))$ pairs such that $v \wedge v' \leq n/K$, the first term of (4.28) is $(1 - o(1))(E[\widetilde{\mathcal{N}}_n])^2$. The proof will be concluded once we prove that the second term is $o(1)(E[\widetilde{\mathcal{N}}_n])^2$.

Guided by the dichotomy of scales, we do not lose by dropping the constraints on m smaller than the branching scales for one of the v's since they are identical. The second term is smaller than

$$\sum_{r=2}^{K} \sum_{\substack{v,v' \in V_n \\ \frac{(r-1)n}{K} < v \wedge v' \leq \frac{rn}{K}}} \prod_{m=2}^{K} P(\mathcal{E}_m(v)) \prod_{m \geq r+1} P(\mathcal{E}_m(v')),$$

where we use the independence of the increments after the branching scales. Since there are at most $2^n(2^{n-\frac{(r-1)n}{K}})$ pairs $v \wedge v'$ such that $\frac{(r-1)n}{K} < v \wedge v' \leq \frac{rn}{K}$, we have that the above is smaller than

$$(E[\widetilde{\mathcal{N}}_n])^2 \sum_{r=2}^{K} \left(2^{\frac{(r-1)n}{K}} \prod_{m=2}^{r} P(\mathcal{E}_m(v))\right)^{-1} \leq (E[\widetilde{\mathcal{N}}_n])^2 \sum_{r=2}^{\infty} \left(2^{(r-1)nK^{-1}(1-(1+\varepsilon)^2 E^2/c^2)}\right)^{-1}$$

where we used the Gaussian estimate with variance n/K for each variable W_m. The sum is geometric. It converges because of the choice $(1+\varepsilon)E < c$. Thus the second term of (4.28) is smaller than

$$2(E[\widetilde{\mathcal{N}}_n])^2 \, 2^{-nK^{-1}(1-(1+\varepsilon)^2 E^2/c^2)} = o(1)(E[\widetilde{\mathcal{N}}_n])^2,$$

for n large enough. This proves (4.27) and concludes the proof of the theorem.

4.3.2 Subleading Order of the Maximum

We now turn to the proof of the subleading order of the maximum for a Gaussian branching random walk. The basic idea goes back to the seminal work of Bramson who showed the convergence of the law of the maximum of branching Brownian motion. It consists in observing that the increments of the maximum must satisfy a linear constraint at each scale. This is consistent with the fact that the first order is linear in the scales and that at each scale l, the variables $X_v(l)$ form themselves a BRW with 2^l points. To prove convergence, more precise estimates are needed than for the subleading order. We follow here the treatment of [46] with the additional use of the *ballot theorem* as in [5], cf. Theorem 4.3.4 below.

Theorem 4.3.3 *Let $X(n) = \{X_v(n) : v \in V_n\}$ be a branching random walk on a binary tree with n levels as in (4.23) with Gaussian increments of variance σ^2. Then*

$$\lim_{n \to \infty} \frac{\max_{v \in V_n} X_v(n) - cn}{\log n} = -\frac{3}{2} \quad \text{in probability,}$$

where $c = \sqrt{2\log 2}\,\sigma$.

For simplicity, we define the deterministic displacement of the maximum as

$$m_n = cn - \frac{3\sigma^2}{2c}\log n, \text{ and more generally, } m_n(\varepsilon) = m_n + \varepsilon \log n. \quad (4.29)$$

The first observation is to realize that the expectation of exceedances at the level m_n diverges! Indeed, using the Gaussian estimate (4.1) and linearity

$$E[\mathcal{N}_n(m_n)] = E[\#\{v \in V_n : X_v(n) > m_n\}] = O(n).$$

This is the first sign that the branching structure matters for the subleading order. The divergence of the expectation comes from atypical events that inflate the expectation. Therefore, it is necessary to restrict the exceedances to the typical behavior values. Since the first order is linear in the scales and the variables $\{X_v(l) : v \in V_n\}$ form a BRW on 2^l values, it is reasonable to expect that $X_l(v) = \sum_{k=1}^{l} Y_k(v) \leq cl + B$ for some appropriate choice of *barrier B*. Keeping this in mind, we introduce a *modified number of exceedances*

$$\widetilde{\mathcal{N}}_n = \#\{v \in V_n : X_v(n) > m_n,\ X_v(l) \leq cl + B\ \forall l \leq n\}.$$

It turns out that $E[\widetilde{\mathcal{N}}_n] = O(1)$, because the probability of a random walk (in this instance $(X_v(l), l \leq n)$) to stay below a barrier is exactly of the order $1/n$. This is the content of the ballot theorem that we now state precisely. The reader is referred to [1,2,24] for more details. The result quoted here is the version appearing in [7].

Theorem 4.3.4 *Let $(S_n)_{n\geq 0}$ be a Gaussian random walk with increments of mean 0 and variance $\sigma^2 > 0$, with $S_0 = 0$. Let $\delta > 0$. There is a constant $C = C(\sigma, \delta)$ such that for all $B > 0$, $b \leq B - \delta$ and $n \geq 1$*

$$P\left[X_n \in (b, b+\delta) \text{ and } X_k \leq B \text{ for } 0 < k < n\right] \leq C \frac{(1+B)(1+B-b)}{n^{3/2}}. \quad (4.30)$$

Moreover, if $\delta < 1$,

$$P\left[X_n \in (0, \delta) \text{ and } X_k \leq 1 \text{ for } 0 < k < n\right] \geq \frac{1}{Cn^{3/2}}. \quad (4.31)$$

With this new tool in hand, we are ready to prove the subleading order.

Proof of Theorem 4.3.3. Upper bound: We want to show that for $\varepsilon > 0$

$$\lim_{n \to \infty} P(\max_{v \in V_n} X_v(n) \geq m_n(\varepsilon)) = 0.$$

As mentioned previously, the idea is to first show that we can constrain the random walks to stay below a linear barrier. More precisely, we prove that for $B = \log^2 n$,

$$\lim_{n \to \infty} P(\exists v \in V_n : X_v(n) \geq m_n(\varepsilon), X_v(l) > cl + B \text{ for some } l \leq n) = 0. \quad (4.32)$$

The proof is by a Markov's inequality at each scale. Indeed, we can bound the above by

$$\sum_{l=1}^{n} P(\exists v : X_v(l) > cl + B) \leq \sum_{l=1}^{n} 2^l \exp\left(-\frac{(cl+B)^2}{2\sigma^2 l}\right) = n \exp(-Bc/\sigma^2) = o(1),$$

where the first inequality is by the Gaussian estimate and the last is by the choice of $B = \log^2 n$.

To conclude the proof of the upper bound, consider the event

$$\mathcal{E}^+(v) = \{X_v(n) \geq m_n(\varepsilon), X_v(l) \leq cl + B \,\forall l \leq n\}.$$

By (4.32), it follows that

$$P(\max_{v \in V_n} X_v(n) \geq m_n(\varepsilon)) = P\left(\sum_{v \in V_n} 1_{\mathcal{E}^+(v)} \geq 1\right) + o(1).$$

Therefore, it remains to show that the first probability goes to 0. By Markov's inequality, this is simply

$$P\left(\sum_{v \in V_n} 1_{\mathcal{E}^+(v)} \geq 1\right) \leq 2^n P(\mathcal{E}^+(v)).$$

To prove that this goes to 0, it is useful to make a change of measure to make the end value $m_n(\varepsilon)$ typical. To this aim, consider the measure Q defined from P through the density

$$\frac{dQ}{dP} = \prod_e \frac{e^{\lambda Y_e}}{e^{\varphi(\lambda)}}, \qquad \varphi(\lambda) = \log E[e^{\lambda Y_e}] = \frac{\lambda^2 \sigma^2}{2},$$

where the product is over all the edges in the binary tree. Note that by definition

$$\varphi'(\lambda) = E_Q[Y_e] = \lambda \sigma^2.$$

In particular, to make $m_n(\varepsilon)$ typical, we choose

$$\lambda = \frac{m_n(\varepsilon)}{\sigma^2 n}.$$

Under this change of measure, we can write the probability as

$$P(\mathcal{E}^+(v)) = e^{-(\lambda m_n(\varepsilon) - \varphi(\lambda))} E_Q[e^{-\lambda(X_v(n) - m_n(\varepsilon))} 1_{\mathcal{E}^+(v)}]$$
$$\leq e^{-\frac{(m_n(\varepsilon))^2}{2\sigma^2 n}} Q(\mathcal{E}^+(v)),$$
(4.33)

where we use the choice of λ and the fact that $X_v(n) - m_n(\varepsilon) \geq 0$ on the event $\mathcal{E}^+(v)$. Since

$$2^n \exp\left(-\frac{(m_n(\varepsilon))^2}{2\sigma^2 n}\right) = O(n^{-\varepsilon c/\sigma^2}) n^{3/2},$$

the upper bound will be proved if

$$n^{3/2} Q(\mathcal{E}^+(v)) = o(n^{\varepsilon c/\sigma^2}).$$

But this follows from the ballot theorem (Theorem 4.3.4) with $B = (\log n)^2$, since $(Y_l(v) - m_n(\varepsilon)/n, l \leq n)$ under Q is a random walk of finite variance.

Lower bound: As for the leading order, to get a matching lower bound, it is necessary to create independence by cutting off the increments at low scales. Of course, we cannot truncate as many scales. We choose

$$r = \log \log n,$$

and consider the truncated walks $X_v(n) - X_v(r)$. We do not lose much by dropping these since, for $\varepsilon > 0$,

$$P(\exists v : X_v(n) < m_n(-2\varepsilon), X_v(n) - X_r(v) \geq m_{n-r}(-\varepsilon)) \leq P(\exists v : X_v(r) < -10cr),$$
(4.34)

where the inequality holds for n large enough depending on ε. By a union bound and the Gaussian estimate (4.1) (note that the variables are symmetric!),

$$P(\exists v : X_v(r) < -10cr) \leq 2^r \exp(-100c^2r^2/(2\sigma^2)) = 2^{-99r} \to 0.$$

In view of (4.34) and by redefining $\varepsilon > 0$, we reduced the proof to showing

$$P(\exists v : X_v(n) - X_r(v) \geq m_{n-r}(-\varepsilon)) \to 1, \; n \to \infty. \tag{4.35}$$

For this purpose, consider

$$\mu = \frac{m_{n-r}(-\varepsilon)}{n-r} \qquad \frac{\mu^2}{2\sigma^2} = \log 2 - \left(\frac{3}{2} - \frac{c}{\sigma^2}\varepsilon\right)\frac{\log(n-r)}{n-r} + o(n^{-1}). \tag{4.36}$$

Denote the truncated walk and its recentering by

$$X_v(r,n) = X_v(n) - X_r(v), \qquad \overline{X}_v(r,n) = X_v(n) - X_r(v) - \mu(n-r).$$

With this notation, the event of exceedance with the barrier $B = 1$ is

$$\mathcal{E}^-(v) = \{v \in V_n : \overline{X}_v(r,n) \in [0,\delta], \overline{X}_v(l,n) \leq 1 \; \forall l \geq r+1\},$$

where $\delta > 0$ is arbitrary. The relevant number of exceedances is

$$\mathcal{N}^-(v) = \sum_{v \in V_n} 1_{\mathcal{E}^-(v)}.$$

By the Paley-Zygmund inequality, we have

$$P(\mathcal{N}^-(v) \geq 1) \geq \frac{(E[\mathcal{N}^-(v)])^2}{E[(\mathcal{N}^-(v))^2]}.$$

To prove (4.35), it remains to show that

$$E[(\mathcal{N}^-(v))^2] = (1 - o(1))(E[\mathcal{N}^-(v)])^2. \tag{4.37}$$

The second moment can be split as a sum over $v \wedge v'$. We get

$$E[(\mathcal{N}^-(v))^2] = \sum_{v,v': v \wedge v' \leq r} P(\mathcal{E}^-(v) \cap \mathcal{E}^-(v'))$$

$$+ \sum_{m=r+1}^{n} \sum_{v,v': v \wedge v' = m} P(\mathcal{E}^-(v) \cap \mathcal{E}^-(v')). \tag{4.38}$$

For the first term, note that $\mathcal{E}^-(v)$ is independent of $\mathcal{E}^-(v')$ because $v \wedge v' \leq r$ so the walks share no increments. Thus, the first term is

$$\sum_{v,v': v \wedge v' \leq r} P(\mathcal{E}^-(v) \cap \mathcal{E}^-(v')) = (1 - o(1))(E[\mathcal{N}^-(v)])^2 \tag{4.39}$$

since there are $2^{2n}(1 - o(1))$ pairs v, v' with $v \wedge v' \leq r$. We show that the second term is $o(1)(E[\mathcal{N}^-(v)])^2$ to conclude the proof.

Observe that, by proceeding as (4.33) with the measure Q (for $\lambda = \mu/\sigma^2$) and the Gaussian estimate (4.1), the expected number of exceedances satisfy

$$E[\mathcal{N}^-(v)] \geq C2^n e^{-\lambda \delta} e^{-\frac{\mu^2(n-r)}{2\sigma^2}} Q(\mathcal{E}^-(v))$$
$$\geq C2^r e^{-\lambda \delta} (n-r)^{\frac{c}{\sigma^2}\varepsilon} (n-r)^{3/2} Q(\mathcal{E}^-(v)) \qquad (4.40)$$
$$\geq C2^r (n-r)^{\frac{c}{\sigma^2}\varepsilon} e^{-\lambda \delta}$$

for some constant $C > 0$. Here we used the estimate on $\frac{\mu^2}{2\sigma^2}$ in (4.36). The last inequality is a consequence of the ballot theorem (Theorem 4.3.4). For $v \wedge v' = m$, we decompose the probability $P(\mathcal{E}^-(v) \cap \mathcal{E}^-(v'))$ on the events $\{\overline{X}_v(r,m) \in (-q, -q+1]\}$ for $q \geq 0$. Note that $X_v(m) = X_{v'}(m)$! We thus get

$$P(\mathcal{E}^-(v) \cap \mathcal{E}^-(v'))$$
$$= \sum_{q=0}^{\infty} P(\mathcal{E}^-(v) \cap \mathcal{E}^-(v') \cap \{\overline{X}_v(m,n) \in (-q, -q+1]\})$$
$$\leq \sum_{q=0}^{\infty} P(\overline{X}_v(r,m) \in (-q, -q+1], \overline{X}_v(r,l) \leq 1 \; \forall l = r+1, \ldots, m)$$
$$\times \left(P(\overline{X}_v(m,n) \geq q - 1, \overline{X}_v(m,l) \leq 1 + q \; \forall l = m+1, \ldots, n)\right)^2,$$

where we used the bound on $X_v(m)$, and the square of the probability comes from the independence of increments after the branching scale m. We evaluate these two probabilities. Using the change of measure with $\lambda = \mu/\sigma^2$ and the bounds on $\overline{X}_v(m,n)$, the first one is smaller than

$$e^{\lambda q} e^{-\frac{\mu^2(n-r)}{2\sigma^2}} Q(\overline{X}_v(r,m) \in (-q, -q+1], \overline{X}_v(r,l) \leq 1 \; \forall l = r+1, \ldots, m).$$

By the ballot theorem, this is smaller than

$$e^{\lambda q} e^{-\frac{\mu^2(n-r)}{2\sigma^2}} \frac{q+1}{(m-r)^{3/2}}.$$

An identical treatment of the square of the second probability yields that it is smaller than

$$e^{-2\lambda q} e^{-\frac{\mu^2(m-r)}{\sigma^2}} \frac{q+1}{(n-m)^3}.$$

Putting all the above together, we finally obtained that

$$\sum_{m=r+1}^{n} \sum_{v,v':v\wedge v'=m} P(\mathcal{E}^-(v)\cap\mathcal{E}^-(v'))$$

$$\leq \sum_{m=r+1}^{n} 2^{2n-m} \frac{e^{-\frac{\mu^2}{2\sigma^2}(2n-m-r)}}{(m-r)^{3/2}(n-m)^3} \sum_{q=0}^{\infty} (q+1)^2 e^{-\lambda q},$$

since there are at most 2^{2n-m} pairs with $v\wedge v'=m$. The last sum is finite. Moreover, using the estimate in (4.36), the first sum is smaller than

$$C 2^r (n-r)^{2\varepsilon\frac{c}{\sigma^2}} \sum_{m=r+1}^{n} \frac{(n-r)^{\frac{3}{2}(2-\frac{m-r}{n-r})}}{(m-r)^{3/2}(n-m)^3}.$$

It is not hard to show by an integral test that the series is bounded for any n. Thus

$$\sum_{m=r+1}^{n} \sum_{v,v':v\wedge v'=m} P(\mathcal{E}^-(v)\cap\mathcal{E}^-(v')) \leq C 2^r (n-r)^{2\varepsilon\frac{c}{\sigma^2}}. \qquad (4.41)$$

Using (4.41), (4.40), and (4.39) in (4.38), we conclude that

$$P(\mathcal{N}^-(v)\geq 1) \geq \frac{1}{1+C 2^r e^{2\lambda\delta}},$$

which goes to 0 by taking $n\to\infty$ ($r=\log\log n$) then $\delta\to 0$. This proves the theorem. $\square$

4.4 Universality Classes of Log-correlated Fields

The heuristics developed in Section 4.3 from branching random walks turns out to be applicable to a wide class of stochastic processes. It is a research topic of current interest to extend these techniques to other non-Gaussian log-correlated models with the purpose of proving results similar to Conjecture 4.1.2. One interesting problem that we will not have time to consider is the cover time of the discrete random walk on the two-dimensional torus, which, perhaps surprisingly, can be estimated using such methods. We refer the reader to [15, 29] for results on this question. We focus our attention here to two open problems: the large values of the Riemann zeta function and of the characteristic polynomials of random matrices. The behavior of the large values of these two models was conjectured by Fydorov, Hiary and Keating [39, 40] to mimic that of a Gaussian log-correlated model. We shall explain here this connection in terms of branching random walks.

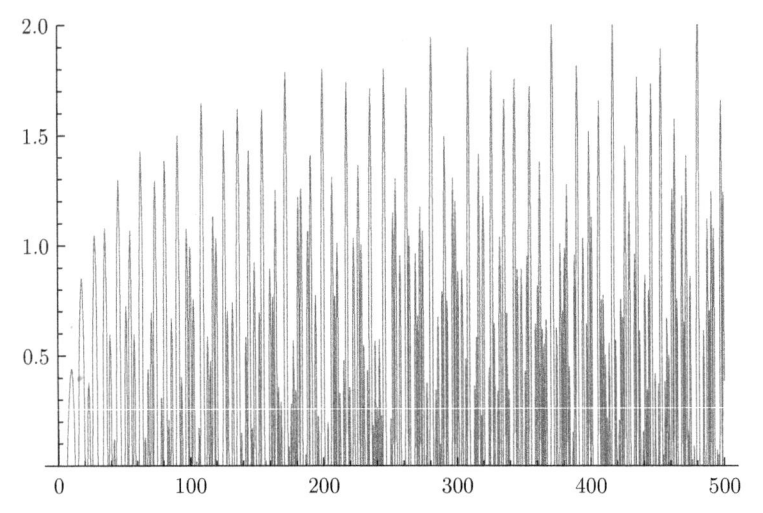

Figure 4.5. The function $\log|\zeta(s)|$ for $s = 1/2 + it$ for $t \in [0, 500]$.

4.4.1 Maximum of the Riemann Zeta Function on an Interval

Let $s \in \mathbb{C}$. If Re $s > 1/2$, the Riemann zeta function is defined by

$$\zeta(s) = \sum_{n=1}^{\infty} \frac{1}{n^s} = \prod_{p \text{ primes}} (1 - p^{-s})^{-1}. \tag{4.42}$$

This definition can be analytically continued using the functional equation

$$\zeta(s) = \chi(s)\zeta(1-s), \qquad \chi(s) = 2^s \pi^{s-1} \sin\left(\frac{\pi}{2}s\right) \Gamma(1-s),$$

to the whole complex plane with a pole at $s = 1$. The distribution of the primes is famously linked to information on the zeros of the function. There are trivial zeros (coming from the functional equation) at the negative even integers. The Riemann hypothesis states that all other zeros lie on the critical line Re $s = 1/2$ suggested by the symmetry of the functional equation. See Figure 4.5 for the plot of the modulus of ζ on the critical line.

We shall be interested instead in the large values of the modulus of ζ in an interval. In other words, what can be said about the local maxima of zeta? The global behavior is much more intricate, see the Lindelöf hypothesis.

Fyodorov, Hiary and Keating made an astonishing prediction for the local maximum of ζ on an interval, say $[0, 1]$, of the critical line. Their claim is based on numerics and on the connections between ζ and random matrices.

Conjecture 4.4.1 (Fyodorov–Hiary–Keating [39, 40]) *For τ sampled uniformly from $[0,T]$,*

$$\lim_{T\to\infty} \frac{\max_{h\in[0,1]} \log|\zeta(1/2+i(\tau+h))| - \log\log T}{\log\log\log T} = -\frac{3}{4} \text{ in probability.} \quad (4.43)$$

In fact, their conjecture is even more precise, suggesting an explicit distribution for the fluctuation. Roughly speaking, the above means that the maximum in a typical interval of the critical line is

$$\max_{h\in[0,1]} \log|\zeta(1/2+i(\tau+h))| \approx \log\log T - \frac{3}{4}\log\log\log T + O(1).$$

The connection with Conjecture 4.1.2 is made by taking $\log T = 2^n$ and $\sigma^2 = \frac{\log 2}{2}$. Observe that it is a perfect match! As we will see, the relation with log-correlated models goes beyond the numerological curiosity. It also proposes a starting point to prove the conjecture. In fact, the conjecture can be proved to the subleading order for a random model of zeta, cf. Theorem 4.4.2.

Suppose first the Euler product of the Riemann zeta function extended to the critical line. By expanding the logarithm, we would have

$$\log|\zeta(1/2+it)| = -\text{Re}\sum_p \log(1-p^{-(1/2+it)}) = \sum_{k=1}^\infty \frac{1}{k} \sum_p \text{Re}\frac{1}{p^{k(1/2+it)}}$$

$$= \sum_p \frac{\text{Re } p^{-it}}{p^{1/2}} + O(1). \quad (4.44)$$

This expansion can be made partially rigorous if the sum over primes is cut off at T, (see Proposition 1 of [44] based on [56]). Now replace t by a random $\tau + h$ where $h \in [0,1]$ and τ is sampled uniformly from $[0,T]$. The randomness entirely comes from the random variables ($p^{-i\tau}$, p primes). It is an easy exercise by moment computations to verify that, since the values of $\log p$ are linearly independent for distinct primes, the process ($p^{-i\tau}$, p primes) converges in the sense of finite-dimensional distribution as $T \to \infty$ to independent random variables distributed uniformly on the unit circle. This suggests the following model for zeta: let (U_p, p primes) be IID uniform random variables on the unit circle indexed by the primes and take

$$\left(\sum_{p\leq T} \frac{\text{Re}(U_p p^{-ih})}{p^{1/2}}, h \in [0,1]\right). \quad (4.45)$$

For this model, we can prove the conjecture up to subleading order.

Theorem 4.4.2 ([7]) *Let $(U_p, p \text{ primes})$ be independent random variables distributed uniformly on the unit circle. Then*

$$\lim_{T \to \infty} \frac{\max_{h \in [0,1]} \sum_{p \leq T} \frac{\operatorname{Re}(U_p p^{-ih})}{p^{1/2}} - \log \log T}{\log \log \log T} = -\frac{3}{4} \text{ in probability.} \quad (4.46)$$

The proof is based on finding an approximate branching random walk embedded in zeta. What is this approximate branching random walk? In particular, what is the multiscale decomposition and does it exhibit the usual self-similarity and dichotomy of scales of log-correlated models? It is tempting to think of the decomposition $\sum_p \frac{\operatorname{Re} p^{-it}}{p^{1/2}}$ as the right multiscale decomposition since the variables $\operatorname{Re} p^{-it}$ decorrelate in the limit. However, for the scales to be self-similar, it is necessary to consider a coarse-graining of the sum. For the sake of clarity, we take $\log T = 2^n$. For the random model (4.45), the increment at scale l for $l = 1, \ldots, n$ is

$$Y_h(l) = \sum_{2^{l-1} < \log p \leq 2^l} \frac{\operatorname{Re}(U_p p^{-ih})}{p^{1/2}}, \quad h \in [0,1]. \quad (4.47)$$

The multiscale decomposition is therefore

$$\sum_{p \leq T} \frac{\operatorname{Re}(U_p p^{-ih})}{p^{1/2}} = \sum_{l=1}^{n} Y_h(l).$$

For the random model, the increments are independent by construction. For zeta, they are only independent in the limit which is a substantial difficulty. We verify that the $Y(l)$'s are self-similar and exhibit a dichotomy. The variance is easily calculated for the random model

$$E[(Y_h(k))^2] = \sum_{2^{k-1} < \log p \leq 2^k} \frac{1}{p} E[(\operatorname{Re} U_p)^2] = \frac{1}{2} \sum_{2^{k-1} < \log p \leq 2^k} \frac{1}{p},$$

where we used the fact that $E[U_p^2] = 0$ and $E[U_p \overline{U}_p] = 0$ where $\bar{z}$ is the complex conjugate of z. The sum over primes can be evaluated using the *prime number theorem*, (see e.g., [53]), which states that

$$\#\{p \leq x : p \text{ prime}\} = \int_2^x \frac{1}{\log u} du + \mathcal{O}(x e^{-c\sqrt{\log x}}). \quad (4.48)$$

We get

$$E[(Y_h(l))^2] = \frac{\log 2}{2} + \mathcal{O}(e^{-c\sqrt{2^l}}).$$

Thus the variances of the increments are approximately equal. Moreover, the parameter σ^2 should be taken to be $\log 2/2$. This motivates the above coarse-graining to get *self-similarity*. For the *dichotomy*, we need the equivalent of the branching scale. For $h, h' \in [0, 1]$, take

$$h \wedge h' = \log_2 |h - h'|^{-1}. \tag{4.49}$$

The covariance between increments at scale l is

$$E[Y_h(k)Y_{h'}(k)] = \frac{1}{2} \sum_{2^{k-1} < \log p \leq 2^k} \frac{\cos(|h - h'|\log p)}{p}.$$

Again, the sum can be evaluated using (4.48). This time, the oscillating nature of cosine will produce a dichotomy depending on the branching scale between h and h'. In fact, it is not hard to show that, (see Lemma 2.1 in [7]),

$$E[Y_k(h)Y_k(h')] = \begin{cases} \frac{\log 2}{2} + O\left(2^{-2(h \wedge h' - k)}\right) & \text{if } k \leq h \wedge h', \\ O\left(2^{-(k - h \wedge h')}\right) & \text{if } k > h \wedge h'. \end{cases} \tag{4.50}$$

The dichotomy is not as clean as for BRW. On the other hand, the increments couple and decouple exponentially fast in the scales. This turns out to be sufficient for the purpose of the subleading order.

To prove Theorem 4.4.2, there are additional difficulties. Firstly, the process is continuous and not discrete like BRW. Therefore, an argument is needed to show that a discrete set of 2^n points capture the order of the maximum. This is based on a chaining argument. Secondly, the process here is not Gaussian so the estimate (4.1) cannot be used. It is replaced by precise large deviation principles obtained from estimates on the exponential moments of the sum.

Of course, there are tremendous technical obstacles to extending the method of proof of Theorem 4.4.2 to prove Conjecture 4.4.1 on the actual zeta function. For one, large deviation estimates are much harder to get. Moreover, the contribution to ζ of primes larger than T needs to be addressed. Despite these hurdles, the branching random walk heuristic gives a promising path to prove at least the leading order of the conjecture.

4.4.2 Maximum of the Characteristic Polynomial of Random Unitary Matrices

In this final section, we return to the problem we alluded to in Section 4.1.1. Let U_N be a random $N \times N$ unitary matrix sampled uniformly from the unitary group. This is often referred to as the *Circular Unitary Ensemble* (CUE). We consider the modulus of the characteristic polynomial on the unit circle

$$\left|P_{U_N}(\theta)\right| = \left|\det(e^{i\theta} - U_N)\right| = \prod_{j=1}^{2^{10}} |e^{i\theta} - e^{i\lambda_j}|.$$

where the eigenvalues of U_N are denoted by $(\lambda_j, j \leq N)$ and lie on the unit circle. A realization of this process was given for $N = 2^{10}$ in Figure 4.2. The prediction for the order of the maximum of this process is as follows.

Conjecture 4.4.3 (Fyodorov–Hiary–Keating [39, 40]) *For $N \in \mathbb{N}$, let U_N be a random matrix sampled uniformly from the group of $N \times N$ unitary matrices. Then*

$$\lim_{N \to \infty} \frac{\max_{\theta \in [0, 2\pi]} \log |P_{U_N}(\theta)| - \log N}{\log \log N} = -\frac{3}{4} \quad \text{in probability.} \tag{4.51}$$

In other words, it it expected that

$$\max_{\theta \in [0, 2\pi]} \log |P_{U_N}(\theta)| \approx \log N - \frac{3}{4} \log \log N + O(1).$$

The conjecture is well motivated by numerics and precise computations by Fyodorov, Hiary and Keating of moments of the partition function of the models. They infer from the expression for the moments the order of magnitude of the moments as well as a prediction for the fluctuations of the maximum, assuming that the system undergoes a *freezing transition* that is similar to the random energy model. Since there is strong empirical evidence that the characteristic polynomial of CUE is a good model for the Riemann zeta function locally, the authors use in part Conjecture 4.4.3 to motivate Conjecture 4.4.1.

As for the Riemann zeta function, the conjecture is exactly what would be expected for the maximum of a log-correlated model with $N = 2^n$ random variables with variance $\sigma^2 n = \frac{\log 2}{2} n$, cf. Conjecture 4.1.2. It turns out that for a given θ, it was shown in [45] that $\log |P_{U_N}(\theta)|$ normalized by $\sqrt{\frac{1}{2} \log N}$ converges in distribution to a standard Gaussian. Therefore, the choice of σ^2 is already consistent with this result. Kistler's multiscale refinement of the second moment method, as described in Section 4.3, can be adapted to prove the leading order of the conjecture.

Theorem 4.4.4 ([6]) *For $N \in \mathbb{N}$, let U_N be a random matrix sampled uniformly from the group of $N \times N$ unitary matrices. Write $P_N(\theta)$, $\theta \in [0, 2\pi]$, for its characteristic polynomial on the unit circle. Then*

$$\lim_{N \to \infty} \frac{\max_{\theta \in [0, 2\pi]} \log |P_N(\theta)|}{\log N} = 1 \quad \text{in probability.} \tag{4.52}$$

In this section, we explain the connection to an approximate branching random walk which illustrates the approach to prove the theorem and the general conjecture.

To see what plays the role of the *multiscale decomposition*, write the characteristic polynomial as

$$\log|P_{U_N}(\theta)| = \sum_{j=1}^{N} \log|1 - e^{i(\lambda_j - \theta)}|.$$

By expanding the logarithm, we get

$$\log|P_{U_N}(\theta)| = \sum_{j=1}^{N}\sum_{k=1}^{\infty} -\frac{\mathrm{Re}(e^{ik(\lambda_j - \theta)})}{j} = \sum_{k=1}^{\infty} -\frac{\mathrm{Re}(e^{-ik\theta}\mathrm{Tr}\,U_N^k)}{k}, \quad (4.53)$$

where Tr stands for the trace. It turns out that an integrable periodic function always has a pointwise convergent Fourier series wherever it is differentiable, so the above series makes sense for θ away from the eigenvalues. This already seems a good candidate for the *multiscale decomposition*. However, as was the case for zeta, it will be necessary to group the traces of powers to get the equivalent of increments at each scale. In other words, the traces of the powers play the role of the primes.

To see this, we will need the seminal result of Diaconis and Shahshahani [34] (see also [33]) who proved that

$$\mathbb{E}\left[\mathrm{Tr}\,U_N^j \,\overline{\mathrm{Tr}\,U_N^k}\right] = \delta_{kj}\min(k,N), \quad (4.54)$$

where $\mathbb{E}$ stands for the expectation under the uniform measure on the unitary group. It is also easy to see by rotation invariance of the uniform measure that $\mathbb{E}\left[\mathrm{Tr}\,U_N^j\mathrm{Tr}\,U_N^k\right] = 0$. This shows that the traces of powers are *uncorrelated*. However, and this is one of the main issues, they are not independent for fixed N. The formula (4.54) is useful to our heuristic in many ways. First, an easy calculation shows that the variance of the powers less than N is

$$E\left[\left(\sum_{k=1}^{N} -\frac{\mathrm{Re}(e^{-ik\theta}\mathrm{Tr}\,U_N^k)}{k}\right)^2\right] = \frac{1}{2}\sum_{k\le N}\frac{1}{k} = \frac{1}{2}\log N + O(1).$$

In particular, if we take $N = 2^n$ to emphasize the analogy with BRW, it leads us to define the *increment at scale l* for $l = 1,\ldots,n$ as

$$Y_\theta(l) = \sum_{2^{l-1} < k \le 2^l} -\frac{\mathrm{Re}(e^{-ik\theta}\mathrm{Tr}\,U_N^k)}{k}.$$

This gives
$$\mathbb{E}[(Y_\theta(l))^2] = \frac{\log 2}{2} + o(1),$$
where $o(1)$ is summable in l. From (4.53), the candidate for the multiscale decomposition is
$$\log|P_{U_N}(\theta)| = \sum_{l=1}^{n} Y_\theta(l) + Z_\theta(n), \tag{4.55}$$

where $Z_\theta(n)$ stands for the sum of powers greater than n. Moreover, equation (4.54) suggests that the contribution of $Z_\theta(n)$ to $\log|P_N(\theta)|$ in (4.53) should be of order 1 since $\sum_{j>N} \frac{N}{j^2} = O(1)$. We stress that, in this decomposition, the increments are uncorrelated though not independent. They are however asymptotically independent Gaussians, cf. [34]. Finally, they are almost *self-similar* in that regard since their variances are almost identical.

It remains to show the *dichotomy of scales* to complete the connection with BRW. A simple calculation using (4.54) gives that the covariance between increments for $\theta, \theta' \in [0, 2\pi]$ is
$$\mathbb{E}[Y_\theta(l)Y_{\theta'}(l)] = \sum_{2^{l-1} < k \le 2^l} \frac{\cos(j\|\theta - \theta'\|)}{2k}, \tag{4.56}$$

where $\|\theta - \theta'\|$ stands for the periodic distance on $[0, 2\pi]$. As was the case for the Riemann zeta function, the presence of the cosine is responsible for the dichotomy. To see this, define the *branching scale*
$$\theta \wedge \theta' = -\log_2 \|\theta - \theta'\|. \tag{4.57}$$

For j, such that $j\|h - h'\|$ is small, the cosine is essentially 1, and for j, such that $j\|\theta - \theta'\|$ is large, the oscillation of the cosine leads to cancellation. This can be proved by Taylor expansion of cosine in the first case and summation by parts in the other. This argument shows
$$\mathbb{E}[Y_\theta(l)Y_{\theta'}(l)] = \begin{cases} \frac{1}{2} + O(2^{l-\theta\wedge\theta'}) & \text{if } l \le \theta \wedge \theta', \\ O(2^{-2(l-\theta\wedge\theta')}) & \text{if } l > \theta \wedge \theta'. \end{cases} \tag{4.58}$$

In particular, this shows that the sum of traces of powers less than N are *log-correlated*:
$$\mathbb{E}\left[\left(\sum_{l=1}^{n} Y_\theta(l)\right)\left(\sum_{l=1}^{n} Y_{\theta'}(l)\right)\right] = \frac{\theta \wedge \theta'}{2} + O(1) = -\frac{1}{2}\log\|\theta - \theta'\| + O(1).$$

There are substantial difficulties in implementing the method of Section 4.3 to prove Theorem 4.4.4, and, more generally, Conjecture 4.4.3. First, it is necessary to control the contribution of the high powers, that is $Z_\theta(n)$ in (4.55). This turns out to be technically very challenging from a random matrix standpoint. Second, as opposed to BRW, the increments are not independent. This makes it much harder to get good large deviation estimates on the sum of the increments. Finally, since we are dealing with a continuous process on $[0, 2\pi]$ with singularities at the eigenvalues, one needs to justify that taking the maximum on $N = 2^n$ discrete points is enough to capture the order of the maximum.

Acknowledgments

I am grateful to Jean-Philippe Bouchaud, Pierluigi Contucci, Cristian Giardinà, Pierre Nolin, Vincent Vargas and Vladas Sidoravicius for the organization of the trimester *Disordered Systems, Random Spatial Processes and Some Applications* at the the Institut Henri Poincaré in the spring 2015. I also thank the Centre International de Rencontres Mathématiques and its staff for the hospitality and the organization during the Introductory School. This work would not have been possible without the numerous discussions with my collaborators on the subject: David Belius, Paul Bourgade, Anton Bovier, Adam J. Harper, Nicola Kistler, Olivier Zindy, and my students Samuel April, Frédéric Ouimet, Roberto Persechino and Jean-Sébastien Turcotte.

References

[1] L. Addario-Berry and B. Reed, *Minima in branching random walks,* Ann. Probab. 37(3), 1044–1079 (2009). MR2537549 (2011b:60338).

[2] L. Addario-Berry and B. A. Reed, *Ballot theorems, old and new,* Horizons of Combinatorics. 9–35 (2008).

[3] E. Aïdékon, *Convergence in law of the minimum of a branching random walk,* Ann. Probab. 41(3A), 1362–1426 (2013). MR3098680.

[4] E. Aïdékon, J. Berestycki, É. Brunet and Z. Shi, *Branching Brownian motion seen from its tip,* Probab. Theory Related Fields. 157(1–2), 405–451 (2013). MR3101852.

[5] E. Aïdékon, J. Berestycki, É. Brunet and Z. Shi, *Branching Brownian motion seen from its tip,* Probab. Theory Related Fields. 157(1–2), 405–451 (2013). MR3101852 (2011g:60153).

[6] L.-P. Arguin, D. Belius and P. Bourgade, *Maximum of the characteristic polynomial of random unitary matrices.* (2015). *Preprint* arXiv:1511.07399. http://link.springer.com/article/10.1007/s00220-016-2740-6

[7] L.-P. Arguin, D. Belius and A. J. Harper, *Maxima of a randomized Riemann zeta function, and branching random walks.* (2015). *Preprint* arXiv:1506.00629.

[8] L.-P. Arguin, A. Bovier and N. Kistler, *Genealogy of extremal particles of branching Brownian motion,* Comm. Pure Appl. Math. 64(12), 1647–1676 (2011). MR2838339 (2012g:60269).

[9] L.-P. Arguin and A. Bovier, *Poissonian statistics in the extremal process of branching Brownian motion*, Ann. Appl. Probab. 22(4), 1693–1711 (2012). MR2985174.

[10] L.-P. Arguin and A. Bovier, *The extremal process of branching Brownian motion*, Probab. Theory Related Fields. 157(3–4), 535–574 (2013). MR3129797.

[11] L.-P. Arguin and F. Ouimet, *Extremes of the two-dimensional Gaussian free field with scale-dependent variance.* (2015). *Preprint* arXiv:1508.06253. http://alea.impa.br/articles/v13/13-31.pd

[12] L.-P. Arguin and O. Zindy, *Poisson-Dirichlet statistics for the extremes of a log-correlated Gaussian field*, Ann. Appl. Probab. 24(4), 1446–1481 (2014). MR3211001.

[13] L.-P. Arguin and O. Zindy, *Poisson-Dirichlet statistics for the extremes of the two-dimensional discrete Gaussian free field*, Electron. J. Probab. 20(59), 19 (2015). MR3354619.

[14] M. Bachmann, *Limit theorems for the minimal position in a branching random walk with independent logconcave displacements*, Adv. in Appl. Probab. 32(1), 159–176 (2000). MR1765165 (2001m:60189).

[15] D. Belius and N. Kistler, *The subleading order of two dimensional cover times.* (2014). *Preprint* arXiv:1405.0888.

[16] J. D. Biggins, *The first- and last-birth problems for a multitype age-dependent branching process*, Advances in Appl. Probability 8(3), 446–459 (1976).

[17] M. Biskup and O. Louidor, *Extreme local extrema of two-dimensional discrete Gaussian free field.* (2013). *Preprint* arXiv:1306.2602.

[18] E. Bolthausen, J.-D. Deuschel and G. Giacomin, *Entropic repulsion and the maximum of the two-dimensional harmonic crystal*, Ann. Probab. 29(4), 1670–1692 (2001). MR1880237 (2003a:82028).

[19] A. Bovier, *Extremes*, Online lecture notes (2006). Available at http://wt.iam.uni-bonn.de/fileadmin/WT/Inhalt/people/Anton_Bovier/lecture-notes/extreme.pdf.

[20] A. Bovier, *From spin glasses to branching Brownian motion – and back?*, in *Random walks, random fields, and disordered systems* (A. Bovier et al. eds.). Berlin: Springer, 1–64 (2015).

[21] A. Bovier and L. Hartung, *The extremal process of two-speed branching Brownian motion*, Elect. J. Probab. 19(18), 1–28 (2014).

[22] A. Bovier and L. Hartung, *Variable speed branching Brownian motion 1. Extremal processes in the weak correlation regime*, ALEA Lat. Am. J. Probab. Math. Stat. 12(1), 261–291 (2015). MR3351476.

[23] M. Bramson, J. Ding and O. Zeitouni, *Convergence in law of the maximum of the two-dimensional discrete Gaussian free field.* (2013). *Preprint* arXiv:1301.6669.

[24] M. Bramson, J. Ding and O. Zeitouni, *Convergence in law of the maximum of nonlattice branching random walk.* (2014). *Preprint* arXiv: 1404.3423.

[25] M. D. Bramson, *Maximal displacement of branching Brownian motion*, Comm. Pure Appl. Math. 31(5), 531–581 (1978). MR0494541 (58 #13382).

[26] D. Carpentier and P. Le Doussal, *Glass transition of a particle in a random potential, front selection in nonlinear renormalization group, and entropic phenomena in Liouville and sinh-Gordon models*, Phys. Rev. E. 63, 026110 (2001Jan).

[27] O. Daviaud, *Extremes of the discrete two-dimensional Gaussian free field*, Ann. Probab. 34(3), 962–986 (2006).

[28] L. de Haan and A. Ferreira, *Extreme value theory: an introduction.* New York: Springer, (2006). MR2234156 (2007g:62008).

[29] A. Dembo, Y. Peres, J. Rosen and O. Zeitouni, *Cover times for Brownian motion and random walks in two dimensions*, Ann. of Math. 160(2), 433–464 (2004). MR2123929 (2005k:60261).

[30] B. Derrida, *Random-energy model: an exactly solvable model of disordered systems*, Phys. Rev. B. 24(5), 2613–2626 (1981). MR627810 (83a:82018).

[31] B. Derrida, *A generalisation of the random energy model that includes correlations between the energies*, J. Phys. Lett. 46(3), 401–407 (1985).

[32] B. Derrida and H. Spohn, *Polymers on disordered trees, spin glasses, and traveling waves*, J. Statist. Phys. 51(5–6), 817–840 (1988). MR971033 (90i:82045).

[33] P. Diaconis and S. N. Evans, *Linear functionals of eigenvalues of random matrices*, Trans. Amer. Math. Soc. 353(7), 2615–2633 (2001). MR1828463 (2002d:60003).

[34] P. Diaconis and M. Shahshahani, *On the eigenvalues of random matrices*, J. Appl. Probab. Studies in Applied Probability. 31A, 49–62 (1994). MR1274717 (95m:60011).

[35] J. Ding and Roy R. and O. Zeitouni, *Convergence of the centered maximum of log-correlated gaussian fields*. (2015). Preprint arXiv:1503.04588.

[36] M. Fang and O. Zeitouni, *Branching random walks in time-inhomogeneous environments*, Electron. J. Probab. 17(67), 18 (2012).

[37] M. Fang and O. Zeitouni, *Slowdown for time inhomogeneous branching Brownian motion*, J. Stat. Phys. 149(1), 1–9 (2012). MR2981635.

[38] Y. V. Fyodorov and J.-P. Bouchaud, *Freezing and extreme-value statistics in a random energy model with logarithmically correlated potential*, J. Phys. A. 41(37), 372001 (2008). MR2430565 (2010g:82036).

[39] Y. V. Fyodorov, G. A. Hiary and J. P. Keating, *Freezing transition, characteristic polynomials of random matrices, and the Riemann zeta function*, Phys. Rev. Lett. 108, 170601 (2012).

[40] Y. V. Fyodorov and J. P. Keating, *Freezing transitions and extreme values: random matrix theory, and disordered landscapes*, Philos. Trans. R. Soc. Lond. Ser. A Math. Phys. Eng. Sci. 372(2007), 20120503 (2014). MR3151088.

[41] Y. V. Fyodorov, P. Le Doussal and A. Rosso, *Statistical mechanics of logarithmic REM: duality, freezing and extreme value statistics of 1/f noises generated by Gaussian free fields*, J. Stat. Mech. Theory Exp. 10, 10005 (2009).

[42] B. Gnedenko, *Sur la distribution limite du terme maximum d'une sèerie aléeatoire*, Ann. of Math. 44, 423–453 (1943). MR0008655 (5,41b).

[43] E. J. Gumbel, *Statistics of extremes*. New York: Columbia University Press, (1958). MR0096342 (20 #2826).

[44] A. J. Harper, *A note on the maximum of the Riemann zeta function, and log-correlated random variables*. (2013). Preprint arXiv:1304.0677.

[45] J. P. Keating and N. C. Snaith, *Random matrix theory and $\zeta(1/2+it)$*, Comm. Math. Phys. 214(1), 57–89 (2000). MR1794265 (2002c:11107).

[46] N. Kistler, *Derrida's random energy models*, in *Correlated random systems: five different methods* (V. Gayrard and N. Kistler (eds.). Berlin: Springer, 71–120 (2015).

[47] S. P. Lalley and T. Sellke, *A conditional limit theorem for the frontier of a branching Brownian motion*, Ann. Probab. 15(3), 10521061 (1987). MR893913 (88h:60161).

[48] G. F. Lawler and V. Limic, *Random walk: a modern introduction*, Cambridge Studies in Advanced Mathematics, (vol. 123). Cambridge: Cambridge University Press, (2010).

[49] M. R. Leadbetter, G. Lindgren and H. Rootzéen, *Extremes and related properties of random sequences and processes*, Springer Series in Statistics. New York: Springer-Verlag, (1983). MR691492 (84h:60050).

[50] T. Madaule, *Maximum of a log-correlated Gaussian field.* (2014). *Preprint* arXiv:1307.1365.
[51] T. Madaule, *Convergence in law for the branching random walk seen from its tip*, Journal of Theoretical Probability. 29(111), 1–37 (2015).
[52] P. Maillard and O. Zeitouni, *Slowdown in branching brownian motion with inhomogeneous variance.* (2015). arXiv:1307.3583.
[53] H. L. Montgomery and R. C. Vaughan, *Multiplicative number theory. I. Classical theory*, Cambridge Studies in Advanced Mathematics, (vol. 97). Cambridge: Cambridge University Press, (2007). MR2378655 (2009b:11001).
[54] S. I. Resnick, *Extreme values, regular variation and point processes*, Springer Series in Operations Research and Financial Engineering. New York: Springer, (2006). MR2364939 (2008h:60002).
[55] R. Rhodes and V. Vargas, *Gaussian multiplicative chaos and applications: a review*, Probab. Surveys. 11, 315–392 (2014).
[56] K. Soundararajan, *Moments of the Riemann zeta function*, Ann. of Math. 170(2), 981–993 (2009). MR2552116 (2010i:11132).
[57] O. Zeitouni, *Branching random walks and Gaussian fields*, Online lecture notes (2013). Available at http://www.wisdom.weizmann.ac.il/ zeitouni/notesGauss.pdf.

5
Scaling Limits, Brownian Loops, and Conformal Fields

FEDERICO CAMIA

5.1 Introduction
5.1.1 Critical Scaling Limits

One of the main goals of both probability theory and statistical physics is to understand and describe the behavior of random systems consisting of a very large number of components (atoms, molecules, pixels, individuals, etc.) where the effect of each single component is negligible and the behavior of the system as a whole is determined by the combined effect of all its components. One usually wishes to study the behavior of such systems via some *observables*, suitably defined quantities that can be of an analytic or geometric nature. The asymptotic (in the number of system components) behavior of these quantities is often deterministic, but in some interesting cases it turns out to be random. This type of macroscopic randomness can be observed in *critical systems*, i.e., systems at a continuous phase transition point (the *critical point*).

In the physical theory of critical systems, it is usually assumed that, when a system approaches the critical point, it is characterized by a single length scale (the *correlation length*) in terms of which all other lengths should be measured. When combined with the experimental observation that the correlation length diverges at the critical point, this simple but strong assumption, known as the *scaling hypothesis*, leads to the belief that a critical system has no characteristic length, and is therefore invariant under scale transformations. This implies that all thermodynamic functions at criticality are homogeneous functions, and predicts the appearance of power laws.

It also suggests that, for models of critical systems realized on a lattice, one can attempt to take a *continuum scaling limit* in which the mesh of the lattice is sent to zero while focus is kept on macroscopic observables that capture the large scale behavior. In the limit, the discrete model should converge to a continuum one that encodes the large scale properties of the original model, containing at the same time more symmetry. In many cases, this allows for the derivation of additional insight by combining methods of discrete mathematics with considerations inspired by the continuum limit picture. The simplest example of such a continuum random model is Brownian motion, which is the scaling limit of the simple random walk. In general, though, the complexity of the discrete model makes it impossible even to guess the nature of the scaling limit, unless some additional feature can be used to pin down properties of the continuum limit. Two-dimensional critical systems belong to the class of models for which this can be done, and the additional feature is *conformal invariance.*

Indeed, thanks to the work of Polyakov [41] and others [7, 8], it has been understood by physicists since the early seventies that critical statistical mechanical models should possess continuum scaling limits with a global conformal invariance that goes beyond simple scale invariance (as long as the discrete models have "enough" rotation invariance), but this remains a conjecture for most models describing critical systems.

The appearance of conformal invariance in the scaling limit, besides being an interesting phenomenon in its own right, poses constraints on the possible continuum limits. However, since the conformal group is in general a finite dimensional Lie group, the resulting constraints are limited in number and provide only limited information. The phenomenon becomes particularly interesting in two dimensions, where every analytic function defines a conformal transformation (at points where its derivative is non-vanishing), and the conformal group is infinite-dimensional.

After this observation was made, a large number of critical problems in two dimensions were analyzed using conformal methods, which were applied, among others, to Ising and Potts models, Brownian motion, Self-Avoiding Walk (SAW), percolation, and Diffusion Limited Aggregation (DLA). The resulting body of knowledge and techniques, starting with the work of Belavin, Polyakov and Zamolodchikov [7, 8] in the early eighties, goes under the name of Conformal Field Theory (CFT).

At the hands of theoretical physicists, CFT has been very successful in producing many interesting results which until recently remained beyond any rigorous mathematical justification. This has changed in the last fifteen years, with the emergence in the mathematics literature of important developments in the area of two-dimensional critical systems which have followed a completely new direction, providing new tools and a new way of looking at critical systems and related conformal field theories.

These developments came on the heels of interesting results on the scaling limits of discrete models (see, e.g., the work of Aizenman [2,3], Benjamini and Schramm [9], Aizenman and Burchard [4], Aizenman, Burchard, Newman and Wilson [5],

Kenyon [31, 32] and Aizenman, Duplantier and Aharony [6]), but they differ greatly from those because they are based on a radically new approach whose main tool is the Stochastic Loewner Evolution (SLE), or Schramm-Loewner Evolution, as it is also known, introduced by Schramm [42]. The new approach, which is probabilistic in nature, focuses directly on non-local structures that characterize a given system, such as cluster boundaries in Ising, Potts and percolation models, or loops in the $O(n)$ model. At criticality, these non-local objects become, in the continuum limit, random curves whose distributions can be uniquely identified thanks to their conformal invariance and a certain Markovian property. There is a one-parameter family of SLEs, indexed by a positive real number κ, and they appear to be essentially the only possible candidates for the scaling limits of interfaces of two-dimensional critical systems that are believed to be conformally invariant.

The identification of the scaling limit of interfaces of critical lattice models with SLE curves has led to tremendous progress in recent years. The main power of SLE stems from the fact that it allows different quantities to be computed; for example, percolation crossing probabilities and various percolation critical exponents. Therefore, relating the scaling limit of a critical lattice model to SLE allows for a rigorous determination of some aspects of the large scale behavior of the lattice model. For the mathematician, the biggest advantage of SLE over CFT possibly lies in its mathematical rigor. But many physicists working on critical phenomena and CFT have promptly recognized the importance of SLE and added it to their toolbox.

In the context of the Ising, Potts and $O(n)$ models, as well as percolation, an SLE curve is believed to describe the scaling limit of a single interface, which can be obtained by imposing special boundary conditions. A single SLE curve is therefore not in itself sufficient to immediately describe the scaling limit of the unconstrained model without boundary conditions in the whole plane (or in domains with boundary conditions that do not determine a single interface), and contains only limited information concerning the connectivity properties of the model.

A more complete description can be obtained in terms of loops, corresponding to the scaling limit of cluster boundaries. Such loops should also be random and have a conformally invariant distribution, closely related to SLE. This observation led to the definition of Conformal Loop Ensembles (CLEs) [44, 45, 49], which are, roughly speaking, random collections of fractal loops with a certain "conformal restriction property." As for SLE, there is a one parameter family of CLEs.

Several interesting models of statistical mechanics, such as percolation and the Ising and Potts models, can be described in terms of clusters. In two dimensions and at the critical point, the scaling limit geometry of the boundaries of such clusters is known (see [19, 20, 22, 46]) or conjectured (see [30]) to be described by some member of the one-parameter family of Schramm-Loewner Evolutions (SLE_κ with $\kappa > 0$) and related Conformal Loop Ensembles (CLE_κ with $8/3 < \kappa < 8$). SLEs can be used to describe the scaling limit of single interfaces; CLEs are collections of loops and are therefore suitable to describe the scaling limit of the collection of all macroscopic boundaries at once. For example, the scaling limit of the critical percolation exploration path is SLE_6 [20, 46], and the scaling limit of the collection of all critical percolation interfaces in a bounded domain is CLE_6 [19, 21]. For $8/3 < \kappa \leq 4$, CLE_κ can be obtained [45] from the Brownian loop soup, as introduced by Lawler and Werner [36].

5.1.2 Near-critical Scaling Limits

A meaningful continuum scaling limit that differs from the critical one can usually be obtained by considering a system near the critical point, the so-called *off-critical* regime. This is done by adjusting some parameter of the model and sending it to the critical value at a specific rate while taking the scaling limit in such a way that the correlation length, in macroscopic units, stays bounded away from 0 and ∞. Depending on the context, this situation is described as *near-critical*, *off-critical* or *massive* scaling limit. The term *massive* refers to the persistence of a macroscopic correlation length, which should give rise to what is known in the physics literature as a *massive* field theory. In Euclidean field theory, the term *massless* is used to describe a scale-invariant system, while the term *massive* refers to a system with exponential decay of correlations.

Near-critical scaling limits are in general not expected to be conformally invariant, since the persistence of a macroscopic correlation length implies the absence of scale invariance, and are instead expected to have exponentially decaying correlations, while "resembling" their critical counterparts at distances smaller than the correlation length. (For a rigorous example of this behavior, see [17].) The lack of conformal invariance implies that the geometry of near-critical scaling limits cannot be described by an SLE or CLE. Indeed, much less has been proved about the geometry of off-critical models than about that of critical ones.

5.1.3 Random Walk Loop Soups and Brownian Loop Soups

Symanzik, in his seminal work on Euclidean quantum field theory [47], introduced a representation of the ϕ^4 Euclidean field as a "gas" of weakly interacting random paths. The use of random paths in the analysis of Euclidean field theories and statistical mechanical models was subsequently developed by various authors, most notably Brydges, Fröhlich, Spencer and Sokal [12, 13], and Aizenman [1], proving extremely useful (see [26] for a comprehensive account). The probabilistic analysis of Brownian and random walk paths and associated local times was carried out by Dynkin [24, 25]. More recently, "gases" or "soups" (i.e., Poissonian ensembles) of Brownian and random walk loops have been extensively studied in connection with SLE and the Gaussian free field (see, e.g., [36–38, 48–50]).

In this chapter, we provide a prototypical example, amenable to rigorous mathematical analysis, of critical and off-critical behavior by studying the scaling limit of an "ideal gas" of loops, called *random walk loop soup*. The loops are weighed according to a random walk measure on the square lattice with killing rates $\{k_x\}_{x\in\mathbb{Z}^2}$. Lawler and Trujillo Ferreras [35] have shown that, in the absence of killing, the random walk loop soup converges in the scaling limit to the *Brownian loop soup* introduced by Lawler and Werner [36].

The Brownian loop soup is, roughly speaking, a Poisson point process with intensity measure $\lambda\mu$, where λ is a positive constant and μ is the *Brownian loop measure* studied in [51]. μ is uniquely determined (up to a multiplicative constant) by its *conformal restriction property*, a combination of conformal invariance and the property that the measure in a subdomain is the original measure restricted to loops that stay in that subdomain. A realization of the loop soup consists of a countable collection of loops. Given a bounded domain D, there is an infinite number of loops that stay in D; however, the number of loops in D of diameter at least $\varepsilon > 0$ is finite. A consequence of conformal invariance is scale invariance: if $\mathcal{A}$ is a realization of the Brownian loop soup and each loop is scaled in space by $1/N$ and in time by $1/N^2$, the resulting configuration also has the distribution of the Brownian loop soup.

The Brownian loop soup exhibits a connectivity phase transition in the parameter $\lambda > 0$ that multiplies the intensity measure. When $\lambda \leq 1/2$, the loop soup in D is composed of disjoint clusters of loops [45, 49, 50] (where a cluster is a maximal collection of loops that intersect each other). When $\lambda > 1/2$, there is a unique cluster [45, 49, 50] and the set of points not surrounded by a loop is totally disconnected (see [11]). Furthermore, when $\lambda \leq 1/2$, the outer boundaries of the loop soup clusters are distributed like Conformal Loop Ensembles (CLE$_\kappa$) [44, 45, 49] with $8/3 < \kappa \leq 4$. As mentioned earlier, the latter are conjectured to describe the scaling limit of cluster boundaries in various critical models of statistical mechanics, such as the critical Potts models for $q \in [1, 4]$.

More precisely, if $8/3 < \kappa \leq 4$, then $0 < \frac{(3\kappa-8)(6-\kappa)}{4\kappa} \leq 1/2$ and the collection of all outer boundaries of the clusters of the Brownian loop soup with intensity parameter $\lambda = \frac{(3\kappa-8)(6-\kappa)}{4\kappa}$ is distributed like CLE$_\kappa$ [45]. For example, the continuum scaling limit of the collection of all macroscopic boundaries of critical Ising spin clusters is conjectured to correspond to CLE$_3$ and to a Brownian loop soup with $\lambda = 1/4$.

We note that there is some confusion in most of the existing literature regarding the critical intensity corresponding to the connectivity phase transition in the Brownian loop soup, as well as regarding the relation between λ and κ. This is due to the fact that the Brownian loop measure is an infinite measure. As a consequence, a choice of normalization is required when defining the Brownian loop soup. This choice is then reflected in the relation between λ and κ. Our choice of normalization is made explicit in Section 5.2.3 and coincides with that of [36].

The intensity λ of the Brownian loop soup is related to the central charge c of the corresponding statistical mechanical model. A discussion of the central charge of the Brownian loop soup can be found in Section 6 of [15]. Here we will just mention that, in some vague sense, λ determines how much the system feels a change in the shape of the domain. To understand this, suppose that $\mathcal{A}(\lambda, D)$ is a realization of the Brownian loop soup in D with intensity λ, and consider a subdomain $D' \subset D$. By removing from $\mathcal{A}(\lambda, D)$ all loops that are not contained in D', one obtains the loop soup $\mathcal{A}(\lambda, D')$ in D' with the same intensity λ. (This property of the Brownian loop soup follows from its Poissonian nature and the conformal restriction property of the Brownian loop measure μ.) The number of loops removed is stochastically increasing in λ, so that larger values of λ imply that the system is more sensitive to changes in the shape of the domain. In some sense, the loops can be seen as mediators of correlations from the boundary of the domain. (A precise formulation of this observation is presented in Section 5.2.2, in the context of the discrete Gaussian free field.) We note that the change from $\mathcal{A}(\lambda, D)$ to $\mathcal{A}(\lambda, D')$ has a non-local effect since the loops that are removed are extended objects, and that even a small local change to the shape of the domain can have an effect very far away, due to the scale invariance of the Brownian loop soup. This is a manifestation of the criticality of the system.

If one considers random walk loop soups with killing, the situation is very different. If one scales space by $1/N$ and time by $1/N^2$ keeping the killing rates constant, the resulting scaling limit is trivial, in the sense that it does not contain any loops larger than one point. This is so because, under the random walk loop measure, only loops of duration of order at least N^2 have diameter of order at least N with non-negligible probability as $N \to \infty$, and are therefore "macroscopic" in the scaling limit. It is then clear that, in order to obtain a nontrivial scaling limit, the killing rates need to be rescaled as well. When that is done appropriately, the scaling limit yields a *massive* (non-scale-invariant) version of the Brownian loop soup [14]. This *massive Brownian loop soup* is perhaps the simplest modification of the massless Brownian loop soup of Lawler and Werner combining several properties that are considered typical of off-critical systems, including exponential decay of correlations and a weaker form of conformal symmetry called *conformal covariance*. The mechanism by which it arises from the Brownian loop soup of

Lawler and Werner is analogous to that appearing in the standard Brownian motion representation of the continuum Gaussian free field when a mass term is present. Moreover, it is the only possible nontrivial, near-critical scaling limit of a random walk loop soup with killing.

The random walk loop soups studied in this chapter are shown to conform to the picture described in Sections 5.1.1 and 5.1.2, providing explicit examples of critical and near-critical scaling limits satisfying the general features typically associated with them. Because of their simplicity and amenability to a full analysis, the loop soups studied in this chapter can be considered a prototypical example of a system with critical and off-critical behavior. Moreover, the Brownian loop soup is deeply related to SLE and the critical scaling limit of models of statistical mechanics, as explained above, while its massive counterpart could potentially play a role in the description of various near-critical scaling limits.

5.1.4 Conformal Correlation Functions in the Brownian Loop Soup

There is a rich but still not fully understood connection between the conformal field theories studied by physicists and conformally invariant stochastic models, such as SLE and the Brownian loop soup. Physicists are often interested in conformal field theories defined by a Lagrangian and focus on the correlation functions (i.e., expectations of products) of local "observables" or "operators." In field theories related to the scaling limit of lattice models, the local observables typically considered are sums of (products of) local (random) variables over some macroscopic region, for example, the local magnetization in the Ising model. In the scaling limit, if the lattice model is critical, such observables may converge to "conformal fields." Mathematically, these "fields" are random generalized functions (distributions in the sense of Schwartz).

By contrast, conformal stochastic models are often defined and studied using very different methods and with different goals. In particular, the Brownian loop soup and SLE describe the behavior of non-local observables, macroscopic random curves such as Ising cluster boundaries.

In the last part of this chapter, based on [15], we use the Brownian loop soup to construct families of functions that behave like correlation functions of conformal fields, and explore some mathematical questions related to the scaling limit of critical correlation functions and to conformal fields. We expect that these correlation functions may define a full-fledged conformal field theory (one that could be related to the physics of de Sitter space-time and to eternal inflation). While this has not been fully established, progress has been made [18] in the direction of constructing random generalized functions with the appropriate correlation functions. The putative field theory would have several novel features, such as a periodic spectrum of conformal dimensions [15].

5.2 Loop Soups

5.2.1 Random Walk Loop Soups

Let $k_x \geq 0$ for every $x \in \mathbb{Z}^2$ and define $p_{x,y} = 1/(k_x+4)$ if $|x-y|=1$ and $p_{x,y}=0$ otherwise. If $k_x=0$ for all x, $\{p_{x,y}\}_{y \in \mathbb{Z}^2}$ is the collection of transition probabilities for the simple symmetric random walk on $\mathbb{Z}^2$. If $k_x \neq 0$, then $p_{x,y} = \frac{1}{4}(1+\frac{k_x}{4})^{-1} < \frac{1}{4}$ and one can interpret $\{p_{x,y}\}_{y \in \mathbb{Z}^2}$ as the collection of transition probabilities for a random walker "killed" at x with probability $1-(1+\frac{k_x}{4})^{-1} = \frac{k_x}{k_x+4}$. (Equivalently, one can introduce a "cemetery" state Δ not in $\mathbb{Z}^2$ to which the random walker jumps from $x \in \mathbb{Z}^2$ with probability $\frac{k_x}{k_x+4}$, and where it stays forever once it reaches it.) Because of this interpretation, we will refer to the collection $\mathbf{k} = \{k_x\}_{x \in \mathbb{Z}^2}$ as *killing rates*.

Given a $(2n+1)$-tuple $(x_0, x_1, \ldots, x_{2n})$ with $x_0 = x_{2n}$ and $|x_i - x_{i-1}|=1$ for $i=1,\ldots,2n$, we call *rooted lattice loop* the continuous path $\tilde{\gamma}: [0, 2n] \to \mathbb{C}$ with $\tilde{\gamma}(i) = x_i$ for integer $i=0,\ldots,2n$ and $\tilde{\gamma}(t)$ obtained by linear interpolation for other t. We call x_0 the *root* of the loop and denote by $|\tilde{\gamma}| = 2n$ the *length* or *duration* of the loop.

Now let D denote either $\mathbb{C}$ or a connected subset of $\mathbb{C}$. Following Lawler and Trujillo Ferreras [35], but within the more general framework of the previous paragraph, we introduce the *rooted random walk loop measure* $v_D^{r,\mathbf{k}}$ which assigns the loop $\tilde{\gamma}$ of length $|\tilde{\gamma}|$, with root x_0, weight

$$|\tilde{\gamma}|^{-1} p_{x_0, x_1} p_{x_1, x_2} \cdots p_{x_{|\tilde{\gamma}|-1}, x_0} = |\tilde{\gamma}|^{-1} \prod_{i=0}^{|\tilde{\gamma}|-1} (k_{x_i}+4) = |\tilde{\gamma}|^{-1} \prod_{x \in \tilde{\gamma}} (k_x+4)^{-n(x,\tilde{\gamma})}$$

if $x_0, \ldots, x_{|\tilde{\gamma}|-1} \in D$ and 0 otherwise, where $x \in \tilde{\gamma}$ means that x is visited by $\tilde{\gamma}$ and $n(x, \tilde{\gamma})$ denotes the number of times $\tilde{\gamma}$ makes a jump from x. (Note that $n(x, \tilde{\gamma})$ equals the number of visits of $\tilde{\gamma}$ to x for all vertices visited by $\tilde{\gamma}$ except the root x_0. For the root x_0, $n(x, \tilde{\gamma})$ is the number of visits minus one since the walk ends at x_0, so the last visit doesn't count.)

The *unrooted random walk loop measure* $v_D^{u,\mathbf{k}}$ is obtained from the rooted one by "forgetting the root." More precisely, if $\tilde{\gamma}$ is a rooted lattice loop and j a positive integer, $\theta_j \tilde{\gamma} : t \mapsto \tilde{\gamma}(j+t \mod |\tilde{\gamma}|)$ is again a rooted loop. This defines an equivalence relation between rooted loops; an *unrooted lattice loop* is an equivalence class of rooted lattice loops under that relation. By a slight abuse of notation, in the rest of the paper we will use $\tilde{\gamma}$ to denote unrooted lattice loops and $\tilde{\gamma}(\cdot)$ to denote any rooted lattice loop in the equivalence class of $\tilde{\gamma}$. The $v_D^{u,\mathbf{k}}$-measure of the unrooted loop $\tilde{\gamma}$ is the sum of the $v_D^{r,\mathbf{k}}$-measures of the rooted loops in the equivalence class of $\tilde{\gamma}$. The *length* or *duration*, $|\tilde{\gamma}|$, of an unrooted loop $\tilde{\gamma}$ is the length of any one of the rooted loops in the equivalence class $\tilde{\gamma}$.

Definition 5.2.1 *A random walk loop soup in D with intensity λ is a Poisson realization from $\lambda v_D^{u,\mathbf{k}}$.*

A realization of the random walk loop soup in D is a multiset (i.e., a set whose elements can occur multiple times) of unrooted loops. If we denote by $N_{\tilde{\gamma}}$ the multiplicity of $\tilde{\gamma}$ in a loop soup with intensity λ, $\{N_{\tilde{\gamma}}\}$ is a collection of independent Poisson random variables with parameters $\lambda v_D^{u,\mathbf{k}}(\tilde{\gamma})$. Therefore, the probability that a realization of the random walk loop soup in D with intensity λ contains each loop $\tilde{\gamma}$ in D with multiplicity $n_{\tilde{\gamma}} \geq 0$ is equal to

$$\prod_{\tilde{\gamma}} \exp\left(-\lambda v_D^{u,\mathbf{k}}(\tilde{\gamma})\right) \frac{1}{n_{\tilde{\gamma}}!} \left(\lambda v_D^{u,\mathbf{k}}(\tilde{\gamma})\right)^{n_{\tilde{\gamma}}} = \frac{1}{\mathcal{Z}_D^{\lambda,\mathbf{k}}} \prod_{\tilde{\gamma}} \frac{1}{n_{\tilde{\gamma}}!} \left(\lambda v_D^{u,\mathbf{k}}(\tilde{\gamma})\right)^{n_{\tilde{\gamma}}}, \quad (5.1)$$

where the product $\prod_{\tilde{\gamma}}$ is over all unrooted lattice loops in D and

$$\mathcal{Z}_D^{\lambda,\mathbf{k}} := \exp\left(\lambda \sum_{\tilde{\gamma}} v_D^{u,\mathbf{k}}(\tilde{\gamma})\right) = 1 + \sum_{n=1}^{\infty} \frac{1}{n!} \sum_{(\tilde{\gamma}_1,\ldots,\tilde{\gamma}_n)} \prod_{i=1}^{n} \lambda v_D^{u,\mathbf{k}}(\tilde{\gamma}_i), \quad (5.2)$$

where the sum over $(\tilde{\gamma}_1,\ldots,\tilde{\gamma}_n)$ is over all ordered configurations of n loops, not necessarily distinct. From a statistical mechanical viewpoint, $\mathcal{Z}_D^{\lambda,\mathbf{k}}$ can be interpreted as the grand canonical partition function of a "gas" of loops, and one can think of the random walk loop soup as describing a grand canonical ensemble of non-interacting loops (an "ideal gas") with the killing rates $\{k_x\}$ and the intensity λ as free "parameters." (For more on the statistical mechanical interpretation of the model, see Section 6.4 of [10].)

Exercise 5.2.2 *Compute the probability $p_\ell(A)$ that a random walk loop soup configuration contains ℓ loops from a given set A and use the result to compute the probability generating function $\sum_{\ell=0}^{\infty} p_\ell x^\ell$.*

When $k_x = 0 \ \forall x \in D \cap \mathbb{Z}^2$, we use v_D^u to denote the unrooted random walk loop measure in D; for reasons that will be clear when we talk about scaling limits, later in this section, a random walk loop soup obtained using such a measure will be called *critical*.

Now let $m : \mathbb{C} \to \mathbb{R}$ be a nonnegative function; we say that a random walk loop soup has *mass (function) m* if $k_x = 4(e^{m^2(x)} - 1)$ for all $x \in D \cap \mathbb{Z}^2$, and call *massive* a random walk loop soup with mass m that is not identically zero on $D \cap \mathbb{Z}^2$. For a massive random walk loop soup in D with intensity λ and mass m we use the notation $\tilde{\mathcal{A}}(\lambda,m,D)$.

The next proposition gives a construction for generating a massive random walk loop soup from a critical one, establishing a useful probabilistic coupling between the two (i.e., a way to construct the two loop soups on the same probability space).

Proposition 5.2.3 *A random walk loop soup in D with intensity λ and mass function m can be realized in the following way.*

1. *Take a realization of the* critical *random walk loop soup in D with intensity λ.*
2. *Assign to each loop $\tilde{\gamma}$ an independent, mean-one, exponential random variable $T_{\tilde{\gamma}}$.*
3. *Remove from the soup the loop $\tilde{\gamma}$ of length $|\tilde{\gamma}|$ if*

$$\sum_{i=0}^{|\tilde{\gamma}|-1} m^2(\tilde{\gamma}(i)) > T_{\tilde{\gamma}}. \tag{5.3}$$

Remark 5.2.4 *Note that equation (5.3) requires choosing a rooted loop from the equivalence class $\tilde{\gamma}$ but is independent of the choice.*

Proof. The proof is analogous to that of Proposition 5.2.9, so we leave the details to the reader. Below we just compare the expected number of loops in D generated by the construction of Proposition 5.2.3 with that of the massive random walk loop soup, to verify the relation between killing rates $\{k_x\}$ and mass function m. Since different loops are independent, we can compare the expected numbers of individual loops.

Writing $p_{x,y} = \frac{1}{k_x+4} = \frac{1}{4}\frac{4}{k_x+4}$ when $|x-y|=1$, for the massive random walk soup and loop $\tilde{\gamma}$, we have

$$\lambda v_D^{u,\mathbf{k}}(\tilde{\gamma}) = \lambda v_D^u(\tilde{\gamma}) \prod_{i=0}^{|\tilde{\gamma}|-1} \frac{4}{k_{x_i}+4}, \tag{5.4}$$

where, in the right hand side of the equation, we have chosen a representative for $\tilde{\gamma}$ such that $\tilde{\gamma}(i) = x_i$ (but the equation is independent of the choice of representative).

The expected number of loops $\tilde{\gamma}$ resulting from the construction of Proposition 5.2.3 is

$$\lambda v_D^u(\tilde{\gamma}) \int_0^\infty e^{-t} \mathbb{1}_{\{\sum_{i=0}^{|\tilde{\gamma}|-1} m^2(\tilde{\gamma}(i)) < t\}} dt = \lambda v_D^u(\tilde{\gamma}) \prod_{i=0}^{|\tilde{\gamma}|-1} e^{-m^2(x_i)}, \tag{5.5}$$

where we have chosen the same representative with $\tilde{\gamma}(i) = x_i$ as before and the result is again independent of the choice. Comparing equations (5.4) and (5.5), we see that the two expected numbers are indeed the same when $e^{-m^2(x)} = \frac{4}{k_x+4} = (1+k_x/4)^{-1}$ or $k_x = 4(e^{m^2(x)} - 1)$. □

5.2.2 Boundary Correlations in the Discrete Gaussian Free Field

In this section we discuss some interesting relations between the random walk loop soups of the previous section and the discrete Gaussian free field. We will use the setup of the previous section, but we need some additional notation and definitions.

Let D be a bounded subset of $\mathbb{C}$, define $D^{\#} := D \cap \mathbb{Z}^2$ and let $\Phi_D^{\mathbf{k}} = \{\phi_x\}_{x \in D^{\#}}$ denote a collection of mean-zero Gaussian random variables with covariance $\mathbb{E}_D^{\mathbf{k}}(\phi_x \phi_y) = G_D^{\mathbf{k}}(x,y)$, where $G_D^{\mathbf{k}}(x,y)$ denotes the Green function of the random walk introduced at the beginning of Section 5.2, with killing rates $\mathbf{k} = \{k_x\}_{x \in D^{\#}}$ and killed upon exiting the domain D (i.e., if the random walker attempts to leave D, it is sent to the cemetery Δ, where it stays forever). The lattice field $\Phi_D^{\mathbf{k}}$ is the *discrete Gaussian free field* in D with zero (Dirichlet) boundary condition. If $k_x = 0 \, \forall x \in D^{\#}$, the field is called *massless*, otherwise we will call it *massive*. (If the nature of the field is not specified, it means that it can be either massless or massive.)

The distribution of $\Phi_D^{\mathbf{k}}$ has density with respect to the Lebesgue measure on $\mathbb{R}^{D^{\#}}$ given by

$$\frac{1}{Z_D^{\mathbf{k}}} \exp\left(-H_D^{\mathbf{k}}(\varphi)\right),$$

where

$$Z_D^{\mathbf{k}} = \int_{\mathbb{R}^{D^{\#}}} \exp\left(-H_D^{\mathbf{k}}(\varphi)\right) \prod_{x \in D^{\#}} d\varphi_x$$

is a normalizing constant (the *partition function* of the model) and the Hamiltonian $H_D^{\mathbf{k}}$ is defined as follows:

$$H_D^{\mathbf{k}}(\varphi) := \frac{1}{4} \sum_{x,y \in D^{\#}: x \sim y} (\varphi_y - \varphi_x)^2 + \frac{1}{2} \sum_{x \in D^{\#}} k_x \varphi_x^2 + \frac{1}{2} \sum_{x \in \partial D^{\#}} \sum_{y \notin D^{\#}: x \sim y} \varphi_x^2$$

$$= -\frac{1}{2} \sum_{x,y \in D^{\#}: x \sim y} \varphi_x \varphi_y + \frac{1}{2} \sum_{x \in D^{\#}} (k_x + 4) \varphi_x^2,$$

where the first sum is over all ordered pairs $x,y \in D^{\#}$ such that $|x-y| = 1$ (denoted by $x \sim y$), and $\partial D^{\#}$ is the set $\{x \in D^{\#} : \exists y \notin D^{\#} \text{ such that } |x-y| = 1\}$.

In the first expression for $H_D^{\mathbf{k}}$, the second sum accounts for the massive nature of the field, while the third sum accounts for the Dirichlet boundary condition. (To understand the third sum, note that one can extend the field $\Phi_D^{\mathbf{k}}$ on $D^{\#}$ to a field $\Phi^{\mathbf{k}} = \{\phi_x\}_{x \in \mathbb{Z}^2}$ on $\mathbb{Z}^2$ by setting $\phi_x = 0 \, \forall x \notin D^{\#}$.)

Exercise 5.2.5 *Use Lemma 1.2 of [13] and Gaussian integration to show that the partition function of the discrete Gaussian free field in D can be expressed as*

$$Z_D^{\mathbf{k}} = \left(\prod_{x \in D^{\#}} \frac{2\pi}{k_x + 4} \right)^{1/2} \mathcal{Z}_D^{1/2,\mathbf{k}}.$$

The next result shows that the probability that the value of the field at a point inside the domain is affected by a change in the shape of the domain can be computed using the random walk loop soup with intensity $\lambda = 1/2$.

Proposition 5.2.6 *Let m be a nonnegative function (possibly identically zero), D and D' be bounded subsets of $\mathbb{C}$ containing $x_0 \in \mathbb{Z}^2$, with $D' \subset D$, and $\mathbf{k}$ denote the collection $\{4(e^{m^2(x)} - 1)\}_{x \in D \cap \mathbb{Z}^2}$. There exist versions of $\Phi_D^{\mathbf{k}} = \{\phi_x\}_{x \in D \cap \mathbb{Z}^2}$ and $\Phi_{D'}^{\mathbf{k}} = \{\phi_x'\}_{x \in D' \cap \mathbb{Z}^2}$, defined on the same probability space, such that*

$$P(\phi_{x_0} \neq \phi_{x_0}') = P_{1/2,m}(\text{there is a loop through } x_0 \text{ that intersects } D \setminus D'),$$

where P denotes the joint probability distribution of $\Phi_D^{\mathbf{k}}$ and $\Phi_{D'}^{\mathbf{k}}$, and $P_{1/2,m}$ is the law of the random walk loop soup in D with intensity $\lambda = 1/2$ and mass function m.

The proof of Proposition 5.2.6 will follow from a probabilistic coupling that allows us to define the random walk loop soup in D with intensity $1/2$ and the Gaussian free field in D with Dirichlet boundary condition on the same probability space. The coupling is given in Lemma 5.2.7 below, but first we need some additional notation.

We say that $x \in \mathbb{Z}^2$ is *touched* by the unrooted loop $\tilde{\gamma}$, and write $x \in \tilde{\gamma}$, if $\tilde{\gamma}(i) = x$ for some $i \in \{0, \ldots, |\tilde{\gamma}| - 1\}$ and some representative $\tilde{\gamma}(\cdot)$ of $\tilde{\gamma}$. If $\tilde{\gamma}(\cdot)$ is any rooted version of $\tilde{\gamma}$, the number of indices in $\{0, \ldots, |\tilde{\gamma}| - 1\}$ such that $\tilde{\gamma}(i) = x$ is denoted by $n(x, \tilde{\gamma})$ (note that the notation makes sense because $n(x, \tilde{\gamma})$ is independent of the choice of representative $\tilde{\gamma}(\cdot)$). To each $x \in \mathbb{Z}^2$ touched by $\tilde{\gamma}$, we associate $n(x, \tilde{\gamma})$ independent, exponentially distributed random variables with mean one, denoted by $\{\tau_x^i(\tilde{\gamma})\}_{i=1,\ldots,n(x,\tilde{\gamma})}$. We call the quantity

$$T_x(\tilde{\gamma}) := \begin{cases} \sum_{i=1}^{n(x,\tilde{\gamma})} \frac{\tau_x^i(\tilde{\gamma})}{k_x + 4} & \text{if } x \in \tilde{\gamma} \\ 0 & \text{if } x \notin \tilde{\gamma} \end{cases}$$

the *occupation time* at x associated to $\tilde{\gamma}$.

If $\tilde{\mathcal{A}}_{\lambda,m}$ is a realization of the random walk loop soup, we define the *occupation field* at x associated to $\tilde{\mathcal{A}}_{\lambda,m}$ as

$$L_x(\tilde{\mathcal{A}}_{\lambda,m}) := \sum_{\tilde{\gamma} \in \tilde{\mathcal{A}}_{\lambda,m}} T_x(\tilde{\gamma}) + \frac{\tau_x^0/2}{k_x + 4},$$

where $\{\tau_x^0\}_{x\in\mathbb{Z}^2}$ is an additional collection of independent, exponential random variables with mean one. We denote by $\mathbf{L}_D^\mathbf{k}$ the collection $\{L_x\}_{x\in D^\#}$.

The next result provides the coupling needed to prove Proposition 5.2.6. It says that, if $\{S_x\}$ are random variables with the Ising-type distribution (5.6) below, where the $\{L_x\}$ are distributed like the components of the occupation field of the random walk loop soup in D with intensity $1/2$ and mass function m, then the random variables $\psi_x = \sqrt{2L_x}S_x$ are equidistributed with the components of the discrete Gaussian free field in D with $k_x = 4(e^{m^2(x)} - 1)$ and Dirichlet boundary condition. More formally, one has the following proposition, where $\hat{P}_{1/2,m}$ denotes the joint distribution of the random walk loop soup in D with intensity $1/2$ and mass function m, and the collection of all exponential random variables needed to define the occupation field. The proof of the proposition follows easily from Theorem A.1 in Appendix A, which is a version of a recent result of Le Jan [37] (see also [38] and Theorem 4.5 of [48]).

Lemma 5.2.7 *Let $\mathbf{L}_D^\mathbf{k} = \{L_x\}_{x\in D^\#}$, denote the occupation field of the random walk loop soup in D with intensity $1/2$ and mass function m. Let $\mathbf{S} = \{S_x\}_{x\in D^\#}$ be (± 1)-valued random variables with (random) distribution*

$$P_{\mathbf{L}_D^\mathbf{k}}(S_x = \sigma_x \; \forall x \in D^\#)$$

$$= \frac{1}{Z}\exp\left(\sum_{x,y\in D^\#: x\sim y} \sqrt{L_x L_y}\sigma_x\sigma_y\right), \qquad (5.6)$$

where $\sigma_x = 1$ or -1 and Z is a normalization constant. Let $\psi_x = \sqrt{2L_x}S_x$ for all $x \in D^\#$; then under $\hat{P}_{1/2,m}\otimes P_{\mathbf{L}_D^\mathbf{k}}$, $\{\psi_x\}_{x\in D^\#}$ is distributed like the discrete Gaussian free field $\Phi_D^\mathbf{k} = \{\phi_x\}_{x\in D^\#}$ in D with $k_x = 4(e^{m^2(x)} - 1)$ and Dirichlet boundary condition.

Proof. According to Theorem A.1, ϕ_x^2 and $\psi_x^2 = 2L_x$ have the same distribution for every $x \in D^\#$. The lemma follows immediately from this fact and the observation that, letting $\sigma_x = \text{sgn}(\varphi_x)$, one can write

$$H_D^\mathbf{k}(\varphi) = -\frac{1}{2}\sum_{x,y\in D^\#: x\sim y}\sqrt{\varphi_x^2\varphi_y^2}\sigma_x\sigma_y + \frac{1}{2}\sum_{x\in D^\#}(k_x+4)\varphi_x^2. \qquad \square$$

Proof of Proposition 5.2.6 Let $\tilde{\mathcal{A}}_{1/2,m}$ be a realization of the random walk loop soup in D with intensity $\lambda = 1/2$ and mass function m, and let $\tilde{\mathcal{A}}'_{1/2,m}$ denote the collection of loops obtained by removing from $\tilde{\mathcal{A}}_{1/2,m}$ all loops that intersect $D\setminus D'$. $\tilde{\mathcal{A}}'_{1/2,m}$ is a random walk loop soup in D' with intensity $1/2$ and mass function m.

If x_0 is touched by a loop from $\tilde{\mathcal{A}}_{1/2,m}$ that intersects $D \setminus D'$, use $\tilde{\mathcal{A}}_{1/2,m}$ and Lemma 5.2.7 to generate a collection $\{\psi_x\}_{x \in D \cap \mathbb{Z}^2}$ distributed like the discrete Gaussian free field in D with Dirichlet boundary condition, and $\tilde{\mathcal{A}}'_{1/2,m}$ and Lemma 5.2.7 to generate a collection $\{\psi'_x\}_{x \in D' \cap \mathbb{Z}^2}$ distributed like the free field in D'. Then, with probability one,

$$\psi_{x_0}^2 = 2L_{x_0}(\tilde{\mathcal{A}}_{1/2,m}) > 2L_{x_0}(\tilde{\mathcal{A}}'_{1/2,m}) = (\psi'_{x_0})^2.$$

If x_0 is not touched by any loop from $\tilde{\mathcal{A}}_{1/2,m}$ that intersects $D \setminus D'$, use $\tilde{\mathcal{A}}_{1/2,m}$ and Lemma 5.2.7 to generate a $\{\psi_x\}_{x \in D \cap \mathbb{Z}^2}$ distributed like the free field in D. Because of symmetry, $\psi_{x_0} > 0$ with probability $1/2$.

- If $\psi_{x_0} > 0$, use $\tilde{\mathcal{A}}'_{1/2,m}$ and Lemma 5.2.7 to generate a collection $\{\psi'_x\}_{x \in D' \cap \mathbb{Z}^2}$, conditioned on $\psi'_{x_0} > 0$.
- If $\psi_{x_0} < 0$, use $\tilde{\mathcal{A}}'_{1/2,m}$ and Lemma 5.2.7 to generate a collection $\{\psi'_x\}_{x \in D' \cap \mathbb{Z}^2}$, conditioned on $\psi'_{x_0} < 0$.

Because of the $\pm$ symmetry of the Gaussian free field, it follows immediately that $\{\psi'_x\}_{x \in D' \cap \mathbb{Z}^2}$ is distributed like a Gaussian free field in D', and that $\psi'_{x_0} = \psi_{x_0}$. □

5.2.3 Brownian Loop Soups

A *rooted loop* $\gamma : [0, t_\gamma] \to \mathbb{C}$ is a continuous function with $\gamma(0) = \gamma(t_\gamma)$. We will consider only loops with $t_\gamma \in (0, \infty)$. The *Brownian bridge measure* μ^{br} is the probability measure on rooted loops of duration 1 with $\gamma(0) = 0$ induced by the Brownian bridge $B_t := W_t - tW_1, t \in [0,1]$, where W_t is standard, two-dimensional Brownian motion. A measure $\mu^{br}_{z,t}$ on loops rooted at $z \in \mathbb{C}$ (i.e., with $\gamma(0) = z$) of duration t is obtained from μ^{br} by Brownian scaling, using the map

$$(\gamma, z, t) \mapsto z + t^{1/2}\gamma(s/t), \ s \in [0,t].$$

More precisely, we let

$$\mu^{br}_{z,t}(\cdot) := \mu^{br}(\Phi_{z,t}^{-1}(\cdot)), \tag{5.7}$$

where

$$\Phi_{z,t} : \gamma(s), s \in [0,1] \mapsto z + t^{1/2}\gamma(s/t), s \in [0,t]. \tag{5.8}$$

The *rooted Brownian loop measure* is defined as

$$\mu_r := \int_{\mathbb{C}} \int_0^\infty \frac{1}{2\pi t^2} \mu^{br}_{z,t} \, dt \, d\mathbf{A}(z), \tag{5.9}$$

where $\mathbf{A}$ denotes area.

The (*unrooted*) *Brownian loop measure* μ is obtained from the rooted one by "forgetting the root." More precisely, if γ is a rooted loop, $\theta_u \gamma : t \mapsto \gamma(u + t \mod t_\gamma)$

is again a rooted loop. This defines an equivalence relation between rooted loops, whose equivalence classes we refer to as (*unrooted*) *loops*; $\mu(\gamma)$ is the μ_r-measure of the equivalence class γ. With a slight abuse of notation, in the rest of the paper we will use γ to denote an unrooted loop and $\gamma(\cdot)$ to denote any representative of the equivalence class γ.

The *massive (unrooted) Brownian loop measure* μ^m is defined by the relation

$$d\mu^m(\gamma) = \exp(-R_m(\gamma))d\mu(\gamma), \tag{5.10}$$

where $m : \mathbb{C} \to \mathbb{R}$ is a nonnegative *mass function* and

$$R_m(\gamma) := \int_0^{t_\gamma} m^2(\gamma(t))dt$$

for any rooted loop $\gamma(t)$ in the equivalence class of the unrooted loop γ. (Analogously, one can also define a massive *rooted* Brownian loop measure: $d\mu_r^m(\gamma) := \exp(-R_m(\gamma))d\mu_r(\gamma)$.)

If D is a subset of $\mathbb{C}$, we let μ_D (respectively, μ_D^m) denote μ (resp., μ^m) restricted to loops that lie in D. The family of measures $\{\mu_D\}_D$ (resp., $\{\mu_D^m\}_D$), indexed by $D \subset \mathbb{C}$, satisfies the *restriction property*, i.e., if $D' \subset D$, then $\mu_{D'}$ (resp., $\mu_{D'}^m$) is μ_D (resp., μ_D^m) restricted to loops lying in D'.

An equivalent characterization of the Brownian loop measure μ is as follows (see [51]). Given a conformal map $f : D \to D'$, let $f \circ \gamma(s)$ denote the loop $f(\gamma(t))$ in D' with parametrization

$$s = s(t) = \int_0^t |f'(\gamma(u))|^2 du. \tag{5.11}$$

Given a subset A of the space of loops in D, let $f \circ A = \{\hat{\gamma} = f \circ \gamma \text{ with } \gamma \in A\}$. Up to a multiplicative constant, μ_D is the unique measure satisfying the following two properties, collectively known as the *conformal restriction property*.

- For any conformal map $f : D \to D'$,

$$\mu_{D'}(f \circ A) = \mu_D(A). \tag{5.12}$$

- If $D' \subset D$, $\mu_{D'}$ is μ_D restricted to loops that stay in D'.

The conformal invariance of the Brownian loop measure and the result just mentioned are discussed in Appendix B.

As a consequence of the conformal invariance of $\{\mu_D\}_D$, the family of massive measures $\{\mu_D^m\}_D$ satisfies a property called *conformal covariance*, defined below. Given a conformal map $f : D \to D'$, let $\tilde{m}$ be defined by the map

$$m(z) \stackrel{f}{\mapsto} \tilde{m}(w) = \left|f'(f^{-1}(w))\right|^{-1} m(f^{-1}(w)), \tag{5.13}$$

where $w = f(z)$. This definition, combined with (5.11) and Brownian scaling, implies that $m^2 dt = \tilde{m}^2 ds$. From this and (5.12), it follows that

$$\mu_{D'}^{\tilde{m}}(f \circ A) = \mu_D^m(A), \tag{5.14}$$

where A and $f \circ A$ have the same meaning as in equation (5.12). We call this property conformal covariance and say that the massive Brownian loop measure μ^m is *conformally covariant*.

Definition 5.2.8 *A Brownian loop soup in D with intensity λ is a Poissonian realization from $\lambda \mu_D$. A massive Brownian loop soup in D with intensity λ and mass function m is a Poissonian realization from $\lambda \mu_D^m$.*

The (massive) Brownian loop soup "inherits" the property of conformal invariance (covariance) from the (massive) Brownian loop measure. Note that in a homogeneous massive Brownian loop soup, that is, if m is constant, loops are exponentially suppressed at a rate proportional to their time duration. We will sometimes call the conformally invariant Brownian loop soup introduced by Lawler and Werner *critical*, to distinguish it from the *massive* Brownian loop soup defined above.

The definition of the massive Brownian loop soup has a nice interpretation in terms of "killed" Brownian motion. For a given function f on the space of loops, one can write

$$\int f(\gamma) e^{-R_m(\gamma)} d\mu(\gamma) = \int \mathbb{E}_{T_\gamma} f(\gamma) \mathbb{1}_{\{R_m(\gamma) < T_\gamma\}} d\mu(\gamma),$$

where $\mathbb{E}_{T_\gamma}$ denotes expectation with respect to the law of the mean-one, exponential random variable T_γ, and $\mathbb{1}_{\{\cdot\}}$ denotes the indicator function. In view of this, one can think of the Brownian loop γ under the measure μ^m defined in (5.10) as being "killed" at rate $m^2(\gamma(t))$. More precisely, one has the following alternative and useful characterization.

Proposition 5.2.9 *A massive Brownian loop soup in D with intensity λ and mass function m can be realized in the following way.*

1. *Take a realization of the* critical *Brownian loop soup in D with intensity λ.*
2. *Assign to each loop γ of duration t_γ an independent, mean-one, exponential random variable, T_γ.*
3. *Remove from the soup all loops γ such that*

$$\int_0^{t_\gamma} m^2(\gamma(t)) dt > T_\gamma. \tag{5.15}$$

Remark 5.2.10 *Note that equation (5.15) requires choosing a time parametrization for the loop γ but is independent of the choice.*

Proof. Let $\mathcal{L}^D$ denote the set of loops contained in D and define $\mathcal{L}^D_{>\varepsilon} := \{\gamma \in \mathcal{L}^D : \text{diam}(\gamma) > \varepsilon\}$ and $\mathcal{L}^D_{>\varepsilon,r} := \{\gamma \in \mathcal{L}^D_{>\varepsilon} : R_m(\gamma) = r\}$. For a subset A of $\mathcal{L}^D_{>\varepsilon}$, let $A_r = A \cap \mathcal{L}^D_{>\varepsilon,r}$. For every $\varepsilon > 0$, the restriction to loops of diameter larger than ε of the massive Brownian loop soup in D with mass function m is a Poisson point process on $\mathcal{L}^D_{>\varepsilon}$ such that the expected number of loops in $A \subset \mathcal{L}^D_{>\varepsilon}$ at level $\lambda > 0$ is

$$\lambda \mu^m(A) = \lambda \int_A e^{-R_m(\gamma)} d\mu(\gamma)$$
$$= \lambda \int_0^\infty \int_{A_r} e^{-r} d\mu(\gamma) dr$$
$$= \lambda \int_0^\infty e^{-r} \mu(A_r) dr.$$

We will now show that, when attention is restricted to loops of diameter larger than ε, the construction of Proposition 5.2.9 produces a Poisson point process on $\mathcal{L}^D_{>\varepsilon}$ with the same expected number of loops at level $\lambda > 0$.

Let $N_\lambda(A)$ denote the number of loops in A obtained from the construction of Proposition 5.2.9. Because the Brownian loop soup is a Poisson point process and loops are removed independently, for every $A \subset \mathcal{L}^D_{>\varepsilon}$ we have that

(i) $N_0(A) = 0$,
(ii) $\forall \lambda, \delta > 0$ and $0 \leq \ell \leq \lambda$, $N_{\lambda+\delta}(A) - N_\lambda(A)$ is independent of $N_\ell(A)$,
(iii) $\forall \lambda, \delta > 0$, $\Pr(N_{\lambda+\delta}(A) - N_\lambda(A) \geq 2) = o(\delta)$,
(iv) $\forall \lambda, \delta > 0$, $\Pr(N_{\lambda+\delta}(A) - N_\lambda(A) = 1) = \mu^m(A)\delta + o(\delta)$,

where (iv) follows from the fact that, conditioned on the event $N_{\lambda+\delta}(A) - N_\lambda(A) = 1$, the additional point (i.e., loop) that appears going from λ to $\lambda + \delta$ is distributed according to the density $\frac{\mu(A_r)dr}{\mu(A)}$ on A. Conditions (i)–(iv) ensure that the point process is Poisson.

In order to identify the Poisson point process generated by the construction of Proposition 5.2.9 with the massive Brownian loop soup, it remains to compute the expected number of loops in A at level λ. For every $\varepsilon > 0$ and $A \subset \mathcal{L}^D_{>\varepsilon}$, this is given by

$$\int_0^\infty e^{-r} \lambda \mu(A_r) dr = \lambda \mu^m(A),$$

which concludes the proof. $\square$

5.2.4 Some Properties of the Massive Brownian Loop Soup

Let $\mathcal{A}(\lambda, m, D)$ denote a massive Brownian loop soup in $D \subset \mathbb{C}$ with mass function m and intensity λ. We say that two loops are *adjacent* if they intersect; this adjacency relation defines *clusters* of loops, denoted by $\mathcal{C}$. (Note that clusters can be nested.) For each cluster $\mathcal{C}$, we write $\overline{\mathcal{C}}$ for the closure of the union of all the loops in $\mathcal{C}$; furthermore, we write $\hat{\mathcal{C}}$ for the *filling* of $\mathcal{C}$, i.e., the complement of the unbounded connected component of $\mathbb{C} \setminus \overline{\mathcal{C}}$. With a slight abuse of notation, we call $\hat{\mathcal{C}}$ a *cluster* and denote by $\hat{\mathcal{C}}_z$ the cluster containing z. We set $\hat{\mathcal{C}}_z = \emptyset$ if z is not contained in any cluster $\hat{\mathcal{C}}$, and call the set $\{z \in D : \hat{\mathcal{C}}_z = \emptyset\}$ the *carpet* (or *gasket*). (Informally, the carpet is the complement of the "filled-in" clusters.)

It is shown in [45] that, in the *critical* case ($m = 0$), if D is bounded, the set of outer boundaries of the clusters $\hat{\mathcal{C}}$ that are not surrounded by other outer boundaries are distributed like a conformal loop ensemble in D, as explained in the introduction.

Theorem 5.2.11 *Let $\mathcal{A}(\lambda, m, D)$ be a massive Brownian loop soup in D with intensity λ and mass function m, and denote by $\mathbb{P}_{\lambda,m}$ the distribution of $\mathcal{A}(\lambda, m, \mathbb{C})$.*

- *If $\lambda > 1$, m is bounded and D is bounded, with probability one the vacant set of $\mathcal{A}(\lambda, m, D)$ is totally disconnected.*
- *If $\lambda \leq 1$ and m is bounded away from zero, the vacant set of $\mathcal{A}(\lambda, m, \mathbb{C})$ contains a unique infinite connected component. Moreover, there is a $\xi < \infty$ such that, for any $z \in \mathbb{C}$ and all $L > 0$,*

$$\mathbb{P}_{\lambda,m}(\mathrm{diam}(\hat{\mathcal{C}}_z) \geq L) \leq e^{-L/\xi}. \tag{5.16}$$

Note that, although in a massive loop soup individual large loops are exponentially suppressed, equation (5.16) is far from obvious, and in fact false when $\lambda > 1$, since in that equation the exponential decay refers to clusters of loops.

Theorem 5.2.12 *For any bounded domain $D \subset \mathbb{C}$ and any $m : D \to \mathbb{R}$ nonnegative and bounded, the carpet of the* massive *Brownian loop soup in D with mass function m and intensity λ has the same Hausdorff dimension as the carpet of the* critical *Brownian loop soup in D with the same intensity.*

It is expected that certain features of a near-critical scaling limit are the same as for the critical scaling limit. One of these features is the Hausdorff dimension of certain geometric objects. For instance, it is proved in [40] that the almost sure Hausdorff dimension of near-critical percolation interfaces in the scaling limit is 7/4, exactly as in the critical case. In view of the results in Section 5.3, Theorem 5.2.12 can be interpreted in the same spirit.

In the rest of the section, we present the proofs of the two theorems. To prove Theorem 5.2.11, we will use the following lemma, where, according to the notation of the theorem, $\mathbb{P}_{\lambda,m}$ denotes the distribution of the massive Brownian loop soup in $\mathbb{C}$ with intensity λ and mass function m.

Lemma 5.2.13 *Let $D \subset \mathbb{C}$ be a bounded domain with $\mathrm{diam}(D) > 1$ and m a positive function bounded away from zero. There exist constants $c < \infty$ and $m_0 > 0$, independent of D, such that, for every $\lambda > 0$ and $\ell_0 > 1$,*

$$\mathbb{P}_{\lambda,m}(\nexists \gamma \text{ with } \mathrm{diam}(\gamma) > \ell_0 \text{ and } \gamma \cap D \neq \emptyset) \geq 1 - \lambda c (\mathrm{diam}(D) + 2\ell_0)^2 e^{-m_0 \ell_0}.$$

Proof. Let $D_\ell := \cup_{z \in D} B_\ell(z)$, where $B_\ell(z)$ denotes the disc of radius ℓ centered at z. We define several sets of loops, namely,

$$\mathcal{L}_\ell := \{\text{loops } \gamma \text{ with } \mathrm{diam}(\gamma) = \ell\}$$
$$\mathcal{L}'_\ell := \{\text{loops } \gamma \text{ with } \mathrm{diam}(\gamma) = \ell \text{ and duration } t_\gamma \geq \ell\}$$
$$\mathcal{L}''_\ell := \{\text{loops } \gamma \text{ with } \mathrm{diam}(\gamma) = \ell \text{ and duration } t_\gamma < \ell\}$$
$$\mathcal{L}^D_\ell := \{\text{loops } \gamma \text{ with } \mathrm{diam}(\gamma) = \ell \text{ and } \gamma \cap D \neq \emptyset\}$$
$$\mathcal{L}^D_{>\ell_0} := \cup_{\ell > \ell_0} \mathcal{L}^D_\ell = \{\text{loops } \gamma \text{ with } \mathrm{diam}(\gamma) > \ell_0 \text{ and } \gamma \cap D \neq \emptyset\}.$$

We note that

$$\mathbb{P}_{\lambda,m}(\nexists \gamma \text{ with } \mathrm{diam}(\gamma) > \ell_0 \text{ and } \gamma \cap D \neq \emptyset) = \exp[-\lambda \mu^m(\mathcal{L}^D_{>\ell_0})]$$
$$\geq 1 - \lambda \mu^m(\mathcal{L}^D_{>\ell_0}) \quad (5.17)$$

where μ^m denotes the massive Brownian loop measure with mass function m. Thus, in order to prove the lemma, we look for an upper bound for $\mu^m(\mathcal{L}^D_{>\ell_0})$.

Denoting by $\mu^m_{D_\ell}$ the restriction of μ^m to D_ℓ, we can write

$$\mu^m(\mathcal{L}^D_{>\ell_0}) \leq \int_{\ell > \ell_0} d\mu^m_{D_\ell}(\mathcal{L}_\ell) = \int_{\ell > \ell_0} d\mu^m_{D_\ell}(\mathcal{L}'_\ell) + \int_{\ell > \ell_0} d\mu^m_{D_\ell}(\mathcal{L}''_\ell). \quad (5.18)$$

From the definition of the massive Brownian loop measure (see equations (5.10) and (5.9)), we have that

$$\int_{\ell > \ell_0} d\mu^m_{D_\ell}(\mathcal{L}'_\ell) \leq \int_{\ell_0}^\infty \frac{\pi}{4} \frac{\mathrm{diam}^2(D_\ell)}{2\pi \ell^2} \exp\left(-\ell \inf_{z \in D_\ell} m(z)\right) d\ell$$
$$< \frac{1}{8} \int_{\ell_0}^\infty \mathrm{diam}^2(D_\ell) \exp\left(-\ell \inf_{z \in D_\ell} m(z)\right) d\ell. \quad (5.19)$$

To bound the second term of the r.h.s. of (5.18), we observe that, if a loop γ of duration t_γ has $\mathrm{diam}(\gamma) \geq \ell$, for any time parametrization of the loop, there exists a $t_0 \in (0, t_\gamma)$ such that $|\gamma(t_0) - \gamma(0)| \geq \ell/2$. The image $\hat{\gamma}$ of γ under Φ^{-1} must then satisfy $|\hat{\gamma}(t_0/t_\gamma)| \geq \ell/(2\sqrt{t_\gamma})$ (see equation (5.8)).

Let W_s and $B_s := W_s - sW_1$ denote standard, two-dimensional Brownian motion and Brownian bridge, respectively, with $s \in [0,1]$. Let W^1_s denote standard, one-dimensional Brownian motion. Noting that $|B_s| \leq |W_s| + |W_1| \leq 2|W_s| + |W_1 - W_s|$, and using the reflection principle and known properties of the complementary error function (erfc), the above observation gives the bound

$$\mu_{z,t}^{br}(\gamma : \text{diam}(\gamma) \geq \ell)$$
$$\leq \mu_{z,t}^{br}(\exists s \in (0,t) : |\gamma(s) - z| \geq \ell/2)$$
$$= \mu^{br}\left(\exists s \in (0,1) : |\hat{\gamma}(s)| \geq \frac{\ell}{2\sqrt{t}}\right)$$
$$\leq \Pr\left(\exists s \in (0,1) : |W_s| + |W_1 - W_s| \geq \frac{\ell}{4\sqrt{t}}\right)$$
$$\leq \Pr\left(\exists s \in (0,1) : |W_s| \geq \frac{\ell}{12\sqrt{t}}\right)$$
$$\leq 4\Pr\left(\sup_{0 \leq s \leq 1} W_s^1 \geq \frac{\ell}{12\sqrt{t}}\right)$$
$$= 8\Pr\left(W_1^1 \geq \frac{\ell}{12\sqrt{t}}\right)$$
$$= 4\,\text{erfc}\left(\frac{\ell}{12\sqrt{t}}\right) \leq 4e^{-\ell^2/288t}. \qquad (5.20)$$

Using (5.20), we have that, for $\ell_0 > 1$,

$$\int_{\ell > \ell_0} d\mu_{D_\ell}^m(\mathcal{L}_\ell'') \leq \int_{\ell_0}^\infty \frac{\pi}{4} \text{diam}^2(D_\ell) \int_0^\ell \frac{1}{2\pi t^2} \mu_{0,t}^{br}(\gamma : \text{diam}(\gamma) \geq \ell) \, dt \, d\ell$$
$$\leq \int_{\ell_0}^\infty \frac{\pi}{4} \text{diam}^2(D_\ell) \int_0^\ell \frac{2}{\pi t^2} e^{-\ell^2/288t} \, dt \, d\ell$$
$$\leq \tilde{c} \int_{\ell_0}^\infty \text{diam}^2(D_\ell) e^{-\ell/288} \, d\ell,$$

where $\tilde{c} < \infty$ is a suitably chosen constant that does not depend on D, ℓ_0 or m.

Combining this bound with the bound (5.19), we can write

$$\mu^m(\mathcal{L}_{>\ell_0}^D) \leq \hat{c} \int_{\ell_0}^\infty \text{diam}^2(D_\ell) e^{-m_0 \ell} \, d\ell,$$

where $m_0 > 0$ is any positive number smaller than $\min(1/288, \inf_{z \in \mathbb{C}} m(z))$ and $\hat{c} < \infty$ is a suitably chosen constant independent of D, ℓ_0 and m. From this, a simple calculation leads to

$$\mu^m(\mathcal{L}_{>\ell_0}^D) \leq \hat{c} \int_{\ell_0}^\infty (\text{diam}(D) + 2\ell)^2 e^{-m_0 \ell} \, d\ell \leq c(\text{diam}(D) + 2\ell_0)^2 e^{-m_0 \ell_0},$$

where $c < \infty$ is a constant that does not depend on D and ℓ_0. The proof is concluded using inequality (5.17).

Proof of Theorem 5.2.11. We first prove the statement in the first bullet. Let $\overline{m} := \sup_{z \in \mathbb{C}} m(z)$; since m is bounded, $\overline{m} < \infty$. Let $\tau > 0$ be so small that $\lambda' := e^{-\overline{m}^2 \tau} \lambda > 1$ and denote by $\mathcal{A}(\lambda, 0, D)$ a critical loop soup in D with intensity λ obtained from the full-plane loop soup $\mathcal{A}(\lambda, 0, \mathbb{C})$. Let $\mathcal{A}(\lambda, m, D)$ be a massive soup obtained from $\mathcal{A}(\lambda, 0, D)$ via the construction of Proposition 5.2.9, and let $\overline{\mathcal{A}}(\lambda', \tau, D)$ denote the collection of loops obtained from $\mathcal{A}(\lambda, 0, D)$ by removing all loops of duration $> \tau$ with probability one, and other loops with probability $1 - e^{-\overline{m}^2 \tau}$, so that $\overline{\mathcal{A}}(\lambda', \tau, D)$ is a loop soup with intensity $\lambda' = e^{-\overline{m}^2 \tau} \lambda$, restricted to loops of duration $\leq \tau$. If we use the same exponential random variables to generate $\mathcal{A}(\lambda, m, D)$ and $\overline{\mathcal{A}}(\lambda', \tau, D)$ from $\mathcal{A}(\lambda, 0, D)$, the resulting loop soups are coupled in such a way that $\mathcal{A}(\lambda, m, D)$ contains all the loops that are contained in $\overline{\mathcal{A}}(\lambda', \tau, D)$, and the vacant set of $\mathcal{A}(\lambda, m, D)$ is a subset of the vacant set of $\overline{\mathcal{A}}(\lambda', \tau, D)$.

We will now show that, for every $S \subset D$ at positive distance from the boundary of D, the intersection with S of the vacant set of $\overline{\mathcal{A}}(\lambda', \tau, D)$ is totally disconnected. The same statement is then true for the intersection with S of the vacant set of $\mathcal{A}(\lambda, m, D)$, which concludes the proof of the first bullet, since the presence of a connected component larger than one point in the vacant set of $\mathcal{A}(\lambda, m, D)$ would lead to a contradiction.

For a given $S \subset D$ and any $\varepsilon > 0$, take $\tau = \tau(S, \varepsilon)$ so small that $e^{-\overline{m}^2 \tau} \lambda > 1$ and the probability that a loop from $\mathcal{A}(\lambda, 0, \mathbb{C})$ of duration $\leq \tau$ intersects both S and the complement of D is less than ε. (This is possible because the μ-measure of the set of loops that intersect both S and the complement of D is finite.) If that event does not happen, the intersection between S and the vacant set of $\overline{\mathcal{A}}(\lambda', \tau, D)$ coincides with the intersection between S and the vacant set of the full-plane loop soup $\overline{\mathcal{A}}(\lambda', \tau, \mathbb{C})$ with cutoff τ on the duration of loops. The latter intersection is a totally disconnected set with probability one by an application of Theorem 2.5 of [11]. (Note that the result does not follow directly from Theorem 2.5 of [11], which deals with full-space soups with a cutoff on the *diameter* of loops, but can be easily obtained from it, for example with a coupling between $\overline{\mathcal{A}}(\lambda', \tau, \mathbb{C})$ and a full-plane soup with a cutoff on the diameter of loops chosen to be much smaller than $\sqrt{\tau}$, and using arguments along the lines of those in the proof of Lemma 5.2.13. We leave the details to the interested reader.) Since τ can be chosen arbitrarily small, this shows that the intersection with any S of the vacant set of $\overline{\mathcal{A}}(\lambda', \tau, D)$ is totally disconnected with probability one.

To prove the statement in the second bullet we need some definitions. Let $R_l := [0, 3l] \times [0, l]$ and denote by A_l the event that the vacant set of a loop soup contains a crossing of R_l in the long direction, i.e., that it contains a connected component which stays in R_l and intersects both $\{0\} \times [0, l]$ and $\{3l\} \times [0, l]$. Furthermore, let $E_l^\ell := \{\nexists \gamma \text{ with } \operatorname{diam}(\gamma) > \ell \text{ and } \gamma \cap R_l \neq \emptyset\}$ and denote by $\overline{\mathbb{P}}_{\lambda,\ell}$ the distribution of the critical Brownian loop soup in $\mathbb{C}$ with intensity λ and cutoff ℓ on the diameter of the loops (i.e., with all the loops of diameter $> \ell$ removed).

We have that, for any $\ell_0 > 0$ and $n \in \mathbb{N}$,

$$\begin{aligned}
\mathbb{P}_{\lambda,m}(A_{3^n}) &\geq \mathbb{P}_{\lambda,m}(A_{3^n} \cap E_{3^n}^{n\ell_0}) \\
&= \mathbb{P}_{\lambda,m}(A_{3^n} | E_{3^n}^{n\ell_0}) \mathbb{P}_{\lambda,m}(E_{3^n}^{n\ell_0}) \\
&\geq \overline{\mathbb{P}}_{\lambda,n\ell_0}(A_{3^n}) \mathbb{P}_{\lambda,m}(E_{3^n}^{n\ell_0}) \\
&= \overline{\mathbb{P}}_{\lambda,3^{-n}n\ell_0}(A_1) \mathbb{P}_{\lambda,m}(E_{3^n}^{n\ell_0}),
\end{aligned}$$

where we have used the Poissonian nature of the loop soup in the second inequality, and the last equality follows from scale invariance.

Now consider a sequence $\{\overline{\mathcal{A}}(\lambda, 3^{-n}n\ell_0, \mathbb{C})\}_{n \geq 1}$ of full-plane soups with cutoffs $\{3^{-n}n\ell_0\}_{n \geq 1}$, obtained from the same critical Brownian loop soup $\mathcal{A}(\lambda, 0, \mathbb{C})$ by removing all loops of diameter larger than the cutoff. The soups are then coupled in such a way that their vacant sets form an increasing (in the sense of inclusion of sets) sequence of sets. Therefore, by Kolmogorov's zero-one law, $\lim_{n \to \infty} \overline{\mathbb{P}}_{\lambda,3^{-n}n\ell_0}(A_1)$ is either 0 or 1. (Note that this limit can be seen as the probability of the union over $n \geq 1$ of the events that the rectangle R_1 is crossed in the long direction by the vacant set of the soup with cutoff $3^{-n}n\ell_0$.) Since $\overline{\mathbb{P}}_{\lambda,3^{-n}n\ell_0}(A_1)$ is strictly positive for $n = 1$ (see Section 3 of [11]) and clearly increasing in n, we conclude that $\lim_{n \to \infty} \overline{\mathbb{P}}_{\lambda,3^{-n}n\ell_0}(A_1) = 1$.

Moreover, it follows from Lemma 5.2.13 that

$$\mathbb{P}_{\lambda,m}(E_{3^n}^{n\ell_0}) \geq 1 - \lambda c(\sqrt{10} + 2\ell_0)^2 e^{(2\log 3 - m_0 \ell_0)n},$$

for some constants $c < \infty$ and $m_0 > 0$ independent of ℓ_0 and n. Choosing $\ell_0 > (2\log 3)/m_0$, we have that $\mathbb{P}_{\lambda,m}(E_{3^n}^{n\ell_0}) \xrightarrow{n \to \infty} 1$, which implies that $\lim_{n \to \infty} \mathbb{P}_{\lambda,m}(A_{3^n}) = 1$. Note that the result does not depend on the position and orientation of the rectangles R_l chosen to define the event A_l.

Crossing events for the vacant set are decreasing in λ and are therefore positively correlated (see, e.g., Lemma 2.2 of [29]). (An event A is decreasing if $\mathcal{A} \notin A$ implies $\mathcal{A}' \notin A$ whenever $\mathcal{A}$ and $\mathcal{A}'$ are two soup realizations such that $\mathcal{A}'$ contains all the loops contained in $\mathcal{A}$.) Let $\mathbb{A}_n := [-3^{n+1}/2, 3^{n+1}/2] \times [-3^{n+1}/2, 3^{n+1}/2] \setminus [-3^n/2, 3^n/2] \times [-3^n/2, 3^n/2]$ and C_n denote the event that a connected component of the vacant set makes a circuit inside $\mathbb{A}_n$ surrounding $[-3^n/2, 3^n/2] \times [-3^n/2, 3^n/2]$. Using the positive correlation of crossing events, and the fact that the circuit described above can be obtained by "pasting" together four crossings of rectangles, we conclude that $\lim_{n \to \infty} \mathbb{P}_{\lambda,m}(C_n) = 1$.

The existence of a unique unbounded component in the vacant set now follows from standard arguments (see, e.g., the proof of Theorem 3.2 of [11]).

The exponential decay of loop soup clusters also follows immediately, since the occurrence of C_n prevents the cluster of the origin from extending beyond the square $[-3^{n+1}/2, 3^{n+1}/2] \times [-3^{n+1}/2, 3^{n+1}/2]$. □

Proof of Theorem 5.2.12. Let $\overline{m}_D := \sup_{z \in D} m(z)$; since m is bounded, $\overline{m}_D < \infty$. Fix $\tau \in (0, \infty)$ and define $\lambda' := e^{-\overline{m}_D^2 \tau} \lambda < \lambda$. Denote by $\mathcal{A}(\lambda, 0, D)$ a critical loop soup in D with intensity λ. Let $\mathcal{A}(\lambda, m, D)$ be a massive loop soup obtained from $\mathcal{A}(\lambda, 0, D)$ via the construction of Proposition 5.2.9, and let $\overline{\mathcal{A}}(\lambda', \tau, D)$ denote the set of loops obtained from $\mathcal{A}(\lambda, 0, D)$ by removing all loops of duration $> \tau$ with probability one, and other loops with probability $1 - e^{-\overline{m}_D^2 \tau}$, so that $\overline{\mathcal{A}}(\lambda', \tau, D)$ is a loop soup with intensity $\lambda' = e^{-\overline{m}_D^2 \tau} \lambda$, restricted to loops of duration $\leq \tau$. If we use the same exponential random variables to generate $\mathcal{A}(\lambda, m, D)$ and $\overline{\mathcal{A}}(\lambda', \tau, D)$ from $\mathcal{A}(\lambda, 0, D)$, the resulting loop soups are coupled in such a way that $\mathcal{A}(\lambda, m, D)$ contains all the loops that are contained in $\overline{\mathcal{A}}(\lambda', \tau, D)$, and the vacant set of $\mathcal{A}(\lambda, m, D)$ is a subset of the vacant set of $\overline{\mathcal{A}}(\lambda', \tau, D)$. Let $\mathbb{P}$ denote the probability distribution corresponding to the coupling between soups described above.

Note that, if we denote carpets by $G[\,\cdot\,]$, we have that

$$G[\overline{\mathcal{A}}(\lambda', \tau, D)] \supset G[\mathcal{A}(\lambda, m, D)] \supset G[\mathcal{A}(\lambda, 0, D)]. \tag{5.21}$$

Moreover, letting

$$h(\ell) := \frac{187 - 7\ell + \sqrt{25 + \ell^2 - 26\ell}}{96},$$

denoting the Hausdorff dimension of a set S by $H(S)$, and combining the computation of the expectation dimension for carpets/gaskets of Conformal Loop Ensembles [43] with the results of [39] (see, in particular, Section 4.5 of [39]), we have that, for any $\ell \in [0, 1]$, with probability one,

$$H(G(\mathcal{A}(\ell, 0, D))) = h(\ell).$$

We will now show that the almost sure Hausdorff dimension of $\overline{\mathcal{A}}(\lambda', \tau, D)$ equals $h(\lambda')$. To do this, consider the event E that $\mathcal{A}(\lambda, 0, D)$ contains no loop of duration $> \tau$. Note that $\mathbb{P}(E) > 0$, since the set of loops of duration $> \tau$ that stay in D has finite mass for the Brownian loop measure μ. Note also that, on the event E, $\overline{\mathcal{A}}(\lambda', \tau, D)$ coincides with $\mathcal{A}(\lambda', 0, D)$. Thus, since the sets of loops of duration $> \tau$ and $\leq \tau$ are disjoint, the Poissonian nature of the loop soups implies that

$$\mathbb{P}(H(G[\overline{\mathcal{A}}(\lambda', \tau, D)]) = h(\lambda')) = \mathbb{P}(H(G[\mathcal{A}(\lambda', 0, D)]) = h(\lambda') | E) = 1.$$

From this and (5.21), it follows that, with probability one,

$$h(\lambda') = H(G[\overline{\mathcal{A}}(\lambda',\tau,D)]) \geq H(G[\mathcal{A}(\lambda,m,D)]) \geq H(G[\mathcal{A}(\lambda,0,D)]) = h(\lambda).$$

Since h is continuous, letting $\tau \to 0$ (so that $\lambda' \to \lambda$) concludes the proof. □

5.3 Scaling Limits

We are now going to consider scaling limits for the random walk loop soup defined in the previous section.

5.3.1 The Critical Case

Consider a critical, full-plane, random walk loop soup $\tilde{\mathcal{A}}_\lambda \equiv \tilde{\mathcal{A}}(\lambda, 0, \mathbb{Z}^2)$. Following [35], for each integer $N \geq 2$, we define the *rescaled random walk loop soup*

$$\tilde{\mathcal{A}}_\lambda^N := \{\tilde{\Phi}_N \tilde{\gamma} : \tilde{\gamma} \in \tilde{\mathcal{A}}_\lambda\} \text{ with } \tilde{\Phi}_N \tilde{\gamma}(t) := N^{-1}\tilde{\gamma}(2N^2 t). \quad (5.22)$$

$\tilde{\Phi}_N \tilde{\gamma}$ is a lattice loop of duration $t_{\tilde{\gamma}} := |\tilde{\gamma}|/(2N^2)$ on the rescaled lattice $\frac{1}{N}\mathbb{Z}^2$ and so $\tilde{\mathcal{A}}_\lambda^N$ is a random walk loop soup on $\frac{1}{N}\mathbb{Z}^2$, with rescaled time.

For each positive integer N we also define the *rescaled Brownian loop soup*

$$\mathcal{A}_\lambda^N := \{\Phi_N \gamma : \gamma \in \mathcal{A}_\lambda\} \text{ with } \Phi_N \gamma(t) := N^{-1}\gamma(N^2 t). \quad (5.23)$$

$\Phi_N \gamma$ is a lattice loop of duration $t_\gamma/(N^2)$ on the rescaled lattice $\frac{1}{N}\mathbb{Z}^2$ and so $\mathcal{A}_\lambda^N$ is a random walk loop soup on $\frac{1}{N}\mathbb{Z}^2$, with rescaled time.

It is shown in [35] that, as $N \to \infty$, $\tilde{\mathcal{A}}_\lambda^N$ converges to the Brownian loop soup of [36] in an appropriate sense. In this section we give a very brief sketch of the proof of this convergence result, which implies that the critical random walk loop soup has a conformally invariant scaling limit (the Brownian loop soup) and explains our use of the term *critical*.

Theorem 5.3.1 ([35]) *There exist two sequences $\{\mathcal{A}_\lambda^N\}_{N \geq 2}$ and $\{\tilde{\mathcal{A}}_\lambda^N\}_{N \geq 2}$ of loop soups, defined on the same probability space, such that the following holds.*

- *For each $\lambda > 0$, $\mathcal{A}_\lambda^N$ is a Brownian loop soup in $\mathbb{C}$ with intensity λ; the realizations of the loop soup are increasing in λ.*
- *For each $\lambda > 0$, $\tilde{\mathcal{A}}_\lambda^N$ is a random walk loop soup on $\frac{1}{N}\mathbb{Z}^2$ with intensity λ and time scaled as in (5.23); the realizations of the loop soup are increasing in λ.*
- *For every bounded $D \subset \mathbb{C}$, with probability going to one as $N \to \infty$, loops from $\mathcal{A}_\lambda^N$ and $\tilde{\mathcal{A}}^N 2_\lambda$ that are contained in D and have duration at least $N^{-1/6}$ can be put in a one-to-one correspondence with the following property. If $\gamma \in \mathcal{A}_\lambda^N$ and*

$\tilde\gamma \in \tilde{\mathcal{A}}_\lambda^N$ are paired in that correspondence and t_γ and $t_{\tilde\gamma}$ denote their respective durations, then

$$|t_\gamma - t_{\tilde\gamma}| \le \frac{5}{8} N^{-2}$$

$$\sup_{0\le s\le 1} |\gamma(st_\gamma) - \tilde\gamma(st_{\tilde\gamma})| \le cN^{-1}\log N$$

for some constant $c < \infty$.

Sketch of the proof. For simplicity we consider the *rooted* random walk loop soup; this suffices since one can obtain an unrooted loop soup by starting with a rooted one and forgetting the root. Let $\mathcal{L}_n^z$ be the set of loops of length $2n$ rooted at z. If $\tilde\gamma$ has length $2n$, then $\nu^r(\tilde\gamma) = \frac{1}{2n}(\frac{1}{4})^{2n}$, which implies that

$$\tilde q_n := \lambda \nu^r(\mathcal{L}_n^z) = \lambda \nu^r(\mathcal{L}_n^0) = \frac{\lambda}{2n}\left[\left(\frac{1}{4}\right)^n \binom{2n}{n}\right]^2$$

$$= \frac{\lambda}{2n}\left[\frac{1}{\pi n} - \frac{1}{4\pi n^2} + O\left(\frac{1}{n^3}\right)\right],$$

where we have used the fact that $n! = \sqrt{2\pi}\, n^{n+1/2} e^{-n}\left[1 + \frac{1}{12n} + O\left(\frac{1}{n^2}\right)\right]$. Thus, the number of loops rooted at z with length $2n$ is a Poisson random variable with parameter $\tilde q_n = \frac{\lambda}{2\pi n^2} - \frac{\lambda}{8\pi n^3} + O\left(\frac{1}{n^4}\right)$.

A realization of the random walk loop soup can be obtained by drawing a Poisson random variable $\tilde N_n^z$ with parameter $\tilde q_n = \frac{\lambda}{2\pi n^2} - \frac{\lambda}{8\pi n^3} + O\left(\frac{1}{n^4}\right)$ for each $n \in \mathbb{N}$ and each vertex $z \in \mathbb{Z}^2$, and rooting $\tilde N_n^z$ loops of length $2n$ at z, with the loops chosen independently according to the uniform probability measure on $\mathcal{L}_n^z$.

If we now consider the *rooted* Brownian loop soup with intensity measure

$$\lambda \mu_r = \lambda \int_{\mathbb{C}} \int_0^\infty \frac{1}{2\pi t^2} \mu_{z,t}^{br}\, dt\, d\mathbf{A}(z),$$

we see that the number N_n^z of rooted Brownian loops with root in the unit square centered at $z \in \mathbb{Z}^2$ with duration between $n - \frac{3}{8}$ and $n + \frac{5}{8}$ is a Poisson random variable with parameter

$$q_n := \lambda \int_{n-3/8}^{n+5/8} \frac{dt}{2\pi t^2} = \frac{\lambda}{2\pi[(n+5/8) - (n-3/8)]} = \frac{\lambda}{2\pi n^2} - \frac{\lambda}{8\pi n^3} + O\left(\frac{1}{n^4}\right),$$

so that $q_n - \tilde q_n = O\left(\frac{1}{n^4}\right)$.

A realization of the Brownian loop soup can be obtained by drawing a Poisson random variable N_n^z with parameter $q_n = \frac{\lambda}{2\pi n^2} - \frac{\lambda}{8\pi n^3} + O\left(\frac{1}{n^4}\right)$ for each $n \in \mathbb{N}$ and each vertex $z \in \mathbb{Z}^2$, and rooting N_n^z loops of duration between $n - 3/8$ and $n + 5/8$ uniformly in the unit square centered at z, with the loops chosen independently according to the Brownian bridge measure of duration $T_n^z(i), i = 1, \ldots, N_n^z$, where the $T_n^z(i)$'s are independent random variables with density

$$\frac{(n + \frac{5}{8})(n - \frac{3}{8})}{s^2}, \quad n - \frac{3}{8} \leq s \leq n + \frac{5}{8}.$$

Finally, in order to get a full soup, one needs to add loops of duration less than 5/8. We will not attempt to couple these short loops with random walk loops.

Now let $N(n, z; t)$, $n \in \mathbb{N}$, $z \in \mathbb{Z}^2$, be independent Poisson processes (in the time variable t) with parameter 1, and let $\tilde{N}_n^z = N(n, z; \tilde{q}_n)$ and $N_n^z = N(n, z; q_n)$. In order to conclude the proof, Lawler and Trujillo Ferreras use a version of a coupling lemma due to Komlós, Major and Tusnády [33] which shows that the random walk bridge and the Brownian bridge can be coupled very closely. Using this lemma to draw coupled versions of the random walk loops and Brownian loops needed in the constructions of the random walk and Brownian loop soups described above, one obtains coupled versions of the soups that are very close, except possibly for small loops. □

5.3.2 The Near-critical Case

If we rescale a massive random walk loop soup with constant mass function $m > 0$ in the same way as discussed in Section 5.3.1, the resulting scaling limit is trivial, in the sense that it does not contain any loops larger than one point. This is because, under the random walk loop measure, only loops of duration of order at least N^2 have a diameter of order of at least N with non-negligible probability as $N \to \infty$, and are therefore "macroscopic" in the scaling limit. It is then clear that, in order to obtain a nontrivial scaling limit, the mass function needs to be rescaled while taking the scaling limit.

Suppose, for simplicity, that the mass function m is constant, and let m_N denote the rescaled mass function. When m_N tends to zero, $k_x \approx 4m_N^2$ and one has the following dichotomy.

- If $\lim_{N \to \infty} Nm_N = 0$, loops with a number of steps of the order of N^2 or smaller are not affected by the killing in the scaling limit and one recovers the critical Brownian loop soup.
- If $\lim_{N \to \infty} Nm_N = \infty$, all loops with a number of steps of the order of N^2 or more are removed from the soup in the scaling limit and no "macroscopic" loop (larger than one point) is left.

In view of this observation, a *near-critical* scaling limit, that is, a nontrivial scaling limit that differs from the critical one, can only be obtained if the mass function m is rescaled by $O(1/N)$. This leads us to considering the loop soup $\tilde{\mathcal{A}}^N_{\lambda,m}$ defined as a random walk loop soup on the rescaled lattice $\frac{1}{N}\mathbb{Z}^2$ with mass function $m/(\sqrt{2}N)$ and rescaled time as in (5.23). Such a soup can be obtained from $\tilde{\mathcal{A}}^N_\lambda$ using the construction in Proposition 5.2.3, replacing $m^2(\tilde{\gamma}(i))$ with $\frac{1}{2N^2}m^2(\tilde{\gamma}(i)/N)$ in equation (5.3).

Theorem 5.3.2 *Let m be a nonnegative function such that m^2 is Lipschitz continuous. There exist two sequences $\{\mathcal{A}^N_{\lambda,m}\}_{N\geq 2}$ and $\{\tilde{\mathcal{A}}^N_{\lambda,m}\}_{N\geq 2}$ of loop soups, defined on the same probability space, such that the following holds.*

- *For each $\lambda > 0$, $\mathcal{A}^N_{\lambda,m}$ is a massive Brownian loop soup in $\mathbb{C}$ with intensity λ and mass m; the realizations of the loop soup are increasing in λ.*
- *For each $\lambda > 0$, $\tilde{\mathcal{A}}^N_{\lambda,m}$ is a massive random walk loop soup on $\frac{1}{N}\mathbb{Z}^2$ with intensity λ, mass $m/(\sqrt{2}N)$ and time scaled as in (5.23); the realizations of the loop soup are increasing in λ.*
- *For every bounded $D \subset \mathbb{C}$, with probability going to one as $N \to \infty$, loops from $\mathcal{A}^N_{\lambda,m}$ and $\tilde{\mathcal{A}}^N_{\lambda,m}$ that are contained in D and have duration at least $N^{-1/6}$ can be put in a one-to-one correspondence with the following property. If $\gamma \in \mathcal{A}^N_{\lambda,m}$ and $\tilde{\gamma} \in \tilde{\mathcal{A}}^N_{\lambda,m}$ are paired in that correspondence and t_γ and $t_{\tilde{\gamma}}$ denote their respective durations, then*

$$|t_\gamma - t_{\tilde{\gamma}}| \leq \frac{5}{8}N^{-2}$$

$$\sup_{0\leq s\leq 1} |\gamma(st_\gamma) - \tilde{\gamma}(st_{\tilde{\gamma}})| \leq c_1 N^{-1} \log N$$

for some constant $c_1 < \infty$.

Proof. Let $\mathcal{A}_\lambda$ be a critical Brownian loop soup in $\mathbb{C}$ with intensity λ and $\tilde{\mathcal{A}}_\lambda$ a critical random walk loop soup on $\mathbb{Z}^2$ with intensity λ, coupled as in Theorem 5.3.1. Consider the scaled loop soups $\mathcal{A}^N_\lambda$ and $\tilde{\mathcal{A}}^N_\lambda$, where $\tilde{\mathcal{A}}^N_\lambda$ is defined in (5.23) and $\mathcal{A}^N_\lambda := \{\Phi_N \gamma : \gamma \in \mathcal{A}_\lambda\}$ with $\Phi_N \gamma(t) = N^{-1}\gamma(N^2 t)$ for $0 \leq t \leq t_\gamma/N^2$. Note that, because of scale invariance, $\mathcal{A}^N_\lambda$ is a critical Brownian loop soup in $\mathbb{C}$ with parameter λ.

It follows from Theorem 5.3.1 that, if one considers only loops of duration greater than $N^{-1/6}$, loops from $\mathcal{A}^N_\lambda$ and $\tilde{\mathcal{A}}^N_\lambda$ contained in D can be put in a one-to-one correspondence with the properties described in Theorem 5.3.2, except perhaps on an event of probability going to zero as $N \to \infty$. For simplicity, in the rest of the proof we will call *macroscopic* the loops of duration greater than $N^{-1/6}$.

On the event that such a one-to-one correspondence between macroscopic loops in D exists, we construct the massive loop soups $\mathcal{A}^N_{\lambda,m}$ and $\tilde{\mathcal{A}}^N_{\lambda,m}$ in the following way. To each pair of macroscopic loops $\gamma \in \mathcal{A}_\lambda^N$ and $\tilde{\gamma} \in \tilde{\mathcal{A}}_\lambda^N$, paired in the correspondence of Theorem 5.3.1, we assign an independent, mean-one, exponential random variable T_γ. We let t_γ denote the (rescaled) duration of γ and $t_{\tilde{\gamma}}$ the (rescaled) duration of $\tilde{\gamma}$, and let $M = 2N^2 t_{\tilde{\gamma}}$ denote the number of steps of the lattice loop $\tilde{\gamma}$. As in the constructions described in Propositions 5.2.9 and 5.2.3, we remove γ from the Brownian loop soup if $\int_0^{t_\gamma} m^2(\gamma(s))ds > T_\gamma$ and remove $\tilde{\gamma}$ from the random walk loop soup if $\frac{1}{2N^2}\sum_{k=0}^{M-1} m^2(\tilde{\gamma}(\frac{k}{2N^2})) > T_\gamma$. The resulting loop soups, $\mathcal{A}_{\lambda,m}^N$ and $\tilde{\mathcal{A}}_{\lambda,m}^N$, are defined on the same probability space and are distributed like a massive Brownian loop soup with mass function m and a random walk loop soup with mass function $m/(\sqrt{2}N)$, respectively. We use $\mathbb{P}$ to denote the joint distribution of $\mathcal{A}_{\lambda,m}^N$, $\tilde{\mathcal{A}}_{\lambda,m}^N$ and the collection $\{T_\gamma\}$.

For loops that are not macroscopic, the removal of loops is done independently for the Brownian loop soup and the random walk loop soup. If there is no one-to-one correspondence between macroscopic loops in D, the removal is done independently for all loops, including the macroscopic ones.

We want to show that, on the event that there is a one-to-one correspondence between macroscopic loops in D, the one-to-one correspondence survives the removal of loops described above with probability going to one as $N \to \infty$. For that purpose, we need to compare $\int_0^{t_\gamma} m^2(\gamma(s))ds$ and $\frac{1}{2N^2}\sum_{k=0}^{M-1} m^2(\tilde{\gamma}(\frac{k}{2N^2}))$ for loops γ and $\tilde{\gamma}$ paired in the above correspondence. In order to do that, we write

$$\int_0^{t_\gamma} m^2(\gamma(s))ds = t_\gamma \int_0^1 m^2(\gamma(t_\gamma u))du$$

$$= \lim_{n \to \infty} \frac{t_\gamma}{n t_{\tilde{\gamma}}} \sum_{i=0}^{\lfloor n t_{\tilde{\gamma}} \rfloor} m^2\left(\gamma\left(\frac{t_\gamma i}{n t_{\tilde{\gamma}}}\right)\right)$$

$$= \lim_{q \to \infty} \frac{t_\gamma/t_{\tilde{\gamma}}}{4qN^2} \sum_{i=0}^{2qM-1} m^2\left(\gamma\left(\frac{i}{2qM}t_\gamma\right)\right),$$

where $t_{\tilde{\gamma}} = \frac{M}{2N^2}$ and the last expression is obtained by letting $n = 4qN^2$, with $q \in \mathbb{N}$. Thus, for fixed N and γ, the quantity

$$\Omega(N,q;\gamma) := \left| \int_0^{t_\gamma} m^2(\gamma(s))ds - \frac{t_\gamma/t_{\tilde{\gamma}}}{4qN^2} \sum_{i=0}^{2qM-1} m^2\left(\gamma\left(\frac{i}{2qM}t_\gamma\right)\right) \right|$$

can be made arbitrarily small by choosing q sufficiently large.

Define the sets of indexes $I_0 = \{i : 0 \leq i < q\} \cup \{i : (2M-1)q \leq i < 2qM\}$ and $I_k = \{i : (2k-1)q \leq i < (2k+1)q\}$ for $1 \leq k \leq M-1$. For $i \in I_k$, $0 \leq k \leq M-1$, we have that

$$\left| \gamma\left(\frac{i}{2qM}t_\gamma\right) - \tilde{\gamma}\left(\frac{k}{M}t_{\tilde{\gamma}}\right) \right|$$

$$\leq \left| \gamma\left(\frac{i}{2qM}t_\gamma\right) - \tilde{\gamma}\left(\frac{i}{2qM}t_{\tilde{\gamma}}\right) \right| + \left| \tilde{\gamma}\left(\frac{i}{2qM}t_{\tilde{\gamma}}\right) - \tilde{\gamma}\left(\frac{k}{M}t_{\tilde{\gamma}}\right) \right|$$

$$\leq \frac{c_1 \log N}{N} + \frac{\sqrt{2}}{N}$$

for some constant c_1, where the first term in the last line comes from Theorem 5.3.1 and the second term comes from the fact that $\tilde{\gamma}(s)$ is defined by interpolation, and that

- if $i \in I_0$, either $0 \leq \frac{i}{2qM}t_{\tilde{\gamma}} < \frac{1}{2M}t_{\tilde{\gamma}}$ so that $\tilde{\gamma}(\frac{i}{2qM}t_{\tilde{\gamma}})$ falls on the edge of $\frac{1}{N}\mathbb{Z}^2$ between $\tilde{\gamma}(0)$ and $\tilde{\gamma}(\frac{t_{\tilde{\gamma}}}{M}) = \tilde{\gamma}(\frac{1}{2N^2})$, or $(1 - \frac{1}{2M})t_{\tilde{\gamma}} \leq \frac{i}{2qM}t_{\tilde{\gamma}} < t_{\tilde{\gamma}}$ so that $\tilde{\gamma}(\frac{i}{2qM}t_{\tilde{\gamma}})$ falls on the edge between $\tilde{\gamma}(t_{\tilde{\gamma}} - \frac{t_{\tilde{\gamma}}}{M}) = \tilde{\gamma}(t_{\tilde{\gamma}} - \frac{1}{2N^2})$ and $\tilde{\gamma}(t_{\tilde{\gamma}}) = \tilde{\gamma}(0)$,
- if $i \in I_k$ with $1 \leq k \leq M - 1$, $\frac{k}{M}t_{\tilde{\gamma}} - \frac{t_{\tilde{\gamma}}}{2M} \leq \frac{i}{2qM}t_{\tilde{\gamma}} < \frac{k}{M}t_{\tilde{\gamma}} + \frac{t_{\tilde{\gamma}}}{2M}$ so that $\tilde{\gamma}(\frac{i}{2qM}t_{\tilde{\gamma}})$ falls either on the edge of $\frac{1}{N}\mathbb{Z}^2$ between $\tilde{\gamma}(\frac{k-1}{M}t_{\tilde{\gamma}}) = \tilde{\gamma}(\frac{k-1}{2N^2})$ and $\tilde{\gamma}(\frac{k}{2N^2})$, or on the edge between $\tilde{\gamma}(\frac{k}{2N^2})$ and $\tilde{\gamma}(\frac{k+1}{M}t_{\tilde{\gamma}}) = \tilde{\gamma}(\frac{k+1}{2N^2})$.

Since m^2 is Lipschitz continuous, for each $i \in I_k$, $0 \leq k \leq M - 1$, we have that

$$\left| m^2\left(\gamma\left(\frac{i}{2qM}t_\gamma\right)\right) - m^2\left(\tilde{\gamma}\left(\frac{k}{M}t_{\tilde{\gamma}}\right)\right) \right| \leq \frac{c_2 \log N}{N}$$

for some constant $c_2 < \infty$ and all $N \geq 2$. We let $\overline{m}_D^2 := \sup_{x \in D} m^2(x)$ and observe that, since $t_{\tilde{\gamma}} \geq N^{-1/6}$, the inequality $|t_\gamma - t_{\tilde{\gamma}}| < \frac{5}{8}N^{-2}$ from Theorem 5.3.1 implies that $t_\gamma / t_{\tilde{\gamma}} < 1 + \frac{5}{8}N^{-11/6}$ and that $M = 2N^2 t_{\tilde{\gamma}} < 2N^2 t_\gamma + \frac{5}{4}$. It then follows that

$$\left| \frac{1}{2N^2}\sum_{k=0}^{M-1} m^2\left(\tilde{\gamma}\left(\frac{k}{2N^2}\right)\right) - \frac{t_\gamma/t_{\tilde{\gamma}}}{4qN^2}\sum_{i=0}^{2qM-1} m^2\left(\gamma\left(\frac{i}{2qM}t_\gamma\right)\right) \right|$$

$$\leq \frac{1}{2N^2}\sum_{k=0}^{M-1}\left| m^2\left(\tilde{\gamma}\left(\frac{k}{2N^2}\right)\right) - \frac{t_\gamma/t_{\tilde{\gamma}}}{2q}\sum_{i \in I_k} m^2\left(\gamma\left(\frac{i}{2qM}t_\gamma\right)\right) \right|$$

$$\leq \frac{1}{2N^2}\sum_{k=0}^{M-1}\frac{1}{2q}\sum_{i \in I_k}\left| m^2\left(\tilde{\gamma}\left(\frac{k}{2N^2}\right)\right) - \frac{t_\gamma}{t_{\tilde{\gamma}}}m^2\left(\gamma\left(\frac{i}{2qM}t_\gamma\right)\right) \right|$$

$$\leq \frac{|1-t_\gamma/t_{\tilde\gamma}|}{2N^2} \sum_{k=0}^{M-1} m^2\left(\tilde\gamma\left(\frac{k}{2N^2}\right)\right)$$

$$+ \frac{t_\gamma/t_{\tilde\gamma}}{4qN^2} \sum_{k=0}^{M-1} \sum_{i\in I_k} \left| m^2\left(\tilde\gamma\left(\frac{k}{2N^2}\right)\right) - m^2\left(\gamma\left(\frac{i}{2qM}t_\gamma\right)\right) \right|$$

$$\leq \frac{|1-t_\gamma/t_{\tilde\gamma}|}{2N^2} \sum_{k=0}^{M-1} m^2\left(\tilde\gamma\left(\frac{k}{2N^2}\right)\right) + t_\gamma \frac{c_2 \log N}{N}$$

$$\leq \frac{|1-t_\gamma/t_{\tilde\gamma}|}{2N^2} M \overline{m}_D^2 + t_\gamma \frac{c_2 \log N}{N} < t_\gamma \frac{c_3' \log N}{N},$$

for some positive constant $c_3' = c_3'(D,m) < \infty$ independent of γ and $\tilde\gamma$.

Therefore, for fixed N and pair of macroscopic loops, γ and $\tilde\gamma$, and for any $q \in \mathbb{N}$,

$$\left| \int_0^{t_\gamma} m^2(\gamma(s))ds - \frac{1}{2N^2} \sum_{k=0}^{M-1} m^2\left(\tilde\gamma\left(\frac{k}{2N^2}\right)\right) \right| < t_\gamma \frac{c_3' \log N}{N} + \Omega(N,q;\gamma).$$

For fixed N and γ, one can choose q^* so large that

$$\Omega(N,q^*;\gamma) < t_\gamma \frac{c_3' \log N}{N}.$$

Hence, there is a positive constant $c_3 = 2c_3'$ such that, for every $N \geq 2$ and every pair of macroscopic loops, γ and $\tilde\gamma$, paired in the correspondence of Theorem 5.3.1,

$$\left| \int_0^{t_\gamma} m^2(\gamma(s))ds - \frac{1}{2N^2} \sum_{k=0}^{M-1} m^2\left(\tilde\gamma\left(\frac{k}{2N^2}\right)\right) \right| < t_\gamma \frac{c_3 \log N}{N}.$$

We now need to estimate the number of macroscopic loops contained in D. For that purpose, we note that, using the rooted Brownian loop measure (5.9), the mean number, $\mathcal{M}$, of macroscopic loops contained in D can be bounded above by

$$\mathcal{M} = \lambda \int_D \int_{N^{-1/6}}^\infty \frac{1}{2\pi t^2} \mu_{z,t}^{br}(\gamma : \gamma \subset D) \, dt \, dA(z) \leq \frac{\lambda \operatorname{diam}^2(D)}{8} N^{1/6}. \qquad (5.24)$$

Let $\mathcal{A}^N(\lambda, m; D)$ (respectively, $\tilde{\mathcal{A}}^N(\lambda, m; D)$) denote the massive Brownian (resp., random walk) loop soup in D, i.e., the set of loops from $\mathcal{A}^N_{\lambda,m}$ (respectively, $\tilde{\mathcal{A}}^N_{\lambda,m}$) contained in D. For the critical soups, we use the same notation omitting the m.

Let A_N denote the event that there is a one-to-one correspondence between macroscopic loops from $\mathcal{A}^N(\lambda; D)$ and $\tilde{\mathcal{A}}^N(\lambda; D)$, and let A_N^m denote the event that there is a one-to-one correspondence between macroscopic loops from $\mathcal{A}^N(\lambda,m;D)$ and $\tilde{\mathcal{A}}^N(\lambda,m;D)$. Furthermore, we denote by X the number of macroscopic loops in $\mathcal{A}^N(\lambda;D)$, and by T a mean-one exponential random variable. We have that, for any $c_4, \theta > 0$ and for all N sufficiently large,

$$\mathbb{P}(A_N^m) \geq \mathbb{P}(A_N^m \cap A_N \cap \{X \leq c_4 N^{1/6}\} \cap \{\nexists \gamma \in \mathcal{A}^N(\lambda;D) : t_\gamma \geq \theta\})$$
$$= \mathbb{P}(A_N^m | A_N \cap \{X \leq c_4 N^{1/6}\} \cap \{\nexists \gamma \in \mathcal{A}^N(\lambda;D) : t_\gamma \geq \theta\})$$
$$\mathbb{P}(A_N \cap \{X \leq c_4 N^{1/6}\} \cap \{\nexists \gamma \in \mathcal{A}^N(\lambda;D) : t_\gamma \geq \theta\})$$
$$\geq \left[1 - \sup_{x \geq 0} \Pr\left(x \leq T \leq x + \frac{c_3 \theta \log N}{N}\right)\right]^{c_4 N^{1/6}}$$
$$\mathbb{P}(A_N \cap \{X \leq c_4 N^{1/6}\} \cap \{\nexists \gamma \in \mathcal{A}^N(\lambda;D) : t_\gamma \geq \theta\})$$
$$= \exp\left(-\frac{c_5 \theta \log N}{N^{5/6}}\right)$$
$$\mathbb{P}(A_N \cap \{X \leq c_4 N^{1/6}\} \cap \{\nexists \gamma \in \mathcal{A}^N(\lambda;D) : t_\gamma \geq \theta\}),$$

where $c_5 = c_3 c_4$.

Since $\exp\left(-\frac{c_5 \theta \log N}{N^{5/6}}\right) \to 1$ as $N \to \infty$ for any $c_5, \theta > 0$, in order to conclude the proof, it suffices to show that $\mathbb{P}(A_N \cap \{X \leq c_4 N^{1/6}\} \cap \{\nexists \gamma \in \mathcal{A}^N(\lambda;D) : t_\gamma \geq \theta\})$ can be made arbitrarily close to one for some choice of c_4 and θ, and N sufficiently large. But by Theorem 5.3.1, $\mathbb{P}(A_N) \geq 1 - c(\lambda+1)\operatorname{diam}^2(D)N^{-7/2} \to 1$ as $N \to \infty$; moreover, if $c_4 > \frac{\lambda \operatorname{diam}^2(D)}{8}$, by equation (5.24), $c_4 N^{1/6}$ is larger than the mean number of macroscopic loops in D. Since X is a Poisson random variable with parameter equal to the mean number $\mathcal{M}$ of macroscopic loops in D, the latter fact (together with a Chernoff bound argument) implies that

$$\mathbb{P}(X > c_4 N^{1/6}) \leq \frac{e^{-\mathcal{M}}(e\mathcal{M})^{c_4 N^{1/6}}}{(c_4 N^{1/6})^{c_4 N^{1/6}}} \leq \left(\frac{e\lambda \operatorname{diam}^2(D)}{8c_4}\right)^{c_4 N^{1/6}}.$$

This shows that, if $c_4 > e\lambda \operatorname{diam}^2(D)/8$, $\mathbb{P}(X \leq c_4 N^{1/6}) \to 1$ as $N \to \infty$.

To find a lower bound for $\mathbb{P}(\nexists \gamma \in \mathcal{A}^N(\lambda;D) : t_\gamma \geq \theta)$, we define

$$\mathcal{L}_{\theta,D} := \{\text{loops } \gamma \text{ with } t_\gamma \geq \theta \text{ that stay in } D\}.$$

We then have

$$\mathbb{P}(\nexists \gamma \in \mathcal{A}^N(\lambda;D) : t_\gamma \geq \theta) = \exp\left[-\lambda \mu_D(\mathcal{L}_{\theta,D})\right]$$
$$\geq 1 - \lambda \mu_D(\mathcal{L}_{\theta,D})$$
$$\geq 1 - \frac{\lambda \operatorname{diam}^2(D)}{\theta}, \quad (5.25)$$

where the last line follows from the bound

$$\mu_D(\mathcal{L}_{\theta,D}) = \int_D \int_\theta^\infty \frac{1}{2\pi t^2} \mu_{z,t}^{br}(\gamma : \gamma \text{ stays in } D) \, dt \, d\mathbf{A}(z) \leq \frac{\text{diam}^2(D)}{\theta}.$$

The lower bound (5.25), together with the previous observations, shows that $\mathbb{P}(A_N^m)$ can be made arbitrarily close to one by choosing $c_4 > e\lambda \text{diam}^2(D)/8$, θ sufficiently large, depending on D, and then N sufficiently large, depending on the values of c_4 and θ. $\square$

5.4 Conformal Correlation Functions in the Brownian Loop Soup

In the rest of this chapter, we will focus on the *critical (massless)* Brownian loop soup and analyze certain correlation functions that characterize aspects of its distribution. (One could perform a similar analysis using the *massive* Brownian loop soup, but we will not do that here.) We will show that these correlation functions need to be defined via a regularization procedure that requires introducing a cutoff, which is then removed via a limiting procedure. The limiting procedure produces a nontrivial limit only if the correlation functions with a cutoff are scaled by an appropriate power of the cutoff. In that case, the limiting functions have multiplicative scaling behavior under conformal transformations – a type of behavior that characterizes correlation functions of *primary fields* in conformal field theory. The models described in this section are interesting in their own right and are studied in [15]. (One of them was inspired by [27].) Here, we use them as "toy" models for the behavior of lattice correlation functions in the scaling limit.

5.4.1 Motivation and Summary of Results

The study of lattice models involves local lattice "observables." A typical example of such local observables are the spin variables in the Ising model, or, more generally, sums of (products of) spin variables. At the critical point, the scaling limit of such observables can lead to conformal fields, which are not defined pointwise (they are not functions), but can be defined as generalized functions (distributions in the sense of Schwartz). For example, the scaling limit of the critical Ising magnetization in two dimensions leads to a conformally covariant magnetization field [16].

Correlation functions (or *n-point functions*), that is, expectations of products of local lattice observables, are important tools in analyzing the behavior of both lattice models and field theories. In general, lattice correlation functions do not possess a finite, nontrivial scaling limit. However, the theory of *renormalization* (based on the study of exactly solved models, as well as the perturbative analysis of cutoff quantum field theories) suggests that certain correlation functions are *multiplicatively renormalizable*; that is, they possess a scaling limit when multiplied by appropriate powers of the lattice spacing. When that happens, the limiting functions are expected to have multiplicative scaling behavior under conformal transformations.

From the point of view of field theory, a lattice can be seen as a way to avoid divergencies, with the lattice spacing playing the role of an *ultraviolet cutoff*, preventing infinities due to small scales. Below, we will define and study correlation functions from the Brownian loop soup. We will see that, in the absence of a lattice, we will need to impose a different ultraviolet cutoff $\delta > 0$. The behavior of the correlation functions described below is conceptually the same as that of lattice correlation functions, with the cutoff δ playing the role of the lattice spacing. We will analyze the correlation functions for two different types of "fields," which we now describe informally.

Winding operator. Let $N_w(z)$ count the total winding number of loops around a point z. We call N_w the *winding operator*. Since the loops in a loop soup have an orientation, winding numbers can be positive or negative (see Figure 5.1). We are interested in the n-point correlation functions of the "field" N_w. However, since the Brownian loop soup is scale invariant, any given z is almost surely "surrounded" by infinitely many loops and $N_w(z)$ is almost surely *infinite*, so the "field" N_w is not well defined. We will need to restrict the soup to a bounded domain and to introduce a cutoff $\delta > 0$ and count only the winding of loops of diameter at least δ.

Layering operator. For each loop, define the interior as the set of points inside the outermost edge of the loop. Points in this set are "covered" by the "filled-in" loop. Furthermore, declare each loop to be of type 1 or type 2 with equal probability, independently of all other loops. For a point z, let $N_j(z)$ be the number of distinct

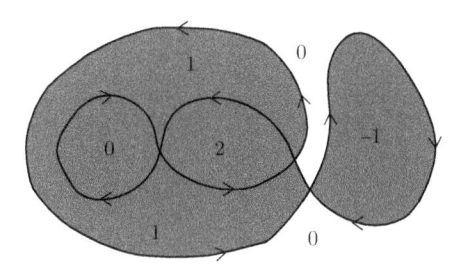

Figure 5.1 A stylized Brownian loop. The numbers indicate the winding numbers of the loop that contribute additively to N_w, while the shaded region is the interior of the loop (the set of points disconnected from infinity by the loop) that contributes ± 1 (where the sign is a Boolean variable assigned randomly to each loop) to the layering number N_ℓ.

loops of type $j = 1, 2$ that cover z, and let $N_\ell(z) = N_1(z) - N_2(z)$. We'll call $N_\ell(z)$ the "layering number" at z, and N_ℓ the *layering operator*. Because the Brownian loop soup is scale invariant, any given point of the plane is covered by infinitely many loops with probability one, so the layering "field" N_ℓ is not well defined. As before, we will need to restrict to bounded domains, as well as introduce a cutoff $\delta > 0$, and count only the contribution from loops of diameter at least δ.

The first model described above appears more natural; the advantage of the second model is that it is simpler to analyze and it can be easily defined on the full plane, as we will see later. Both models present logarithmic divergences (like massless fields in two dimensions) as one tries to remove the cutoff by letting $\delta \to 0$. For this reason, we will focus on exponentials of the winding and layering numbers times imaginary coefficients. We will denote these *exponential operators* as

$$V_\beta(z) = e^{i\beta N(z)},$$

where N can be either a layering or winding number, and will prove the following results.

- For both models in finite domains D, *correlators* of $n \in \mathbb{N}$ exponential operators

$$\left\langle \prod_{j=1}^n V_{\beta_j}(z_j) \right\rangle_{\delta,D} = \left\langle e^{i \sum_{j=1}^n \beta_j N(z_j)} \right\rangle_{\delta,D} \quad (5.26)$$

exist as long as a cutoff $\delta > 0$ on the the diameters of the loops is imposed, and

$$\lim_{\delta \to 0} \frac{\left\langle \prod_{j=1}^n V_{\beta_j}(z_j) \right\rangle_{\delta,D}}{\prod_{j=1}^n \delta^{2\Delta(\beta_j)}} =: \phi_D(z_1, \ldots, z_n; \beta_1, \ldots, \beta_n) \equiv \phi_D(z; \boldsymbol{\beta})$$

exists and is finite. Moreover, if D' is another finite domain and $f : D \to D'$ is a conformal map such that $z'_1 = f(z_1), \ldots, z'_n = f(z_n)$, then

$$\phi_{D'}(z'; \boldsymbol{\beta}) = \prod_{j=1}^n |f'(z_j)|^{-2\Delta(\beta_j)} \phi_D(z; \boldsymbol{\beta}),$$

where the $\Delta(\beta)$ is defined below. This is the behavior expected for a conformal primary operator/field. (In the conformal field theory literature, the terms field and operator are sometimes used interchangeably.)

- For both versions in infinite volume, correlators of n exponential operators

$$\left\langle \prod_j V_{\beta_j}(z_j) \right\rangle_\delta = \left\langle e^{i \sum_j \beta_j N(z_j)} \right\rangle_\delta \quad (5.27)$$

vanish. However, in the case of the layering model, one can let $\delta \to 0$ and still obtain a nontrivial limit by imposing the following condition, satisfied mod 2π,

$$\sum_j \beta_j = 2\pi k, \ k \in \mathbb{Z}. \tag{5.28}$$

- The correlators (5.27) of the layering model in the plane are finite and non-zero when (5.28) is satisfied, so long as the loop soup is cut off at short scales (no other cutoff is necessary).
- In the case of two points, assuming (5.28), the $\delta \to 0$ limit of the renormalized correlators of the layering model in the plane (5.27) can be explicitly computed up to an overall multiplicative constant. The result is

$$\phi_{\mathbb{C}}(z_1, z_2; \beta_1, \beta_2) = C_2 \left| \left(\frac{1}{z_1 - z_2} \right)^{\Delta(\beta_1)+\Delta(\beta_2)} \right|^2,$$

where C_2 is a constant (see Section 5.4.2.4).
- The $\delta \to 0$ limit of the renormalized 3-point function for the layering model in the plane, assuming (5.28), is

$$\phi_{\mathbb{C}}(z_1, z_2, z_3; \beta_1, \beta_2, \beta_3) =$$

$$C_3 \left| \left(\frac{1}{|z_1 - z_2|} \right)^{\Delta(\beta_1)+\Delta(\beta_2)-\Delta(\beta_3)} \left(\frac{1}{|z_1 - z_3|} \right)^{\Delta(\beta_1)+\Delta(\beta_3)-\Delta(\beta_2)} \right.$$

$$\left. \left(\frac{1}{|z_2 - z_3|} \right)^{\Delta(\beta_2)+\Delta(\beta_3)-\Delta(\beta_1)} \right|^2,$$

where C_3 is a constant (Corollary 5.4.9).
- The exponents $\Delta(\beta)$ are called *conformal dimensions* and differ for the two models. For the layering model,

$$\Delta_l(\beta) = \frac{\lambda}{10}(1 - \cos \beta).$$

For the winding model,

$$\Delta_w(\beta) = \lambda \beta (2\pi - \beta)/8\pi^2,$$

where this formula applies for $0 \le \beta < 2\pi$, and $\Delta_w(\beta)$ is periodic under $\beta \to \beta + 2\pi$.

5.4.2 Correlation Functions of the Layering and Winding Models

5.4.2.1 Correlation Functions of the Layering Model

As already discussed, the layering model is defined by randomly assigning a Boolean variable to each loop in the Brownian loop soup. Alternatively, one can think of this as two independent Brownian loop soups, each with a Poisson distribution $P_{\lambda_{+(-)},\mu}$ with intensity measure $\lambda_{+(-)}\mu$, where we take $\lambda_+ = \lambda_- = \lambda/2$. (This follows from the fact that the collection of all loops from a Brownian loop soup of intensity λ_+ and an independent one of intensity λ_- is distributed like a Brownian loop soup with intensity $\lambda_+ + \lambda_-$.)

Denote by $N_{+(-)}(z)$ the number of loops γ in the first (respectively, second) class such that the point $z \in \mathbb{C}$ is separated from infinity by the image of γ in $\mathbb{C}$. If $\bar{\gamma}$ is the "filled in" loop γ, then this condition becomes $z \in \bar{\gamma}$, or z is covered by γ. We are interested in the layering field N_ℓ, with $N_\ell(z) = N_+(z) - N_-(z)$. This is purely formal as both $N_{+(-)}(z)$ are infinite with probability one for any z. They are infinite for two reasons: both because there are infinitely many large loops surrounding z (infrared, or IR, divergence), and because there are infinitely many small loops around z (ultraviolet, or UV, divergence).

We will consider correlators of the exponential operator $V_\beta = e^{i\beta N_\ell(z)}$, and show that there are choices of β that remove the IR divergence and a normalization which removes the UV divergence. Specifically, we are interested in the correlators $V_{\boldsymbol{\beta}}(z_1,\ldots,z_n) := \prod_{j=1}^n V_{\beta_j}(z_j) = e^{i\sum_{j=1}^n \beta_j N_\ell(z_j)}$ and their moments

$$\langle V_{\boldsymbol{\beta}}(z_1,\ldots,z_n)\rangle := \mathbb{E}_\lambda(V_{\boldsymbol{\beta}}(z_1,\ldots,z_n)),$$

where $z_j \in \mathbb{C}$, $\boldsymbol{\beta} = (\beta_1,\ldots,\beta_n) \in \mathbb{R}^n$, and the expected value $\mathbb{E}_\lambda$ is taken with respect to two independent copies of the Brownian loop soup with equal intensities $\lambda/2$.

As the field N_ℓ has both IR and UV divergences, we get a meaningful definition by introducing cutoffs which restrict the loops to have diameter within some δ and $R \in \mathbb{R}^+$, $\delta < R$: let $\mu_{\delta,R}(\cdot) = \mu(\cdot \cap \{\gamma : \delta \leq \mathrm{diam}(\gamma) < R\})$ and consider the correlators

$$\langle V_{\boldsymbol{\beta}}(z_1,\ldots,z_n)\rangle_{\delta,R} := \mathbb{E}_{\lambda,\delta,R}\left(e^{i\sum_{j=1}^n \beta_j N_\ell(z_j)}\right),$$

where the expectation $\mathbb{E}_{\lambda,\delta,R}$ is with respect to the distribution $P_{\lambda,\mu_{\delta,R}} \otimes P_{1/2}$, where $P_{\lambda,\mu_{\delta,R}}$ is the Poisson distribution with intensity measure $\lambda\mu_{\delta,R}$ and $P_{1/2}$ is the distribution of a countable sequence of independent Bernoulli random variables with parameter $1/2$ (remember that each loop belongs to one of two classes with equal probability). It is for these correlators that a suitable choice of $\boldsymbol{\beta}$ and a suitable normalization will allow us to remove both IR and UV cutoffs.

5.4.2.2 The 1-point Function in the Layering Model

In this section we explicitly compute the 1-point function in the presence of IR and UV cutoffs. Replacing the area of a filled Brownian loop of time length 1 with that of a disc of radius 1, the result reproduces the 1-point function in the disc model of [27].

Lemma 5.4.1 *For all $z \in \mathbb{C}$, we have that*

$$\langle V_\beta(z) \rangle_{\delta,R} = \left(\frac{R}{\delta}\right)^{-\frac{\lambda}{5}(1-\cos\beta)}.$$

Proof. With IR and UV cutoffs in place, the field $N_\ell(z)$ can be realized as follows. Let η be a realization of loops, and let $\{X_\gamma\}_{\gamma \in \eta}$ be a collection of Bernoulli symmetric random variables taking values in $\{-1, 1\}$. The quantity

$$N_\ell(z) = \sum_{\gamma \in \eta, z \in \bar{\gamma}, \delta \leq \mathrm{diam}(\gamma) < R} X_\gamma =: \sum\nolimits^* X_\gamma$$

is finite $P_{\lambda,\mu}$ almost surely, since $\mu\{\gamma : z \in \bar{\gamma}, \delta \leq \mathrm{diam}(\gamma) < R\} = \mu_{\delta,R}\{\gamma : z \in \bar{\gamma}\} < \infty$ (see Appendix B and [51]). Now,

$$\langle V_\beta(z) \rangle_{\delta,R} = \mathbb{E}_{\lambda,\delta,R}\left(e^{i\beta N_\ell(z)}\right)$$

$$= \sum_{k=0}^{\infty} \mathbb{E}_{\lambda,\delta,R}\left(e^{i\beta N_\ell(z)} | \mathcal{L}_k\right) P_{\lambda,\mu_{\delta,R}}(\mathcal{L}_k),$$

where $\mathcal{L}_k = \{\eta : |\{\gamma \in \eta : z \in \bar{\gamma}, \delta \leq \mathrm{diam}(\gamma) < R\}| = k\}$. If X denotes a (± 1)-valued symmetric random variable,

$$\mathbb{E}_{\lambda,\delta,R}\left(e^{i\beta \sum^* X_\gamma} | \mathcal{L}_k\right) = \left(E\left(e^{i\beta X}\right)\right)^k = (\cos\beta)^k.$$

Therefore, for $\alpha_{z,\delta,R} = \mu_{\delta,R}(\gamma : z \in \bar{\gamma})$, we have that

$$\langle V_\beta(z) \rangle_{\delta,R} = \sum_{k=0}^{\infty} (\cos\beta)^k \frac{(\lambda \alpha_{z,\delta,R})^k}{k!} e^{-\lambda \alpha_{z,\delta,R}}$$

$$= e^{-\lambda \alpha_{z,\delta,R}(1-\cos\beta)}.$$

Moreover, by Lemma B.8,

$$\alpha_{z,\delta,R} = \frac{1}{5} \log \frac{R}{\delta},$$

which implies

$$\langle V_\beta(z) \rangle_{\delta,R} = \left(\frac{R}{\delta}\right)^{-\frac{\lambda}{5}(1-\cos(\beta))},$$

as claimed. □

5.4.2.3 The 1-point Function in the Winding Model

To define the second model, let $N_w(z)$ denote the total winding number about the point z of all loops in a Brownian loop soup; as for the layering operators, this is a formal definition as, in general, $N_w(z)$ is infinite. Consider again the correlators $V_\beta(z_1,\ldots,z_n) = e^{i\sum_{j=1}^n \beta_j N_w(z_j)}$ and their moments $\langle V_\beta(z_1,\ldots,z_n)\rangle = E(V_\beta(z_1,\ldots,z_n))$ where $z_j \in \mathbb{C}$, $\boldsymbol{\beta} = (\beta_1,\ldots,\beta_n) \in \mathbb{R}^n$, and the expected value is taken with respect to the Brownian loop soup distribution. Denoting by $P_{\lambda,\mu}$ the Poisson distribution with intensity measure $\lambda\mu$, and restricting the loops to have diameter between some δ and $R \in \mathbb{R}^+$, with $\delta < R$, we let $\mu_{\delta,R}(\cdot) = \mu(\cdot \cap \{\gamma : \delta \leq \mathrm{diam}(\gamma) < R\})$ and consider the correlators

$$\langle V_\beta(z_1,\ldots,z_n)\rangle_{\delta,R} := \mathbb{E}_{\lambda,\delta,R}\left(e^{i\sum_{j=1}^n \beta_j N_w(z_j)}\right).$$

We now explicitly compute the 1-point function in the presence of IR and UV cutoffs for the winding model, using the following result, which we leave as an exercise.

Exercise 5.4.2 *Show that*

$$\sum_{m=1}^\infty \frac{1}{m^2}(1-\cos(m\beta)) = \frac{1}{4}\beta(2\pi-\beta),$$

where, on the right hand side of the equation, β is to be interpreted modulo 2π. (Hint: the left hand side is a convergent series representing a periodic function and can therefore be written as a Fourier series.)

Lemma 5.4.3 *For all $z \in \mathbb{C}$, we have that*

$$\langle V_\beta(z)\rangle_{\delta,R} = \left(\frac{R}{\delta}\right)^{-\lambda\frac{\beta(2\pi-\beta)}{4\pi^2}}, \quad (5.29)$$

where the formula is valid for $\beta \in [0,2\pi)$, and for $\beta \notin [0,2\pi)$, in the right hand side, β should be replaced by $(\beta \mod 2\pi)$.

Proof. For a point z and a loop γ, let $\theta_\gamma(z)$ indicate the winding number of γ around z. Moreover, for $k \in (\mathbb{N} \cup \{0\})^\mathbb{N}$ let

$$\mathcal{L}_k = \{\eta : |\{\gamma \in \eta : z \in \bar{\gamma}, \delta \leq \mathrm{diam}(\gamma) < R, |\theta_\gamma(z)| = m\}| = k_m \text{ for all } m \in \mathbb{N}\}.$$

If a loop γ has $|\theta_\gamma(z)| = m$, then the winding number $\theta_\gamma(z)$ is $\pm m$ with equal probability under $P_{\lambda,\mu}$. Finally, for $m \geq 1$ we have

$$\begin{aligned}
\alpha_{z,\delta,R,m} &:= \mu_{\delta,R}(\gamma : z \in \bar{\gamma}, |\theta_\gamma(z)| = m) \\
&= \mu(\gamma \in \eta : z \in \bar{\gamma}, \delta \leq \mathrm{diam}(\gamma) < R, |\theta_\gamma(z)| = m) \\
&= \mu(\gamma : z \in \bar{\gamma}, \gamma \not\subset B_{z,\delta}, \gamma \subset B_{z,R}, |\theta_\gamma(z)| = m) \\
&= \frac{1}{\pi^2 m^2} \log \frac{R}{\delta},
\end{aligned}$$

where, in the last equality, we have used Lemma B.9.

For all $k \in (\mathbb{N} \cup \{0\})^{\mathbb{N}}$ we have

$$P_{\lambda,\mu_{\delta,R}}(\mathcal{L}_k) = \prod_{m=1}^{\infty} \frac{(\lambda \alpha_{z,\delta,R,m})^{k_m}}{k_m!} e^{-\lambda \alpha_{z,\delta,R,m}},$$

as for different m's the sets of loops with those winding numbers are disjoint. Hence with IR and UV cutoffs in place, denoting by $E_{\lambda,\delta,R}$ the expectation with respect to the Poisson distribution $P_{\lambda,\mu_{\delta,R}}$ with intensity measure $\mu_{\delta,R}$, we have, for all z,

$$\begin{aligned}
\langle V_\beta(z) \rangle_{\delta,R} &= E_{\lambda,\delta,R}\left(e^{i\beta N_w(z)}\right) \\
&= \sum_{k \in (\mathbb{N}\cup\{0\})^{\mathbb{N}}} E_{\lambda,\delta,R}\left(e^{i\beta N_w(z)} | \mathcal{L}_k\right) P_{\lambda,\mu_{\delta,R}}(\mathcal{L}_k) \\
&= \sum_{k \in (\mathbb{N}\cup\{0\})^{\mathbb{N}}} \prod_{m=1}^{\infty} (\cos(m\beta))^{k_m} \frac{(\lambda \alpha_{z,\delta,R,m})^{k_m}}{k_m!} e^{-\lambda \alpha_{z,\delta,R,m}} \\
&= \prod_{m=1}^{\infty} e^{-\lambda \alpha_{z,\delta,R,m}(1-\cos(m\beta))} \\
&= \left(\frac{R}{\delta}\right)^{-\lambda \sum_{m=1}^{\infty} \frac{1}{\pi^2 m^2}(1-\cos(m\beta))} = \left(\frac{R}{\delta}\right)^{-\lambda \frac{\beta(2\pi-\beta)}{4\pi^2}},
\end{aligned}$$

where the β on the r.h.s. of the last equality is to be interpreted modulo 2π. □

5.4.2.4 The 2-point Function in the Layering Model

We now analyze the 2-point function when the IR cutoff is removed by the charge conservation condition (5.28).

Theorem 5.4.4 *If $\beta_1 + \beta_2 = 2k\pi$ with $k \in \mathbb{Z}$, there is a positive constant $C_2 < \infty$ such that, for all $z_1 \neq z_2$,*

$$\lim_{R \to \infty} \langle V_{\beta_1}(z_1) V_{\beta_2}(z_2) \rangle_{\delta,R} = C_2 \left(\frac{|z_1 - z_2|}{\delta} \right)^{-\frac{\lambda}{5}(2 - \cos\beta_1 - \cos\beta_2)}.$$

As a consequence,

$$\lim_{\delta \to 0} \lim_{R \to \infty} \frac{\langle V_{\beta_1}(z_1) V_{\beta_2}(z_2) \rangle_{\delta,R}}{\delta^{\frac{\lambda}{5}(2 - \cos\beta_1 - \cos\beta_2)}} = C_2 |z_1 - z_2|^{-\frac{\lambda}{5}(2 - \cos\beta_1 - \cos\beta_2)}.$$

Proof. Letting $d := |z_1 - z_2|$, for given β_1 and β_2, and $d \geq \delta$, we have that

$$\langle V_{\beta_1}(z_1) V_{\beta_2}(z_2) \rangle_{\delta,R} = \left\langle e^{i(\beta_1 N_\ell(z_1) + \beta_2 N_\ell(z_2))} \right\rangle_{\delta,R}$$

$$= \left\langle e^{i(\beta_1 + \beta_2) N_{12}} \right\rangle_{d,R} \left\langle e^{i\beta_1 N_1} \right\rangle_{\delta,R} \left\langle e^{i\beta_2 N_2} \right\rangle_{\delta,R},$$

where N_{12} is the number of loops that cover both z_1 and z_2, and N_1 (N_2) is the number of loops that cover z_1 but not z_2 (z_2 but not z_1, respectively). The 2-point function factorizes because the sets of loops contributing to N_{12}, N_1 and N_2 are disjoint; the δ is replaced by d in the first factor in the second line because a loop covering both z_1 and z_2 must have diameter at least d.

As in the 1-point function calculation, we can write

$$\left\langle e^{i(\beta_1 + \beta_2) N_{12}} \right\rangle_{d,R} = \sum_{n=0}^{\infty} (\cos(\beta_1 + \beta_2))^n P_{\lambda, \mu_{d,R}^{loop}}(N_{12} = n)$$

$$= e^{-\lambda \alpha_{d,R}(z_1, z_2)(1 - \cos(\beta_1 + \beta_2))},$$

where $\alpha_{d,R}(z_1, z_2) \equiv \mu_{d,R}^{loop}(\gamma : z_1, z_2 \in \bar{\gamma})$. Similarly, if $\alpha_{\delta,R}(z_1, \neg z_2) = \mu_{\delta,R}^{loop}(\gamma : z_1 \in \bar{\gamma}, z_2 \notin \bar{\gamma})$ and $\alpha_{\delta,R}(\neg z_1, z_2)$ is correspondingly defined, then

$$\left\langle e^{i\beta_1 N_1} \right\rangle_{\delta,R} = e^{-\lambda \alpha_{\delta,R}(z_1, \neg z_2)(1 - \cos\beta_1)}$$

and

$$\left\langle e^{i\beta_2 N_2} \right\rangle_{\delta,R} = e^{-\lambda \alpha_{\delta,R}(\neg z_1, z_2)(1 - \cos\beta_2)}.$$

Combining the three terms we obtain

$$\left\langle e^{i(\beta_1 N_\ell(z_1) + \beta_2 N_\ell(z_2))} \right\rangle_{\delta,R}$$

$$= e^{-\lambda \alpha_{d,R}(z_1, z_2)(1 - \cos(\beta_1 + \beta_2))} e^{-\lambda \alpha_{\delta,R}(z_1, \neg z_2)(1 - \cos\beta_1)} e^{-\lambda \alpha_{\delta,R}(\neg z_1, z_2)(1 - \cos\beta_2)}.$$

It is easy to see that $\lim_{R \to \infty} \alpha_{d,R}(z_1, z_2) = \infty$. (This follows from the scale invariance of μ^{loop} by considering an increasing – in size – sequence of disjoint, concentric annuli around z_1 and z_2 that are scaled versions of each other.) Hence, in order to remove the IR cutoff, we must impose (5.28) and set $\beta_1 + \beta_2 = 2k\pi$, so that $1 - \cos(\beta_1 + \beta_2) = 0$.

Assuming that $\beta_1 + \beta_2 = 2k\pi$, we are left with

$$\left\langle e^{i(\beta_1 N_\ell(z_1) + \beta_2 N_\ell(z_2))} \right\rangle_{\delta, R} = e^{-\lambda \alpha_{\delta, R}(z_1, \neg z_2)(1 - \cos \beta_1) - \lambda \alpha_{\delta, R}(\neg z_1, z_2)(1 - \cos \beta_2)}. \quad (5.30)$$

To remove the infrared cutoff, we use the fact that the Brownian loop soup is *thin*: if $z_1 \neq z_2$, $\mu^{loop}(\gamma : z_1 \in \bar{\gamma}, z_2 \notin \bar{\gamma}, \text{diam}(\gamma) \geq \delta) < \infty$ for any $\delta > 0$ (see [39], Lemma 4). By the obvious monotonicity of $\alpha_{\delta, R}(z_1, \neg z_2)$ in R, this implies that

$$\lim_{R \to \infty} \alpha_{\delta, R}(z_1, \neg z_2) = \mu^{loop}\{\gamma \in \eta : z_1 \in \bar{\gamma}, z_2 \notin \bar{\gamma}, \text{diam}(\gamma) \geq \delta\} \equiv \alpha_\delta(z_1, \neg z_2).$$

By scale, rotation and translation invariance of the Brownian loop measure μ^{loop}, $\alpha_\delta(z_1, \neg z_2)$ can only depend on the ratio $x = d/\delta$, so we can introduce the notation $\alpha(x) \equiv \alpha_\delta(z_1, \neg z_2)$. The function α has the following properties, which are also immediate consequences of the scale, rotation and translation invariance of the Brownian loop measure.

- $\alpha(x) = \alpha_\delta(0, \neg z)$ for any z such that $|z| = d$.
- For $\sigma \geq 1$, if $\delta < d$, letting $\alpha_{\delta, R}(z) \equiv \alpha_{z, \delta, R} = \mu_{\delta, R}(\gamma : z \in \bar{\gamma})$,

$$\alpha(\sigma x) = \alpha_\delta(0, \neg \sigma z)$$
$$= \alpha_{\sigma \delta}(0, \neg \sigma z) + \alpha_{\delta, \sigma \delta}(0, \neg \sigma z)$$
$$= \alpha(x) + \alpha_{\delta, \sigma \delta}(0) = \alpha(x) + \alpha_{1, \sigma}(0). \quad (5.31)$$

Now let

$$G(x) := \left\langle e^{i(\beta_1 N_\ell(z_1) + \beta_2 N_\ell(z_2))} \right\rangle_\delta := \lim_{R \to \infty} \left\langle e^{i(\beta_1 N_\ell(z_1) + \beta_2 N_\ell(z_2))} \right\rangle_{\delta, R};$$

using (5.30) and the definition of the function α, we can write

$$G(x) = e^{-\lambda \alpha(x)(2 - \cos \beta_1 - \cos \beta_2)}.$$

Then, for $\sigma \geq 1$,

$$G(\sigma x) = e^{-\lambda \alpha_{1, \sigma}(0)(2 - \cos \beta_1 - \cos \beta_2)} G(x).$$

Using Lemma B.8, we have that

$$\alpha_{1, \sigma}(0) = \frac{1}{5} \log \sigma.$$

It then follows that, for $\sigma \geq 1$,

$$G(\sigma x) = \sigma^{-\frac{\lambda}{5}(2 - \cos \beta_1 - \cos \beta_2)} G(x). \quad (5.32)$$

For $0 < \sigma < 1$, (5.31) implies

$$\alpha(\sigma x) = \alpha(x) - \alpha_{1,1/\sigma}(0).$$

But since

$$\alpha_{1,1/\sigma}(0) = -\frac{1}{5}\log\sigma,$$

equation (5.32) is unchanged when $0 < \sigma < 1$.

The fact that (5.32) is valid for all $\sigma > 0$ immediately implies that

$$G(x) = C_2 x^{-\frac{\lambda}{5}(2-\cos\beta_1 - \cos\beta_2)}$$

for some constant $C_2 > 0$. □

5.4.3 Conformal Covariance of the n-point Functions

We now analyze the n-point functions for general $n \geq 1$ and their conformal invariance properties. In bounded domains $D \subset \mathbb{C}$, we show, for both models, how to remove the UV cutoff $\delta > 0$ by dividing by $\delta^{2\sum_{j=1}^{n} \Delta_j}$, with the appropriate Δ_j's. We also show that this procedure leads to conformally covariant functions of the domain D. The scaling with δ originates from the fact that loops with diameter less than δ can only wind around a single point in the limit $\delta \to 0$, and so for these small loops the n-point function reduces to the product of 1-point functions.

In Section 5.4.3.3, we deal with the layering model in the full plane, $\mathbb{C}$, and show that, together with the UV cutoff $\delta > 0$, we can also remove the IR cutoff $R < \infty$, provided we impose the condition $\sum_{j=1}^{n} \beta_j \in 2\pi\mathbb{Z}$ (see (5.28)). We refer to this condition as "charge conservation" because – apart from the periodicity – it is reminiscent of momentum or charge conservation for the vertex operators of the free boson.

In the layering model the IR convergence (given "charge conservation") is due to the finiteness of the total mass of the loops which cover some points but not others; this is basically the property that the soup of outer boundaries of a Brownian loop soup is *thin* in the language of Nacu and Werner [39].

5.4.3.1 The Layering Model in Finite Domains

In the theorem below, we let $\left\langle \prod_{j=1}^{n} V_{\beta_j}(z_j) \right\rangle_{\delta,D} = \mathbb{E}_{\lambda,\delta,D}\left(\prod_{j=1}^{n} e^{i\beta_j N_\ell(z_j)}\right)$ denote the expectation of the product $\prod_{j=1}^{n} e^{i\beta_j N_\ell(z_j)}$ with respect to a loop soup in D with intensity $\lambda > 0$ containing only loops of diameter at least $\delta > 0$, that is, with respect to the distribution $P_{\lambda,\mu_{\delta,D}} \otimes P_{1/2}$, where $P_{\lambda,\mu_{\delta,D}}$ is the Poisson distribution with intensity measure $\mu_{\delta,D} = \mu_D \mathbb{1}_{\{\text{diam}(\gamma) \geq \delta\}} = \mu \mathbb{1}_{\{\gamma \subset D, \text{diam}(\gamma) \geq \delta\}}$ and $P_{1/2}$ is the distribution of a countable sequence of independent Bernoulli random variables with parameter 1/2 (remember that each loop belongs to one of two classes with equal probability).

Theorem 5.4.5 *If $n \in \mathbb{N}$, $D \subset \mathbb{C}$ is bounded and $\boldsymbol{\beta} = (\beta_1, \ldots, \beta_n)$, then*

$$\lim_{\delta \to 0} \frac{\left\langle \prod_{j=1}^n V_{\beta_j}(z_j) \right\rangle_{\delta, D}}{\delta^{\frac{\lambda}{5} \sum_{j=1}^n (1 - \cos \beta_j)}} =: \phi_D(z_1, \ldots, z_n; \boldsymbol{\beta})$$

exists and is finite and real. Moreover, if D' is another bounded subset of $\mathbb{C}$ and $f : D \to D'$ is a conformal map, such that $z'_1 = f(z_1), \ldots, z'_n = f(z_n)$, then

$$\phi_{D'}(z'_1, \ldots, z'_n; \boldsymbol{\beta}) = \prod_{j=1}^n |f'(z_j)|^{-\frac{\lambda}{5}(1 - \cos \beta_j)} \phi_D(z_1, \ldots, z_n; \boldsymbol{\beta}).$$

The proof of the theorem will make use of the following lemma, where $B_\delta(z)$ denotes the disc of radius δ centered at z, $\bar{\gamma}$ denotes the complement of the unique unbounded component of $\mathbb{C} \setminus \gamma$.

Lemma 5.4.6 *Let $D, D' \subset \mathbb{C}$ and let $f : D \to D'$ be a conformal map. For $n \geq 1$, assume that $z_1, \ldots, z_n \in D$ are distinct and that $z'_1 = f(z_1), \ldots, z'_n = f(z_n)$, and let $s_j = |f'(z_j)|$ for $j = 1, \ldots, n$. Then we have that, for each $j = 1, \ldots, n$,*

$$\mu_{D'}(\gamma : z'_j \in \bar{\gamma}, \bar{\gamma} \not\subset f(B_\delta(z_j)), z'_k \notin \bar{\gamma} \; \forall k \neq j)$$
$$- \mu_{D'}(\gamma : z'_j \in \bar{\gamma}, \bar{\gamma} \not\subset B_{s_j \delta}(z'_j), z'_k \notin \bar{\gamma} \; \forall k \neq j) = o(1) \text{ as } \delta \to 0.$$

Proof. Let $B_{in}(z'_j)$ denote the largest (open) disc centered at z'_j contained inside $f(B_\delta(z_j)) \cap B_{s_j \delta}(z'_j)$, and $B_{out}(z'_j)$ denote the smallest disc centered at z'_j containing $f(B_\delta(z_j)) \cup B_{s_j \delta}(z'_j)$. A moment of thought reveals that, for δ sufficiently small,

$$|\mu_{D'}(\gamma : z'_j \in \bar{\gamma}, \bar{\gamma} \not\subset f(B_\delta(z_j)), z'_k \notin \bar{\gamma} \; \forall k \neq j)$$
$$- \mu_{D'}(\gamma : z'_j \in \bar{\gamma}, \bar{\gamma} \not\subset B_{s_j \delta}(z'_j), z'_k \notin \bar{\gamma} \; \forall k \neq j)|$$
$$= \mu_{D'}(\gamma : z'_j \in \bar{\gamma}, \bar{\gamma} \not\subset f(B_\delta(z_j)) \cap s_j B_\delta(z_j), \bar{\gamma} \subset f(B_\delta(z_j)) \cup s_j B_\delta(z_j), z'_k \notin \bar{\gamma} \; \forall k \neq j)$$
$$\leq \mu_{D'}(\gamma : z'_j \in \bar{\gamma}, \bar{\gamma} \not\subset B_{in}(z'_j), \bar{\gamma} \subset B_{out}(z'_j))$$
$$= \frac{1}{5} \log \frac{\text{diam}(B_{out}(z'_j))}{\text{diam}(B_{in}(z'_j))}, \tag{5.33}$$

where we have used equation (B.9) from Theorem B.4 in Appendix B. Note that, when $D' = \mathbb{C}$, the quantities above involving $\mu_\mathbb{C}$ are bounded because of the fact that the Brownian loop soup is *thin* (see [39]).

Since f is analytic, for every $w \in \partial B_\delta(z_j)$, we have that

$$|f(w) - z'_j| = s_j \delta + O(\delta^2),$$

which implies that
$$\lim_{\delta \to 0} \frac{\text{diam}(B_{out}(z'_j))}{\text{diam}(B_{in}(z'_j))} = 1.$$

In view of (5.33), this concludes the proof of the lemma. $\square$

Proof of Theorem 5.4.5. We first show that the limit is finite. We let η denote a realization of loops and $\{X_\gamma\}_{\gamma \in \eta}$ a collection of independent, Bernoulli, symmetric random variables taking values in $\{-1, 1\}$. Moreover, let $[n] \equiv \{1, \ldots, n\}$, let $\mathcal{K}$ denote the space of assignments of a nonnegative integer to each non-empty subset S of $\{z_1, \ldots, z_n\}$, and for $S \subset \{z_1, \ldots, z_n\}$, let $I_S \subset [n]$ be the set of indices such that $k \in I_S$ if and only if $z_k \in S$. We have that

$$\left\langle \prod_{j=1}^n V_{\beta_j}(z_j) \right\rangle_{\delta,D} = \mathbb{E}_{\lambda,\delta,D}\left(e^{i\sum_{j=1}^n \beta_j N_\ell(z_j)}\right)$$

$$= \sum_{k \in \mathcal{K}} \mathbb{E}_{\lambda,\delta,D}\left(e^{i\sum_{j=1}^n \beta_j N_\ell(z_j)} | \mathcal{L}_k\right) P_{\lambda,\mu_{\delta,D}}(\mathcal{L}_k)$$

where $\mathcal{L}_k = \{\eta : \forall S \subset \{z_1, \ldots, z_n\}, S \neq \emptyset, |\{\gamma \in \eta : z_j \in \bar{\gamma} \ \forall j \in I_S, z_j \notin \bar{\gamma} \ \forall j \notin I_S\}| = k(S)\}$. With probability one with respect to $P_{\lambda,\mu_{\delta,D}}$ we have that, for each $j = 1, \ldots, n$,

$$N_\ell(z_j) = \sum_{\gamma: z_j \in \bar{\gamma}, \text{diam}(\gamma) \geq \delta} X_\gamma = \sum_{S \subset \{z_1,\ldots,z_n\}: z_j \in S} \sum_{\gamma: S \subset \bar{\gamma}, S^c \subset \bar{\gamma}^c, \text{diam}(\gamma) \geq \delta} X_\gamma.$$

With the notation $\sum^S := \sum_{\gamma: S \subset \bar{\gamma}, S^c \subset \bar{\gamma}^c, \text{diam}(\gamma) \geq \delta}$, and letting X denote a $(\pm)$-valued, symmetric random variable, we have that

$$\mathbb{E}_{\lambda,\delta,R}\left(e^{i\sum_{j=1}^n \beta_j N_\ell(z_j)} | \mathcal{L}_k\right) = \mathbb{E}_{\lambda,\delta,R}\left(e^{i\sum_{j=1}^n \beta_j \sum_{S \subset \{z_1,\ldots,z_n\}: z_j \in S} \sum^S X_\gamma} | \mathcal{L}_k\right)$$

$$= \mathbb{E}_{\lambda,\delta,R}\left(e^{i\sum_{S \subset \{z_1,\ldots,z_n\}} \sum^S (\sum_{j \in I_S} \beta_j) X_\gamma} | \mathcal{L}_k\right)$$

$$= \prod_{S \subset \{z_1,\ldots,z_n\}, S \neq \emptyset} \left(E\left(e^{i(\sum_{j \in I_S} \beta_j) X}\right)\right)^{k(S)}$$

$$= \prod_{S \subset \{z_1,\ldots,z_n\}, S \neq \emptyset} \left(\cos\left(\sum_{j \in I_S} \beta_j\right)\right)^{k(S)}.$$

Next, given $S \subset \{z_1, \ldots, z_n\}$ with $|S| \geq 2$, let $\alpha_D(S) := \mu_D(\gamma : S \subset \bar{\gamma}, S^c \subset \bar{\gamma}^c)$ and $\alpha_{\delta,D}(z_j) := \mu_D(\gamma : \text{diam}(\gamma) \geq \delta, z_j \in \bar{\gamma}, z_k \notin \bar{\gamma} \ \forall k \neq j)$. Furthermore, let

$m = \min_{i,j: i \neq j} |z_i - z_j| \wedge \min_i \text{dist}(z_i, \partial D)$ and note that, when $\delta < m$, we can write

$$\left\langle \prod_{j=1}^n V_{\beta_j}(z_j) \right\rangle_{\delta, D} = \sum_{k \in \mathcal{K}} \prod_{S \subset \{z_1, \ldots, z_n\}, |S| > 1} \left(\cos\left(\sum_{k \in I_S} \beta_k\right) \right)^{k(S)} \frac{(\lambda \alpha_D(S))^{k(S)}}{(k(S))!} e^{-\lambda \alpha_D(S)}$$

$$\prod_{j=1}^n \left(\cos \beta_j \right)^{k(z_j)} \frac{(\lambda \alpha_{\delta, D}(z_j))^{k(z_j)}}{(k(z_j))!} e^{-\lambda \alpha_{\delta, D}(z_j)}$$

$$= \prod_{S \subset \{z_1, \ldots, z_n\}, |S| > 1} \exp\left[-\lambda \alpha_D(S) \left(1 - \cos\left(\sum_{k \in I_S} \beta_k\right) \right) \right]$$

$$\prod_{j=1}^n \exp\left[-\lambda \alpha_{\delta, D}(z_j)(1 - \cos \beta_j) \right]. \tag{5.34}$$

For every $j = 1, \ldots, n$, using Lemma B.8, we have that

$$\alpha_{\delta, D}(z_j) = \mu_D(\gamma : \text{diam}(\gamma) \geq \delta, z_j \in \bar{\gamma}, z_k \notin \bar{\gamma} \; \forall k \neq j)$$
$$= \mu_D(\gamma : m > \text{diam}(\gamma) \geq \delta, z_j \in \bar{\gamma})$$
$$+ \mu_D(\gamma : \text{diam}(\gamma) \geq m, z_j \in \bar{\gamma}, z_k \notin \bar{\gamma} \; \forall k \neq j)$$
$$= \frac{1}{5} \log \frac{m}{\delta} + \alpha_{m, D}(z_j).$$

Therefore, we obtain

$$\lim_{\delta \to 0} \frac{\left\langle \prod_{j=1}^n V_{\beta_j}(z_j) \right\rangle_{\delta, D}}{\delta^{\frac{\lambda}{5} \sum_{j=1}^n (1 - \cos(\beta_j))}} = \prod_{S \subset \{z_1, \ldots, z_n\}, |S| > 1} \exp\left[-\lambda \alpha_D(S) \left(1 - \cos\left(\sum_{k \in I_S} \beta_k\right) \right) \right]$$

$$m^{-\frac{\lambda}{5} \sum_{j=1}^n (1 - \cos \beta_j)} e^{-\sum_{j=1}^n \lambda \alpha_{m, D}(z_j)(1 - \cos \beta_j)}$$

$$= m^{-\frac{\lambda}{5} \sum_{j=1}^n (1 - \cos \beta_j)} \exp\left[-\lambda \sum_{j=1}^n \alpha_{m, D}(z_j)(1 - \cos \beta_j) \right]$$

$$\exp\left[-\lambda \sum_{S \subset \{z_1, \ldots, z_n\}, |S| > 1} \alpha_D(S) \left(1 - \cos\left(\sum_{k \in I_S} \beta_k\right) \right) \right]$$

$$=: \phi_D(z_1, \ldots, z_n; \boldsymbol{\beta}).$$

This concludes the first part of the proof.

To prove the second part of the theorem, using (5.34), we write

$$\left\langle \prod_{j=1}^{n} V_{\beta_j}(z_j) \right\rangle_{\delta,D} = \exp\left[-\lambda \sum_{S \subset \{z_1,\ldots,z_n\}, |S| \geq 2} \alpha_D(S)\left(1 - \cos\left(\sum_{k \in I_S} \beta_k\right)\right)\right]$$

$$\prod_{j=1}^{n} \exp\left[-\lambda \alpha_{\delta,D}(z_j)(1 - \cos \beta_j)\right]. \tag{5.35}$$

For each $S \subset \{z_1,\ldots,z_n\}$ with $|S| \geq 2$, $\alpha_D(S)$ is invariant under conformal transformations; that is, if $f : D \to D'$ is a conformal map from D to another bounded domain D', and $S' = \{z'_1,\ldots,z'_n\}$, where $z'_1 = f(z_1),\ldots,z'_n = f(z_n)$, then $\alpha_{D'}(S') = \alpha_D(S)$. Therefore, the first exponential term in (5.35) is also invariant under conformal transformations. This implies that, for δ sufficiently small,

$$\frac{\left\langle \prod_{j=1}^{n} V_{\beta_j}(z_j) \right\rangle_{\delta,D}}{\left\langle \prod_{j=1}^{n} V_{\beta_j}(z'_j) \right\rangle_{\delta,D'}} = \prod_{j=1}^{n} \exp\left\{-\lambda \left[\alpha_{\delta,D}(z_j) - \alpha_{\delta,D'}(z'_j)\right](1 - \cos \beta_j)\right\}.$$

Writing

$$\alpha_{\delta,D}(z_j) = \mu_D(\gamma : \text{diam}(\gamma) \geq \delta, z_j \in \bar{\gamma}, \bar{\gamma} \subset B_\delta(z_j))$$
$$+ \mu_D(\gamma : z_j \in \bar{\gamma}, \bar{\gamma} \not\subset B_\delta(z_j), z_k \notin \bar{\gamma} \ \forall k \neq j)$$

and noticing that $\mu_D(\gamma : \text{diam}(\gamma) \geq \delta, z_j \in \bar{\gamma}, \bar{\gamma} \subset B_\delta(z_j)) = \mu_{D'}(\gamma : \text{diam}(\gamma) \geq \delta, z'_j \in \bar{\gamma}, \bar{\gamma} \subset B_\delta(z'_j))$ (where we have assumed, without loss of generality, that δ is so small that $B_\delta(z_j) \subset D$ and $B_\delta(z'_j) \subset D'$), we have that

$$\alpha_{\delta,D}(z_j) - \alpha_{\delta,D'}(z'_j) = \mu_D(\gamma : z_j \in \bar{\gamma}, \bar{\gamma} \not\subset B_\delta(z_j), z_k \notin \bar{\gamma} \ \forall k \neq j)$$
$$- \mu_{D'}(\gamma : z'_j \in \bar{\gamma}, \bar{\gamma} \not\subset B_\delta(z'_j), z'_k \notin \bar{\gamma} \ \forall k \neq j).$$

To evaluate this difference, using conformal invariance, we write

$$\mu_D(\gamma : z_j \in \bar{\gamma}, \bar{\gamma} \not\subset B_\delta(z_j), z_k \notin \bar{\gamma} \ \forall k \neq j)$$
$$= \mu_{D'}(\gamma : z'_j \in \bar{\gamma}, \bar{\gamma} \not\subset f(B_\delta(z_j)), z'_k \notin \bar{\gamma} \ \forall k \neq j).$$

Thus, letting $s_j = |f'(z_j)|$ and using Lemmas B.6 and B.8 from Appendix B, we can write

$$\alpha_{\delta,D}(z_j) - \alpha_{\delta,D'}(z'_j) = \mu_{D'}(\gamma : z'_j \in \bar{\gamma}, \bar{\gamma} \not\subset f(B_\delta(z_j)), z'_k \notin \bar{\gamma} \; \forall k \neq j)$$
$$-\mu_{D'}(\gamma : z'_j \in \bar{\gamma}, \bar{\gamma} \not\subset B_\delta(z'_j), z'_k \notin \bar{\gamma} \; \forall k \neq j)$$
$$= \mu_{D'}(\gamma : z'_j \in \bar{\gamma}, \bar{\gamma} \not\subset f(B_\delta(z_j)), z'_k \notin \bar{\gamma} \; \forall k \neq j)$$
$$-\mu_{D'}(\gamma : z'_j \in \bar{\gamma}, \bar{\gamma} \not\subset B_{s_j\delta}(z'_j), z'_k \notin \bar{\gamma} \; \forall k \neq j)$$
$$-[\mu_{D'}(\gamma : z'_j \in \bar{\gamma}, \bar{\gamma} \not\subset B_\delta(z'_j), z'_k \notin \bar{\gamma} \; \forall k \neq j)$$
$$-\mu_{D'}(\gamma : z'_j \in \bar{\gamma}, \bar{\gamma} \not\subset B_{s_j\delta}(z'_j), z'_k \notin \bar{\gamma} \; \forall k \neq j)]$$
$$= \mu_{D'}(\gamma : z'_j \in \bar{\gamma}, \bar{\gamma} \not\subset f(B_\delta(z_j)), z'_k \notin \bar{\gamma} \; \forall k \neq j)$$
$$-\mu_{D'}(\gamma : z'_j \in \bar{\gamma}, \bar{\gamma} \not\subset B_{s_j\delta}(z'_j), z'_k \notin \bar{\gamma} \; \forall k \neq j)$$
$$-\frac{1}{5}\log s_j.$$

Using Lemma 5.4.6, we obtain that $\alpha_{\delta,D}(z_j) - \alpha_{\delta,D'}(z'_j) = -\frac{1}{5}\log|f'(z_j)| + o(1)$ as $\delta \to 0$, which gives

$$\frac{\langle V_\beta(z_1,\ldots,z_n)\rangle_{\delta,D}}{\langle V_\beta(z'_1,\ldots,z'_n)\rangle_{\delta,D'}} = e^{-o(1)} \prod_{j=1}^{n} |f'(z_j)|^{\frac{\lambda}{5}(1-\cos\beta_j)} \text{ as } \delta \to 0.$$

Letting $\delta \to 0$ concludes the proof. □

5.4.3.2 The Winding Model in Finite Domains

As above, let $N_w(z)$ denote the total number of windings of all loops of a given soup around $z \in \mathbb{C}$. We have the following theorem.

Theorem 5.4.7 *If $n \in \mathbb{N}$, $D \subset \mathbb{C}$ is bounded and $\boldsymbol{\beta} = (\beta_1,\ldots,\beta_n)$, then*

$$\lim_{\delta \to 0} \frac{\langle e^{i\beta_1 N_w(z_1)}\ldots e^{i\beta_n N_w(z_n)}\rangle_{\delta,D}}{\delta^{\frac{\lambda}{4\pi^2}\sum_{j=1}^{n}\beta_j(2\pi-\beta_j)}} =: \psi_D(z_1,\ldots,z_n;\boldsymbol{\beta})$$

exists and is finite and real. Moreover, if D' is another bounded subset of $\mathbb{C}$ and $f : D \to D'$ is a conformal map, such that $z'_1 = f(z_1),\ldots,z'_n = f(z_n)$, then

$$\psi_{D'}(z'_1,\ldots,z'_n;\boldsymbol{\beta}) = \prod_{j=1}^{n} |f'(z_j)|^{-\lambda\frac{\beta_j(2\pi-\beta_j)}{4\pi^2}} \psi_D(z_1,\ldots,z_n;\boldsymbol{\beta}),$$

where, in the exponent, the β_j's are to be interpreted modulo 2π.

Proof. The proof is analogous to that of Theorem 5.4.5. Let $\alpha_D(S; k_{i_1}, \ldots, k_{i_l}) := \mu_D(\gamma : N_\gamma(z_{i_j}) = k_{i_j}$ for each $z_{i_j} \in S$ and $S^c \subset \bar{\gamma}^c$), and $\alpha_{\delta,D}(z_j;k) := \mu_D(\gamma : z_j \in \bar{\gamma}, \theta_\gamma(z_j) = k, z_k \notin \bar{\gamma} \ \forall k \neq j)$. With this notation we can write

$$\left\langle e^{\beta_1 N_w(z_1)} \ldots e^{\beta_n N_w(z_n)} \right\rangle_{\delta,D}$$

$$= \exp\Big[-\lambda \sum_{l=2}^{n} \sum_{\substack{S \subset \{z_1,\ldots,z_n\} \\ |S|=l}} \sum_{k_{i_1},\ldots,k_{i_l}=-\infty}^{\infty} \alpha_D(S; k_{i_1},\ldots,k_{i_l})$$

$$\left(1 - \cos(k_{i_1}\beta_{i_1} + \ldots + k_{i_l}\beta_{i_l})\right)\Big]$$

$$\prod_{j=1}^{n} \exp\Big[-\lambda \sum_{k=-\infty}^{\infty} \alpha_{\delta,D}(z_j;k)(1 - \cos(k\beta_j))\Big],$$

For each $S \subset \{z_1, \ldots, z_n\}$ with $|S| = l \geq 2$, $\alpha_D(S; k_{i_1}, \ldots, k_{i_l})$ is invariant under conformal transformations; therefore,

$$\frac{\left\langle e^{i\beta_1 N_w(z_1)} \ldots e^{i\beta_n N_w(z_n)} \right\rangle_{\delta,D}}{\left\langle e^{i\beta_1 N_w(z_1')} \ldots e^{i\beta_n N_w(z_n')} \right\rangle_{\delta,D'}}$$

$$= \prod_{j=1}^{n} \exp\Big\{-\lambda \sum_{k=-\infty}^{\infty} \Big[\alpha_{\delta,D}(z_j;k) - \alpha_{\delta,D'}(z_j';k)\Big](1 - \cos(k\beta_j))\Big\}.$$

Proceeding as in the proof of Theorem 5.4.5, but using Lemma B.9 instead of Lemma B.8, gives

$$\alpha_{\delta,D}(z_j;k) - \alpha_{\delta,D'}(z_j';k) = -c_k \log|f'(z_j)| + o(1) \text{ as } \delta \to 0,$$

where $c_k = \frac{1}{2\pi^2 k^2}$ for $k \in \mathbb{Z} \setminus \{0\}$ and $c_0 = 1/30$.

This, together with the observation, already used at the end of Section 5.4.2.3, that $\sum_{k=-\infty}^{\infty} c_k(1 - \cos(k\beta)) = \frac{\beta(2\pi - \beta)}{4\pi^2}$ (where, in the right hand side, β should be interpreted modulo 2π), readily implies the statement of the theorem. $\square$

5.4.3.3 The Layering Model in the Plane

In the theorem below, we let $\left\langle \prod_{j=1}^{n} V_{\beta_j}(z_j) \right\rangle_{\delta,R}$ denote the expectation of the product $\prod_{j=1}^{n} e^{i\beta_j N_\ell(z_j)}$ with respect to a loop soup in $\mathbb{C}$ with intensity $\lambda > 0$ containing only loops γ of diameter $0 < \delta \leq \text{diam}(\gamma) < R < \infty$.

Theorem 5.4.8 *If $n \in \mathbb{N}$ and $\boldsymbol{\beta} = (\beta_1, \ldots, \beta_n)$ with $|\boldsymbol{\beta}| = \sum_{j=1}^{n} \beta_j \in 2\pi\mathbb{Z}$, then*

$$\lim_{\delta \to 0, R \to \infty} \frac{\left\langle \prod_{j=1}^{n} V_{\beta_j}(z_j) \right\rangle_{\delta,R}}{\delta^{\frac{\lambda}{5}\sum_{j=1}^{n}(1-\cos\beta_j)}} =: \phi_{\mathbb{C}}(z_1, \ldots, z_n; \boldsymbol{\beta})$$

exists and is finite and real. Moreover, if $f : \mathbb{C} \to \mathbb{C}$ *is a conformal map, such that* $z'_1 = f(z_1), \ldots, z'_n = f(z_n)$, *then*

$$\phi_{\mathbb{C}}(z'_1, \ldots, z'_n; \boldsymbol{\beta}) = \prod_{j=1}^{n} |f'(z_j)|^{-\frac{\lambda}{5}(1-\cos\beta_j)} \phi_{\mathbb{C}}(z_1, \ldots, z_n; \boldsymbol{\beta}).$$

Proof sketch. The beginning of the proof proceeds like that of Theorem 5.4.5 until equation (5.34), leading to the following equation:

$$\left\langle \prod_{j=1}^{n} V_{\beta_j}(z_j) \right\rangle_{\delta,R} = \prod_{S \subset \{z_1,\ldots,z_n\}, 1 < |S| < n} \exp\left[-\lambda \alpha_R(S)\left(1 - \cos\left(\sum_{k \in I_S} \beta_k\right)\right)\right]$$
$$\prod_{j=1}^{n} \exp\left[-\lambda \alpha_{\delta,R}(z_j)(1 - \cos\beta_j)\right],$$

where $\alpha_R(S) := \mu_{\mathbb{C}}(\gamma : S \subset \bar{\gamma}, S^c \subset \bar{\gamma}^c, \text{diam}(\gamma) < R)$, for $S \subset \{z_1, \ldots, z_n\}$ with $2 \leq |S| < n$, and $\alpha_{\delta,R}(z_j) := \mu_{\mathbb{C}}(\gamma : \delta \leq \text{diam}(\gamma) < R, z_j \in \bar{\gamma}, z_k \notin \bar{\gamma} \ \forall k \neq j)$, and where I_S denotes the set of indices such that $k \in I_S$ if and only if $z_k \in S$.

Note, in the equation above, the condition $|S| < n$ in the first product on the right hand side; this condition comes from the fact that the term $-\lambda \alpha_R(S)$ with $S = \{z_1, \ldots, z_n\}$ is multiplied by $1 - \cos(\sum_{k=1}^{n} \beta_k) = 0$, where we have used the "charge conservation" condition $|\boldsymbol{\beta}| = \sum_{j=1}^{n} \beta_j \in 2\pi\mathbb{Z}$.

For every $j = 1, \ldots, n$, using Lemma B.8, we have that

$$\alpha_{\delta,R}(z_j) = \mu_{\mathbb{C}}(\gamma : \delta \leq \text{diam}(\gamma) < R, z_j \in \bar{\gamma}, z_k \notin \bar{\gamma} \ \forall k \neq j)$$
$$= \mu_{\mathbb{C}}(\gamma : m > \text{diam}(\gamma) \geq \delta, z_j \in \bar{\gamma})$$
$$+ \mu_D(\gamma : m \leq \text{diam}(\gamma) < R, z_j \in \bar{\gamma}, z_k \notin \bar{\gamma} \ \forall k \neq j)$$
$$= \frac{1}{5} \log \frac{m}{\delta} + \alpha_{m,R}(z_j).$$

Now note that monotonicity and the fact that the Brownian loop soup is *thin* [39] imply that $\alpha_{m,\mathbb{C}}(z_j) := \lim_{R \to \infty} \alpha_{m,R}(z_j)$ and $\alpha_{\mathbb{C}}(S) := \lim_{R \to \infty} \alpha_R(S)$, for $S \subset \{z_1, \ldots, z_n\}$ with $2 \leq |S| < n$, exist and are bounded. After letting $R \to \infty$, the proof proceeds like that of Theorem 5.4.5, with $D = D' = \mathbb{C}$. □

We have already seen the behavior of the 2-point function in the layering model in Section 5.4.2.4; the corollary below deals with the 3-point function.

Corollary 5.4.9 *Let $z_1, z_2, z_3 \in \mathbb{C}$ be three distinct points, then we have that*

$$\phi_\mathbb{C}(z_1,z_2,z_3;\beta_1,\beta_2,\beta_3) = C_3 \left| \left(\frac{1}{|z_1-z_2|}\right)^{\Delta_l(\beta_1)+\Delta_l(\beta_2)-\Delta_l(\beta_3)} \left(\frac{1}{|z_1-z_3|}\right)^{\Delta_l(\beta_1)+\Delta_l(\beta_3)-\Delta_l(\beta_2)} \left(\frac{1}{|z_2-z_3|}\right)^{\Delta_l(\beta_2)+\Delta_l(\beta_3)-\Delta_l(\beta_1)} \right|^2$$

for some constant C_3.

Proof. Theorem 5.4.8 implies that the 3-point function in the full plane transforms covariantly under conformal maps. This immediately implies the corollary following standard argument (see, e.g., [23]). We briefly sketch those arguments below for the reader's convenience.

Scale invariance, rotation invariance and translation invariance immediately imply that there are constants C_{abc} such that

$$\phi_\mathbb{C}(z_1,z_2,z_3;\beta_1,\beta_2,\beta_3) = \sum C_{abc} z_{12}^{-a} z_{13}^{-b} z_{23}^{-c}, \tag{5.36}$$

where $z_{ij} = |z_i - z_j|$ and the sum is over all triplets $a,b,c \geq 0$ satisfying $a+b+c = 2(\Delta_l(\beta_1) + \Delta_l(\beta_2) + \Delta_l(\beta_3))$ (the constraint on the exponents a,b,c follows from Theorem 5.4.8 applied to scale transformations).

Now let f be a conformal transformation from $\mathbb{C}$ to $\mathbb{C}$; f is then a Möbius transformation and has the form $f(z) = \frac{Az+B}{Cz+D}$, with $f'(z) = \frac{AD-BC}{(Cz+D)^2}$. Letting $\gamma_j := |f'(z_j)|^{-1}$, if $\tilde{z} = f(z)$, it is easy to check that $\tilde{z}_{ij} = \gamma_i^{-1/2} \gamma_j^{-1/2} z_{ij}$. Using this fact and Theorem 5.4.8, we have that

$$\phi_\mathbb{C}(\tilde{z}_1,\tilde{z}_2,\tilde{z}_3;\beta_1,\beta_2,\beta_3) = \left(\gamma_1^{\Delta_l(\beta_1)} \gamma_2^{\Delta_l(\beta_2)} \gamma_3^{\Delta_l(\beta_3)}\right)^2 \sum C_{abc} z_{12}^{-a} z_{13}^{-b} z_{23}^{-c}$$

$$= \left(\gamma_1^{\Delta_l(\beta_1)} \gamma_2^{\Delta_l(\beta_2)} \gamma_3^{\Delta_l(\beta_3)}\right)^2 \sum C_{abc} \frac{\tilde{z}_{12}^{-a} \tilde{z}_{13}^{-b} \tilde{z}_{23}^{-c}}{\gamma_1^{a/2+b/2} \gamma_2^{a/2+c/2} \gamma_3^{b/2+c/2}}.$$

For this last expression to be of the correct form (5.36), the γ's need to cancel; this immediately leads to the relations $a = 2(\Delta_l(\beta_2) + \Delta_l(\beta_3) - \Delta_l(\beta_1))$, $b = 2(\Delta_l(\beta_1) + \Delta_l(\beta_3) - \Delta_l(\beta_2))$, $c = 2(\Delta_l(\beta_1) + \Delta_l(\beta_2) - \Delta_l(\beta_3))$. □

Appendix A: Occupation Field and Gaussian Free Field

The theorem used in the proof of Lemma 5.2.7, stated below, is a version of a recent result of Le Jan [37] (see also [38] and Theorem 4.5 of [48]). In this appendix we give a self-contained proof, both for completeness and because the occupation field $\{L_x\}$ in Theorem A.1 below is not exactly the same as the occupation field of [37,38] and [48]. Indeed, the loop soup that appears in [37,38] and in Theorem 4.5 of [48] is not the same as the loop soup studied in this chapter, although the two are closely related. In this appendix we use the notation of Sections 5.2.1 and 5.2.2.

Theorem A.1 *Let m be a nonnegative function and $\mathbf{k}$ denote the collection $\{4(e^{m^2(x)} - 1)\}_{x \in D^\#}$. Then the occupation field $\{L_x(\tilde{\mathcal{A}}_{1/2,m})\}_{x \in D^\#}$ has the same distribution as the field $\{\frac{1}{2}\phi_x^2\}_{x \in D^\#}$, where $\{\phi_x\}_{x \in D^\#}$ is the Gaussian free field with covariance $\mathbb{E}(\phi_x \phi_y) = G_D^{\mathbf{k}}(x,y)$.*

Proof. We will show that the Laplace transform of the occupation field is the same as that of the field $\{\frac{1}{2}\phi_x^2\}_{x \in D^\#}$. For that purpose, it is crucial to notice that $L_x(\mathcal{A}_{\lambda,m})$ has the gamma distribution with parameters $\sum_{\tilde{\gamma}} N_{\tilde{\gamma}} n(x, \tilde{\gamma}) + 1/2$ and $\frac{1}{k_x+4}$, and consequently, for any real number v and any collection $\{n_{\tilde{\gamma}}\}$ of nonnegative numbers,

$$\mathbb{E}[\exp(-vL_x) | \{N_{\tilde{\gamma}}\} = \{n_{\tilde{\gamma}}\}] = \left(1 + \frac{v}{k_x+4}\right)^{-1/2} \prod_{\tilde{\gamma}} \left(1 + \frac{v}{k_x+4}\right)^{-n_{\tilde{\gamma}} n(x, \tilde{\gamma})},$$

where the product is over all unrooted lattice loops in D.

Let $\{v_x\}_{x \in D^\#}$ be a collection of real numbers, the Laplace transform of the occupation field is given by the following expectation, where the sum $\sum_{\{n_{\tilde{\gamma}}\}}$ is over all collections of possible multiplicities for the lattice loops $\tilde{\gamma}$ in D,

$$\mathbb{E}\left[\exp\left(-\sum_{x \in D^\#} v_x L_x\right)\right]$$

$$= \sum_{\{n_{\tilde{\gamma}}\}} \mathbb{E}\left[\exp\left(-\sum_{x \in D^\#} v_x L_x\right) \bigg| \{N_{\tilde{\gamma}}\} = \{n_{\tilde{\gamma}}\}\right] \mathbb{P}(\{N_{\tilde{\gamma}}\} = \{n_{\tilde{\gamma}}\}).$$

Recalling (5.1), we can write

$$\mathbb{E}\left[\exp\left(-\sum_{x \in D^\#} v_x L_x\right)\right]$$

$$= \sum_{\{n_{\tilde{\gamma}}\}} \left[\prod_{x \in D^\#}\left(1+\frac{v_x}{k_x+4}\right)^{-1/2} \prod_{\tilde{\gamma}}\left(1+\frac{v_x}{k_x+4}\right)^{-n_{\tilde{\gamma}} n(x,\tilde{\gamma})}\right]$$

$$\prod_{\tilde{\gamma}} \exp\left(-\lambda v_D^{u,\mathbf{k}}(\tilde{\gamma})\right) \frac{1}{n_{\tilde{\gamma}}!} \left(\lambda v_D^{u,\mathbf{k}}(\tilde{\gamma})\right)^{n_{\tilde{\gamma}}}$$

$$= \left(\prod_{x \in D^\#} (k_x+4)^{1/2}\right) \left(\prod_{\tilde{\gamma}} \exp\left(-\lambda v_D^{u,\mathbf{k}}(\tilde{\gamma})\right)\right) \left(\prod_{x \in D^\#} (v_x+k_x+4)^{-1/2}\right)$$

$$\sum_{\{n_{\tilde{\gamma}}\}} \prod_{\tilde{\gamma}} \frac{1}{n_{\tilde{\gamma}}!} \left[\lambda v_D^{u,\mathbf{k}}(\tilde{\gamma}) \prod_{x \in D^\#} \left(\frac{v_x+k_x+4}{k_x+4}\right)^{-n(x,\tilde{\gamma})}\right]^{n_{\tilde{\gamma}}}.$$

Now let $\rho_{\tilde{\gamma}}$ denote the number of rooted loops in the equivalence class $\tilde{\gamma}$, then we can write

$$v_D^{u,\mathbf{k}}(\tilde{\gamma}) = \frac{\rho_{\tilde{\gamma}}}{|\tilde{\gamma}|} \prod_{x \in D^\#} (k_x+4)^{-n(x,\tilde{\gamma})}$$

and, letting $\mathbf{k}+\mathbf{v}$ denote the collection $\{k_x+v_v\}_{x \in D^\#}$, it is easy to see that

$$v_D^{u,\mathbf{k}}(\tilde{\gamma}) \prod_{x \in D^\#} \left(\frac{v_x+k_x+4}{k_x+4}\right)^{-n(x,\tilde{\gamma})} = v_D^{u,\mathbf{k}+\mathbf{v}}(\tilde{\gamma}).$$

Using this fact, the fact that (5.1) defines a probability distribution, and equation (5.2), one has that

$$\sum_{\{n_{\tilde{\gamma}}\}} \prod_{\tilde{\gamma}} \frac{1}{n_{\tilde{\gamma}}!} \left[\lambda v_D^{u,\mathbf{k}}(\tilde{\gamma}) \prod_{x \in D^\#} \left(\frac{v_x+k_x+4}{k_x+4}\right)^{-n(x,\tilde{\gamma})}\right]^{n_{\tilde{\gamma}}}$$

$$= \sum_{\{n_{\tilde{\gamma}}\}} \prod_{\tilde{\gamma}} \frac{1}{n_{\tilde{\gamma}}!} \left(\lambda v_D^{u,\mathbf{k}+\mathbf{v}}(\tilde{\gamma})\right)^{n_{\tilde{\gamma}}}$$

$$= \mathcal{Z}_{\lambda,\mathbf{k}+\mathbf{v}} = \exp\left(\lambda \sum_{\tilde{\gamma}} v_D^{u,\mathbf{k}+\mathbf{v}}(\tilde{\gamma})\right).$$

Then the Laplace transform of the occupation field can be written as

$$\mathbb{E}\left[\exp\left(-\sum_{x\in D^\#} v_x L_x\right)\right]$$

$$=\left(\prod_{x\in D^\#}(k_x+4)\right)^{1/2}\exp\left(-\lambda\sum_{\tilde{\gamma}} v_D^{u,\mathbf{k}}(\tilde{\gamma})\right)$$

$$\left(\prod_{x\in D^\#}(v_x+k_x+4)\right)^{-1/2}\exp\left(\lambda\sum_{\tilde{\gamma}} v_D^{u,\mathbf{k}+\mathbf{v}}(\tilde{\gamma})\right)$$

$$=\left(\prod_{x\in D^\#}(k_x+4)\right)^{1/2}\exp\left(-\lambda\sum_{\tilde{\gamma}}\frac{\rho_{\tilde{\gamma}}}{|\tilde{\gamma}|}\prod_{x\in D^\#}(k_x+4)^{-n(x,\tilde{\gamma})}\right)$$

$$\left(\prod_{x\in D^\#}(v_x+k_x+4)\right)^{-1/2}\exp\left(\lambda\sum_{\tilde{\gamma}}\frac{\rho_{\tilde{\gamma}}}{|\tilde{\gamma}|}\prod_{x\in D^\#}(v_x+k_x+4)^{-n(x,\tilde{\gamma})}\right).$$

If $\lambda = 1/2$, using Lemma 1.2 of [13] and a standard Gaussian integration formula, the last expression can be written as

$$\left(\int_{\mathbb{R}^{D^\#}} e^{-H_D^{\mathbf{k}}(\varphi)}\prod_{x\in D^\#} d\varphi_x\right)^{-1}\int_{\mathbb{R}^{D^\#}}\exp\left(-H_D^{\mathbf{k}}(\varphi)-\frac{1}{2}\sum_{x\in D^\#} v_x\varphi_x^2\right)\prod_{x\in D^\#} d\varphi_x$$

$$=\mathbb{E}_D^{\mathbf{k}}\left[\exp\left(-\sum_{x\in D^\#} v_x\frac{\phi_x^2}{2}\right)\right],$$

which concludes the proof. □

Appendix B: The Brownian Loop Measure

The Brownian loop measure studied in [51] plays an important role in this chapter as the intensity measure used to define the Brownian loop soup. Indeed, the conformal invariance properties of the latter are a consequence of those of the Brownian loop measure. The purpose of this appendix is to discuss those properties, starting with some considerations about Brownian motion and Brownian bridges.

A two-dimensional, or complex, standard Brownian motion $W \equiv (W_t, t \geq 0)$ is the process $W_t = W_t^1 + iW_t^2$, where $(W_t^1, t \geq 0)$ and $(W_t^2, t \geq 0)$ are two independent, one-dimensional, standard Brownian motions. (*Standard* here means that it starts at the origin and that the quadratic variation is 1.)

Lemma B.1 (Conformal invariance of Brownian motion) *Let $f : D \to \mathbb{C}$ be a conformal map. Let W be a complex Brownian motion started from $z \in D$ and stopped at*

$$\tau_D := \inf\{t \geq 0 : W_t \notin D\}.$$

Define $S(t) := \int_0^t |f'(W_u)|^2 du$, $0 \leq t \leq \tau_D$ and let $\sigma(t)$ be such that $\int_0^{\sigma(t)} |f'(W_u)|^2 du = t$. Then $Y_s = \left(f(W_{\sigma(s)}), 0 \leq s \leq S(\tau_D)\right)$ is a complex Brownian motion started at $f(z)$ and stopped $\tau_{f(D)}$.

Proof. Since f is a conformal map, we can write $f = u + iv$, where u and v are harmonic functions, such that $\partial_x u = \partial_y v$ and $\partial_y u = -\partial_x v$. Now note that

$$du(W_t) = \partial_x u(W_t) dW_t^1 + \partial_y u(W_t) dW_t^2 + \frac{1}{2}\left(\partial_{xx} u(W_t) + \partial_{yy} u(W_t)\right) dt$$

$$= \partial_x u(W_t) dW_t^1 + \partial_y u(W_t) dW_t^2.$$

Since the dt term vanishes, $u(W_t)$ is a local martingale; the same applies to $v(W_t)$. Moreover, the quadratic variations are

$$[u(W)]_t = [v(W)]_t = \int_0^t \left[(\partial_x u(W_s))^2 + (\partial_y u(W_s))^2\right] ds$$

$$= \int_0^t |f'(W_s)|^2 ds = S(t),$$

implying that Y is a complex Brownian motion. The remaining properties follow immediately from the definition of Y. □

Following Lawler and Werner [34, 36], let $\mathcal{K}$ be the set of parametrized, continuous, planar curves γ defined on a time interval $[0, t_\gamma]$, endowed with the metric

$$d_\mathcal{K}(\gamma, \gamma') := \inf_\theta \left[\sup_{0 \leq s \leq t_\gamma} |s - \theta(s)| + |\gamma(s) - \gamma'(\theta(s))|\right],$$

where the infimum is over all increasing homeomorphisms $\theta : [0, t_\gamma] \to [0, t_\gamma]$. Let $\mathcal{K}(D)$ be the subset of $\mathcal{K}$ consisting of those curves that stay in D. If $z, w \in \mathbb{C}$, denote by $\mathcal{K}_z$ (respectively, $\mathcal{K}^w$) the set of $\gamma \in \mathcal{K}$ with $\gamma(0) = z$ (resp., $\gamma(t_\gamma) = w$). We let $\mathcal{K}_z^w = \mathcal{K}_z \cup \mathcal{K}^w$ and define $\mathcal{K}_z(D)$, $\mathcal{K}^w(D)$ and $\mathcal{K}_z^w(D)$ similarly.

Let $\tilde{\mu}(z, \cdot; t)$ denote the law of standard complex Brownian motion $(W_s, 0 \leq s \leq t)$ with $W_0 = z$. We can write

$$\tilde{\mu}(z, \cdot; t) = \int_\mathbb{C} \tilde{\mu}(z, w; t) d\mathbf{A}(w),$$

where **A** denotes area and $\tilde{\mu}(z,w;t)$ is a measure supported on $\gamma \in \mathcal{K}_z^w$ with $t_\gamma = t$. The total mass of $\tilde{\mu}(z,w;t)$ is $|\tilde{\mu}(z,w;t)| = (2\pi t)^{-1} \exp(-|z-w|^2/(2t))$. The normalized (probability) measure $\mu^{br}(z,w;t) := \tilde{\mu}(z,w;t)/|\tilde{\mu}(z,w;t)|$ is the law of the *Brownian bridge* from z to w in time t.

The σ-finite measure $\tilde{\mu}(z,w)$ is defined by

$$\tilde{\mu}(z,w) := \int_0^\infty \tilde{\mu}(z,w;t)dt = \int_0^\infty \frac{1}{2\pi t}\mu^{br}(z,w;t)e^{-|z-w|^2/2t}dt.$$

Note that the integral explodes at infinity so that the total mass of large loops is infinite. (When $z = w$, the integral also diverges at zero).

If $w = z$, one has $\tilde{\mu}(z,z;t) = (2\pi t)^{-1}\mu^{br}_{z,t}$, where $\mu^{br}_{z,t} := \mu^{br}(z,z;t)$ is the law of the Brownian bridge of time duration t started and ended at z. The measure

$$\tilde{\mu}(z,z) = \int_0^\infty \frac{1}{2\pi t}\mu^{br}_{z,t}dt,$$

is an infinite measure on Brownian loops that start and end at z.

If $D \subset \mathbb{C}$ is a domain and $z,w \in D$, we define $\tilde{\mu}_D(z,w)$ to be $\tilde{\mu}(z,w)$ restricted to $\mathcal{K}(D)$. If $z \neq w$ and D is such that Brownian motion started in D eventually exits D, then $|\tilde{\mu}_D(z,w)| < \infty$. More precisely, $|\tilde{\mu}_D(z,w)| = \frac{1}{\pi}G_D(z,w)$, where $G_D(z,w)$ is Green's function normalized so that, for the unit disc $\mathbb{D}$, $G_\mathbb{D}(0,z) = -\log|z|$.

Suppose that $f : D \to D'$ is a conformal transformation and $\gamma \in \mathcal{K}$; let

$$s(t) := \int_0^t |f'(\gamma(u))|^2 du. \qquad (B.1)$$

If $s(t) < \infty$ for all $t < t_\gamma$, define $f \circ \gamma$ by

$$(f \circ \gamma)(s(t)) := f(\gamma(t)).$$

Moreover, if $A \in \mathcal{K}$, let $f \circ A = \{\hat{\gamma} = f \circ \gamma \text{ with } \gamma \in A\}$. Then

$$\tilde{\mu}_{f(D)}(f(z),f(w))(f \circ A) = \tilde{\mu}_D(z,w)(A). \qquad (B.2)$$

The conformal invariance of $\mu_D(z,w)$ expressed by equation (B.2) is an immediate consequence of two classical results: the conformal invariance of Brownian motion, which follows from Lemma B.1, and the fact that $G_{f(D)}(f(z),f(w)) = G_D(z,w)$. The conformal invariance of $\mu(z,z)$ follows by letting $w \to z$.

Now let $\tilde{\mathcal{K}}$ be the set of *loops*, i.e., the set of $\gamma \in \mathcal{K}$ with $\gamma(0) = \gamma(t_\gamma)$. Such a γ can also be considered as a function with domain $(-\infty,\infty)$ satisfying $\gamma(s) = \gamma(s+t_\gamma)$. For $r \in \mathbb{R}$, define the shift operator $\theta_r : \tilde{\mathcal{K}} \to \tilde{\mathcal{K}}$ by $t_{\theta_r \gamma} = t_\gamma$ and $\theta_r \gamma(s) = \gamma(s+r)$. We say that two loops γ and γ' are equivalent if for some r, $\gamma' = \theta_r \gamma$.

We write $[\gamma]$ for the equivalent class of γ. Let $\tilde{\mathcal{K}}_U$ be the set of *unrooted loops*, i.e., the equivalence classes in $\tilde{\mathcal{K}}$. Note that $\tilde{\mathcal{K}}_U$ is a metric space under the metric

$$d_U(\gamma,\gamma') := \inf_{r\in[0,t_\gamma]} d_{\mathcal{K}}(\theta_r\gamma,\gamma'),$$

and that a measure supported on $\tilde{\mathcal{K}}$ gives a measure on $\tilde{\mathcal{K}}_U$ by "forgetting the root," i.e., by considering the map $\gamma \mapsto [\gamma]$. If D is a domain, define $\tilde{\mathcal{K}}_U(D)$ to be the set of unrooted loops that lie entirely in D, i.e., $\gamma[0,t_\gamma] \subset D$.

The *Brownian loop measure* μ on $\tilde{\mathcal{K}}_U$ is defined by

$$\mu := \int_{\mathbb{C}} \frac{1}{t_\gamma} \tilde{\mu}(z,z) d\mathbf{A}(z) = \int_{\mathbb{C}} \int_0^\infty \frac{1}{2\pi t^2} \mu_{z,t}^{br} dt\, d\mathbf{A}(z), \tag{B.3}$$

where dA denotes the Lebesgue measure on $\mathbb{C}$. Informally, one can say that μ is the measure on unrooted loops obtained by averaging μ over the root (starting point):

$$"\mu(\gamma) = \frac{1}{t_\gamma}\int_0^{t_\gamma} \tilde{\mu}(\gamma(s),\gamma(s))ds." \tag{B.4}$$

If D is a domain of the complex plane, we define μ_D to be μ restricted to curves in $\tilde{\mathcal{K}}(D)$, i.e., μ_D is given by (B.3) with $\mathbb{C}$ replaced by D and $\tilde{\mu}(z,z)$ replaced by $\tilde{\mu}_D(z,z)$. The measure μ_D is conformally invariant in the sense that, given a conformal map $f: D \to D'$ and a set $A \in \tilde{\mathcal{K}}_U(D)$,

$$\mu_{D'}(f \circ A) = \mu_D(A). \tag{B.5}$$

$$\mu_{D'}(f \circ A) = \int_{D'} \frac{1}{t_{\hat{\gamma}}} \tilde{\mu}_{D'}(w,w)(f \circ A) d\mathbf{A}(w)$$

$$= \int_D \frac{1}{t_{f\circ\gamma}} \tilde{\mu}_D(z,z)(A)|f'(z)|^2 d\mathbf{A}(z) \tag{B.6}$$

$$= \int_D \frac{|f'(z)|^2}{t_{f\circ\gamma}} \tilde{\mu}_D(z,z)(A) d\mathbf{A}(z) \tag{B.7}$$

$$= \int_D \frac{1}{t_\gamma} \tilde{\mu}_D(z,z)(A) d\mathbf{A}(z) = \mu_D(A), \tag{B.8}$$

where in (B.6) we have used the conformal invariance on $\tilde{\mu}_D$ and in (B.8) we have used the fact that

$$\int_0^{t_\gamma} \frac{|f'(\gamma(u))|^2}{t_{f\circ\gamma}} du = 1.$$

(To understand the calculation above, it may help to remember (B.1) and (B.4) and to think of (B.7) as follows:

$$\text{``} \sum_{\gamma \in A} \int_0^{t_\gamma} \frac{|f'(\gamma(u))|^2}{t_{f \circ \gamma}} \tilde{\mu}(\gamma(u), \gamma(u)) \, du. \text{''}$$

Because the measure μ_D is the restriction of the measure μ to loops that stay in D, equation (B.5) is called *conformal restriction*. The Brownian loop measure is fully characterized by its conformal restriction property, in a way that we are now going to make precise. In order to do that, we need the following definitions.

Definition B.2 *A bounded open set is an* annular region *in the plane if it is conformally equivalent to some annulus $\{z : 1 < |z| < R\}$. In other words, A is a bounded open set such that $\mathbb{C} \setminus A$ has two connected components (that are not singletons).*

Definition B.3 *A measure on the set of self-avoiding loops in the plane is* nontrivial *if for some $0 < \delta < \Delta < \infty$, the mass of loops of diameter at least δ that stay in some disc of radius Δ is neither zero nor infinite. The bounded connected component of $\mathbb{C} \setminus A$ will be called the* hole *of A.*

For each such annular region A, define the set $\mathcal{U}_A$ of self-avoiding loops that stay in A and that have a non-zero index around the hole of A (or in other words, that disconnects the two connected components of ∂A). We will denote by $\mathcal{G}$ the σ-field generated by this family. This is the σ-field that we will work with.

The next theorem follows from the restriction property of the Brownian loop measure and [51].

Theorem B.4 *Up to a multiplicative constant, the Brownian loop measure μ is the only nontrivial measure on the set of self-avoiding loops in the plane satisfying conformal restriction. Furthermore, for any simply connected sets $\tilde{D} \subset D$ and any $z \in \tilde{D}$,*

$$\mu(\{\gamma : \gamma \subset D, \gamma \not\subset \tilde{D}, \gamma \text{ disconnects } z \text{ from } \partial D\}) = \frac{1}{5} \log f'(z) \qquad \text{(B.9)}$$

where f is the conformal map from $\tilde{D}$ to D such that $f(z) = z$ and $f'(z)$ is real and positive.

The proof of the theorem contains several parts, so we split it into several lemmas.

Lemma B.5 *Up to a multiplicative constant, there exists at most one nontrivial measure μ^* on the set of self-avoiding loops in the plane satisfying conformal* *restriction. Furthermore, for any simply connected sets $\tilde{D} \subset D$ and any $z \in \tilde{D}$,*

$$\mu^*(\{\gamma : \gamma \subset D, \gamma \not\subset \tilde{D}, \gamma \text{ disconnects } z \text{ from } \partial D\}) = c \log f'(z) \quad \text{(B.10)}$$

where f is the conformal map from $\tilde{D}$ to D such that $f(z) = z$, $f'(z)$ is real and positive, and $0 < c < \infty$.

Sketch of the proof. This lemma is essentially Proposition 3 of [51]; here we omit some of the technical details of the proof, but discuss the main ideas. Let $\tilde{f}$ be the unique conformal map from D to the the unit disc $\mathbb{U}$ such that $\tilde{f}(z) = 0$ and $\tilde{f}'(z)$ is real and positive. Let $U := \tilde{f}(\tilde{D}) \subset \mathbb{U}$ and note that U contains the origin 0. In the rest of the proof, we will work with simply connected subsets U of the unit disc $\mathbb{U}$ containing the origin, and we will prove that conformal restriction implies a version of (B.10) for such sets. The conformal invariance of μ^* then implies (B.10).

Let f_U be the unique conformal map from U to $\mathbb{U}$ such that $f_U(0) = 0$ and $f'_U(0) > 0$. Since $U \subset \mathbb{U}, f'_U(0) \geq 1$. Since $U = f^{-1}(\mathbb{U})$, the map f_U describes U fully. Let

$$A(f_U) = \mu^*(\gamma : 0 \in \bar{\gamma}, \gamma \subset \mathbb{U}, \gamma \not\subset U),$$

where $0 \in \bar{\gamma}$ means that γ disconnects the origin from infinity, as in the rest of this chapter. It is easy to convince oneself that

$$A(f_V \circ f_U) = \mu^*(\gamma : 0 \in \bar{\gamma}, \gamma \text{ intersects } \mathbb{U} \setminus U \text{ or } f_U^{-1}(\mathbb{U} \setminus V))$$
$$= \mu^*(\{\gamma : 0 \in \bar{\gamma}, \gamma \subset \mathbb{U}, \gamma \not\subset U\} \cup \{\gamma : 0 \in \bar{\gamma}, \gamma \subset U, \gamma \text{ int.} f_U^{-1}(\mathbb{U} \setminus V)\})$$
$$= A(f_U) + A(f_V).$$

Let's now consider a special class of domains: $U_t := \mathbb{U} \setminus [r_t, 1)$, where r_t is the positive real number such that $f'_{U_t}(0) = e^t$. (We won't need the exact expression for r_t.) It is a well-know fact that, if Z is a planar Brownian motion that started from the origin and stopped at its first-exit time T of U_t, then $\log f'_{U_t}(0) = -E \log(|Z_T|)$. This implies that $t = E(-\log |Z_T|) \in (0, \infty)$.

Note that $f_{U_t} \circ f_{U_s} = f_{U_\tau}$ for some $\tau \in (0, \infty)$; but since $(f_{U_t} \circ f_{U_s})' = e^{t+s}$, it follows that $f_{U_t} \circ f_{U_s} = f_{U_{t+s}}$, so the family $(f_{U_t})_{t \geq 0}$ is a semi-group. Hence, $t \mapsto A(f_{U_t})$ is a non-decreasing function from $(0, \infty)$ to $(0, \infty)$ such that

$$A(f_{U_{t+s}}) = A(f_{U_t} \circ f_{U_s}) = A(f_{U_t}) + A(f_{U_s}).$$

This implies that $A(f_{U_t}) = ct = c \log f'_{U_t}(0)$ for some constant $0 < c < \infty$.

For each $\theta \in [0, 2\pi)$ and $t > 0$, let $U_{t,\theta} := \setminus [r_t e^{i\theta}, e^{i\theta})$; clearly, $f'_{U_{t,\theta}}(0) = e^t$ and $A(f_{U_{t,\theta}}) = A(f_{U_t}) = ct$. Let S denote the semi-group of conformal maps generated by the family $(f_{U_{t,\theta}}, t > 0, \theta \in [0, 2\pi))$ (i.e., the set of finite compositions of those maps). If we take $f = f_{U_{t,\theta}} \circ f_{U_{s,\alpha}} \in S$, then $f'(0) = f'_{U_{t,\theta}}(0)f'_{U_{s,\alpha}}(0) = e^{t+s}$, that is, $c \log f'(0) = c(t+s)$. On the other hand, $A(f) = A(f_{U_{t,\theta}}) + A(f_{U_{s,\alpha}}) = ct + cs$, therefore $A(f) = c \log f'(0)$.

The theory developed by Loewner (for the proof of the Bieberbach conjecture in the first highly nontrivial case of the third coefficient) implies that S is "dense" in the class of conformal maps f_U from some simply connected domain $U \subset \mathbb{U}$ onto $\mathbb{U}$ in the sense that, for any f_U, one can find a sequence f_{U_n} in S such that U_n is an increasing in n (in the sense of inclusion) family of domains, such that $\cup_n U_n = U$. Therefore, $f'_{U_n}(0) \to f'_U(0)$ as $n \to \infty$. Furthermore, because a loop is a compact subset of the complex plane, it exits U if and only if it exits U_n for every n. Thus, for any simply connected subset U of $\mathbb{U}$ containing the origin, we have that

$$A(f_U) = \lim_{n \to \infty} A(f_{U_n}) = c \lim_{n \to \infty} \log f'_{U_n}(0) = c \log f'_U(0).$$

The conformal invariance of μ^* concludes the proof of (B.10). The uniqueness of μ^* follows from rather simple measure-theoretical considerations that we omit. (The interested reader can find the details in [51].) □

In the next lemmas, as in the rest of this chapter, we let $\bar{\gamma}$ denote the "filling" of the loop γ, $B_{z,a}$ be a disc of radius a around z and $\gamma \not\subset B_{z,\delta}$ indicate that the image of γ is not fully contained in $B_{z,\delta}$.

Lemma B.6 *Let $z \in \mathbb{C}$, then*

$$\mu(\gamma : z \in \bar{\gamma}, \delta \leq \mathrm{diam}(\gamma) < R) = \mu(\gamma : z \in \bar{\gamma}, \gamma \not\subset B_{z,\delta}, \gamma \subset B_{z,R}).$$

Proof. Since $\mu(\gamma : \mathrm{diam}(\gamma) = R) = 0$, we have that

$$\mu(\gamma : z \in \bar{\gamma}, \delta \leq \mathrm{diam}(\gamma) < R) - \mu(\gamma : z \in \bar{\gamma}, \gamma \not\subset B_{z,\delta}, \gamma \subset B_{z,R})$$
$$= \mu(\gamma : z \in \bar{\gamma}, \mathrm{diam}(\gamma) \geq \delta, \gamma \subset B_{z,\delta}) - \mu(\gamma : z \in \bar{\gamma}, \mathrm{diam}(\gamma) \geq R, \gamma \subset B_{z,R}),$$

where the last two terms are identical because of the scale invariance of μ. □

Lemma B.7 *For any $z \in \mathbb{C}$ and any $r > 0$,*

$$\mu\left(\gamma : z \in \bar{\gamma}, 1 \leq \mathrm{diam}(\gamma) < e^r\right) = \mu\left(\gamma : z \in \bar{\gamma}, 1 \leq t_\gamma < e^{2r}\right) = \frac{r}{5}.$$

Proof. Part of the lemma follows from a simple computation:

$$\mu(\gamma : 0 \in \bar{\gamma}, 1 \leq t_\gamma < e^{2r}) = \int_{\mathbb{C}} \int_1^{e^{2r}} \frac{1}{2\pi t^2} \mu_{z,t}^{br}(\{\gamma : 0 \in \bar{\gamma}\}) \, dt \, dA(z)$$

$$= \int_{\mathbb{C}} \int_1^{e^{2r}} \frac{1}{2\pi t^2} \mu_{0,t}^{br}(\{\gamma : z \in \bar{\gamma}\}) \, dt \, dA(z)$$

$$= \int_1^{e^{2r}} \frac{1}{2\pi t^2} \mathbb{E}_{0,t}^{br} \left(\int_{\mathbb{C}} \mathbb{1}_{\{\gamma : z \in \bar{\gamma}\}} \, dA(z) \right) dt$$

$$= \int_1^{e^{2r}} \frac{1}{2\pi t} \mathbb{E}_{0,1}^{br} \left(\int_{\mathbb{C}} \mathbb{1}_{\{\gamma : z \in \bar{\gamma}\}} \, dA(z) \right) dt,$$

where $\mathbb{E}_{0,t}^{br}$ denotes expectation with respect to a complex Brownian bridge of time length t started at the origin, and where, in the last equality, we have used the fact that

$$\mathbb{E}_{0,t}^{br} \left(\int_{\mathbb{C}} \mathbb{1}_{\{\gamma : z \in \bar{\gamma}\}} \, dA(z) \right) = t \, \mathbb{E}_{0,1}^{br} \left(\int_{\mathbb{C}} \mathbb{1}_{\{\gamma : z \in \bar{\gamma}\}} \, dA(z) \right)$$

because of scaling. The expected area of a "filled-in" Brownian bridge, computed in [28], is

$$\mathbb{E}_{0,1}^{br} \left(\int_{\mathbb{C}} \mathbb{1}_{\{\gamma : z \in \bar{\gamma}\}} \, dA(z) \right) = \frac{\pi}{5},$$

so that

$$\mu(\gamma : z \in \bar{\gamma}, 1 \leq t_\gamma < e^{2r}) = \frac{r}{5}. \tag{B.11}$$

Combined with equation (B.10), equation (B.11) implies that

$$\frac{\mu(\gamma : z \in \bar{\gamma}, 1 \leq \mathrm{diam}(\gamma) < e^r)}{\mu(\gamma : z \in \bar{\gamma}, 1 \leq t_\gamma < e^{2r})} = 5c,$$

independently of r. Therefore, to conclude the proof of the lemma, it suffices to show that

$$\lim_{r \to \infty} \frac{\mu(\gamma : z \in \bar{\gamma}, 1 \leq \mathrm{diam}(\gamma) < e^r)}{\mu(\gamma : z \in \bar{\gamma}, 1 \leq t_\gamma < e^{2r})} = 1.$$

Let $A := \{\gamma : z \in \bar{\gamma}, 1 \leq \mathrm{diam}(\gamma) < e^r\}$, $B \equiv \{\gamma : z \in \bar{\gamma}, 1 \leq t_\gamma < e^{2r}\}$, and define the disjoint sets

- $B_1 := \{\gamma : z \in \bar{\gamma}, \mathrm{diam}(\gamma) < e^{-r}, 1 \leq t_\gamma < e^{2r}\} \subset B$,
- $B_2 := \{\gamma : z \in \bar{\gamma}, e^{-r} \leq \mathrm{diam}(\gamma) < 1, 1 \leq t_\gamma < e^{2r}\} \subset B$,
- $B_3 := \{\gamma : z \in \bar{\gamma}, 1 \leq \mathrm{diam}(\gamma) < e^r, 1 \leq t_\gamma < e^{2r}\} = A \cap B$,
- $B_4 := \{\gamma : z \in \bar{\gamma}, e^r \leq \mathrm{diam}(\gamma) < e^{2r}, 1 \leq t_\gamma < e^{2r}\} \subset B$,
- $B_5 := \{\gamma : z \in \bar{\gamma}, \mathrm{diam}(\gamma) \geq e^{2r}, 1 \leq t_\gamma < e^{2r}\} \subset B$,

and the disjoint sets

- $A_1 := \{\gamma : z \in \bar{\gamma}, 1 \leq \operatorname{diam}(\gamma) < e^r, t_\gamma < e^{-2r}\} \subset A,$
- $A_2 := \{\gamma : z \in \bar{\gamma}, 1 \leq \operatorname{diam}(\gamma) < e^r, e^{-2r} \leq t_\gamma < 1\} \subset A,$
- $A_3 := \{\gamma : z \in \bar{\gamma}, 1 \leq \operatorname{diam}(\gamma) < e^r, 1 \leq t_\gamma < e^{2r}\} = A \cap B,$
- $A_4 := \{\gamma : z \in \bar{\gamma}, 1 \leq \operatorname{diam}(\gamma) < e^r, e^{2r} \leq t_\gamma < e^{4r}\} \subset A,$
- $A_5 := \{\gamma : z \in \bar{\gamma}, 1 \leq \operatorname{diam}(\gamma) < e^r, t_\gamma \geq e^{4r}\} \subset A.$

We clearly have that $\mu(A) = \sum_{i=1}^{5} \mu(A_i)$ and $\mu(B) = \sum_{i=1}^{5} \mu(B_i)$.

Note that A_4 can be obtained from B_2 by scaling each loop γ in B_2 by a factor of e^r and scaling t_γ by a factor e^{2r}. Similarly, A_2 can be obtained from B_4 by scaling each loop γ in B_4 by a factor of e^{-r} and scaling t_γ by a factor e^{-2r}. Because of the scaling properties of Brownian motion, those transformations are μ-measure preserving; thus $\mu(A_4) = \mu(B_2)$ and $\mu(A_2) = \mu(B_4)$.

To conclude the proof, we show that the μ-measures of B_1, B_5, A_1, A_5 go to zero as $r \to \infty$. For $\mu(B_5)$, we have the upper bound

$$\mu(B_5) = \int_{\mathbb{C}} \int_1^{e^{2r}} \frac{1}{2\pi t^2} \mu_{z,t}^{br}(\gamma : 0 \in \bar{\gamma}, \operatorname{diam}(\gamma) \geq e^{2r}) \, dt \, d\mathbf{A}(z)$$

$$\leq \pi e^{4r} \int_1^{e^{2r}} \frac{1}{2\pi t^2} \mu_{0,t}^{br}\left(\gamma : \operatorname{diam}(\gamma) \geq e^{2r}\right) dt$$

$$+ \sum_{n=1}^{\infty} \pi \left[\left(e^{2r} + n\right)^2 - \left(e^{2r} + n - 1\right)^2\right]$$

$$\int_1^{e^{2r}} \frac{1}{2\pi t^2} \mu_{0,t}^{br}\left(\gamma : \operatorname{diam}(\gamma) \geq e^{2r} + n - 1\right) dt,$$

where the first term in the upper bound comes from rooted loops with root within the disc of radius e^{2r} centered at the origin, and the other terms come from rooted loops with roots inside one of a family of concentric annuli around that disc.

Plugging (5.20) from the proof of Lemma 5.2.13 into the upper bound for $\mu(B_5)$ gives

$$\mu(B_5) \leq 2e^{4r} \int_1^{e^{2r}} t^{-2} \exp\left(-\frac{e^{4r}}{288t}\right) dt$$

$$+ 2 \sum_{n=1}^{\infty} (2e^{2r} + 2n - 1) \int_1^{e^{2r}} t^{-2} \exp\left(-\frac{(e^{2r} + n - 1)^2}{288t}\right) dt$$

$$\leq \operatorname{const} \exp\left(-\frac{e^{2r}}{288}\right) \to 0 \text{ as } r \to \infty.$$

If we now let W_s and $B_s := W_s - sW_t$, $s \in [0,t]$, denote standard two-dimensional Brownian motion and Brownian bridge, respectively, we have that $\max_{s\in[0,t]} |B_s| \geq |W_{1/2} - \frac{1}{2}W_t|$. Combining this observation with the Markov property of Brownian motion and the Gaussian distribution of its increments leads to the following upper bound for $\mu(B_1)$:

$$\mu(B_1) = \int_{\mathbb{C}} \int_1^{e^{2r}} \frac{1}{2\pi t^2} \mu_{z,t}^{br}\left(\gamma : 0 \in \bar{\gamma}, \mathrm{diam}(\gamma) < e^{-r}\right) dt\, d\mathbf{A}(z)$$

$$\leq \pi e^{-2r} \int_1^{e^{2r}} \Pr\left(|W_t - 2W_{1/2}| \leq e^{-r}\right) dt$$

$$\leq \mathrm{const}\, e^{-2r} \to 0 \text{ as } r \to \infty.$$

Finally, let $B_5' \equiv \{\gamma : z \in \bar{\gamma}, e^{2r} \leq \mathrm{diam}(\gamma) < e^{3r}, t_\gamma < e^{2r}\} \subset B_5$ and $B_1' \equiv \{\gamma : z \in \bar{\gamma}, e^{-2r} \leq \mathrm{diam}(\gamma) < e^{-r}, t_\gamma \geq 1\} \subset B_1$. Note that B_5' can be obtained from A_1 by scaling each loop γ in A_1 by a factor of e^{2r} and scaling t_γ by a factor e^{4r}. Similarly, B_1' can be obtained from A_5 by scaling each loop γ in A_5 by a factor of e^{-2r} and scaling t_γ by a factor e^{-4r}. Because of the scaling properties of Brownian motion, those transformations are μ-measure preserving; thus $\mu(A_1) = \mu(B_5') \leq \mu(B_5)$ and $\mu(A_5) = \mu(B_1') \leq \mu(B_1)$, concluding the proof of the lemma.

Proof of Theorem B.4. We have already seen that the Brownian loop measure μ satisfies conformal restriction; Lemma B.5 then implies that μ is unique, up to a multiplicative constant, and satisfies (B.10) for some constant $0 < c < \infty$. Combining (B.10) with Lemmas B.6 and B.7 gives

$$cr = \mu(\gamma : z \in \bar{\gamma}, \gamma \not\subset B_{z,1}, \gamma \subset B_{z,e^r})$$
$$= \mu\left(\gamma : z \in \bar{\gamma}, 1 \leq \mathrm{diam}(\gamma) < e^r\right)$$
$$= \mu\left(\gamma : z \in \bar{\gamma}, 1 \leq t_\gamma < e^{2r}\right) = \frac{r}{5}. \qquad \square$$

We conclude this appendix with two simple but useful lemmas.

Lemma B.8 *Let $z \in \mathbb{C}$, then*

$$\mu(\gamma : z \in \bar{\gamma}, \delta \leq \mathrm{diam}(\gamma) < R) = \frac{1}{5} \log \frac{R}{\delta}.$$

Proof. This is an immediate consequence of Lemma B.6 and Theorem B.4. $\qquad \square$

Lemma B.9 *Let $z \in \mathbb{C}$ and $k \in \mathbb{Z} \setminus \{0\}$, then*

$$\mu^{loop}(\gamma : \gamma \text{ has winding number } k \text{ around } z, \delta \leq \mathrm{diam}(\gamma) < R) = \frac{1}{2\pi^2 k^2} \log \frac{R}{\delta}.$$

Proof. It is easy to check that the measure on loops surrounding the origin induced by μ, but restricted to loops that wind k times around a given $z \in \mathbb{C}$, satisfies the conformal restriction property. Therefore, following the proofs of lemmas B.6 and B.5, we obtain

$$\mu(\gamma : \gamma \text{ has winding number } k \text{ around } z, \delta \leq \operatorname{diam}(\gamma) < R)$$
$$= \mu(\gamma : \gamma \text{ has winding number } k \text{ around } z, \gamma \not\subset B_{z,\delta}, \gamma \subset B_{z,R})$$
$$= c_k \log \frac{R}{\delta},$$

for some positive constant $c_k < \infty$. In order to find the constants c_k, we can proceed as in the proof of Theorem B.4, using the fact that the expected area of a "filled-in" Brownian loop winding k times around the origin was computed in [28] and is equal to $1/2\pi k^2$ for $k \in \mathbb{Z} \setminus \{0\}$ (and $\pi/30$ for $k = 0$). $\square$

Acknowledgments

It is a pleasure to thank the organizers of the program *Disordered Systems, Random Spatial Processes and Some Applications* (IHP, Paris, 5 January – 3 April, 2015) for their invitation to give a series of lectures, which resulted in these notes. Part of the content of these notes derives from work conducted with Alberto Gandolfi and Matthew Kleban, and part of it profited from discussions and previous work with Erik Broman; I wish to thank all of them for the enjoyable collaborations. Several discussions with Tim van de Brug and Marcin Lis helped to improve the presentation of the material. My gratitude also goes to Claire Berenger and all the IHP staff for their friendly efficiency, and to an anonymous referee for a useful suggestion. The work of the author was supported in part by the Netherlands Organization for Scientific Research (NWO) through grant Vidi 639.032.916.

References

[1] M. Aizenman, *Geometric analysis of ϕ^4 fields and ising models. Parts I and II*, Comm. Math. Phys. 86, 1–48 (1982).
[2] M. Aizenman, *The geometry of critical percolation and conformal invariance*, in *STATPHYS 19, Proceedings Xiamen 1995* (H. Bai-lin, ed.). World Scientific (1995).
[3] M. Aizenman, *Scaling limit for the incipient spanning clusters*, in *Mathematics of Multiscale Materials; the IMA Volumes in Mathematics and its Applications* (K. Golden, G. Grimmett, R. James, G. Milton and P. Sen, eds.). New York: Springer (1998).
[4] M. Aizenman and A. Burchard, *Hölder regularity and dimension bounds for random curves*, Duke Math. J. 99, 419–453 (1999).

[5] M. Aizenman, A. Burchard, C. M. Newman and D. B. Wilson, *Scaling limits for minimal and random spanning trees in two dimensions*, Ran. Structures Alg. 15, 316–367 (1999).
[6] M. Aizenman, B. Duplantier and A. Aharony, *Connectivity exponents and the external perimeter in 2D independent percolation*, Phys. Rev. Lett. 83, 1359–1362 (1999).
[7] A. A. Belavin, A. M. Polyakov and A. B. Zamolodchikov, *Infinite conformal symmetry of critical fluctuations in two dimensions*, J. Stat. Phys. 34, 763–774 (1984).
[8] A. A. Belavin, A. M. Polyakov and A. B. Zamolodchikov, *Infinite conformal symmetry in two-dimensional quantum field theory*, Nucl. Phys. B. 241, 333–380 (1984).
[9] I. Benjamini and O. Schramm, *Conformal invariance of Voronoi percolation*, Comm. Math. Phys. 197, 75–107 (1998).
[10] M. Bauer and D. Bernard, *2D growth processes: SLE and Loewner chains*, Physics Reports. 432, 115–221 (2006).
[11] E. I. Broman, F. Camia, *Universal behavior of connectivity properties in fractal percolation models*, Electron. J. Probab. 15, 1394–1414 (2010).
[12] D. C. Brydges, J. Fröhlich and A. D. Sokal, *The random-walk representation of classical spin systems and correlation inequalities. II. The skeleton inequalities*, Comm. Math. Phys. 91, 117–139 (1983).
[13] D. C. Brydges, J. Fröhlich and T. Spencer, *The random walk representation of classical spin systems and correlation inequalities*, Comm. Math. Phys. 83, 123–150 (1982).
[14] F. Camia, *Off-Criticality and the Massive Brownian Loop Soup*. (2013). Preprint available at arXiv:1309.6068.
[15] F. Camia, A. Gandolfi and M. Kleban, *Conformal correlation functions in the Brownian loop soup*, Nucl. Phys. B. 902, 483–507 (2016).
[16] F. Camia, C. Garban and C. M. Newman, *Planar Ising magnetization field I: Uniqueness of the critical scaling limit*, Ann. Probab. 43, 528–571 (2015).
[17] F. Camia, M. Joosten and R. Meester, *Trivial, critical and near-critical scaling limits of two-dimensional percolation*, J. Stat. Phys. 137, 57–69 (2009).
[18] T. van der Brug, F. Camia and M. Lis, *Conformal fields from Brownian loops*. In preparation (2017).
[19] F. Camia and C. M. Newman, *Two-dimensional critical percolation: the full scaling limit*, Comm. Math. Phys. 268, 1–38 (2006).
[20] F. Camia and C. M. Newman, *Critical percolation exploration path and SLE_6: a proof of convergence*, Probab. Theory Related Fields. 139, 473–519 (2007).
[21] F. Camia and C. M. Newman, *SLE(6) and CLE(6) from critical percolation*, in *Probability, Geometry and Integrable Systems*, 103–130, Math. Sci. Res. Inst. Publ., Cambridge: Cambridge University Press, 55, (2008).
[22] D. Chelkak, H. Duminil-Copin, C. Hongler, A. Kemppainen and S. Smirnov, *Convergence of Ising interfaces to Schramm's SLE curves*, Comptes Rendu Mathematique. 352, 157–161 (2014).
[23] P. Di Francesco, P. Mathieu and D. Sénéchal, *Conformal field theory: Graduate Texts in Contemporary Physics*, New York: Springer-Verlag, (1997).
[24] E. B. Dynkin, *Markov processes as a tool in field theory*, J. of Funct. Anal. 50, 167–187 (1983).
[25] E. B. Dynkin, *Gaussian and non-Gaussian random fields associated with Markov processes*, J. of Funct. Anal. 55, 344–376 (1984).
[26] R. Fernández, J. Fröhlich and A. D. Sokal, *Random Walks, Critical Phenomena, and Triviality in Quantum Field Theory*. Berlin: Springer-Verlag (1992).
[27] B. Freivogel and M. Kleban, *A conformal field theory of eternal inflation?*, Journal of High Energy Physics, 0912, 019 (2009).

[28] C. Garban and J. A. Trujillo Ferreras, *The expected area of the filled planar Brownian loop is $\pi/5$*, Comm. Math. Phys. 264, 797–810 (2006).

[29] S. Janson, *Bounds of the distribution of extremal values of a scanning process*, Stochastic Processes Applications 18, 313–328 (1984).

[30] W. Kager and B. Nienhuis, *A guide to stochastic löwner evolution and its applications*, J. Phys. A. 115, 1149–1229 (2004).

[31] R. Kenyon, *Conformal invariance of domino tiling*, J. Math. Phys. 41, 1338–1363 (2000).

[32] R. Kenyon, *Conformal invariance of domino tiling*, Ann. Probab. 28, 759–795 (2000).

[33] J. Komlós, P. Major and G. Tusnády, *An approximation of partial sums of independent RV's and the sample DF. I.*, Z. Wahr. 32, 111–131 (1975).

[34] G. F. Lawler, *Conformally Invariant Processes in the Plane*, Mathematical Surveys and Monographs. 114. Providence, RI: American Mathematical Society, (2005).

[35] G. F. Lawler and J. A. Trujillo Ferreras, *Random walk loop soup*, Transactions of the American Mathematical Society. 359, 767–787 (2006).

[36] G .F. Lawler and W. Werner, *The Brownian loop soup*, Probab. Theory Relat. Fields. 128, 565–588 (2004).

[37] Y. Le Jan, *Markov loops and renormalization*, Ann. Probab. 38, 1280–1319 (2010).

[38] Y. Le Jan, *Markov paths, loops and fields*, in Lecture Notes in Mathematics, Volume 2026, Ecole d'Eté de Probabilité de St. Flour. Berlin: Springer, (2012).

[39] S. Nacu and W. Werner, *Random soups, carpets and fractal dimensions*, J. London Math. Soc. 83(3), 789–809 (2011).

[40] P. Nolin and W. Werner, *Asymmetry of near-critical percolation interfaces*, J. Amer. Math. Soc. 22, 797–819 (2009).

[41] A. M. Polyakov, *Conformal symmetry of critical fluctuations*, JETP Letters. 12, 381–383 (1970).

[42] O. Schramm, *Scaling limits of loop-erased random walks and uniform spanning trees*, Israel Journal of Mathematics. 118, 221–288 (2000).

[43] O. Schramm, S. Sheffield, D. B. Wilson, *Conformal radii for conformal loop ensembles*, Comm. Math. Phys. 288, 43–53 (2009).

[44] S. Sheffield, *Exploration trees and conformal loop ensembles*, Duke Math. J. 147, 79–129 (2009).

[45] S. Sheffield and W. Werner, *Conformal loop ensembles: the Markovian characterization and the loop-soup construction*, Ann. Math. 176, 1827–1917 (2012).

[46] S. Smirnov, *Critical percolation in the plane: conformal invariance, Cardy's formula, scaling limits*, C. R. Acad. Sci.–Ser. I–Math. 333, 239–244 (2001).

[47] K. Symanzik, *Euclidean quantum field theory*, in Local Quantum Theory, Proceedings of the International School of Physics "Enrico Fermi," Course 45 (R. Jost ed.), New York: Academic Press, 152–223 (1969).

[48] A. -S. Sznitman, *Topics in Occupation Times and Gaussian Free Field*, Zürich: Lectures in Advanced Mathematics. Zürich: European Mathematical Society Publishing House, (2012).

[49] W. Werner, *SLEs as boundaries of clusters of Brownian loops*, C. R. Acad. Sci.–Ser. I–Math. 337, 481–486 (2003).

[50] W. Werner, *Some recent aspects of random conformally invariant systems*, in Les Houches Scool Proceedings: Session LXXXII, Mathematical Statistical Physics (A. Bovier, F. Dunlop, A. van Enter and J. Dalibard eds.). Oxford: Elsevier, 57–98 (2006).

[51] W. Werner, *The conformally invariant measure on self-avoiding loops*, J. Amer. Math. Soc. 21, 137–169 (2008).

6

The Brownian Web, the Brownian Net, and their Universality

EMMANUEL SCHERTZER, RONGFENG SUN
AND JAN M. SWART

6.1 Introduction

The Brownian web originated from the work of Arratia's Ph.D. thesis [1], where he studied diffusive scaling limits of coalescing random walk paths starting from everywhere on $\mathbb{Z}$, which can be seen as the spatial genealogies of the population in the dual voter model on $\mathbb{Z}$. Arratia showed that the collection of coalescing random walks converge to a collection of coalescing Brownian motions on $\mathbb{R}$, starting from every point on $\mathbb{R}$ at time 0. Subsequently, Arratia [2] attempted to generalize his result by constructing a system of coalescing Brownian motions starting from everywhere in the space-time plane $\mathbb{R}^2$, which would be the scaling limit of coalescing random walk paths starting from everywhere on $\mathbb{Z}$ at every time $t \in \mathbb{R}$. However, the manuscript [2] was never completed, even though fundamental ideas have been laid down. This topic remained dormant until Tóth and Werner [99] discovered a surprising connection between the one-dimensional space-time coalescing Brownian motions that Arratia tried to construct, and an unusual process called the *true self-repelling motion*, which is repelled by its own local time profile. Building on ideas from [2], Tóth and Werner [99] gave a construction of the system of space-time coalescing Brownian motions, and then used it to construct the true self-repelling motion.

On the other hand, Fontes, Isopi, Newman and Stein [37] discovered that this system of space-time coalescing Brownian motions also arises in the study of aging and scaling limits of one-dimensional spin systems. To establish weak convergence of discrete models to the system of coalescing Brownian motions, Fontes et al. [38, 40] introduced a topology that the system of coalescing Brownian motions starting from every space-time point can be realized as a random variable taking values in a Polish space, and they named this random variable *the Brownian web*. An extension to the Brownian web was later introduced by the authors in [86], and independently by Newman, Ravishankar and Schertzer in [73]. This object was named *the Brownian net* in [86], where the coalescing paths in the Brownian web are also allowed to branch. To counter the effect of instantaneous coalescence, the branching occurs at an effectively "infinite" rate.

The Brownian web and net have very interesting properties. Their construction is nontrivial due to the uncountable number of starting points in space-time. Coalescence allows one to reduce the system to a countable number of starting points. In fact, the collection of coalescing paths starting from every point on $\mathbb{R}$ at time 0 immediately becomes locally finite when time becomes positive, similar to the phenomenon of *coming down from infinity* in Kingman's coalescent (see e.g., [10]). In fact, the Brownian web can be regarded as the spatial analogue of Kingman's coalescent, with the former arising as the limit of genealogies of the voter model on $\mathbb{Z}$, and the latter arising as the limit of genealogies of the voter model on the complete graph. The key tool in the analysis of the Brownian web, as well as the Brownian net, is its self-duality, similar to the self-duality of critical bond percolation on $\mathbb{Z}^2$. Duality allows one to show that there exist random space-time points where multiple paths originate, and one can give a complete classification of these points. The Brownian web and net also admit a coupling, where the web can be constructed by sampling paths in the net, and conversely, the net can be constructed from the web by Poisson marking a set of "pivotal" points in the web and turning these into points where paths can branch. The latter construction is similar to the construction of scaling limits of near-critical planar percolation from that of critical percolation [19, 48, 49].

The Brownian web and net give rise to a new universality class. In particular, they are expected to arise as the universal scaling limits of one-dimensional interacting particle systems with coalescence, respectively branching-coalescence. One such class of models are population genetic models with resampling and selection, whose spatial genealogies undergo branching and coalescence. Establishing weak convergence to the Brownian web or net can also help in the study of the discrete particle systems themselves. Related models which have been shown to converge to the Brownian web include coalescing random walks [72], succession lines in Poisson trees [36, 23, 46] and drainage network type models [17, 22, 80]. Interesting connections with the Brownian web and net have also emerged from many unexpected sources, including supercritical oriented percolation on $\mathbb{Z}^{1+1}$ [6], planar aggregation models [76, 77], true self-avoiding random walks on $\mathbb{Z}$ [94, 99], random matrix theory [101, 100], and also one-dimensional random walks in i.i.d. space-time random environments [92]. There are also close parallels between the Brownian web and the scaling limit of critical planar percolation, which are the only known examples of two-dimensional *black noise* [95, 96, 87, 33].

The goal of this article is to give an introduction to the Brownian web and net, their basic properties, and how they arise in the scaling limits of one-dimensional interacting particle systems with branching and coalescence. We will focus on the key ideas, while referring many details to the literature. Our emphasis is naturally biased toward our own research. However, we will also briefly survey related work, including the many interesting connections mentioned above. We have left out many other closely related studies, including diffusion-limited reactions [28, 8] where a dynamic phase transition is observed for branching-coalescing random walks, the propagation of cracks in a sheet [27], rill erosion [31] and the directed Abelian Sandpile Model [25], quantum spin chains [60], etc., which all lie within the general framework of non-equilibrium critical phenomena discussed in the physics surveys [79, 53].

The rest of this article is organized as follows. In Section 6.2, we will construct and give a characterization of the Brownian web and study its properties. In Section 6.3, we do the same for the Brownian net. In Section 6.4, we introduce a coupling between the Brownian web and net and show how one can be constructed from the other. In Section 6.5, we will explain how the Brownian web and net can be used to construct the scaling limits of one-dimensional random walks in i.i.d. space-time random environments. In Section 6.6, we formulate convergence criteria for the Brownian web, which are then applied to coalescing random walks. We will also discuss strategies for proving convergence to the Brownian net. In Section 6.7, we survey other interesting models and results connected to the Brownian web and net. Lastly, in Section 6.8, we conclude with some interesting open questions.

6.2 The Brownian Web

The Brownian web is best motivated by its discrete analogue, the collection of discrete time coalescing simple symmetric random walks on $\mathbb{Z}$, with one walker starting from every site in the space-time lattice $\mathbb{Z}^2_{\text{even}} := \{(x,n) \in \mathbb{Z}^2 : x+n \text{ is even}\}$. The restriction to the sublattice $\mathbb{Z}^2_{\text{even}}$ is necessary due to parity. Figure 6.1 illustrates a graphical construction, where from each $(x,n) \in \mathbb{Z}^2_{\text{even}}$ an independent arrow is drawn from (x,n) to either $(x-1,n+1)$ or $(x+1,n+1)$ with probability $1/2$ each, determining whether the walk starting at x at time n should move to $x-1$ or $x+1$ at time $n+1$. The objects of interest for us are the collection of upward random walk paths (obtained by following the arrows) starting from every space-time lattice point. The question is:

Q.1 What is the diffusive scaling limit of this collection of coalescing random walk paths if space and time are scaled by $1/\sqrt{n}$ and $1/n$ respectively?

Intuitively, it is not difficult to see that the limit should be a collection of coalescing Brownian motions, starting from everywhere in the space-time plane $\mathbb{R}^2$. This is what we will call the *Brownian web*. However, a conceptual difficulty arises, namely that we need to construct the joint realization of *uncountably* many

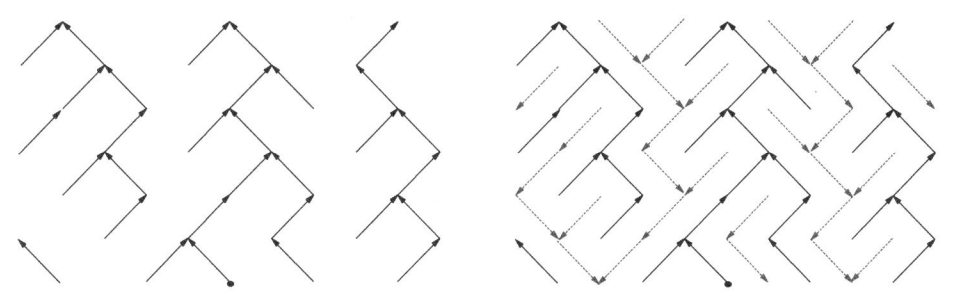

Figure 6.1. Discrete space-time coalescing random walks on $\mathbb{Z}^2_{\text{even}}$, and its dual on $\mathbb{Z}^2_{\text{odd}}$.

Brownian motions. Fortunately it turns out that coalescence allows us to reduce the construction to only a countable collection of coalescing Brownian motions.

Note that in Figure 6.1, we have also drawn a collection of downward arrows connecting points in the odd space-time lattice $\mathbb{Z}^2_{\text{odd}} := \{(x,n) \in \mathbb{Z}^2 : x + n \text{ is odd}\}$, which are dual to the upward arrows by the constraint that the upward and backward arrows do not cross each other. This is the same duality as that for planar bond percolation, and the collection of upward arrows uniquely determines the downward arrows, and vice versa. The collection of downward arrows determines a collection of coalescing random walk paths running backward in time, with one walker starting from each site in $\mathbb{Z}^2_{\text{odd}}$. We may thus strengthen **Q.1** to the following:

Q.2 What is the diffusive scaling limit of the joint realization of the collection of forward and backward coalescing random walk paths?

Observe that the collection of backward coalescing random walk paths has the same distribution as the forward collection, except for a rotation in space-time by 180° and a lattice shift. Therefore, the natural answer to **Q.2** is that the limit consists of two collections of coalescing Brownian motions starting from everywhere in space-time – one running forward in time and the other backward – and the two collections are equally distributed except for a time-reversal. This is what we will call the (forward) *Brownian web* and the *dual* (backward) *Brownian web*.

In the discrete system, we observe that the collection of forward and the collection of backward coalescing random walk paths uniquely determine each other by the constraint that forward and backward paths cannot cross. It is natural to expect the same for their continuum limits, namely that the Brownian web and the dual Brownian web almost surely uniquely determine each other by the constraint that their paths cannot cross.

The heuristic considerations above, based on discrete approximations, outline the key properties that we expect the Brownian web to satisfy and provide a guide for our analysis.

Before proceeding to a proper construction of the Brownian web and establishing its basic properties, we first define a suitable Polish space in which the Brownian web takes its value. This will be essential to prove weak convergence to the Brownian web.

6.2.1 The Space of Compact Sets of Paths

Following Fontes et al. [40], we regard the collection of colaescing Brownian motions as a *set* of space-time paths, which can be shown to be almost surely relatively compact if space and time are suitably compactified. It is known that given a Polish space E (the space of paths in our case), the space of compact subsets of E, equipped with the induced Hausdorff topology, is a Polish space itself. Therefore a natural space for the Brownian web is the space of *compact sets of paths* (after compactifying space and time), with the Brownian web taken to be the almost sure closure of the set of colaescing Brownian motions. This *paths topology* was inspired by a similar topology proposed by Aizenman [3] to study two-dimensional percolation configurations as closed sets of curves, called the *percolation web*, which was then studied rigorously by Aizenman and Burchard in [4]. We now give the details.

We first compactify $\mathbb{R}^2$. Let R_c^2 denote the completion of the space-time plane $\mathbb{R}^2$ w.r.t. the metric

$$\rho((x_1,t_1),(x_2,t_2)) = |\tanh(t_1) - \tanh(t_2)| \vee \left| \frac{\tanh(x_1)}{1+|t_1|} - \frac{\tanh(x_2)}{1+|t_2|} \right|. \qquad (6.1)$$

Note that R_c^2 can be identified with the continuous image of $[-\infty,\infty]^2$ under a map that identifies the line $[-\infty,\infty] \times \{\infty\}$ with a single point $(*,\infty)$, and the line $[-\infty,\infty] \times \{-\infty\}$ with the point $(*,-\infty)$ (see Figure 6.2).

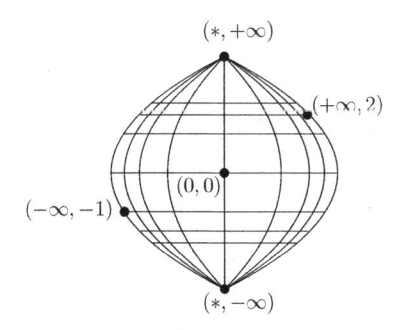

Figure 6.2. The compactification R_c^2 of $\mathbb{R}^2$.

A path π in R_c^2, whose starting time we denote by $\sigma_\pi \in [-\infty, \infty]$, is a mapping $\pi : [\sigma_\pi, \infty] \to [-\infty, \infty] \cup \{*\}$ such that $\pi(\infty) = *$, $\pi(\sigma_\pi) = *$ if $\sigma_\pi = -\infty$, and $t \to (\pi(t), t)$ is a continuous map from $[\sigma_\pi, \infty]$ to (R_c^2, ρ). We then define Π to be the space of all paths in R_c^2 with all possible starting times in $[-\infty, \infty]$. Endowed with the metric

$$d(\pi_1, \pi_2) = \left|\tanh(\sigma_{\pi_1}) - \tanh(\sigma_{\pi_2})\right| \\ \vee \sup_{t \geq \sigma_{\pi_1} \wedge \sigma_{\pi_2}} \left| \frac{\tanh(\pi_1(t \vee \sigma_{\pi_1}))}{1+|t|} - \frac{\tanh(\pi_2(t \vee \sigma_{\pi_2}))}{1+|t|} \right|, \quad (6.2)$$

(Π, d) is a complete separable metric space. Note that convergence in the metric d can be described as locally uniform convergence of paths plus convergence of starting times. (The metric d differs slightly from the original choice in [40], which is somewhat less natural, as explained in the appendix of [86].)

Let $\mathcal{H}$ denote the *space of compact subsets of* (Π, d), equipped with the Hausdorff metric

$$d_{\mathcal{H}}(K_1, K_2) = \sup_{\pi_1 \in K_1} \inf_{\pi_2 \in K_2} d(\pi_1, \pi_2) \vee \sup_{\pi_2 \in K_2} \inf_{\pi_1 \in K_1} d(\pi_1, \pi_2), \quad (6.3)$$

and let $\mathcal{B}_{\mathcal{H}}$ be the Borel σ-algebra associated with $d_{\mathcal{H}}$.

Exercise 6.2.1 *Show that* $(\mathcal{H}, d_{\mathcal{H}})$ *is a complete separable metric space.*

Exercise 6.2.2 *Let* $K \subset \Pi$ *be compact. Show that* $\bar{\mathcal{K}} := \{\bar{A} : A \subset K\}$ *is a compact subset of* $\mathcal{H}$.

For further properties of $(\mathcal{H}, d_{\mathcal{H}})$, such as textcolorreda criterion for the convergence of a sequence of elements in $\mathcal{H}$, or necessary and sufficient conditions for the precompactness of a subset of $\mathcal{H}$, see e.g., [92, Appendix B].

We will construct the Brownian web $\mathcal{W}$ as an $(\mathcal{H}, \mathcal{B}_{\mathcal{H}})$-valued random variable. The following notational convention will be adopted in the rest of this article:

- For $K \in \mathcal{H}$ and $A \subset R_c^2$, $K(A)$ will denote the set of paths in K with starting points in A.
- When $A = \{z\}$ for $z \in R_c^2$, we also write $K(z)$ instead of $K(\{z\})$.

6.2.2 Construction and Characterization of the Brownian Web

The basic ideas in constructing the Brownian web are the following. First we can construct coalescing Brownian motions starting from a deterministic countable dense subset $\mathcal{D}$ of the space-time plane $\mathbb{R}^2$. It is easily seen that coalescence forces paths started at typical points outside $\mathcal{D}$ to be squeezed between coalescing paths started from $\mathcal{D}$. Therefore, to construct paths starting from outside $\mathcal{D}$, we only need to take the closure of the set of paths starting from $\mathcal{D}$. Lastly one shows that the law of the random set of paths constructed does not depend on the choice of $\mathcal{D}$. This construction procedure is effectively contained in the following result from [40, Theorem 2.1], which gives a characterization of the Brownian web $\mathcal{W}$ as an $(\mathcal{H}, \mathcal{B}_{\mathcal{H}})$-valued random variable, i.e., a random compact set of paths.

Theorem 6.2.3 (Characterization of the Brownian web) *There exists an $(\mathcal{H}, \mathcal{B}_{\mathcal{H}})$-valued random variable $\mathcal{W}$, called the standard Brownian web, whose distribution is uniquely determined by the following properties:*

(a) *For each deterministic $z \in \mathbb{R}^2$, almost surely there is a unique path $\pi_z \in \mathcal{W}(z)$.*
(b) *For any finite deterministic set of points $z_1, \ldots, z_k \in \mathbb{R}^2$, the collection $(\pi_{z_1}, \ldots, \pi_{z_k})$ is distributed as coalescing Brownian motions.*
(c) *For any deterministic countable dense subset $\mathcal{D} \subset \mathbb{R}^2$, almost surely, $\mathcal{W}$ is the closure of $\{\pi_z : z \in \mathcal{D}\}$ in (Π, d).*

Proof Sketch. We will sketch the main ideas and ingredients and refer the details to [39, 40]. The main steps are:

(1) Let $\mathcal{D} = \{(x, t) : x, t \in \mathbb{Q}\}$ and construct the collection of coalescing Brownian motions $\mathcal{W}(\mathcal{D}) := \{\pi_z\}_{z \in \mathcal{D}}$, where π_z is the Brownian motion starting at z.
(2) Show that $\mathcal{W}(\mathcal{D})$ is almost surely a precompact set in the space of paths (Π, d), and hence $\mathcal{W} := \overline{\mathcal{W}(\mathcal{D})}$ defines a random compact set, i.e., an $(\mathcal{H}, \mathcal{B}_{\mathcal{H}})$-valued random variable.
(3) Show that properties (a) and (b) hold for $\mathcal{W}$, which can be easily seen to imply that property (c) also holds for $\mathcal{W}$.

The above steps construct a random variable $\mathcal{W}$ satisfying properties (a)–(c). Its law is uniquely determined, since if $\widetilde{\mathcal{W}}$ is another random variable satisfying the same properties, then both $\mathcal{W}(\mathcal{D})$ and $\widetilde{\mathcal{W}}(\mathcal{D})$ are coalescing Brownian motions starting from $\mathcal{D}$, and hence can be coupled to equal almost surely. Property (c) then implies that $\mathcal{W} = \widetilde{\mathcal{W}}$ almost surely under this coupling.

Step (1) Fix an order for points in $\mathcal{D}$, so that $\mathcal{D} = \{z_k\}_{k \in \mathbb{N}}$. Coalescing Brownian motions $(\pi_k)_{k \in \mathbb{N}}$ starting respectively from $(z_k)_{k \in \mathbb{N}}$ can be constructed inductively from independent Brownian motions $(\tilde{\pi}_k)_{k \in \mathbb{N}}$ starting from $(z_k)_{k \in \mathbb{N}}$. First let $\pi_1 := \tilde{\pi}_1$. Assuming that $\pi_1, \ldots, \pi_k$ have already been constructed from $\tilde{\pi}_1, \ldots, \tilde{\pi}_k$, then we define the path π_{k+1} to coincide with the independent Brownian motion $\tilde{\pi}_{k+1}$ until the first time τ when it meets one of the already constructed coalescing paths, say π_j, for some $1 \leq j \leq k$. From time τ onward, we just set π_{k+1} to coincide with π_j. It is not difficult to see that for any $k \in \mathbb{N}$, $(\pi_i)_{1 \leq i \leq k}$ is a collection of coalescing Brownian motions characterized by the property that different paths evolve as independent Brownian motions when they are apart, and evolve as the same Brownian motion from the time when they first meet. Furthermore, any subset of a collection of coalescing Brownian motions is also a collection of coalescing Brownian motions.

Step (2) The main idea is the following. The compactification of space-time as shown in Figure 6.2 allows us to approximate $\mathbb{R}^2$ by a large space-time box $\Lambda_{L,T} := [-L, L] \times [-T, T]$, and proving precompactness of $\mathcal{W}(\mathcal{D})$ can be reduced to proving the equicontinuity of paths in $\mathcal{W}(\mathcal{D})$ restricted to $\Lambda_{L,T}$ (for further details, see [40, Appendix B]). More precisely, it suffices to show that for any $\varepsilon > 0$, almost surely we can choose $\delta > 0$ such that the modulus of continuity

$$\psi_{\mathcal{W}(\mathcal{D}), L, T}(\delta) := \sup\{|\pi_z(t) - \pi_z(s)| : z \in \mathcal{D}, (\pi_z(s), s) \in \Lambda_{L,T}, t \in [s, s + \delta]\} \leq \varepsilon. \tag{6.4}$$

Assuming w.l.o.g. that $\varepsilon, \delta \in \mathbb{Q}$, we will control $\psi_{\mathcal{W}(\mathcal{D}), L, T}(\delta) > \varepsilon$ in terms of the modulus of continuity of coalescing Brownian motions starting from the grid

$$G_{\varepsilon, \delta} := \{(m\varepsilon/4, n\delta) : m, n \in \mathbb{N}\} \cap [-L - \varepsilon, L + \varepsilon] \times [-T - \delta, T] \subset \mathcal{D}.$$

Indeed, $\psi_{\mathcal{W}(\mathcal{D}), L, T}(\delta) > \varepsilon$ means that $|\pi_z(t) - \pi_z(s)| > \varepsilon$ for some $z \in \mathcal{D}$ with $(\pi_z(s), s) \in \Lambda_{L,T}$ and $t \in [s, s + \delta]$. Then there exists a point in the grid $\tilde{z} = (\tilde{x}, \tilde{t}) \in G_{\varepsilon, \delta}$ with $s \in [\tilde{t}, \tilde{t} + \delta)$ and $\tilde{x} \in (\pi(s) \wedge \pi(t) + \varepsilon/4, \pi(s) \wedge \pi(t) - \varepsilon/4)$. Since π_z and $\pi_{\tilde{z}}$ are coalescing Brownian motions, either $\pi_{\tilde{z}}$ coalesces with π_z before time t, or $\pi_{\tilde{z}}$ avoids π_z up to time t. Either way, we must have

$$\sup_{h \in [0, 2\delta]} |\pi_{\tilde{z}}(\tilde{t} + h) - \pi_{\tilde{z}}(\tilde{t})| \geq \varepsilon/4.$$

Denote this event by $E_z^{\varepsilon, \delta}$. Then

$$\mathbb{P}(\psi_{\mathcal{W}(\mathcal{D}), L, T}(\delta) > \varepsilon) \leq \mathbb{P}\Big(\bigcup_{z \in G_{\varepsilon, \delta}} E_z^{\varepsilon, \delta}\Big) \leq \sum_{z \in G_{\varepsilon, \delta}} \mathbb{P}(E_z^{\varepsilon, \delta}) = |G_{\varepsilon, \delta}| \mathbb{P}(\sup_{h \in [0, 2\delta]} |B_h| \geq \varepsilon/4)$$

$$\leq C_{L,T} \varepsilon^{-1} \delta^{-1} e^{-c\varepsilon^2/\delta},$$

where $c, C_{L,T} > 0$, B is a standard Brownian motion, and we have used the reflection principle to bound the tail probability for $\sup |B|$. Since $\mathbb{P}(\psi_{\mathcal{W}(\mathcal{D}), L, T}(\delta) > \varepsilon) \to 0$ as $\delta \downarrow 0$, this implies (6.4). Therefore $\mathcal{W}(\mathcal{D})$ is a.s. precompact, and $\mathcal{W} := \overline{\mathcal{W}(\mathcal{D})}$ defines an $(\mathcal{H}, \mathcal{B}_\mathcal{H})$-valued random variable.

Step (3) We first show that for each $z = (x, t) \in \mathbb{R}^2$, almost surely $\mathcal{W}(z)$, the paths in $\mathcal{W}$ starting at z, contains a unique path. Let $z_n^- = (x - \varepsilon_n, t - \delta_n) \in \mathcal{D}$, $z_n^+ = (x + \varepsilon_n, t - \delta_n) \in \mathcal{D}$, with $\varepsilon_n, \delta_n \downarrow 0$, and let τ_n be the time when $\pi_{z_n^-}$ and $\pi_{z_n^+}$ coalesce. Note that on the event

$$E_n := \Big\{\pi_{z_n^-}(t) < x < \pi_{z_n^+}(t), \tau_n \leq t + \frac{1}{n}\Big\},$$

every path in $\mathcal{W}(z)$ must be enclosed between $\pi_{z_n^-}$ and $\pi_{z_n^+}$, and hence is uniquely determined from time $\tau_n \leq t + \frac{1}{n}$ onward. It is easy to see that we can choose $\varepsilon_n \downarrow 0$ sufficiently fast, and $\delta_n \downarrow 0$ much faster than ε_n^2, such that $\mathbb{P}(E_n) \to 1$ as $n \to \infty$. In particular, almost surely, E_n occurs infinitely often, which implies that the paths in $\mathcal{W}(z)$ all coincide on (t, ∞) and hence $\mathcal{W}(z)$ contains a unique path.

To show that $\mathcal{W}$ satisfies property (b), let us fix $z_1, \ldots, z_k \in \mathbb{R}^2$. For each $1 \leq i \leq k$, let $z_{n,i} \in \mathcal{D}$ with $z_{n,i} \to z_i$ as $n \to \infty$. By the a.s. compactness of $\mathcal{W}$, and the fact that $\mathcal{W}(z_i)$ a.s. contains a unique path π_{z_i} by property (a) that we just verified, we must have $\pi_{z_{n,i}} \to \pi_{z_i}$ in (Π, d) for each $1 \leq i \leq k$. In particular, as a sequence of Π^k-valued random variables, $(\pi_{z_{n,i}})_{1 \leq i \leq k}$ converges in distribution to $(\pi_{z_i})_{1 \leq i \leq k}$. On the other hand, as a subset of $\mathcal{W}(\mathcal{D})$, $(\pi_{z_{n,i}})_{1 \leq i \leq k}$ is a collection of coalescing Brownian motions, and it is easy to show that as their starting points converge, they converge in distribution to a collection of coalescing Brownian motions starting from $(z_i)_{1 \leq i \leq k}$. Therefore $(\pi_{z_i})_{1 \leq i \leq k}$ is distributed as a collection of coalescing Brownian motions.

Lastly, to show that $\mathcal{W}$ satisfies property (c), let $\mathcal{D}'$ be another countable dense subset of $\mathbb{R}^2$. Clearly $\overline{\mathcal{W}(\mathcal{D}')} \subset \mathcal{W}$. To show the converse, $\mathcal{W} \subset \overline{\mathcal{W}(\mathcal{D}')}$, it suffices to show that for each $z \in \mathcal{D}$, $\pi_z \in \overline{\mathcal{W}(\mathcal{D}')}$. This can be seen by taking a sequence $z_n' \in \mathcal{D}'$ with $z_n' \to z$, for which we must have $\pi_{z_n'} \to \pi_z \in \overline{\mathcal{W}(\mathcal{D}')}$ by the compactness of $\overline{\mathcal{W}(\mathcal{D}')} \subset \mathcal{W}$ and the fact that $\mathcal{W}(z) = \{\pi_z\}$. □

6.2.3 The Brownian Web and its Dual

As discussed at the beginning of Section 6.2, similar to the duality between forward and backward coalescing random walks shown in Figure 6.1, we expect the Brownian web $\mathcal{W}$ also to have a dual $\widehat{\mathcal{W}}$. Such a duality provides a powerful tool for analyzing properties of the Brownian web. Since the dual Brownian web $\widehat{\mathcal{W}}$ should be a collection of coalescing paths running backward in time, we first define the space in which $\widehat{\mathcal{W}}$ takes its values.

Given $z = (x,t) \in R_c^2$, which is identified with $[-\infty, \infty]^2$ where $[-\infty, \infty] \times \{\pm\infty\}$ is contracted to a single point $(*, \pm\infty)$, let $-z$ denote $(-x, -t)$. Given a set $A \subset R_c^2$, let $-A$ denote $\{-z : z \in A\}$. Identifying each path $\pi \in \Pi$ with its graph as a subset of R_c^2, $\hat{\pi} := -\pi$ defines a path running backward in time, with starting time $\hat{\sigma}_{\hat{\pi}} = -\sigma_\pi$. Let $\widehat{\Pi} := -\Pi$ denote the set of all such backward paths, equipped with a metric $\hat{d}$ that is inherited from (Π, d) under the mapping $-$. Let $\widehat{\mathcal{H}}$ be the space of compact subsets of $(\widehat{\Pi}, \hat{d})$, equipped with the Hausdorff metric $d_{\widehat{\mathcal{H}}}$ and Borel σ-algebra $\mathcal{B}_{\widehat{\mathcal{H}}}$. For any $K \in \mathcal{H}$, we will let $-K$ denote the set $\{-\pi : \pi \in K\} \in \widehat{\mathcal{H}}$.

The following result characterizes the joint law of the Brownian web $\mathcal{W}$ and its dual $\widehat{\mathcal{W}}$ as a random variable taking values in $\mathcal{H} \times \widehat{\mathcal{H}}$, equipped with the product σ-algebra.

Theorem 6.2.4 (Characterization of the double Brownian web) *There exists an $\mathcal{H} \times \widehat{\mathcal{H}}$-valued random variable $(\mathcal{W}, \widehat{\mathcal{W}})$, called the double Brownian web (with $\widehat{\mathcal{W}}$ called the dual Brownian web), whose distribution is uniquely determined by the following properties:*

(a) *$\mathcal{W}$ and $-\widehat{\mathcal{W}}$ are both distributed as the standard Brownian web.*
(b) *Almost surely, no path $\pi_z \in \mathcal{W}$ crosses any path $\hat{\pi}_{\hat{z}} \in \widehat{\mathcal{W}}$ in the sense that, $z = (x,t)$ and $\hat{z} = (\hat{x}, \hat{t})$ with $t < \hat{t}$, and $(\pi_z(s_1) - \hat{\pi}_{\hat{z}}(s_1))(\pi_z(s_2) - \hat{\pi}_{\hat{z}}(s_2)) < 0$ for some $t < s_1 < s_2 < \hat{t}$.*

Furthermore, for each $z \in \mathbb{R}^2$, $\widehat{\mathcal{W}}(z)$ a.s. consists of a single path $\hat{\pi}_z$ which is the unique path in $\widehat{\Pi}$ that does not cross any path in $\mathcal{W}$, and thus $\widehat{\mathcal{W}}$ is a.s. determined by $\mathcal{W}$ and vice versa.

Proof Sketch. The existence of a double Brownian web $(\mathcal{W}, \widehat{\mathcal{W}})$ satisfying properties (a)–(b) is most easily derived as scaling limits of forward and backward coalescing random walks. We defer this to Section 6.6, after we introduce general criteria for convergence to the Brownian web.

Let us first prove that if $(\mathcal{W}, \widehat{\mathcal{W}})$ satisfies properties (a)–(b), then almost surely $\mathcal{W}$ uniquely determines $\widehat{\mathcal{W}}$. Indeed, fix a deterministic $z = (x,t) \in \mathbb{R}^2$. By the characterization of the Brownian web $\mathcal{W}$, $\mathcal{W}(\mathbb{Q}^2)$ is a collection of coalescing Brownian motions, with a.s. one Brownian motion starting from each point in $\mathbb{Q}^2$. Since Brownian motion has zero probability of hitting a deterministic space-time point, there is zero probability that z lies on $\pi_{z'}$ for any $z' \in \mathbb{Q}^2$. Therefore for any $s \in \mathbb{Q}$ with $s < t$, property (b) implies that for any path $\hat{\pi} \in \widehat{\mathcal{W}}(z)$, we must have
$$\hat{\pi}(s) = \sup\{y \in \mathbb{Q} : \pi_{(y,s)}(t) < x\} = \inf\{y \in \mathbb{Q} : \pi_{(y,s)}(t) > x\}.$$
In other words, $\hat{\pi}$ is uniquely determined at rational times and hence at all times, and $\widehat{\mathcal{W}}(z)$ contains a unique path. It follows that $\widehat{\mathcal{W}}(\mathbb{Q}^2)$ is a.s. uniquely determined by $\mathcal{W}$, and hence so is $\widehat{\mathcal{W}} = \overline{\widehat{\mathcal{W}}(\mathbb{Q}^2)}$.

Lastly we show that the distribution of $(\mathcal{W}, \widehat{\mathcal{W}})$ is uniquely determined by properties (a) and (b). Indeed, if $(\mathcal{W}', \widehat{\mathcal{W}}')$ is another double Brownian web, then $\mathcal{W}$ and $\mathcal{W}'$ can be coupled so that they equal a.s. As we have just shown, $\mathcal{W}$ a.s. uniquely determines $\widehat{\mathcal{W}}$, and $\mathcal{W}'$ determines $\widehat{\mathcal{W}}'$. Therefore $\widehat{\mathcal{W}} = \widehat{\mathcal{W}}'$ a.s., and $(\mathcal{W}', \widehat{\mathcal{W}}')$ has the same distribution as $(\mathcal{W}, \widehat{\mathcal{W}})$. □

Remark 6.2.5 One can characterize the joint law of paths in $(\mathcal{W}, \widehat{\mathcal{W}})$ starting from a finite deterministic set of points. Similar to the construction of coalescing

Brownian motions, we can construct one path at a time. To add a new forward path to an existing collection, we follow an independent Brownian motion until it either meets an existing forward Brownian motion, in which case they coalesce, or it meets an existing dual Brownian motion, in which case it is *Skorohod reflected* by the dual Brownian motion (for further details, see [93]). Extending this pathwise construction to a countable dense set of starting points $\mathcal{D}$ and then taking closure, this gives a direct construction of $(\mathcal{W}, \widehat{\mathcal{W}})$, which is formulated in [41, Theorem 3.7].

Remark 6.2.6 In light of Theorem 6.2.4, one may wonder whether $\mathcal{W}$ a.s. consists of *all* paths in Π which do not cross any path in $\widehat{\mathcal{W}}$, and vice versa. The answer is no, and $\mathcal{W}$ is actually the minimal compact set of paths that do not cross any path in $\widehat{\mathcal{W}}$ while still containing paths starting from every point in $\mathbb{R}^2$. More non-crossing paths can be added to $\mathcal{W}$ by extending paths in $\mathcal{W}$ backward in time, following paths in $\widehat{\mathcal{W}}$ (see [44]). Such paths can be excluded if we impose the further restriction that no path can enter from outside any open region that is enclosed by a pair of paths in $\widehat{\mathcal{W}}$. This is called the *wedge characterization* of the Brownian web, to be discussed in more detail in Remark 6.3.10.

6.2.4 The Coalescing Point Set

The coupling between the Brownian web and its dual given in Theorem 6.2.4 allows one to deduce interesting properties for the Brownian web. The first result is on the density of paths in the Brownian web $\mathcal{W}$ started at time 0.

Given the Brownian web $\mathcal{W}$, and a closed set $A \subset \mathbb{R}$, define the *coalescing point set* by

$$\xi_t^A := \{y \in \mathbb{R} : y = \pi(t) \text{ for some } \pi \in \mathcal{W}(A \times \{0\})\}, \qquad t \geq 0. \tag{6.5}$$

In words, ξ_t^A is the set of points in $\mathbb{R}$ that lie on some path in $\mathcal{W}$ that starts from A at time 0. Note that this process is monotone in the sense that if $A \subset B$, then $\xi_t^A \subset \xi_t^B$ a.s. for all $t \geq 0$.

It turns out that even if started from the whole line, $\xi_t^{\mathbb{R}}$ becomes a.s. locally finite as soon as $t > 0$, as the following density result shows. Such a *coming down from infinity* phenomenon also appears in Kingman's coalescent, see e.g., [10].

Proposition 6.2.7 (Density of the coalescing point set) *Let $\xi_{\cdot}^{\mathbb{R}}$ be the coalescing point set defined from the Brownian web $\mathcal{W}$ as in* (6.5). *Then for all $t > 0$ and $a < b$,*

$$\mathbb{E}[|\xi_t^{\mathbb{R}} \cap [a,b]|] = \frac{b-a}{\sqrt{\pi t}}. \tag{6.6}$$

Proof. Let $\widehat{\mathcal{W}}$ be the dual Brownian web determined a.s. by $\mathcal{W}$, as in Theorem 6.2.4. Observe that by the non-crossing property between paths in $\mathcal{W}$ and $\widehat{\mathcal{W}}$, $\xi_t^\mathbb{R} \cap (a,b) \neq \emptyset$ implies that the paths $\hat{\pi}_a, \hat{\pi}_b \in \widehat{\mathcal{W}}$, starting respectively at (a,t) and (b,t), do not coalesce in the time interval $(0,t)$ (i.e., $\hat{\tau} \leq 0$ if $\hat{\tau}$ denotes the time when $\hat{\pi}_a$ and $\hat{\pi}_b$ coalesce). Conversely, if $\hat{\tau} \leq 0$, then any path in $\mathcal{W}$ started from $[\hat{\pi}_a(0), \hat{\pi}_b(0)]$ at time 0 will hit $[a,b]$ at time t, i.e., $\xi_t^\mathbb{R} \cap [a,b] \neq \emptyset$. Thus,

$$\mathbb{P}(\xi_t^\mathbb{R} \cap (a,b) \neq \emptyset) \leq \mathbb{P}(\hat{\tau} \leq 0) \leq \mathbb{P}(\xi_t^\mathbb{R} \cap [a,b] \neq \emptyset), \tag{6.7}$$

where we observe that

$$\mathbb{P}(\hat{\tau} \leq 0) = \mathbb{P}\Big(\sup_{s \in (0,t)}(B_2(s) - B_1(s)) \leq b-a\Big) = \mathbb{P}\Big(\sup_{s \in (0,t)} B(s) \leq \frac{b-a}{\sqrt{2}}\Big)$$

$$= \frac{1}{\sqrt{2\pi t}} \int_{-\frac{b-a}{\sqrt{2}}}^{\frac{b-a}{\sqrt{2}}} e^{-\frac{x^2}{2t}} dx \sim \frac{b-a}{\sqrt{\pi t}} \quad \text{as } b-a \downarrow 0.$$

By (6.7), this implies that

$$\mathbb{P}(x \in \xi_t^\mathbb{R}) = 0 \quad \text{for all } x \in \mathbb{R},$$

and the inequalities in (6.7) are in fact all equalities.

We can then apply the monotone convergence theorem to obtain

$$\mathbb{E}[|\xi_t^\mathbb{R} \cap [a,b]|] = \lim_{n \to \infty} \mathbb{E}\Big[\Big|\Big\{1 \leq i < (b-a)2^n : \xi_t^\mathbb{R} \cap \Big(a + \frac{i-1}{2^n}, a + \frac{i}{2^n}\Big) \neq \emptyset\Big\}\Big|\Big]$$

$$= \lim_{n \to \infty} (b-a)2^n \mathbb{P}(\xi_t^\mathbb{R} \cap (0, 2^{-n}) \neq \emptyset) = \lim_{n \to \infty} (b-a)2^n \frac{2^{-n}}{\sqrt{\pi t}}$$

$$= \frac{b-a}{\sqrt{\pi t}}.$$

This concludes the proof of the proposition. □

As a corollary of Proposition 6.2.7, we show that when paths in the Brownian web converge, they converge in a strong sense (see e.g., [86, Lemma 3.4]).

Corollary 6.2.8 (Strong convergence of paths in $\mathcal{W}$) *Let $\mathcal{W}$ be the standard Brownian web. Almost surely, for any sequence $\pi_n \in \mathcal{W}$ with $\pi_n \to \pi \in \mathcal{W}$, the time of coalescence τ_n between π_n and π must tend to σ_π as $n \to \infty$.*

Exercise 6.2.9 *Deduce Corollary 6.2.8 from Proposition 6.2.7.*

Remark 6.2.10 Apart from its density, we actually know quite a bit more about the coalescing point set $\xi_t^\mathbb{R}$. It has been shown by Tribe et al. [101, 100] that $\xi_t^\mathbb{R}$

is in fact a *Pfaffian point process*, whose kernel also appears in the real Ginibre random matrix ensemble. Furthermore, $\xi_t^\mathbb{R}$ (and more generally ξ_t^A for any $A \subset \mathbb{R}$) can be shown (see e.g., [52, Appendix C] and [69]) to be negatively associated in the sense that for any $n \in \mathbb{N}$ and any disjoint open intervals $O_1, \cdots, O_n$, we have

$$\mathbb{P}(\cap_{i=1}^n \{\xi_t^\mathbb{R} \cap O_i \neq \emptyset\}) \leq \prod_{i=1}^n \mathbb{P}(\xi_t^\mathbb{R} \cap O_i \neq \emptyset). \tag{6.8}$$

For any $B \subset \mathbb{R}$ with positive Lebesgue measure, we also have

$$\mathbb{P}(|\xi_t^\mathbb{R} \cap B| \geq m+n) \leq \mathbb{P}(|\xi_t^\mathbb{R} \cap B| \geq m)\mathbb{P}(|\xi_t^\mathbb{R} \cap B| \geq n) \quad \forall m,n \in \mathbb{N}. \tag{6.9}$$

On a side note, we remark that when $A \subset \mathbb{R}$ is a finite set, determinantal formulas have also been derived for the distribution of ξ_t^A in [103, Proposition. 9].

6.2.5 Special Points of the Brownian Web

We have seen in Theorem 6.2.3 that for each deterministic $z \in \mathbb{R}^2$, almost surely the Brownian web $\mathcal{W}$ contains a unique path starting from z. However, it is easily seen that there must exist random points $z \in \mathbb{R}^2$ where $\mathcal{W}(z)$ contains multiple paths. Indeed, consider paths in $\mathcal{W}$ starting from $\mathbb{R}$ at time 0. Proposition 6.2.7 shows that these paths coalesce into a locally finite set of points $\xi_t^\mathbb{R}$ at any time $t > 0$. Each point $x_i \in \xi_t^\mathbb{R}$ (with $x_i < x_{i+1}$ for all $i \in \mathbb{Z}$) can be traced back to an interval (u_i, u_{i+1}) at time 0, where all paths starting there pass through the space-time point (x_i, t). At the boundary u_i between two such intervals, we note however that $\mathcal{W}((u_i, 0))$ must contain at least two paths, which are limits of paths in $\mathcal{W}$ starting from (u_{i-1}, u_i), respectively (u_i, u_{i+1}), at time 0. Are there random space-time points where more than two paths originate? It turns out that we can give a complete classification of the type of multiplicity we see almost surely in a Brownian web. The main tool to accomplish this is the self-duality of the Brownian web discussed in Section 6.2.3.

First we give a classification scheme for $z \in \mathbb{R}^2$ according to the multiplicity of paths in $\mathcal{W}$ entering and leaving z. We say a path π enters $z = (x,t)$ if $\sigma_\pi < t$ and $\pi(t) = x$, and π leaves z if $\sigma_\pi \leq t$ and $\pi(t) = x$. Two paths π and π' leaving z are defined to be equivalent, denoted by $\pi \sim_{\text{out}}^z \pi'$, if $\pi = \pi'$ on $[t, \infty)$. Two paths π and π' entering z are defined to be equivalent, denoted by $\pi \sim_{\text{in}}^z \pi'$, if $\pi = \pi'$ on $[t-\varepsilon, \infty)$ for some $\varepsilon > 0$. Note that $\sim_{\text{in}}^z$ and $\sim_{\text{out}}^z$ are equivalence relations.

Let $m_{\text{in}}(z)$, respectively $m_{\text{out}}(z)$, denote the number of equivalence classes of paths in $\mathcal{W}$ entering, respectively leaving, z, and let $\hat{m}_{\text{in}}(z)$ and $\hat{m}_{\text{out}}(z)$ be defined similarly for the dual Brownian web $\hat{\mathcal{W}}$. Given a realization of the Brownian web $\mathcal{W}$, points $z \in \mathbb{R}^2$ are classified according to the value of $(m_{\text{in}}(z), m_{\text{out}}(z))$. We divide points of type (1,2) further into types $(1,2)_l$ and $(1,2)_r$, where the subscript

The Brownian Web and the Brownian Net

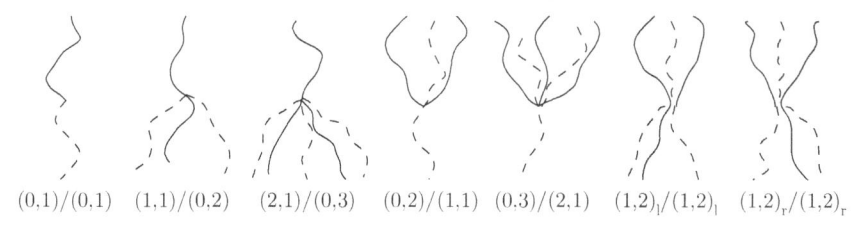

(0,1)/(0,1)   (1,1)/(0,2)   (2,1)/(0,3)   (0,2)/(1,1)   (0,3)/(2,1) $(1,2)_l/(1,2)_l$ $(1,2)_r/(1,2)_r$

Figure 6.3. Special points of the Brownian web.

l (resp. r) indicates that the left (resp. right) of the two outgoing paths is the continuation of the (up to equivalence) unique incoming path. Points in the dual Brownian web $\widehat{\mathcal{W}}$ are labeled according to their type in the Brownian web obtained by rotating the graph of $\widehat{\mathcal{W}}$ in $\mathbb{R}^2$ by 180°.

We are now ready to state the following classification result (see also [99, Proposition 2.4] and [41, Theorems 3.11–3.14]), illustrated in Figure 6.3.

Theorem 6.2.11 (Special points of the Brownian web) *Let $(\mathcal{W}, \widehat{\mathcal{W}})$ be the standard Brownian web and its dual. Then almost surely, each $z \in \mathbb{R}^2$ satisfies*

$$m_{\mathrm{out}}(z) = \hat{m}_{\mathrm{in}}(z) + 1 \quad \text{and} \quad \hat{m}_{\mathrm{out}}(z) = m_{\mathrm{in}}(z) + 1, \tag{6.10}$$

and z is of one of the following seven types according to $(m_{\mathrm{in}}(z), m_{\mathrm{out}}(z))/(\hat{m}_{\mathrm{in}}(z), \hat{m}_{\mathrm{out}}(z))$:

(0,1)/(0,1), (1,1)/(0,2), (0,2)/(1,1), (2,1)/(0,3), (0,3)/(2,1), $(1,2)_l/(1,2)_l$, $(1,2)_r/(1,2)_r$.

Almost surely,

(i) *the set of points of type (0,1)/(0,1) has full Lebesgue measure in $\mathbb{R}^2$;*
(ii) *points of type (1,1)/(0,2) are points in the set $\bigcup_{\pi \in \mathcal{W}}\{(\pi(t),t) : t > \sigma_\pi\}$, excluding points of type (1,2)/(1,2) and (2,1)/(0,3);*
(iii) *the set of points of type (2,1)/(0,3) consists of points at which two paths in $\mathcal{W}$ coalesce and is countable;*
(iv) *points of type $(1,2)_l/(1,2)_l$ are points of intersection between some $\pi \in \mathcal{W}$ and $\hat{\pi} \in \widehat{\mathcal{W}}$, with $\sigma_\pi < t < \hat{\sigma}_{\hat{\pi}}$, $\pi(s) \leq \hat{\pi}(s)$ for all $s \in [\sigma_\pi, \hat{\sigma}_{\hat{\pi}}]$, and π intersects $\hat{\pi}$ at $(\pi(t),t) = (\hat{\pi}(t),t)$.*

Similar statements hold for the remaining three types by symmetry.

Proof. First we will prove relation (6.10). Let $z = (x,t)$, and assume that $\hat{m}_{\mathrm{in}}(z) = k$ for some $k \in \mathbb{N}_0 := \{0\} \cup \mathbb{N}$. Then there exist $\varepsilon > 0$ and k ordered paths $\hat{\pi}_1, \ldots, \hat{\pi}_k \in \widehat{\mathcal{W}}$ starting at time $t + \varepsilon$, such that these paths are disjoint on $(t, t+\varepsilon]$ and coalesce together at time t at position x. Note that the ordered paths $(\hat{\pi}_i)_{1 \leq i \leq k}$ divide the space-time strip $\mathbb{R} \times (t, t+\varepsilon)$ into $k+1$ regions $(I_i)_{1 \leq i \leq k+1}$, where I_1 is the region to the left of $\hat{\pi}_1$, I_i is the region between $\hat{\pi}_{i-1}$ and $\hat{\pi}_i$ for each $2 \leq i \leq k$, and I_{k+1} is the region to the right of $\hat{\pi}_k$. From the interior of each region I_i, we can pick a sequence of starting points $(z_i^n)_{n \in \mathbb{N}}$ with $z_i^n \to z$. Since paths in $\mathcal{W}$ and $\widehat{\mathcal{W}}$ do not cross, as formulated in Theorem 6.2.4, each path $\pi_i^n \in \mathcal{W}(z_i^n)$ must stay confined in $\overline{I_i}$ in the time interval $[t, t+\varepsilon]$, and so must any subsequential limit of $(\pi_i^n)_{n \in \mathbb{N}}$. Such subsequential limits must exist by the almost sure compactness of $\mathcal{W}$, and each subsequential limit is a path $\pi_i \in \mathcal{W}(z)$. Therefore $\mathcal{W}(z)$ must contain at least $k+1$ distinct paths, one contained in $\overline{I_i}$ for each $1 \leq i \leq k+1$. Furthermore, each $\overline{I_i}$ cannot contain more than one path in $\mathcal{W}(z)$. Indeed, if $\overline{I_i}$ contains two distinct paths $\pi, \pi' \in \mathcal{W}(z)$, then any path $\hat{\pi} \in \widehat{\mathcal{W}}$ started strictly between π and π' on the time interval $(t, t+\varepsilon)$ must enter z, and $\hat{\pi}$ is distinct from $(\hat{\pi}_i)_{1 \leq i \leq k}$, which contradicts the assumption that $\hat{m}_{in}(z) = k$. Therefore we must have $m_{out}(z) = k+1 = \hat{m}_{in}(z) + 1$.

Similar considerations as above show that if z is of type $(1,2)_l$ in $\mathcal{W}$, then it must be of the same type in $\widehat{\mathcal{W}}$. The same holds for type $(1,2)_r$.

We now show that the seven types of points listed are all there is. Note that it suffices to show that almost surely $m_{in}(z) + m_{out}(z) \leq 3$ for each $z \in \mathbb{R}^2$. There are four possible cases of $m_{in}(z) + m_{out}(z) > 3$, which we rule out one by one:

(a) For some $z \in \mathbb{R}^2$, $m_{in}(z) \geq 3$ and $m_{out}(z) \geq 1$. Note that Corollary 6.2.8 implies that every path $\pi \in \mathcal{W}$ coincides with some path in $\mathcal{W}(\mathbb{Q}^2)$ on $[\sigma_\pi + \varepsilon, \infty)$, for any given $\varepsilon > 0$. Therefore the event that $m_{in}(z) \geq 3$ for some $z \in \mathbb{R}^2$ is contained in the event that three distinct Brownian motions among $\mathcal{W}(\mathbb{Q}^2)$ coalesce at the same time. Such an event has probability zero, because there are countably many ways of choosing three Brownian motions from $\mathcal{W}(\mathbb{Q}^2)$, and conditioned on two Brownian motions coalescing at a given space-time point, there is zero probability that a third independent Brownian motion (which evolves independently before coalescing) would visit the same space-time point.

(b) For some $z \in \mathbb{R}^2$, $m_{in}(z) \geq 2$ and $m_{out}(z) \geq 2$. In this case, $\hat{m}_{in}(z) \geq 1$, and, again by Corollary 6.2.8, the event we consider is contained in the event that there exist two paths $\pi_1, \pi_2 \in \mathcal{W}(\mathbb{Q}^2)$ and a path $\hat{\pi} \in \widehat{\mathcal{W}}(\mathbb{Q}^2)$, such that $\hat{\pi}$ passes through the point of coalescence between π_1 and π_2. Such an event has probability 0, since conditioned on π_1 and π_2 up to the time of their coalescence, $\hat{\pi}$ is an independent Brownian motion with zero probability of hitting a given space-time point – the point of coalescence between π_1 and π_2.

(c) For some $z \in \mathbb{R}^2$, $m_{in}(z) \geq 1$ and $m_{out}(z) \geq 3$. In this case, $\hat{m}_{out}(z) \geq 2$ and $\hat{m}_{in}(z) \geq 2$, which is equivalent to Case (b) by the symmetry between $\mathcal{W}$ and $\widehat{\mathcal{W}}$.

(d) For some $z \in \mathbb{R}^2$, $m_{\text{out}}(z) \geq 4$. In this case, $\hat{m}_{\text{in}}(z) \geq 3$, which is equivalent to Case (a).

We leave the verification of statements (i)–(iv) as an exercise. □

Exercise 6.2.12 *Verify statements (i)–(iv) in Theorem 6.2.11.*

6.3 The Brownian Net

The Brownian net generalizes the Brownian web by allowing paths to branch. The existence of such an object is again motivated by its discrete analogue, the collection of discrete time branching-coalescing simple symmetric random walks on $\mathbb{Z}$. Figure 6.4 gives an illustration: from each $(x,n) \in \mathbb{Z}^2_{\text{even}}$, an arrow is drawn from (x,n) to either $(x-1, n+1)$ or $(x+1, n+1)$ with probability $(1-\varepsilon)/2$ each, representing whether the walk starting at x at time n should move to $x-1$ or $x+1$ at time $n+1$; and with probability ε, arrows are drawn from (x,n) to both $(x-1, n+1)$ and $(x+1, n+1)$, so that the walk starting at (x,n) branches into two walks, with one moving to $x-1$ and the other to $x+1$ at time $n+1$.

If we consider the collection of all upward random walk paths obtained by following the arrows, then the natural question is: when space-time is scaled diffusively, could this random collection of paths have a nontrivial limit? To have an affirmative answer to this question, it is necessary to choose the branching probability ε to depend suitably on the diffusive scaling parameter. More precisely:

Q.1 If space-time is rescaled by $S_\varepsilon(x,t) := (\varepsilon x, \varepsilon^2 t)$, and the branching probability is chosen to be $b\varepsilon$ for some $b > 0$, then what is the scaling limit of the collection of branching-coalescing random walk paths as $\varepsilon \downarrow 0$?

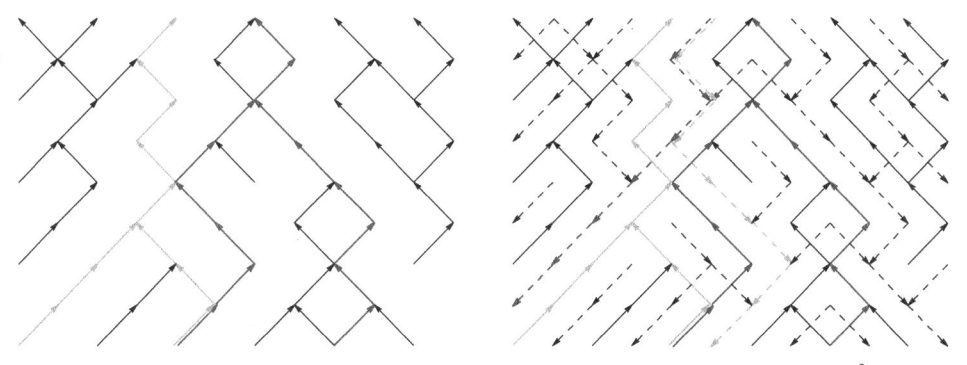

Figure 6.4. Discrete space-time branching-coalescing random walks on $\mathbb{Z}^2_{\text{even}}$, and its dual on $\mathbb{Z}^2_{\text{odd}}$.

The limit will be what we call the *Brownian net* $\mathcal{N}_b$ *with branching parameter* b. For simplicity, we will focus on the case $b = 1$, with the limit called the *standard Brownian net* $\mathcal{N}$. The fact that a nontrivial scaling limit exists with the above choice of the branching probability is hinted by the following observation. Instead of considering the collection of all random walk paths, let us first restrict our attention to two special subsets: the set of *leftmost*, respectively *rightmost*, random walk paths where the random walk always follows the arrow to the left, respectively right, whenever it encounters a branching point (see Figure 6.4). Note that the collection of leftmost paths is a collection of coalescing random walks with drift $-\varepsilon$, which ensures that each path under the diffusive scaling $S_\varepsilon(x,t) = (\varepsilon x, \varepsilon^2 t)$ converges to a Brownian motion with drift -1. Therefore we expect the collection of leftmost paths to converge to a variant of the Brownian web, $\mathcal{W}^l$, which consists of coalescing Brownian motions with drift -1. Similarly, we expect the collection of rightmost paths to converge to a limit $\mathcal{W}^r$, which consists of coalescing Brownian motions with drift $+1$. Of course, $\mathcal{W}^l$ and $\mathcal{W}^r$ are coupled in a nontrivial way.

The above observation explains the choice of the branching probability, and we see that any limit of the branching-coalescing random walk paths must contain the two coupled Brownian webs $(\mathcal{W}^l, \mathcal{W}^r)$. The questions that remain are:

(A) How to characterize the joint law of $(\mathcal{W}^l, \mathcal{W}^r)$?
(B) Can we construct the scaling limit of branching-coalescing random walks from $(\mathcal{W}^l, \mathcal{W}^r)$?

To answer (A), it suffices to characterize the joint distribution $l_{z_1} \in \mathcal{W}^l(z_1), \ldots, l_{z_k} \in \mathcal{W}^l(z_k)$ and $r_{z'_1} \in \mathcal{W}^r(z'_1), \ldots, r_{z'_{k'}} \in \mathcal{W}^r(z'_{k'})$ for a finite collection of $(z_i)_{1 \leq i \leq k}$ and $(z'_i)_{1 \leq i \leq k'}$ in $\mathbb{R}^2$. An examination of their discrete analogue suggests that:

- the paths $(l_{z_1}, \ldots, l_{z_k}, r_{z'_1}, \ldots, r_{z'_{k'}})$ evolve independently when they are apart;
- the *leftmost paths* $(l_{z_1}, \ldots, l_{z_k})$ coalesce when they meet, and the same is true for the *rightmost paths* $(r_{z'_1}, \ldots, r_{z'_{k'}})$;
- a pair of leftmost and rightmost paths $(l_{z_i}, r_{z'_j})$ solves the following pair of stochastic differential equations (SDEs):

$$dL_t = 1_{\{L_t \neq R_t\}} dB_t^l + 1_{\{L_t = R_t\}} dB_t^s - dt,$$
$$dR_t = 1_{\{L_t \neq R_t\}} dB_t^r + 1_{\{L_t = R_t\}} dB_t^s + dt,$$
(6.11)

where the leftmost path L and the rightmost path R are driven by independent Brownian motions B^l and B^r when they are apart, and driven by the same Brownian motion B^s (independent of B^l and B^r) when they coincide; furthermore, L and R are subject to the constraint that $L_t \leq R_t$ for all $t \geq T := \inf\{u \geq \sigma_L \vee \sigma_R : L_u \leq R_u\}$, with σ_L and σ_R being the starting times of L and R.

It turns out that the SDE (6.11) has a unique weak solution, and the above properties uniquely determine the joint law of $(l_{z_1},\ldots,l_{z_k},r_{z'_1},\ldots,r_{z'_{k'}})$, which we will call *left-right coalescing Brownian motions*. Extending the starting points to a countable dense set in $\mathbb{R}^2$, and then taking closure of the resulting set of leftmost, respectively rightmost, paths, a.s. determines $(\mathcal{W}^l,\mathcal{W}^r)$, which we will call the *left-right Brownian web*.

To answer (B), observe that in the discrete case, all random walk paths can be constructed by hopping back and forth between leftmost and rightmost random walk paths. This suggests a similar approach to construct the scaling limit of the set of all branching-coalescing random walk paths, which we will call *the Brownian net* $\mathcal{N}$. More precisely, to construct $\mathcal{N}$ from $(\mathcal{W}^l,\mathcal{W}^r)$, we simply consider the set of all paths that can be obtained by *hopping* a finite number of times between paths in $\mathcal{W}^l$ and $\mathcal{W}^r$, and then take its closure.

The above considerations led to the original construction of the Brownian net $\mathcal{N}$ in [86], called the *hopping construction*.

From Figure 6.4, it is easily seen that the branching-coalescing random walks on $\mathbb{Z}^2_{\text{even}}$ a.s. uniquely determine a dual collection of branching-coalescing random walks on $\mathbb{Z}^2_{\text{odd}}$, running backward in time. Furthermore, the two systems are equally distributed apart from a rotation in space-time by 180° and a lattice shift. Therefore in the scaling limit, we expect the left-right Brownian web $(\mathcal{W}^l,\mathcal{W}^r)$ to have a dual $(\widehat{\mathcal{W}^l},\widehat{\mathcal{W}^r})$, which determines a dual Brownain net $\widehat{\mathcal{N}}$. As for the Brownian web, such a duality provides a powerful tool. In particular, it leads to a second construction of the Brownian net, called the *wedge construction* in [86].

Besides the hopping and wedge constructions of the Brownian net, there are two more constructions, called the *mesh construction*, also developed in [86], and the *marking construction* developed by Newman, Ravishankar and Schertzer in [73], where the Brownian net was conceived independently from [86]. The mesh construction is based on the observation that, given the left-right Brownian web $(\mathcal{W}^l,\mathcal{W}^r)$, there exist space-time regions (called *meshes*) with their left boundaries being rightmost paths, and their right boundaries being leftmost paths. Such unusual configurations make these meshes forbidden regions, where no paths can enter. The mesh construction asserts that the Brownian net consists of all paths which do not enter meshes.

In contrast to the hopping construction, which is an *outside-in* approach where the Brownian net is constructed from its outermost paths – the leftmost and rightmost paths – the *marking construction* developed in [73] is an *inside-out* approach, where one starts from a Brownian web and then constructs the Brownian net by adding branching points. In the discrete setting, this amounts to turning coalescing random walks into branching-coalescing random walks by changing each lattice point independently into a branching point with probability ε. In the continuum setting, this turns out to require *Poisson marking* the set of (1,2) points of the Brownian web (cf. Theorem 6.2.11) and turning them into branching points, so that the incoming Brownian web path can continue along either of the two outgoing Brownian web paths. We will introduce the marking construction in detail in Section 6.4, where we will study couplings between the Brownian web and net.

In the rest of this section, we will define the left-right Brownian web $(\mathcal{W}^l, \mathcal{W}^r)$, give the hopping, wedge and mesh constructions of the Brownian net, and study various properties of the Brownian net, including the branching-coalescing point set, the backbone of the Brownian net, and special points of the Brownian net.

6.3.1 The Left-right Brownian Web and its Dual

The discussions above show that the key object in the construction of the Brownian net $\mathcal{N}$ is the left-right Brownian web $(\mathcal{W}^l, \mathcal{W}^r)$, which should be the diffusive scaling limit of the collections of leftmost and rightmost branching-coalescing random walk paths with branching probability ε. In turn, the key ingredient in the construction of $(\mathcal{W}^l, \mathcal{W}^r)$ is the pair of left-right SDEs in (6.11), which can be shown to be well-posed.

Proposition 6.3.1 (The left-right SDE) *For each initial state $(L_0, R_0) \in \mathbb{R}^2$, there exists a unique weak solution to the SDE (6.11) subject to the constraint that $L_t \leq R_t$ for all $t \geq T := \inf\{s \geq 0 : L_s = R_s\}$. Furthermore, almost surely, if $I := \{t \geq T : L_t = R_t\} \neq \emptyset$, then I is nowhere dense perfect set with positive Lebesgue measure.*

Proof Sketch. We sketch the basic idea and refer to [86, Proposition 2.1 and 3.1] for details. Assume w.l.o.g. that $L_0 = R_0 = 0$. Define

$$T_t := \int_0^t 1_{\{L_s < R_s\}} ds, \qquad S_t := \int_0^t 1_{\{L_s = R_s\}} ds,$$

and

$$\tilde{B}^l_{T_t} := \int_0^t 1_{\{L_s < R_s\}} dB^l_s, \quad \tilde{B}^r_{T_t} := \int_0^t 1_{\{L_s < R_s\}} dB^r_s, \quad \tilde{B}^s_{S_t} := \int_0^t 1_{\{L_s = R_s\}} dB^s_s.$$

Then (L_t, R_t) solves

$$\begin{aligned} L_t &= \tilde{B}^l_{T_t} + \tilde{B}^s_{S_t} - t, \\ R_t &= \tilde{B}^r_{T_t} + \tilde{B}^s_{S_t} + t, \end{aligned} \tag{6.12}$$

and the difference $D_t := R_t - L_t$ satisfies

$$D_t = (\tilde{B}^r - \tilde{B}^r)_{T_t} + 2t = \sqrt{2}\tilde{B}_{T_t} + 2T_t + 2S_t \tag{6.13}$$

with the constraint that D is nonnegative, where $\tilde{B} := \frac{1}{\sqrt{2}}(\tilde{B}^r - \tilde{B}^l)$ is also a standard Brownian motion.

Since $T_t = \int_0^t 1_{\{L_s < R_s\}} ds = \int_0^t 1_{\{D_s > 0\}} ds$, it is easily seen that T_t must be continuous and strictly increasing in t. Therefore $\tau := T_t$ admits an inverse $T^{-1}\tau = t$. Rewriting the equation (6.13) for D with respect to the variable τ, the time D spent at the origin, we obtain

$$\tilde{D}_\tau := \frac{1}{\sqrt{2}} D_{T^{-1}\tau} = \tilde{B}_\tau + \sqrt{2}\tau + \sqrt{2}S_{T^{-1}\tau}, \tag{6.14}$$

where $\tilde{D}_\tau$ can be regarded as a transformation of $\tilde{B}_\tau + \sqrt{2}\tau$ by adding an increasing function $\tilde{S}_\tau := \sqrt{2}S_{T^{-1}\tau}$, which increases only when $\tilde{D}_\tau = 0$ such that $\tilde{D}_\tau$ stays nonnegative. Such an equation is known as a Skorohod equation, with $\tilde{D}_t$ being the Skorohod reflection of $\tilde{B}_\tau + \sqrt{2}\tau$ at the origin. Such a Skorohod equation admits a pathwise unique solution [59, Section 3.6.C], with

$$\tilde{S}_\tau := \sqrt{2}S_{T^{-1}\tau} = -\inf_{0 \leq s \leq \tau}(\tilde{B}_\tau + \sqrt{2}\tau),$$

which is in fact also the local time at the origin for the drifted Brownian motion $\tilde{B}_\tau + \sqrt{2}\tau$ reflected at the origin.

Having determined $S_{T^{-1}\tau}$ and $D_{T^{-1}\tau}$ almost surely from $\tilde{B} = \frac{1}{\sqrt{2}}(\tilde{B}^r - \tilde{B}^l)$, to recover D_t, we only need to make a time change from τ back to $t := T_t + S_t = \tau + S_{T^{-1}\tau}$. Note that this time change has no effect when D is away from 0, but adds positive Lebesgue time when D is at 0. Therefore in contrast to $D_{T^{-1}\tau}$, which is a drifted Brownian motion reflected instantaneously at the origin, D_t is the same Brownian motion *sticky reflected* at the origin (see e.g., [102] and the references therein for further details on sticky reflected Brownian motions). Similarly, from $T_t = \tau$ and $S_t = S_{T^{-1}\tau}$, we can construct (L_t, R_t) in (6.12). From the same arguments, it is also easily seen that $I := \{t \geq 0 : L_t = R_t\}$ is almost surely a nowhere dense perfect set with positive Lebesgue measure. □

Having characterized the interaction of a single pair of leftmost and rightmost paths, we can now construct a collection of *left-right coalescing Brownian motions* $(l_{z_1}, \ldots, l_{z_k}, r_{z'_1}, \ldots, r_{z'_{k'}})$ with the properties that: (1) the paths evolve independently when they do not coincide; (2) $(l_{z_1}, \ldots, l_{z_k})$, respectively $(r_{z'_1}, \ldots, r_{z'_{k'}})$, is distributed as a collection of coalescing Brownian motions with drift -1, respectively $+1$; (3) every pair (l_{z_i}, r_{z_j}) is a weak solution to the left-right SDE (6.11). The construction can be carried out inductively. Assume w.l.o.g. that the paths all start at the same time. Then

- Let the paths evolve independently until the first time two paths meet.
- If this pair of paths are of the same type, then let them coalesce and iterate the construction with one path less than before.
- If this pair of paths are of different types, then let them evolve as a left-right pair solving the SDE (6.11), and let all other paths evolve independently, until the first time two paths (other than paths in the same left-right pair) meet.
- If the two meeting paths are of the same type, then let them coalesce and iterate the construction.
- If the two meeting paths are of different types, then let them form a left-right pair (breaking whatever pair relations they were in), and iterate the construction.

It is easily seen that this iterative construction terminates after a finite number of steps, when either a single path, or a single pair of leftmost and rightmost paths remains.

We are now ready to characterize the *left-right Brownian web* $(\mathcal{W}^l, \mathcal{W}^r)$.

Theorem 6.3.2 (The left-right Brownian web and its dual) *There exists an $\mathcal{H}^2$-valued random variable $(\mathcal{W}^l, \mathcal{W}^r)$, called the (standard) left-right Brownian web, whose distribution is uniquely determined by the following properties:*

(i) *The left Brownian web $\mathcal{W}^l$ (resp. right Brownian web $\mathcal{W}^r$) is distributed as a Brownian web $\mathcal{W}$ tilted with drift -1 (resp. $+1$), i.e., $\mathcal{W}^l$ (resp. $\mathcal{W}^r$) has the same distribution as the image of $\mathcal{W}$ under the space-time transformation $(x,t) \to (x-t, t)$ (resp. $(x,t) \to (x+t, t)$).*

(ii) *For any finite deterministic set of points $z_1, \ldots, z_k, z'_1, \ldots, z'_{k'} \in \mathbb{R}^2$, the collection of paths $(l_{z_1}, \cdots, l_{z_k}; r_{z'_1}, \cdots, r_{z'_{k'}})$ is distributed as a family of left-right coalescing Brownian motions.*

Furthermore, almost surely there exists a dual left-right Brownian web $(\widehat{\mathcal{W}^l}, \widehat{\mathcal{W}^r}) \in \widehat{\mathcal{H}}^2$, such that $(\mathcal{W}^l, \widehat{\mathcal{W}^l})$ (resp. $(\mathcal{W}^r, \widehat{\mathcal{W}^r})$) is distributed as $(\mathcal{W}, \widehat{\mathcal{W}})$ tilted with drift -1 (resp. $+1$), and $(-\widehat{\mathcal{W}^l}, -\widehat{\mathcal{W}^r})$ has the same distribution as $(\mathcal{W}^l, \mathcal{W}^r)$.

Proof Sketch. The existence and uniqueness of a left-right Brownian web satisfying properties (i)–(ii) follow the same argument as for the Brownian web in Theorem 6.2.3. The almost sure existence of a dual left-right Brownian web follows from the duality of the Brownian web. The fact that $(\widehat{\mathcal{W}^l}, \widehat{\mathcal{W}^r})$ has the same distribution as $(\mathcal{W}^l, \mathcal{W}^r)$, except for rotation around the origin in space-time by 180°, can be derived by taking the diffusive scaling limits of their discrete counterparts, where such an equality in distribution is trivial (see [86, Theorem 5.3] for further details). □

Exercise 6.3.3 *Show that almost surely, no path in $\mathcal{W}^l$ can cross any path in $\mathcal{W}^r$ or $\widehat{\mathcal{W}}^r$ from left to right, where π_1 is said to cross π_2 from left to right if there exist $s < t$ such that $\pi_1(s) < \pi_2(s)$ and $\pi_1(t) > \pi_2(t)$. Similarly, paths in $\mathcal{W}^r$ cannot cross paths in $\mathcal{W}^l$ or $\widehat{\mathcal{W}}^l$ from right to left.*

6.3.2 The Hopping Construction of the Brownian Net

We are now ready to construct the Brownian net $\mathcal{N}$ by allowing paths to hop back and forth between paths in the left-right Brownian web $(\mathcal{W}^l, \mathcal{W}^r)$, where a path π obtained by hopping from π_1 to π_2 at time t, with $\pi_1(t) = \pi_2(t)$, is defined by $\pi := \pi_1$ on $[\sigma_{\pi_1}, t]$ and $\pi := \pi_2$ on $[t, \infty)$.

Given π_1 and π_2 with $\pi_1(t) = \pi_2(t)$, which are in the scaling limit $\mathcal{N}$ of branching-coalescing random walk paths, it is not guaranteed that π obtained by hopping from π_1 to π_2 at time t is also in $\mathcal{N}$. Indeed, even though $\pi_1(t) = \pi_2(t)$, π_1 and π_2 may still arise as limits of random walk paths which do not meet, for which hopping is not possible. Therefore π has no approximating analogue among the branching-coalescing random walk paths, and hence π may not be in $\mathcal{N}$. One remedy is to allow hopping from π_1 to π_2 at time t only if the two paths *cross at time t*, i.e., $\sigma_{\pi_1}, \sigma_{\pi_2} < t$, there exist times $t^- < t^+$ with $(\pi_1(t^-) - \pi_2(t^-))(\pi_1(t^+) - \pi_2(t^+)) < 0$, and $t = \inf\{s \in (t^-, t^+) : (\pi_1(t^-) - \pi_2(t^-))(\pi_1(s) - \pi_2(s)) < 0\}$. We call t the *crossing time* between π_1 and π_2. If $\pi_1, \pi_2 \in \mathcal{N}$ and they cross at time t, then it is easily seen by discrete approximations that the path π obtained by hopping from π_1 to π_2 at time t must also be in $\mathcal{N}$.

Given a set of paths K, let $\mathcal{H}_{\text{cross}}(K)$ denote the set of paths obtained by hopping a finite number of times among paths in K at crossing times. The Brownian net $\mathcal{N}$ can then be constructed by setting $\mathcal{N} := \overline{\mathcal{H}_{\text{cross}}(\mathcal{W}^l \cup \mathcal{W}^r)}$. Here is the hopping characterization of the Brownian net from [86, Theorem 1.3].

Theorem 6.3.4 (Hopping characterization of the Brownian net) *There exists an $(\mathcal{H}, \mathcal{B}_\mathcal{H})$-valued random variable $\mathcal{N}$, called the standard Brownian net, whose distribution is uniquely determined by the following properties:*

(i) *For each $z \in \mathbb{R}^2$, $\mathcal{N}(z)$ a.s. contains a unique left-most path l_z and right-most path r_z.*
(ii) *For any finite deterministic set of points $z_1, \ldots, z_k, z'_1, \ldots, z'_{k'} \in \mathbb{R}^2$, the collection of paths $(l_{z_1}, \ldots, l_{z_k}, r_{z'_1}, \ldots, r_{z'_{k'}})$ is distributed as a family of left-right coalescing Brownian motions.*
(iii) *For any deterministic countable dense sets $\mathcal{D}^l, \mathcal{D}^r \subset \mathbb{R}^2$,*

$$\mathcal{N} = \overline{\mathcal{H}_{\text{cross}}(\{l_z : z \in \mathcal{D}^l\} \cup \{r_z : z \in \mathcal{D}^r\})} \quad \text{a.s.} \tag{6.15}$$

Proof. The uniqueness in law of a random variable $\mathcal{N}$ satisfying the above properties is easily verified by the same argument as that for the Brownian web in Theorem 6.2.3. For existence, we can just define $\mathcal{N} := \overline{\mathcal{H}_{\mathrm{cross}}(\mathcal{W}^{\mathrm{l}} \cup \mathcal{W}^{\mathrm{r}})}$, where $(\mathcal{W}^{\mathrm{l}}, \mathcal{W}^{\mathrm{r}})$ is the standard left-right Brownian web.

To show that $\mathcal{N}$ satisfies properties (i)–(iii), first note that $\mathcal{H}_{\mathrm{cross}}(\mathcal{W}^{\mathrm{l}} \cup \mathcal{W}^{\mathrm{r}})$ satisfies properties (i)–(ii), where for each deterministic z, $\mathcal{H}_{\mathrm{cross}}(\mathcal{W}^{\mathrm{l}} \cup \mathcal{W}^{\mathrm{r}})$ contains a leftmost element and a rightmost element, which are just $l_z \in \mathcal{W}^{\mathrm{l}}(z)$ and $r_z \in \mathcal{W}^{\mathrm{r}}(z)$. We claim that taking closure of $\mathcal{H}_{\mathrm{cross}}(\mathcal{W}^{\mathrm{l}} \cup \mathcal{W}^{\mathrm{r}})$ does not change the leftmost and rightmost element starting from any given $z = (x,t)$.

Indeed, if $\mathcal{N}(z)$ contains any path π with $\pi(t') < l_z(t') - \varepsilon$ for some $t' > t$ and $\varepsilon > 0$, then there exists a sequence $\pi_n \in \mathcal{H}_{\mathrm{cross}}(\mathcal{W}^{\mathrm{l}} \cup \mathcal{W}^{\mathrm{r}})$ starting from z_n, such that $\pi_n \to \pi$ and $\pi_n(t') \leq l_z(t') - \varepsilon$ for all n large. Since $l_{z_n} \in \mathcal{W}^{\mathrm{l}}(z_n)$ is the leftmost path in $\mathcal{H}_{\mathrm{cross}}(\mathcal{W}^{\mathrm{l}} \cup \mathcal{W}^{\mathrm{r}})$ starting from z_n, we have $l_{z_n} \leq \pi_n$, and hence also $l_{z_n}(t') \leq l_z(t') - \varepsilon$ for all n large. However, this is impossible because $l_{z_n} \to l_z$, and hence the time of coalescence between l_{z_n} and l_z tends to t as $n \to \infty$ by Corollary 6.2.8.

We can therefore conclude that $\mathcal{N} := \overline{\mathcal{H}_{\mathrm{cross}}(\mathcal{W}^{\mathrm{l}} \cup \mathcal{W}^{\mathrm{r}})}$ also satisfies properties (i)–(ii). Since any path in $\mathcal{W}^{\mathrm{l}} \cup \mathcal{W}^{\mathrm{r}}$ can be approximated by paths in $\mathcal{W}^{\mathrm{l}}(\mathcal{D}^{\mathrm{l}}) \cup \overline{\mathcal{W}^{\mathrm{r}}(\mathcal{D}^{\mathrm{r}})}$ in the strong sense as in Corollary 6.2.8, it is easily seen that $\mathcal{H}_{\mathrm{cross}}(\mathcal{W}^{\mathrm{l}} \cup \mathcal{W}^{\mathrm{r}}) \subset \overline{\mathcal{H}_{\mathrm{cross}}(\mathcal{W}^{\mathrm{l}}(\mathcal{D}^{\mathrm{l}}) \cup \mathcal{W}^{\mathrm{r}}(\mathcal{D}^{\mathrm{r}}))}$, and hence $\mathcal{N}$ also satisfies property (iii).

What we have left out in the proof is the a.s. precompactness of $\mathcal{H}_{\mathrm{cross}}(\mathcal{W}^{\mathrm{l}} \cup \mathcal{W}^{\mathrm{r}})$, which is needed for $\mathcal{N}$ to qualify as an $(\mathcal{H}, \mathcal{B}_{\mathcal{H}})$-valued random variable. We leave this as an exercise. □

Exercise 6.3.5 *Show that almost surely, $\mathcal{H}_{\mathrm{cross}}(\mathcal{W}^{\mathrm{l}} \cup \mathcal{W}^{\mathrm{r}})$ is precompact.*

Exercise 6.3.6 *Show that a.s., no path in $\mathcal{N}$ can cross any path in $\mathcal{W}^{\mathrm{r}}$ or $\widehat{\mathcal{W}^{\mathrm{r}}}$ from left to right, or cross any path in $\mathcal{W}^{\mathrm{l}}$ or $\widehat{\mathcal{W}^{\mathrm{l}}}$ from right to left, where the definition of crossing is, as in Exercise 6.3.3.*

Remark 6.3.7 Of course, we need to justify that the Brownian net characterized in Theorem 6.3.4 is indeed the scaling limit of branching-coalescing random walks, as motivated at the start of Section 6.3. Such a convergence result was established in [86, Section 5.3].

6.3.3 The Wedge Construction of the Brownian Net

The wedge and mesh constructions of the Brownian net $\mathcal{N}$ are both based on the observation that there are certain forbidden regions in space-time where Brownian net paths cannot enter. It turns out that the Brownian net can also be characterized as the set of paths that do not enter these forbidden regions. In the wedge construction, these forbidden regions, called *wedges*, are defined from the dual

The Brownian Web and the Brownian Net 293

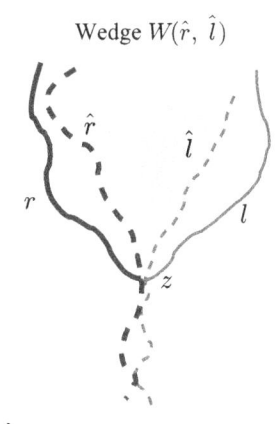

Figure 6.5. A wedge $W(\hat{r},\hat{l})$ with bottom point z.

left-right Brownian web $(\widehat{\mathcal{W}}^l, \widehat{\mathcal{W}}^r)$, while in the mesh construction, these forbidden regions, called *meshes*, are defined from the left-right Brownian web $(\mathcal{W}^l, \mathcal{W}^r)$.

Definition 6.3.8 (Wedges) *Let* $(\mathcal{W}^l, \mathcal{W}^r, \widehat{\mathcal{W}}^l, \widehat{\mathcal{W}}^r)$ *be the standard left-right Brownian web and its dual. For any* $\hat{r} \in \widehat{\mathcal{W}}^r$ *and* $\hat{l} \in \widehat{\mathcal{W}}^l$ *that are ordered with* $\hat{r}(s) < \hat{l}(s)$ *at the time* $s := \hat{\sigma}_{\hat{r}} \wedge \hat{\sigma}_{\hat{l}}$, *let* $T := \sup\{t < s : \hat{r}(t) = \hat{l}(t)\}$ *(possibly equals* $-\infty$*) be the first hitting time of* $\hat{r}$ *and* $\hat{l}$. *We call the open set (see Figure 6.5)*

$$W = W(\hat{r},\hat{l}) := \{(x,u) \in \mathbb{R}^2 : T < u < s, \hat{r}(u) < x < \hat{l}(u)\} \qquad (6.16)$$

a wedge of $(\widehat{\mathcal{W}}^l, \widehat{\mathcal{W}}^r)$ *with left and right boundary* $\hat{r}$ *and* $\hat{l}$ *and bottom point* $z := (\hat{r}(T),T) = (\hat{l}(T),T)$. *A path* π *is said to* enter W *from outside if there exist* $\sigma_\pi \leq s < t$ *such that* $(\pi(s),s) \notin \overline{W}$ *and* $(\pi(t),t) \in W$.

Theorem 6.3.9 (Wedge characterization of the Brownian net) *Let* $(\mathcal{W}^l, \mathcal{W}^r, \widehat{\mathcal{W}}^l, \widehat{\mathcal{W}}^r)$ *be the standard left-right Brownian web and its dual. Then almost surely,*

$$\mathcal{N} = \{\pi \in \Pi : \pi \text{ does not enter any wedge of } (\widehat{\mathcal{W}}^l, \widehat{\mathcal{W}}^r) \text{ from outside}\} \qquad (6.17)$$

is the standard Brownian net associated with $(\mathcal{W}^l, \mathcal{W}^r)$, *i.e.*, $\mathcal{N} = \overline{\mathcal{H}_{\text{cross}}(\mathcal{W}^l \cup \mathcal{W}^r)}$.

Remark 6.3.10 The wedge characterization can also be applied to the Brownian web $\mathcal{W}$ with both $\widehat{\mathcal{W}}^l$ and $\widehat{\mathcal{W}}^r$ replaced by the dual Brownian web $\widehat{\mathcal{W}}$. Indeed, $\mathcal{W}$ can be seen as a degenerate Brownian net $\mathcal{N}_b$ with branching parameter $b = 0$, where $\mathcal{N}_b$ can be constructed in the same way as the standard Brownian net $\mathcal{N}$ with $b = 1$, except that the left-right coalescing Brownian motions in $(\mathcal{W}^l, \mathcal{W}^r)$ now have drift $\mp b$ respectively. For $\mathcal{W}$, the wedge characterization is stronger than requiring paths not to cross any path in $\widehat{\mathcal{W}}$ (cf. Remark 6.2.6), because it also prevents paths from entering a wedge from outside through its bottom point.

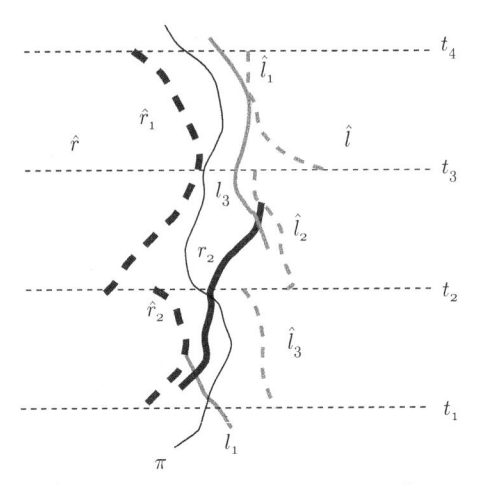

Figure 6.6. Steering a hopping path in the 'fish-trap' $(\hat{r}, \hat{l})$.

Proof of Theorem 6.3.9. First we show that no path in $\overline{\mathcal{H}_{\text{cross}}(\mathcal{W}^l \cup \mathcal{W}^r)}$ can enter any wedge of $(\widehat{\mathcal{W}^l}, \widehat{\mathcal{W}^r})$ from outside. If this is false, then there must be some path $\pi \in \overline{\mathcal{H}_{\text{cross}}(\mathcal{W}^l \cup \mathcal{W}^r)}$ which enters a wedge $W(\hat{r}, \hat{l})$ from outside. There are two possibilities: either π enters W from outside by crossing one of its two boundaries, which is impossible by Exercise 6.3.6; or π enters W from outside through its bottom point z. However, by the same argument as why a point of coalescence between two dual Brownian web paths cannot be hit by a forward Brownian web path (cf. Theorem 6.2.11), no path in $\mathcal{W}^l \cup \mathcal{W}^r$ can enter the bottom point of a wedge W, and hence neither can any path in $\overline{\mathcal{H}_{\text{cross}}(\mathcal{W}^l \cup \mathcal{W}^r)}$. This verifies the desired inclusion.

We now show the converse inclusion that any path not entering wedges from outside must be in $\overline{\mathcal{H}_{\text{cross}}(\mathcal{W}^l \cup \mathcal{W}^r)}$. Let π be such a path. The strategy to approximate π by hopping paths is illustrated in Figure 6.6.

To approximate π in a given time interval, say $[s,t]$, we first partition $[s,t]$ into sub-intervals of equal length, $[t_{i-1}, t_i]$, for $1 \leq i \leq N$. Fix an $\varepsilon > 0$. From the top time t_N, we consider the wedge formed by $\hat{r}_1 \in \widehat{\mathcal{W}^r}$ and $\hat{l}_1 \in \widehat{\mathcal{W}^l}$ starting respectively at $\pi(t_N) - \varepsilon$ and $\pi(t_N) + \varepsilon$ at time t_N. Note that $\hat{r}_1$ and $\hat{l}_1$ cannot meet during the interval $[t_{N-1}, t_N]$, otherwise π would be entering the wedge $W(\hat{r}_1, \hat{l}_1)$ from outside. At time t_{N-1}, we check whether $\hat{r}_1 < \pi(t_{N-1}) - \varepsilon$, and if it is the case, then we start $\hat{r}_2$ at $\pi(t_{N-1}) - \varepsilon$ at time t_{N-1}. Similarly we check whether $\hat{l}_1 > \pi(t_{N-1}) + \varepsilon$, and if it is the case, then start $\hat{l}_2$ at $\pi(t_{N-1}) + \varepsilon$ at time t_{N-1}. In any event, we still have a pair of dual left-right paths which enclose π on the interval $[t_{N-2}, t_{N-1}]$, which are within ε distance of π at the top time t_{N-1}. This procedure is iterated until the time interval $[t_0, t_1]$, and it constructs a 'fish-trap' of dual left-right paths, which can now be used to construct a forward hopping path that stays inside the 'fish-trap'.

Indeed, a forward path $l \in \mathcal{W}^l$ cannot cross dual paths $\hat{l}_i \in \widehat{\mathcal{W}^l}$, which form the right boundary of the 'fish-trap'. When l hits the left-boundary of the 'fish-trap', we can then hop to a path $r \in \mathcal{W}^r$ until it hits the right-boundary of the 'fish-trap'. Iterating this procedure then constructs a hopping path that stays inside the 'fish-trap'. The almost sure equicontinuity of paths in $\mathcal{W}^l \cup \mathcal{W}^r \cup \widehat{\mathcal{W}^l} \cup \widehat{\mathcal{W}^r}$ ensures that only a finite number of hoppings is needed to reach time t_N, and the supnorm distance on the interval $[s,t]$ between π and any path inside the 'fish-trap' can be made arbitrarily small by choosing N large and ε small. Therefore π can be approximated arbitrarily well by paths in $\mathcal{H}_{\text{cross}}(\mathcal{W}^l \cup \mathcal{W}^r)$. For further details, see [86, Lemma 4.7]. □

6.3.4 The Mesh Construction of the Brownian Net

Definition 6.3.11 (Meshes) *Let $(\mathcal{W}^l, \mathcal{W}^r)$ be the standard left-right Brownian web. If for a given $z = (x,t) \in \mathbb{R}^2$, there exist $l \in \mathcal{W}^l(z)$ and $r \in \mathcal{W}^r(z)$ such that $r(s) < l(s)$ on $(t, t + \varepsilon)$ for some $\varepsilon > 0$, then denoting $T := \inf\{s > t : r(s) = l(s)\}$, we call the open set (see Figure 6.7)*

$$M = M(r,l) := \{(y,s) \in \mathbb{R}^2 : t < s < T, \ r(s) < y < l(s)\} \tag{6.18}$$

a mesh of $(\mathcal{W}^l, \mathcal{W}^r)$ with left and right boundaries r and l and bottom point z. A path π is said to enter M *if there exist $\sigma_\pi < s < t$ such that $(\pi(s), s) \notin M$ and $(\pi(t), t) \in M$.*

Theorem 6.3.12 (Mesh characterization of the Brownian net) *Let $(\mathcal{W}^l, \mathcal{W}^r)$ be the standard left-right Brownian web. Then almost surely,*

$$\mathcal{N} = \{\pi \in \Pi : \pi \text{ does not enter any mesh of } (\mathcal{W}^l, \mathcal{W}^r)\} \tag{6.19}$$

is the standard Brownian net associated with $(\mathcal{W}^l, \mathcal{W}^r)$, i.e., $\mathcal{N} = \overline{\mathcal{H}_{\text{cross}}(\mathcal{W}^l \cup \mathcal{W}^r)}$.

Remark 6.3.13 The mesh characterization can also be applied to the Brownian web $\mathcal{W}$, where the bottom point of the mesh must be of either type $(0,2)$, $(1,2)$, or $(0,3)$ in Theorem 6.2.11.

We note that there is a subtle difference between *a path entering a mesh M from outside*, vs *a path entering a mesh M*. In particular, a path entering M (but not entering M from outside) could start inside $\overline{M}$, hit the boundary of M at a later time, and then move inside M. The heart of the proof of Theorem 6.3.12 consists in ruling out such scenarios for Brownian net paths, for which meshes

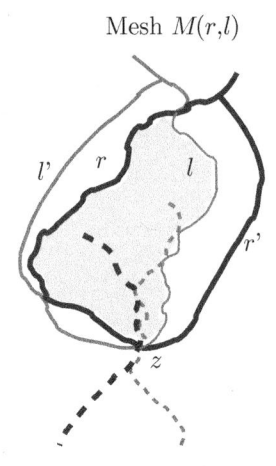

Figure 6.7. A mesh $M(r,l)$ with bottom point z.

play an essential role. As a by-product, one can show the following result (see [86, Proposition 1.8]), which is stronger than the assertions of Exercise 6.3.6.

Proposition 6.3.14 (Containment by left-most and right-most paths) *Let $(\mathcal{W}^l, \mathcal{W}^r)$ be the standard left-right Brownian web, and let $\mathcal{N}$ be the Brownian net associated with it. Then almost surely, there exist no $\pi \in \mathcal{N}$ and $l \in \mathcal{W}^l$ such that $l(s) \leq \pi(s)$ and $\pi(t) < l(t)$ for some $\sigma_\pi \vee \sigma_l < s < t$. An analogue statement holds for right-most paths.*

The proofs of Proposition 6.3.14 and Theorem 6.3.12 are fairly involved and we refer the details to [86, Theorem 1.7 and Proposition 1.8].

The wedge and mesh characterizations of the Brownian net have the following interesting corollary.

Proposition 6.3.15 (Brownian net is closed under hopping) *Let $\mathcal{N}$ be the standard Brownian net. Then:*

(i) *Almost surely for any $\pi_1, \pi_2 \in \mathcal{N}$ with $\pi_1(t) = \pi_2(t)$ and $\sigma_{\pi_1}, \sigma_{\pi_2} < t$ for some t, the path π defined by $\pi := \pi_1$ on $[\sigma_{\pi_1}, t]$ and $\pi := \pi_2$ on $[t, \infty)$ is in $\mathcal{N}$.*
(ii) *For any deterministic t, almost surely for any $\pi_1, \pi_2 \in \mathcal{N}$ with $\pi_1(t) = \pi_2(t)$ and $\sigma_{\pi_1}, \sigma_{\pi_2} \leq t$, the path π defined by $\pi := \pi_1$ on $[\sigma_{\pi_1}, t]$ and $\pi := \pi_2$ on $[t, \infty)$ is in $\mathcal{N}$.*

Exercise 6.3.16 *Use the mesh characterization to prove Proposition 6.3.15 (i), and the use then wedge characterization together with Proposition 6.3.18 below to prove Proposition 6.3.15 (ii).*

Here is an even more striking corollary of the mesh characterization [86, Proposition 1.13].

Proposition 6.3.17 (Image set property) *Let $\mathcal{N}$ be the Brownian net. For $T \in [-\infty, \infty)$, let $\mathcal{N}_T := \{\pi \in \mathcal{N} : \sigma_\pi = T\}$ and $N_T := \{(\pi(t), t) \in \mathbb{R}^2 : \pi \in \mathcal{N}_T, t \in (T, \infty)\}$. Then almost surely for any $T \in [-\infty, \infty)$, any path $\pi \in \Pi$ with $\sigma_\pi = T$ and $\{(\pi(t), t) : t > T\} \subset N_T$ is a path in $\mathcal{N}$.*

Note that for the Brownian web, this property is easily seen to hold.

6.3.5 The Branching-coalescing Point Set

Similar to the definition of the coalescing point set from the Brownian web in Section 6.2.4, we can define the so-called *branching-coalescing point set* from the Brownian net $\mathcal{N}$ as follows.

Given the Brownian net $\mathcal{N}$, and a closed set $A \subset \mathbb{R}$, define the *branching-coalescing point set* by

$$\xi_t^A := \{y \in \mathbb{R} : y = \pi(t) \text{ for some } \pi \in \mathcal{N}(A \times \{0\})\}, \qquad t \geq 0. \tag{6.20}$$

In other words, ξ_t^A is the set of points in $\mathbb{R}$ that lie on some path in $\mathcal{N}$ that start from A at time 0.

Using the wedge characterization of the Brownian net, we can compute the density of $\xi_t^\mathbb{R}$. As $t \downarrow 0$, we see in Proposition 6.3.18 below that the density diverges at the same rate $1/\sqrt{\pi t}$ as for the coalescing point set in Proposition 6.2.7, which indicates that coalescence plays the dominant role for small times, while the density converges to the constant 2 as $t \uparrow \infty$, which results from the balance between branching and coalescence for large times.

Proposition 6.3.18 (Density of branching-coalescing point set) *Let $\xi_\cdot^\mathbb{R}$ be the branching-coalescing point set defined from the Brownian net $\mathcal{N}$ as in (6.20). Then for all $t > 0$ and $a < b$,*

$$\mathbb{E}[|\xi_t^\mathbb{R} \cap [a,b]|] = (b-a)\left(\frac{e^{-t}}{\sqrt{\pi t}} + 2\Phi(\sqrt{2t})\right), \tag{6.21}$$

where $\Phi(x) = \frac{1}{\sqrt{2\pi}} \int_{-\infty}^{x} e^{-\frac{y^2}{2}} dy$.

Exercise 6.3.19 *Prove Proposition 6.3.18 by adapting the proof of Proposition 6.2.7 and showing that, almost surely, $\xi_t^\mathbb{R} \cap (a,b) \neq \emptyset$ if and only if $\hat{r} \in \widehat{\mathcal{W}}^\mathrm{r}$ starting at (a,t), and $\hat{l} \in \widehat{\mathcal{W}}^\mathrm{l}$ starting at (b,t), do not meet above time 0.*

Remark 6.3.20 Surprisingly, we can even identify the law of $\xi_t^\mathbb{R}$ as $t \uparrow \infty$, which is a Poisson point process on $\mathbb{R}$ with intensity 2. Furthermore, $\xi_\cdot$ is reversible with respect to the law of the Poisson point process on $\mathbb{R}$ with intensity 2. Formulated in terms of the Brownian net, this amounts to the statement that $\mathcal{N}(*, -\infty)$, the collection of Brownian net paths started at time $-\infty$ and called the *backbone of the Brownian net* in [86], has the same distribution as $-\mathcal{N}(*, -\infty)$, the set of paths obtained by reflecting the graph of each path in $\mathcal{N}(*, -\infty)$ across the origin in space-time. These results were established in [86, Section 9] by first observing their analogues for a discrete system of branching-coalescing random walks, and then passing to the continuum limit.

Remark 6.3.21 Proposition 6.3.18 implies that for each $t > 0$, almost surely $\xi_t^{\mathbb{R}}$ is a locally finite point set. However, almost surely there exists a dense set of random times at which $\xi_t^{\mathbb{R}}$ contains no isolated point, and is in particular uncountable (see [91, Proposition 3.14]).

Remark 6.3.22 As noted in Remark 6.2.10, started from the whole real line, the coalescing point set forms a Pfaffian point process at each time $t > 0$. It will be interesting to investigate whether the branching-coalescing point set $\xi_t^{\mathbb{R}}$ also admits an explicit characterization as a Pfaffian point process.

6.3.6 Special Points of the Brownian Net

Similar to the classification of special points for the Brownian web formulated in Theorem 6.2.11, we can give an almost sure classification of all points in $\mathbb{R}^2$ according to the configuration of paths in the Brownian net entering and leaving the point. Such an analysis was carried out in [91], where it was shown that a.s. there are 20 types of points, in contrast to the 7 types for the Brownian web.

Since Brownian net paths must be contained between paths in the left Brownian web and right Brownian web, as stated in Proposition 6.3.14, the classification of special points is in fact carried out mainly for $(\mathcal{W}^l, \mathcal{W}^r)$. First we introduce a notion of equivalence between paths entering and leaving a point, which is weaker than that introduced for the Brownian web in Section 6.2.5.

Definition 6.3.23 (Equivalence of paths entering and leaving a point) *We say $\pi_1, \pi_2 \in \Pi$ are equivalent paths entering $z = (x, t) \in \mathbb{R}^2$, denoted by $\pi_1 \sim_{\text{in}}^z \pi_2$, if π_1 and π_2 enter z and $\pi_1(t - \varepsilon_n) = \pi_2(t - \varepsilon_n)$ for a sequence $\varepsilon_n \downarrow 0$. We say π_1, π_2 are equivalent paths leaving z, denoted by $\pi_1 \sim_{\text{out}}^z \pi_2$, if π_1 and π_2 leave z and $\pi_1(t + \varepsilon_n) = \pi_2(t + \varepsilon_n)$ for a sequence $\varepsilon_n \downarrow 0$.*

When applied to paths in the Brownian web, the above notion of equivalence implies the equivalence introduced in Section 6.2.5, which is why we have abused the notation and used the same symbols $\sim_{\text{in}}^z$ and $\sim_{\text{out}}^z$. Although $\sim_{\text{in}}^z$ and $\sim_{\text{out}}^z$ are not equivalence relations on the space of all paths Π, they are easily seen to be equivalence relations on $\mathcal{W}^l \cup \mathcal{W}^r$. This allows us to classify z according to the equivalence classes of paths entering and leaving z, which are necessarily ordered.

To denote the type of a point, we first list the incoming equivalence classes of paths from left to right, and then, separated by a comma, the outgoing equivalence classes of paths from left to right. If an equivalence class contains only paths in $\mathcal{W}^l$, resp. $\mathcal{W}^r$, we will label it by l, resp. r, while if it contains both paths in $\mathcal{W}^l$ and in $\mathcal{W}^r$, we will label it by p, standing for pair. For points with (up to equivalence) one incoming and two outgoing paths, a subscript l, resp. r, means that all incoming paths belong to the left one, resp. right one, of the two outgoing equivalence classes; a subscript s indicates that incoming paths in $\mathcal{W}^l$ belong to the left outgoing equivalence class, while incoming paths in $\mathcal{W}^r$ belong to the right outgoing equivalence class. If at a point there are no incoming paths in $\mathcal{W}^l \cup \mathcal{W}^r$, then we denote this by o or n, where o indicates that there are no incoming paths in the Brownian net $\mathcal{N}$, while n indicates that there are incoming paths in $\mathcal{N}$ (but none in $\mathcal{W}^l \cup \mathcal{W}^r$).

For example, a point is of type (p, lp)$_r$ if at this point there is one equivalence class of incoming paths in $\mathcal{W}^l \cup \mathcal{W}^r$ and there are two outgoing equivalence classes. The incoming equivalence class is of type p while the outgoing equivalence classes are of type l and p, from left to right. All incoming paths in $\mathcal{W}^l \cup \mathcal{W}^r$ continue as paths in the outgoing equivalence class of type p.

Since the dual left-right Brownian web $(\widehat{\mathcal{W}}^l, \widehat{\mathcal{W}}^r)$ can be used to define a dual Brownian net $\widehat{\mathcal{N}}$, the type of z w.r.t. $(\widehat{\mathcal{W}}^l, \widehat{\mathcal{W}}^r)$ and $\widehat{\mathcal{N}}$ can be defined in the same way, after rotating their graphs in $\mathbb{R}^2$ by 180° around the origin.

We now state the classification result from [91, Theorem 1.7], while omitting its proof.

Theorem 6.3.24 (Special points of the Brownian net) *Let $(\mathcal{W}^l, \mathcal{W}^r)$ be the standard left-right Brownian web, let $(\widehat{\mathcal{W}}^l, \widehat{\mathcal{W}}^r)$ be its dual, and let $\mathcal{N}$ and $\widehat{\mathcal{N}}$ be the associated Brownian net and its dual. Then almost surely, each point in $\mathbb{R}^2$ is of one of the following 20 types in $\mathcal{N}/\widehat{\mathcal{N}}$:*

(1) (o, p)/(o, p), (o, pp)/(p, p), (p, p)/(o, pp), (o, ppp)/(pp, p), (pp, p)/(o, ppp), (p, pp)$_l$/(p, pp)$_l$, (p, pp)$_r$/(p, pp)$_r$;
(2) (p, pp)$_s$/(p, pp)$_s$, called *separation points*;
(3) (l, p)/(o, lp), (o, lp)/(l, p), (r, p)/(o, pr), (o, pr)/(r, p);
(4) (l, pp)$_r$/(p, lp)$_r$, (p, lp)$_r$/(l, pp)$_r$, (r, pp)$_l$/(p, pr)$_l$, (p, pr)$_l$/(r, pp)$_l$;
(5) (l, lp)r/(l, lp)$_r$, (r, pr)l/(r, pr)$_l$;
(6) (n, p)/(o, lr), (o, lr)/(n, p);

and all of these types occur. For each deterministic time $t \in \mathbb{R}$, almost surely, each point in $\mathbb{R} \times \{t\}$ is of either type (o, p)/(o, p), (o, pp)/(p, p), *or* (p, p)/(o, pp), *and*

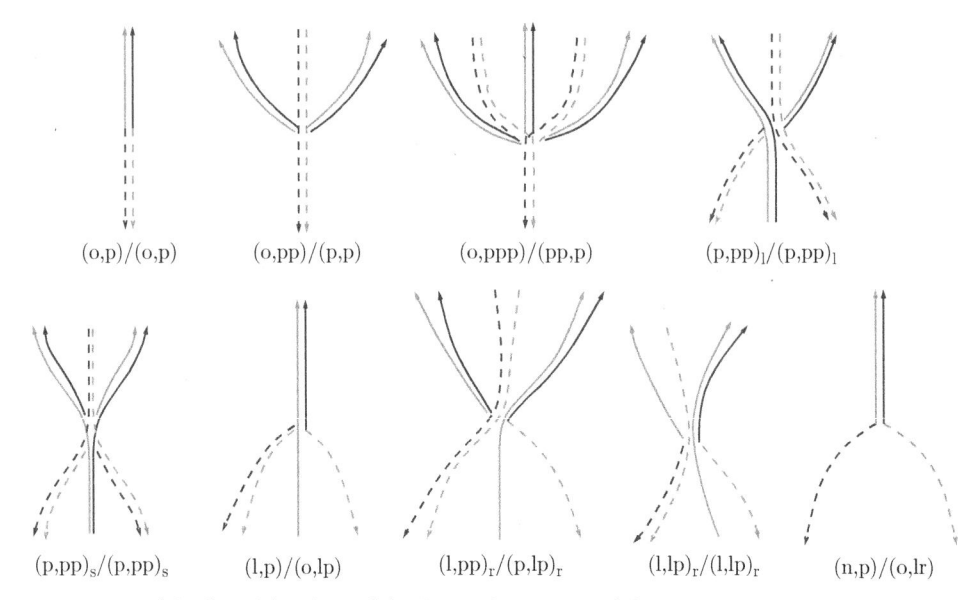

Figure 6.8. Special points of the Brownian net, modulo symmetry.

all of these types occur. A deterministic point $(x,t) \in \mathbb{R}^2$ is almost surely of type $(\mathrm{o},\mathrm{p})/(\mathrm{o},\mathrm{p})$.

Remark 6.3.25 Note that the points listed in item (1) are analogues of the seven types of points of the Brownian web in Figure 6.3, where a path of the Brownian web is replaced by a pair of paths in $(\mathcal{W}^l, \mathcal{W}^r)$. Modulo symmetry, this gives rise to four distinct types of points. The types of points listed within each item from (2)–(6) are related to each other by symmetry. Therefore modulo symmetry, there are nine types of special points for the Brownian net, as illustrated in Figure 6.8.

Remark 6.3.26 A basic ingredient in the proof of Theorem 6.3.24 is the characterization of the interaction between paths in $(\mathcal{W}^l, \mathcal{W}^r)$ and paths in $(\widehat{\mathcal{W}}^l, \widehat{\mathcal{W}}^r)$. The interaction between paths in $\mathcal{W}^l$ and $\widehat{\mathcal{W}}^l$, and similar between paths in $\mathcal{W}^r$ and $\widehat{\mathcal{W}}^r$, is given by Skorohod reflection, as mentioned in Remark 6.2.5. It turns out that paths in $\mathcal{W}^r$ interact with paths in $\widehat{\mathcal{W}}^l$ also via Skorohod reflection, except that given $\hat{l} \in \widehat{\mathcal{W}}^l$, if $r \in \mathcal{W}^r$ initially starts on the left of $\hat{l}$, then it is Skorohod reflected to the left of $\hat{l}$ until the collision local time between r and $\hat{l}$ exceeds an independent exponential random variable, at which time r crosses over to the right of $\hat{l}$ and is Skorohod reflected to the right of $\hat{l}$ from that time on (see [91, Lemma 2.1]).

Remark 6.3.27 Although Theorem 6.3.24 classifies points in $\mathbb{R}^2$ mostly according to the configuration of paths in the left-right Brownian web, it can be used to deduce the configuration of paths in the Brownian net entering and leaving each type of point. It turns out that $\sim_{\text{in}}^{z}$ is in fact an equivalence relation among paths in $\mathcal{N}$, and the same holds for $\sim_{\text{out}}^{z}$ for z of any type other than (o, lr). In particular, when z is not of type (o,lr), each equivalence class of paths in $\mathcal{W}^{\text{l}} \cup \mathcal{W}^{\text{r}}$ entering or leaving z can be enlarged to an equivalence class of paths in $\mathcal{N}$. In particular, by Proposition 6.3.14, all Brownian net paths in an equivalence class of type p, which contains a pair of equivalent paths $(l,r) \in (\mathcal{W}^{\text{l}}, \mathcal{W}^{\text{r}})$, must be bounded between l and r when sufficiently close to z. This applies in particular to points of type $(p, pp)_s$, the *separation points*, as well as points of type (pp,p), which we call the *meeting points*. See [91, Section 1.4] for further details.

Exercise 6.3.28 *Show that the set of separation points of $\mathcal{N}$ is a.s. countable and dense in $\mathbb{R}^2$.*

6.4 Coupling the Brownian Web and Net

We now introduce a coupling between the Brownian web $\mathcal{W}$ and the Brownian net $\mathcal{N}$, which is again best motivated from their discrete analogues, the coalescing and branching-coalescing random walks. In particular, this gives the fourth construction of the Brownian net, the *marking construction* mentioned before Section 6.3.1, which was developed by Newman, Ravishankar and Schertzer in [73].

Given a realization of branching-coalescing random walks as illustrated in Figure 6.9 (a), where each lattice point in $\mathbb{Z}^2_{\text{even}}$ is a branching point with probability ε, we can construct a collection of coalescing random walks by simply forcing the random walk to go either left or right with probability $1/2$ each, independently at each branching point. Interestingly, these branching points have analogues in the continuum limit $\mathcal{N}$, which are the *separation points*, i.e., points of type $(p, pp)_s$ in Theorem 6.3.24. Therefore, given a realization of the Brownian net $\mathcal{N}$, we can sample a Brownian web $\mathcal{W}$ by forcing the Brownian web paths to go either left or right with probability $1/2$ each, independently at each separation point. The complication is that each path in the Brownian net will encounter infinitely many separation points on any finite time interval. But fortunately, the number of separation points that is relevant for determining the path's position at a given time t is almost surely locally finite.

Conversely, given a realization of coalescing random walks as illustrated in Figure 6.9 (b), we can construct a collection of branching-coalescing random walks by turning each lattice point in $\mathbb{Z}^2_{\text{even}}$ independently into a branching point with probability ε. The key observation by Newman et al. in [73] is that each lattice point in $\mathbb{Z}^2_{\text{even}}$ is a point where a random walk path and some dual random walk

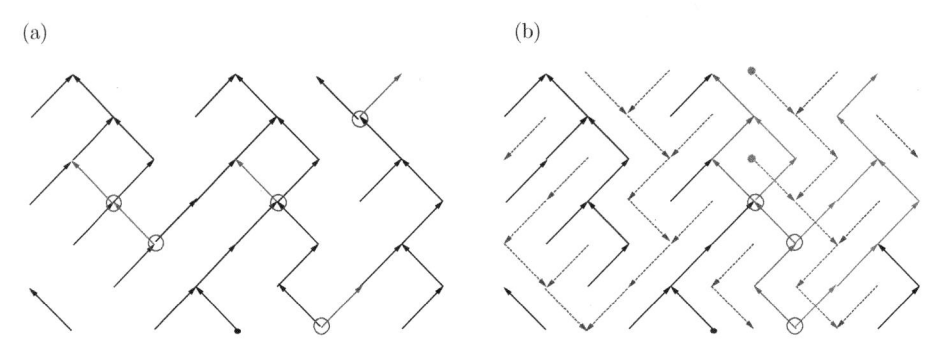

Figure 6.9. (a) Sampling coalescing random walks from branching-coalescing random walks. (b) Turning coalescing random walks into branching-coalescing random walks.

path are 1 unit of distance apart, and turning that lattice point into a branching point adds a new random walk path that crosses the dual random walk path (see Figure 6.9 (b), where the dotted lines are the dual paths, and the branching points are circled). In the diffusive scaling limit, most of the added branching points have no effect because they lead to small excursions from the coalescing random walk paths that vanish in the limit. The branching points that have an effect in the limit are points at which a forward random walk path and a dual random walk path – started macroscopically apart – come within 1 unit of distance from each other. In the scaling limit, these become precisely the (1,2) points of the Brownian web in Theorem 6.2.11, where a Brownian web path meets a dual Brownian web path. In the discrete setting, these points of close encounter between forward and backward random walk paths are turned independently into branching points with probability ε. In the continuum limit, this leads to Poisson marking of the points of collision between Brownian web paths and dual Brownian web paths, with the intensity measure given by the intersection local time measure between the forward and dual paths. The Brownian net can then be constructed by allowing the Brownian web paths to branch at these Poisson marked (1,2) points, which are precisely the separation points in the resulting Brownian net.

To formulate precisely the coupling between the Brownian web and net motivated by the above heuristic discussions, we first introduce the necessary background. By Exercise 6.3.28, the set of separation points S is a.s. a countable set, and it was pointed out in Remark 6.3.27 that any path $\pi \in \mathcal{N}$ entering a separation point $z = (x,t)$ must do so bounded between a pair of equivalent paths $(l,r) \in (\mathcal{W}^l, \mathcal{W}^r)$ entering z, and similarly when leaving z, it must also be enclosed by one of the two pairs of outgoing equivalent paths $(l_1, r_1), (l_2, r_2) \in$

$(\mathcal{W}^l(z), \mathcal{W}^r(z))$, ordered from left to right. We can then define

$$\operatorname{sgn}_\pi(z) := \begin{cases} -1 & \text{if } l_1 \leq \pi \leq r_1 \text{ on } [t, \infty), \\ +1 & \text{if } l_2 \leq \pi \leq r_2 \text{ on } [t, \infty). \end{cases} \quad (6.23)$$

For points of type $(1,2)$ in the Brownian web $\mathcal{W}$, we can similarly define

$$\operatorname{sgn}_\mathcal{W}(z) := \begin{cases} -1 & \text{if } z \text{ is of type } (1,2)_l \text{ in } \mathcal{W}, \\ +1 & \text{if } z \text{ is of type } (1,2)_r \text{ in } \mathcal{W}. \end{cases} \quad (6.24)$$

By setting the sign of paths entering each separation point independently to be ± 1 with probability $1/2$, we will recover the Brownian web as a subset of the Brownian net.

To construct the Brownian net from the Brownian web, a key object is the local time measure on points of intersection between paths in $\mathcal{W}$ and paths in $\widehat{\mathcal{W}}$., i.e., points of type $(1,2)$ in $\mathcal{W}$. Its existence was proved in [73, Proposition 3.1], which we quote below (see also [92, Proposition 3.4]).

Proposition 6.4.1 (Intersection local time) *Let $(\mathcal{W}, \widehat{\mathcal{W}})$ be the Brownian web and its dual. Then a.s. there exists a unique measure ℓ, concentrated on the set of points of type $(1,2)$ in $\mathcal{W}$, such that for each $\pi \in \mathcal{W}$ and $\hat{\pi} \in \widehat{\mathcal{W}}$,*

$$\ell(\{z = (x,t) \in \mathbb{R}^2 : \sigma_\pi < t < \hat{\sigma}_{\hat{\pi}}, \pi(t) = x = \hat{\pi}(t)\}) \\ = \lim_{\varepsilon \downarrow 0} \varepsilon^{-1} |\{t \in \mathbb{R} : \sigma_\pi < t < \hat{\sigma}_{\hat{\pi}}, |\pi(t) - \hat{\pi}(t)| \leq \varepsilon\}|. \quad (6.25)$$

The measure ℓ is a.s. non-atomic and σ-finite. We let ℓ_l and ℓ_r denote the restrictions of ℓ to the sets of points of type $(1,2)_l$ and $(1,2)_r$, respectively.

By Poisson marking $(1,2)$ points of $\mathcal{W}$ with intensity measure $\ell = \ell_l + \ell_r$, we obtain a countable set of $(1,2)$ points. Allowing paths in $\mathcal{W}$ to branch at these points then leads to the Brownian net.

We can now formulate the coupling between the Brownian web and net (see [92, Theorems 4.4 and 4.6]).

Theorem 6.4.2 (Coupling between the Brownian web and net) *Let $\mathcal{W}$ be the standard Brownian web, and $\mathcal{N}$ the standard Brownian net. Let S denote the set of separation points of $\mathcal{N}$. Then there exists a coupling between $\mathcal{W}$ and $\mathcal{N}$ such that:*

(i) *Almost surely $\mathcal{W} \subset \mathcal{N}$, and each separation point $z \in S$ in $\mathcal{N}$ is of type $(1,2)$ in $\mathcal{W}$.*

(ii) *Conditional on $\mathcal{N}$, the random variables* $(\text{sgn}_{\mathcal{W}}(z))_{z \in S}$ *are i.i.d. with* $\mathbb{P}[\text{sgn}_{\mathcal{W}}(z) = \pm 1 \mid \mathcal{N}] = 1/2$, *and a.s.,*

$$\mathcal{W} = \{\pi \in \mathcal{N} : \text{sgn}_{\pi}(z) = \text{sgn}_{\mathcal{W}}(z) \ \forall z \in S \ s.t. \ \pi \ \text{enters} \ z\}. \tag{6.26}$$

(iii) *Conditional on $\mathcal{W}$, the sets* $S_l := \{z \in S : \text{sgn}_{\mathcal{W}}(z) = -1\}$ *and* $S_r := \{z \in S : \text{sgn}_{\mathcal{W}}(z) = +1\}$ *are independent Poisson point sets with intensities ℓ_l and ℓ_r, respectively, and a.s.,*

$$\mathcal{N} = \lim_{\Delta_n \uparrow S} \text{hop}_{\Delta_n}(\mathcal{W}) \tag{6.27}$$

for any sequence of finite sets Δ_n increasing to $S = S_l \cup S_r$, where $\text{hop}_{\Delta_n}(\mathcal{W})$ is the set of paths obtained from $\mathcal{W}$ by allowing paths entering any (1,2) point $z \in \Delta_n$ to continue along either of the two outgoing paths at z.

Remark 6.4.3 Theorem 6.4.2 can be generalized to the case where $\mathcal{W}^l$, $\mathcal{W}$, and $\mathcal{W}^r$ may be tilted with different drifts, as long as the drifts remain ordered (see [92, Theorem 6.15]). We only need to modify the probability $\mathbb{P}(\text{sgn}_{\mathcal{W}}(z) = \pm 1 \mid \mathcal{N})$ in (ii), while the intensity measures ℓ_l and ℓ_r in (iii) need to be multiplied by constants depending on the drifts of $\mathcal{W}^l$, $\mathcal{W}$ and $\mathcal{W}^r$.

Remark 6.4.4 Theorem 6.4.2 (ii) shows how to sample a Brownian web $\mathcal{W}$ from a Brownian net $\mathcal{N}$ by forcing the paths to continue either left or right independently at each separation point. What if we sample independently another Brownian web $\mathcal{W}'$, conditioned on $\mathcal{N}$? How can we characterize the joint distribution of $\mathcal{W}$ and $\mathcal{W}'$? It turns out that $(\mathcal{W}, \mathcal{W}')$ forms a pair of so-called *sticky Brownian webs*, where Brownian motions in $\mathcal{W}$ and $\mathcal{W}'$ undergo sticky interaction. Such an object was first introduced in [56]. For further details, see [92, Section 3.3 and Lemma 6.16].

Proof Sketch. Theorem 6.4.2 is proved in [92, Section 6] via discrete approximation, using the fact that a similar coupling to that in Theorem 6.4.2 (ii) and (iii) holds in the discrete system. The proof is too complex and lengthy to be included here. Instead, we outline below some key ingredients.

The fact that a coupling exists between the Brownian web and net follows by taking the scaling limit of the coupled coalescing and branching-coalescing random walks, where i.i.d. signs are assigned to each branching point to determine the path of the coalescing walks.

To show that such a coupling satisfies Theorem 6.4.2 (ii), the key is to show that the branching-coalescing random walks, together with the branching points, converge to the Brownian net together with its separation points. Furthermore, under such a convergence, we can match the branching points with the separation points and assign them the same i.i.d. signs, such that coalescing random walk paths converge to paths in the Brownian net as long as the discrete and continuum paths follow the same signs at the respective branching and separation points.

The main difficulty with the above approach is that the set of separation points of $\mathcal{N}$ is dense in $\mathbb{R}^2$, so it is unclear in what sense the branching points should converge to the separation points. The solution rests on the observation that, for Brownian net paths starting at some time S, their positions at a later time U depend a.s. only on the signs of these paths at a locally finite set of separation points in the time interval (S, U) (called (S, U)-*relevant separation points*). As we refine our knowledge of the paths at more times in the interval (S, U), more separation points become relevant. Furthermore, the relevant separation points form a locally finite directed graph, called the *finite graph representation* of the Brownian net, which determines a coarse-grained structure of the Brownian net. Therefore a natural notion of convergence is to show that for any $S < U$, the (S, U)-relevant separation points and the associated finite graph representation arise as limits of similar structures in the branching-coalescing random walks. Once such a convergence is verified, Theorem 6.4.2 can then be easily verified.

To show that the coupling between the Brownian web and net satisfies the first statement in Theorem 6.4.2 (iii), the key is to show that when restricted to the set of intersection points between a pair of forward and dual Brownian web paths $(\pi, \hat{\pi})$, the set of separation points is distributed as a Poisson point process with intensity measure given by the intersection local time measure between π and $\hat{\pi}$. These separation points are in fact relevant separation points with respect to the starting times of π and $\hat{\pi}$, and hence must arise as the limit of relevant branching points in the discrete system. On the other hand, the corresponding relevant branching points form a Bernoulli point process with the intensity measure given by the counting measure on the set of space-time points where a pair of forward and dual coalescing random walk paths approximating $(\pi, \hat{\pi})$ come within distance 1. With space-time scaled by $(\varepsilon, \varepsilon^2)$, this counting measure can be shown to converge to the intersection local time measure between π and $\hat{\pi}$, and hence the limit of the branching points (i.e., the set of separation points restricted to the intersections between π and $\hat{\pi}$) must be a Poisson point process with intensity measure given by the intersection local time measure.

To prove the second part of Theorem 6.4.2 (iii), i.e., (6.27), note that under the coupling between the Brownian web and net, the Brownian web is embedded in the Brownian net and the Poisson marked (1,2) points of the web are exactly the separation points of the net. Using the coarse-grained structures of the Brownian net given by the finite graph representations, it can then be shown that turning the separation points into branching points for the Brownian web gives the Brownian net. □

The proof sketch above shows that the key ingredient in the proof of Theorem 6.4.2 is the notion of relevant separation points and the associated finite graph representation of the Brownian net. Since they shed new light on the structure of the Brownian net, we will discuss these notions in detail in the rest of this section.

6.4.1 Relevant Separation Points of the Brownian Net

As noted in Exercise 6.3.28, the set of separation points (i.e., points of type $(p, pp)_s$ in Theorem 6.3.24) is almost surely countable and dense in $\mathbb{R}^2$. In fact almost surely each path in the Brownian net encounters infinitely many separation points in any open time interval. However, for given deterministic times $S < U$, there is only a locally finite set of separation points which are relevant for deciding where paths in the Brownian net started at time S end up at time U. More precisely, as illustrated in Figure 6.10, we define (cf. [91, Section 2.3]):

Definition 6.4.5 (Relevant separation points) *A separation point $z = (x, t)$ of the Brownian net $\mathcal{N}$ is called (S, U)-relevant for some $-\infty \leq S < t < U \leq \infty$, if there exists $\pi \in \mathcal{N}$ such that $\sigma_\pi = S$ and $\pi(t) = x$, and there exist $l \in \mathcal{W}^l(z)$ and $r \in \mathcal{W}^r(z)$ such that $l < r$ on (t, U).*

Note that because z is a separation point, l and r are continuations of incoming paths at z. By Proposition 6.3.15, the paths obtained by hopping from π to either l or r at z are both paths in the Brownian net, which are distinct on the time interval (t, U). We do not require $l(U) < r(U)$ because an (S, U)-relevant separation point defined as above also turns out to be a relevant separation point w.r.t. the dual Brownian net $\widehat{\mathcal{N}}$.

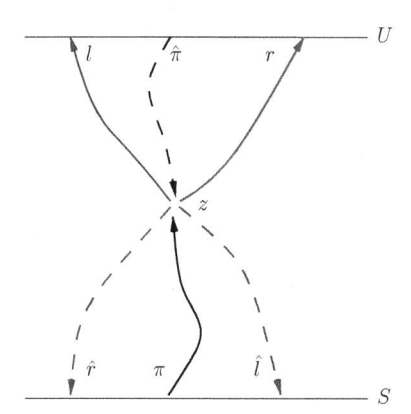

Figure 6.10. An (S, U)-relevant separation point.

Exercise 6.4.6 *Use the wedge characterization of the Brownian net and the steering argument illustrated in Figure 6.6 to show that almost surely, for each $-\infty \leq S < U \leq \infty$, a separation point $z = (x,t)$ with $S < t < U$ is (S,U)-relevant in $\mathcal{N}$ if and only if $-z$ is $(-U,-S)$-relevant in the dual Brownian net $-\widehat{\mathcal{N}}$ rotated 180° around the origin in $\mathbb{R}^2$.*

A crucial property of the (S,U)-relevant separation points is that they are almost surely locally finite, which follows from the following density calculation (cf. [91, Proposition 2.9]).

Proposition 6.4.7 (Density of relevant separation points) *Let $\mathcal{N}$ be a standard Brownian net. Then for each deterministic $-\infty \leq S < U \leq \infty$, if $R_{S,U}$ denotes the set of (S,U)-relevant separation points, then*

$$\mathbb{E}[|R_{S,U} \cap A|] = 2\int_A \Psi(t-S)\Psi(U-t)\,dx\,dt$$

for all Borel-measurable $A \subset \mathbb{R} \times (S,U)$, (6.28)

where

$$\Psi(t) := \frac{e^{-t}}{\sqrt{\pi t}} + 2\Phi(\sqrt{2t}), \quad 0 < t \leq \infty, \quad \text{and} \quad \Phi(x) := \frac{1}{\sqrt{2\pi}}\int_{-\infty}^x e^{-y^2/2}dy.$$
(6.29)

In particular, if $-\infty < S, U < \infty$, then $R_{S,U}$ is a.s. a locally finite subset of $\mathbb{R} \times [S,U]$.

Proof Sketch. For $S < s < u < U$, let

$$E_s := \{\pi(s) : \pi \in \mathcal{N}, \sigma_\pi = S\} \quad \text{and} \quad F_u := \{\hat{\pi}(s) : \hat{\pi} \in \widehat{\mathcal{N}}, \hat{\sigma}_{\hat{\pi}} = U\}.$$

Observe that a.s., each (S,U)-relevant separation point z in the strip $\mathbb{R} \times [s,u]$ can be traced back along some Brownian net path $\pi \in \mathcal{N}$ to a position $x \in E_s$. Furthermore, if $(l,r) \in (\mathcal{W}^l, \mathcal{W}^r)$ is the pair of left-right Brownian web paths starting at (x,s), then we must have $F_u \cap (l(u), r(u)) \neq \emptyset$, and z is bounded between l and r on the time interval $[s,u]$. Therefore each (S,U)-relevant separation point in $\mathbb{R} \times [s,u]$ can be approximated by some (x,s) with x in

$$Q_{s,u} := \{x \in E_s : F_u \cap (l(u), r(u)) \neq \emptyset, \text{ where } (l,r) \in (\mathcal{W}^l(x,s), \mathcal{W}^r(x,s))\}.$$

The density of the set $Q_{s,u}$ can be easily computed. As we partition (S,U) into disjoint intervals $[t_i, t_{i+1})$ of size $1/n$ with $n \uparrow \infty$, each (S,U)-relevant separation point z can be approximated by some (x_n, t_{i_n}) with $x_n \in Q_{t_{i_n}, t_{i_n+1}}$ for some i_n, such that $(x_n, t_{i_n}) \to z$ as $n \to \infty$. The upper bound on the density of $R_{S,U}$ then follows by Fatou's Lemma. For the lower bound, it suffices to consider points in $Q_{t_i,t_{i+1}}$ which are separated from each other by some $\delta > 0$ which is sent to zero after taking the limit $n \to \infty$. □

Exercise 6.4.8 *Prove Proposition 6.4.7 by rigorously implementing the strategy outlined above.*

6.4.2 Finite Graph Representation of the Brownian Net

Having defined (S, U)-relevant separation points and established their local finiteness, we are now ready to introduce the *finite graph representation*, which is a directed graph with the interior vertices given by the (S, U)-relevant separation points, and the directed edges determined by how Brownian net paths go from one relevant separation point to the next. Such a directed graph gives a coarse-grained representation of the Brownian net, and plays a key role in studying the coupling between the Brownian web and net in Theorem 6.4.2.

Let $-\infty < S < U < \infty$ be deterministic times, let $R_{S,U}$ be the set of (S, U)-relevant separation points of $\mathcal{N}$ and set

$$R_S := \mathbb{R} \times \{S\},$$
$$R_U := \{(x, U) : x \in \mathbb{R}, \exists \pi \in \mathcal{N} \text{ with } \sigma_\pi = S \text{ s.t. } \pi(U) = x\}. \tag{6.30}$$

We will make the set $R := R_S \cup R_{S,U} \cup R_U$ into a directed graph, with the directed edges representing the Brownian net paths.

First we identify the directed edges leading out from each $z = (x,t) \in R_{S,U}$. By Remark 6.3.27 and Proposition 6.3.14, any $\pi \in \mathcal{N}$ with $\sigma_\pi = S$ that enters an (S, U)-relevant separation point z must leave z enclosed by one of the two outgoing pairs of equivalent left-right paths $(l_1, r_1), (l_2, r_2) \in (\mathcal{W}^l(z), \mathcal{W}^r(z))$, with $l_1 \sim^z_{\text{out}} r_1$ and $l_2 \sim^z_{\text{out}} r_2$. Take the pair (l_1, r_1) for instance. If $l_1(U) \neq r_1(U)$, then there must be a last separation point z' along the pair (l_1, r_1), which is also an (S, U)-relevant separation point because hopping from π to either l_1 or r_1 at time t still gives a Brownian net path. If $l_1(U) = r_1(U)$, then we just set $z' = (l_1(U), U) = (r_1(U), U) \in R_U$. In either case, we note that all paths that leave z bounded between l_1 and r_1 must continue to do so until they reach z'. Therefore we draw a directed edge from z to z', denoted by $z \to_{l_1, r_1} z'$, representing all Brownian net paths that go from z to z' while bounded between l_1 and r_1. Similarly, paths that leave z while bounded between l_2 and r_2 must all lead to some $z'' \in R_{S,U} \cup R_U$, which is represented by a directed edge $z \to_{l_2, r_2} z''$.

Next we identify the directed edges leading out from $z = (x, S) \in R_S$. By Theorem 6.3.24, at a deterministic time S, almost surely each $z \in R_S$ is of type (o, p), (p, p), or (o, pp). If z is of type (o, p) or (p, p), then a single equivalent pair $l \sim^z_{\text{out}} r$ starts from z, and all Brownian net paths leaving z must be bounded between l and r, and leads to some $z' \in R_{S,U} \cup R_U$, which we represent by a directed edge $z \to_{l,r} z'$. Similarly when z is of type (o, pp), two directed edges start from z.

Given the above directed graph with vertex set $R_S \cup R_{S,U} \cup R_U$, it is not difficult to see that each Brownian net path from time S to U corresponds to a directed path from R_S to R_U, and conversely, each directed path from R_S to R_U can be associated with a family of Brownian net paths from time S to U. We summarize the basic properties below (see [92, Proposition 6.5]).

Proposition 6.4.9 (Finite graph representation) *Let $\mathcal{N}$ be a Brownian net with associated left-right Brownian web $(\mathcal{W}^l, \mathcal{W}^r)$, and let $-\infty < S < U < \infty$ be deterministic times. Let $R := R_S \cup R_{S,U} \cup R_U$ and directed edges $\to_{l,r}$ be defined as above. Then, a.s. (see Figure 6.11):*

(a) *For each $z \in R_S$ that is not of type (o, pp), there exist unique $l \in \mathcal{W}^l(z)$, $r \in \mathcal{W}^r(z)$ and $z' \in R$ such that $z \to_{l,r} z'$.*
(b) *For each $z = (x,t)$ such that either $z \in R_{S,U}$ or $z \in R_S$ is of type (o, pp), there exist unique $l, l' \in \mathcal{W}^l(z)$, $r, r' \in \mathcal{W}^r(z)$ and $z', z'' \in R$ such that $l \leq r' < l' \leq r$ on $(t, t+\varepsilon)$ for some $\varepsilon > 0$, $z \to_{l,r'} z'$ and $z \to_{l',r} z''$. For $z \in R_{S,U}$ one has $z' \neq z''$. For $z \in R_S$ of type (o, pp), one has $z' \neq z''$ if and only if there exists $\hat\pi \in \hat{\mathcal{N}}$ with $\hat\sigma_{\hat\pi} = U$ such that $\hat\pi$ enters z.*
(c) *For each $\pi \in \mathcal{N}$ with $\sigma_\pi = S$, there exist $z_i = (x_i, t_i) \in R$ $(i = 0,\ldots,n)$ and $l_i \in \mathcal{W}^l(z_i)$, $r_i \in \mathcal{W}^l(z_i)$ $(i = 0,\ldots,n-1)$ such that $z_0 \in R_S$, $z_n \in R_U$, $z_i \to_{l_i, r_i} z_{i+1}$ and $l_i \leq \pi \leq r_i$ on $[t_i, t_{i+1}]$ $(i = 0,\ldots,n-1)$.*
(d) *If $z_i = (x_i, t_i) \in R$ $(i = 0,\ldots,n)$ and $l_i \in \mathcal{W}^l(z_i)$, $r_i \in \mathcal{W}^l(z_i)$ $(i = 0,\ldots,n-1)$ satisfy $z_0 \in R_S$, $z_n \in R_U$, and $z_i \to_{l_i,r_i} z_{i+1}$ $(i = 0,\ldots,n-1)$, then there exists a $\pi \in \mathcal{N}$ with $\sigma_\pi = S$ such that $l_i \leq \pi \leq r_i$ on $[t_i, t_{i+1}]$.*

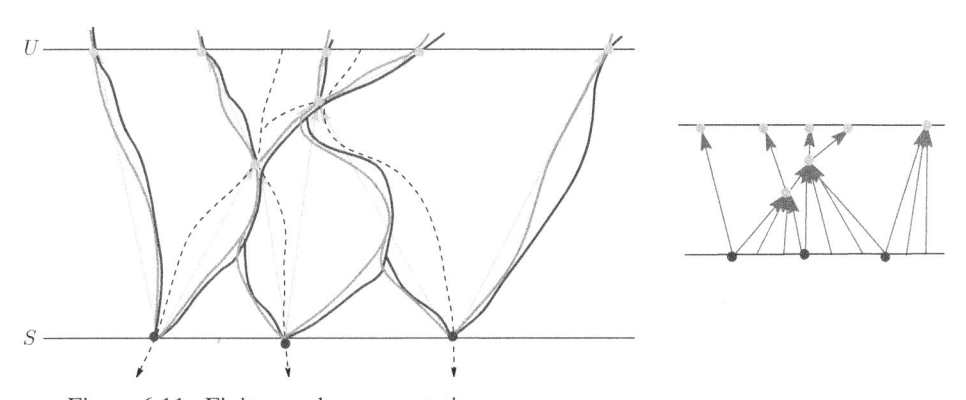

Figure 6.11. Finite graph representation.

Exercise 6.4.10 *Use the finite graph representation to show that: conditioned on the Brownian net $\mathcal{N}$ with the set of separation points S, almost surely for any $(\alpha_z)_{z \in S} \in \{\pm 1\}^S$,*

$$W = \{\pi \in \mathcal{N} : \operatorname{sgn}_\pi(z) = \alpha_z \, \forall z \in S \text{ s.t. } \pi \text{ enters } z\}$$

(cf. (6.26)) is a closed subset of $\mathcal{N}$, containing at least one path starting from each $(x,t) \in \mathbb{R}^2$.

6.5 Scaling Limits of Random Walks in i.i.d. Space-time Environment

We will now show how the Brownian web and net can be used to construct the continuum limits of one-dimensional random walks in i.i.d. random space-time environments.

A random walk X in an i.i.d. random space-time environment is defined as follows. Let $\omega := (\omega_z)_{z \in \mathbb{Z}^2_{\text{even}}}$ be i.i.d. $[0,1]$-valued random variables with common distribution μ. We view ω as a random space-time environment, such that conditioned on ω, if the random walk is at position x at time t, then in the next time step the walk jumps to $x+1$ with probability $\omega_{(x,t)}$ and jumps to $x-1$ with probability $1 - \omega_{(x,t)}$ (see Figure 6.12). Let $P^\omega_{(x,s)}$ denote probability for the walk starting at x at time s in the environment ω, and let $\mathbb{P}$ denote probability for ω.

The question is: if we scale space and time by ε and ε^2 respectively, is it possible to choose a law μ_ε for an environment $\omega^\varepsilon := (\omega^\varepsilon_z)_{z \in \mathbb{Z}^2_{\text{even}}}$ such that the walk in the environment ω^ε converges as $\varepsilon \downarrow 0$ to a limiting random motion in a continuum space-time random environment? Different ways of looking at the random walk in random environment suggest that the answer is yes.

One alternative way of characterizing the law of random walk in random environment is to consider the law of the family of random transition probability kernels

$$K^\omega_{s,t}(x,y) := P^\omega_{(x,s)}(X(t) = y), \qquad (x,s),(y,t) \in \mathbb{Z}^2_{\text{even}}, \ s \leq t. \tag{6.31}$$

Still another way is to specify the *n-point motions*, i.e., the law of n random walks $\vec{X} := (X_1(t), \ldots, X_n(t))_{t \geq 0}$ sampled independently from the same environment ω, and then averaged with respect to the law of ω. Note that $\vec{X}$ is a Markov chain with transition probability kernel

$$K^{(n)}_{0,t}(\vec{x}, \vec{y}) = \int \prod_{i=1}^n K^\omega_{0,t}(x_i, y_i) \mathbb{P}(\mathrm{d}\omega), \quad (x_i, 0), (y_i, t) \in \mathbb{Z}^2_{\text{even}} \text{ for } 1 \leq i \leq n. \tag{6.32}$$

Furthermore, the *n*-point motions are *consistent* in the sense that the law of any k-element subset of $(X_1(\cdot), \ldots, X_n(\cdot))$ is governed by that of the k-point motion.

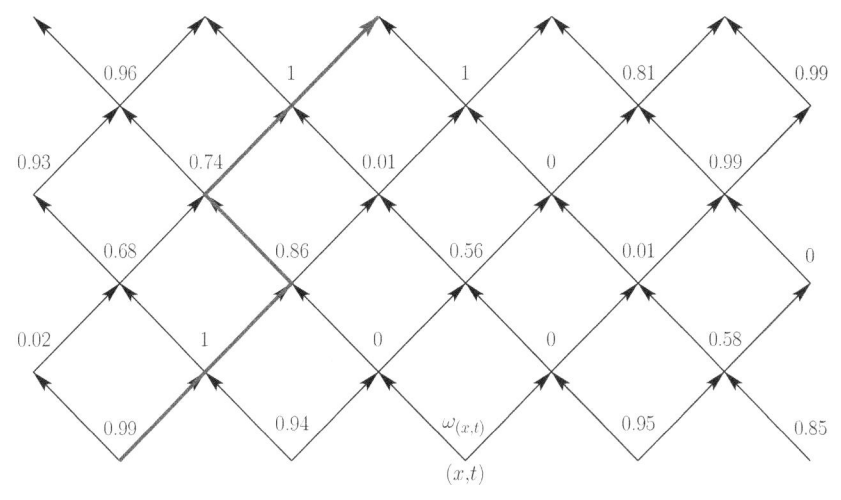

Figure 6.12. Random walk on $\mathbb{Z}^2_{\text{even}}$ in a random environment ω.

Note that the moments of $\omega_{(0,0)}$ are determined by $(K^{(n)})_{n \in \mathbb{N}}$, since

$$\mathbb{E}[\omega^n_{(0,0)}] = K^{(n)}_{0,1}(\vec{0}, \vec{1}).$$

Therefore the law of $\omega_{(0,0)}$, as well as that of ω and K^ω, are uniquely determined by the law of the n-point motions $(K^{(n)})_{n \in \mathbb{N}}$.

Evidence for the existence of a continuum limit for random walks in i.i.d. space-time random environments came from the convergence of the n-point motions, which was first established by Le Jan and Lemaire [62] for i.i.d. Beta-distributed random environments, and subsequently extended to general environments by Howitt and Warren [55]. By the theory of *stochastic flow of kernels* developed by Le Jan and Raimond [63], this implies that the family of random probability kernels K^ω in (6.32) also converges in a suitable sense to a continuum limit. Motivated by these results, it was then shown in [92] that under the same convergence criterion as in [55], not only the n-point motions converge, but also the random environments themselves converge to a continuum space-time limit which can be constructed explicitly from the Brownian web and net.

In Section 6.5.1 below, we will first briefly recall the theory of *stochastic flows of kernels* from [63], and then review Howitt and Warren's convergence result [55] for the n-point motions of random walks in i.i.d. space-time random environments. In Section 6.5.2, we will then show how to construct the limiting continuum space-time environment from the Brownian web and net, which arises naturally if one looks at the discrete environments in the right way. Lastly in Section 6.5.3, we will discuss some properties of the continuum random motion in random environment.

6.5.1 Stochastic Flows of Kernels and the Howitt-Warren Flows

In (6.31) and (6.32), we saw that the family of random transition probability kernels $K_{s,t}^\omega(x,\cdot)$ and the family of consistent n-point motions provide alternative characterizations of a random walk in i.i.d. space-time random environment. It turns out that in general, without knowing the existence of any underlying random environment, there is a correspondence between a consistent family of n-point motions on a Polish space, and a family of random probability kernels called a *stochastic flow of kernels*. This was the main result of Le Jan and Raimond in [63, Theorem 2.1], which motivated the study of concrete examples of consistent n-point motions in [64, 62, 55]. We now recall the notion of a *stochastic flow of kernels*.

For any Polish space E, let $\mathcal{B}(E)$ denote the Borel σ-field on E and let $\mathcal{M}_1(E)$ denote the space of probability measures on E, equipped with the weak topology and Borel σ-algebra. A *random probability kernel*, defined on some probability space $(\Omega, \mathcal{F}, \mathbb{P})$, is a measurable function $K : \Omega \times E \to \mathcal{M}_1(E)$. Two random probability kernels K, K' are said to equal in finite-dimensional distributions if for each $x_1, \ldots, x_n \in E$, the n-tuple of random probability measures $(K(x_1, \cdot), \ldots, K(x_n, \cdot))$ is equally distributed with $(K'(x_1, \cdot), \ldots, K'(x_n, \cdot))$. Two or more random probability kernels are called independent if their finite-dimensional distributions are independent.

Definition 6.5.1 (Stochastic flow of kernels) *A stochastic flow of kernels on a Polish space E is a collection $(K_{s,t})_{s \leq t}$ of random probability kernels on E such that*

(i) *For all $s \leq t \leq u$ and $x \in E$, a.s. $K_{s,s}(x,A) = \delta_x(A)$ and $\int_E K_{s,t}(x, \mathrm{d}y) K_{t,u}(y,A) = K_{s,u}(x,A)$ for all $A \in \mathcal{B}(E)$.*
(ii) *For each $t_0 < \cdots < t_n$, the random probability kernels $(K_{t_{i-1}, t_i})_{i=1,\ldots,n}$ are independent.*
(iii) *$K_{s,t}$ and $K_{s+u,t+u}$ are equal in finite-dimensional distributions for each real $s \leq t$ and u.*

We have omitted two weak continuity conditions on $(K_{s,t}(x,\cdot))_{s \leq t, x \in E}$ from [63, Definition 2.3], which are automatically satisfied by a version of K if the n-point motions defined via (6.32) are a family of Feller processes.

Remark 6.5.2 Although we motivated the notion of stochastic flow of kernels from random walks in i.i.d. space-time random environments, the kernels K in Definition 6.5.1 cannot be associated with an underlying random environment unless there exists a version of K such that Definition 6.5.1 (i) is strengthened to:

(i′)   A.s., $\forall s \leq t \leq u$ and $x \in E$, $K_{s,s}(x,\cdot) = \delta_x(\cdot)$ and $\int_E K_{s,t}(x,\mathrm{d}y) K_{t,u}(y,\cdot)$
$= K_{s,u}(x,\cdot)$.

In general, it is not known whether such a version always exists.

Le Jan and Raimond showed in [63, Theorem 2.1] that every consistent family of Feller n-point motions corresponds to a stochastic flow of kernels. Using Dirichlet form construction of Markov processes, they then constructed as an example in [64] a consistent family of n-point motions on the circle which are a special type of sticky Brownian motions. Subsequently, Le Jan and Lemaire showed in [62] that the n-point motions of random walks in i.i.d. Beta-distributed space-time random environments converge to the n-point motions constructed in [64]. Howitt and Warren [55] then found the general condition for the convergence of n-point motions of random walks in i.i.d. space-time random environments, and they characterized the limiting n-point motions, which are also sticky Brownian motions, in terms of well-posed martingale problems.

We next recall Howitt and Warren's result [55], or rather, a different formulation of their result as presented in [92, Appendix A] with discrete instead of continuous time random walks, and with a reformulation of the martingale problem characterizing the limiting sticky Brownian motions.

Theorem 6.5.3 (Convergence of n-point motions to sticky Brownian motions)
For $\varepsilon > 0$, let μ_ε be the common law of an i.i.d. space-time random environment $(\omega_z^\varepsilon)_{z \in \mathbb{Z}^2_{\mathrm{even}}}$, such that

$$\begin{aligned}&\text{(i)} \quad \varepsilon^{-1} \int (2q-1) \mu_\varepsilon(\mathrm{d}q) \xrightarrow[\varepsilon \to 0]{} \beta, \\ &\text{(ii)} \quad \varepsilon^{-1} q(1-q) \mu_\varepsilon(\mathrm{d}q) \underset{\varepsilon \to 0}{\Longrightarrow} \nu(\mathrm{d}q),\end{aligned} \qquad (6.33)$$

for some $\beta \in \mathbb{R}$ and finite measure ν on $[0,1]$, where $\Rightarrow$ denotes weak convergence. Let $\vec{X}^\varepsilon := (X_1^\varepsilon, \ldots, X_n^\varepsilon)$ be n independent random walks in ω^ε with $\varepsilon \vec{X}^\varepsilon(0) \to \vec{X}(0) = (X_1(0), \ldots, X_n(0)) \in \mathbb{R}^n$ as $\varepsilon \downarrow 0$. Then $(\varepsilon \vec{X}^\varepsilon(\varepsilon^2 t))_{t \geq 0}$ converges weakly to a family of sticky Brownian motions $(\vec{X}(t))_{t \geq 0} = (X_1(t), \ldots, X_n(t))_{t \geq 0}$, whose law is the unique solution of the following Howitt-Warren martingale problem *with drift β and characteristic measure ν:*

(i) *$\vec{X}$ is a continuous, square-integrable semi-martingale with initial condition $\vec{X}(0)$;*

(ii) *The covariance process between X_i and X_j is given by*

$$\langle X_i, X_j \rangle(t) = \int_0^t \mathbf{1}_{\{X_i(s)=X_j(s)\}} ds, \qquad t \geq 0, \ i,j = 1,\ldots,n; \qquad (6.34)$$

(iii) *For each non-empty $\Delta \subset \{1,\ldots,n\}$, let*

$$f_\Delta(\vec{x}) := \max_{i \in \Delta} x_i \quad \text{and} \quad g_\Delta(\vec{x}) := |\{i \in \Delta : x_i = f_\Delta(\vec{x})\}| \qquad (\vec{x} \in \mathbb{R}^n). \qquad (6.35)$$

Then

$$f_\Delta(\vec{X}(t)) - \int_0^t \beta_+(g_\Delta(\vec{X}(s))) ds \qquad (6.36)$$

is a martingale with respect to the filtration generated by $\vec{X}$, where

$$\beta_+(1) := \beta \quad \text{and} \quad \beta_+(m) := \beta + 2 \int v(dq) \sum_{k=0}^{m-2} (1-q)^k \quad \text{for } m \geq 2. \qquad (6.37)$$

Remark 6.5.4 The Howitt-Warren sticky Brownian motions evolve independently when they are apart, and experience sticky interaction when they meet. In particular, when $n = 2$, $X_1(t) - X_2(t)$ is a Brownian motion with stickiness at the origin, which is just a time changed Brownian motion such that its local time at the origin has been turned into real time, modulo a constant multiple that determines the stickiness. More generally, for Howitt-Warren sticky Brownian motions started at $X_1(0) = \cdots = X_n(0)$, the set of times with $X_1(t) = X_2(t) = \cdots = X_n(t)$ is a nowhere dense set with positive Lebesgue measure. The measure v determines a two-parameter family of constants $\theta_{k,l} = \int q^k (1-q)^l \frac{v(dq)}{q(1-q)}$, $k,l \geq 1$, which can be regarded as the rate (in a certain excursion theoretic sense) at which $(X_1, \cdots, X_n)$ split into two groups, $(X_1, \cdots, X_k)$ and $(X_{k+1}, \cdots, X_{k+l})$, with $k + l = n$.

It is easily seen that the n-point motions defined by the Howitt-Warren martingale problem in Theorem 6.5.3 form a consistent family, and it is Feller by [55, Proposition 8.1]. Therefore by the aforementioned result of Le Jan and Raimond [63, Theorem 2.1], there exists a stochastic flow of kernels $(K_{s,t})_{s \leq t}$ on $\mathbb{R}$, unique in finite-dimensional distributions, such that the n-point motions of $(K_{s,t})_{s \leq t}$ in the sense of (6.32) are given by the unique solutions of the Howitt-Warren martingale problem. Therefore we define as follows:

Definition 6.5.5 (Howitt-Warren flow) *We call the stochastic flow of kernels, whose n-point motions solve the Howitt-Warren martingale problem in Theorem 6.5.3 for some $\beta \in \mathbb{R}$ and finite measure v on $[0,1]$, the* Howitt-Warren flow *with* drift β *and* characteristic measure v.

Remark 6.5.6 We single out three special classes of Howitt-Warren flows: (1) $\nu = 0$, for which the n-point motions are coalescing Brownian motions, and the flow is known as the *Arratia flow*; (2) $\nu(dx) = adx$ for some $a > 0$, which we will call the *Le Jan-Raimond flow* since it was first constructed in [64] via Dirichlet forms and subsequently shown in [62] to arise as limits of random walks in i.i.d. Beta-distributed environments; (3) $\nu = a\delta_0 + b\delta_1$ for some $a, b \geq 0$ with $a + b > 0$, called the *erosion flow*, which was studied in [56].

As noted in Remark 6.5.2, it is not known a priori whether there exists a version of the Howitt-Warren flow K such that almost surely, $(K_{s,t}(x,\cdot))_{s<t,x\in\mathbb{R}}$ are truly transition probability kernels of a random motion in a random environment. It is then natural to ask whether such an underlying random environment indeed exists for the Howitt-Warren flows, and if yes, whether the environment can be explicitly characterized as in the discrete case. This is where the Brownian web and Brownian net enter the picture.

6.5.2 The Space-time Random Environment for the Howitt-Warren Flows

How can we construct a continuum space-time random environment such that the Howitt-Warren flow with drift β and characteristic measure ν is indeed the family of transition probability kernels of a random motion in this random environment? The answer again lies in discrete approximation.

Special case: ν satisfies $\int q^{-1}(1-q)^{-1}\nu(dq) < \infty$. In this case, the continuum random environment can be constructed from the Brownian net. For $\varepsilon > 0$, define a probability measure μ_ε on $[0,1]$ by

$$\mu_\varepsilon := b\varepsilon\bar{\nu} + \tfrac{1}{2}(1-(b+c)\varepsilon)\delta_0 + \tfrac{1}{2}(1-(b-c)\varepsilon)\delta_1$$

$$\text{where} \quad b := \int \frac{\nu(dq)}{q(1-q)}, \quad c := \beta - \int (2q-1)\frac{\nu(dq)}{q(1-q)}, \quad \bar{\nu}(dq) := \frac{\nu(dq)}{bq(1-q)}.$$

(6.38)

When ε is sufficiently small, such that $1 - (b+|c|)\varepsilon \geq 0$, μ_ε is a probability measure on $[0,1]$ and is easily seen to satisfy (6.33) as $\varepsilon \downarrow 0$. Therefore by Theorem 6.5.3, μ_ε determines the law of an i.i.d. random environment $\omega^\varepsilon := (\omega_z^\varepsilon)_{z\in\mathbb{Z}^2_{\text{even}}}$, whose associated n-point motions converge to that of the Howitt-Warren flow with drift β and characteristic measure ν.

Note that for small ε, most of the ω_z^ε are either zero or one. We can thus encode ω^ε in two steps. First we identify a collection of branching-coalescing random walks determined by ω^ε, where at each $z = (x,t) \in \mathbb{Z}^2_{\text{even}}$, the walk moves to $(x+1, t+1)$ if $\omega_z^\varepsilon = 1$, moves to $(x-1, t+1)$ if $\omega_z^\varepsilon = 0$, and is a branching point

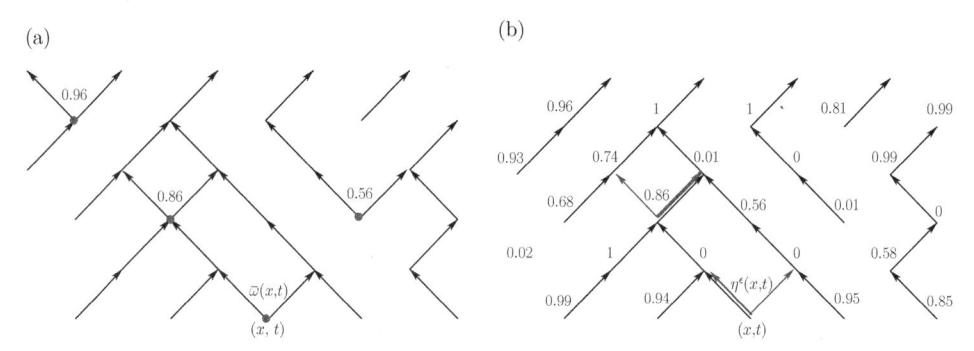

Figure 6.13. Representation of the random environment $(\omega_z^\varepsilon)_{z \in \mathbb{Z}^2_{\text{even}}}$ in terms of: (a) a marked discrete net $(N^\varepsilon, \bar{\omega}^\varepsilon)$; (b) a marked discrete web $(W^\varepsilon, \eta^\varepsilon)$.

if $\omega_z^\varepsilon \in (0,1)$ since the walk has strictly positive probability of moving to either $(x+1,t+1)$ or $(x-1,t+1)$. Given the set of branching-coalescing random walk paths, which we denote by N^ε and call a *discrete net*, we can then specify the value of ω_z^ε at each branching point by sampling i.i.d. random variables $\bar{\omega}_z^\varepsilon$ with common law $\bar{\nu}$ (see Figure 6.13(a)). The pair $(N^\varepsilon, \bar{\omega}^\varepsilon)$ then gives an alternative representation of the random environment ω, wherein a walk must navigate along N^ε, and when it encounters a branching point z, it jumps either left or right with probability $1 - \bar{\omega}_z^\varepsilon$, respectively $\bar{\omega}_z^\varepsilon$.

The above setup is essentially the same as the coupling between branching-coalescing random walks and coalescing random walks discussed in Section 6.4, which corresponds to taking $(\bar{\omega}_z^\varepsilon)_{z \in \mathbb{Z}^2_{\text{even}}}$ to be i.i.d. $\{0,1\}$-valued random variables in the current setting. As we rescale space and time by ε and ε^2 respectively, we note that N^ε converges to a variant of the Brownian net $\mathcal{N}_{\beta_-,\beta_+}$, constructed from a pair of left-right Brownian webs as in Section 6.3 with respective drifts

$$\beta_- = \beta - 2 \int \nu(\mathrm{d}q)(1-q)^{-1}, \qquad \beta_+ = \beta + 2 \int \nu(\mathrm{d}q) q^{-1}. \qquad (6.39)$$

The branching points of N^ε converge to the set of separation points of $\mathcal{N}_{\beta_-,\beta_+}$, denoted by S. Since the law of $\bar{\omega}_z^\varepsilon$ at the branching points is $\bar{\nu}$, independent of ε, we should then assign i.i.d. random variables ω_z with the same law $\bar{\nu}$ to each separation point $z \in S$. The pair $(\mathcal{N}_{\beta_-,\beta_+}, (\omega_z)_{z \in S})$ then gives the desired continuum space-time random environment, wherein a random motion π must navigate along $\mathcal{N}_{\beta_-,\beta_+}$, and independently at each separation point z it encounters, it chooses $\mathrm{sgn}_\pi(z)$ (see (6.23)) to be $+1$ with probability ω_z and -1 with probability $1 - \omega_z$.

We next give a precise formulation of how the subclass of Howitt-Warren flows with $\int q^{-1}(1-q)^{-1} \nu(\mathrm{d}q) < \infty$ can be obtained from the random environment $(\mathcal{N}_{\beta_-,\beta_+}, (\omega_z)_{z \in S})$ (cf. [92, Theorem 4.7]). For its proof, see [92]. To construct a version of the Howitt-Warren flow which a.s. satisfies the Chapman-Kolmogorov equation (condition (i′) in Remark 6.5.2), we will sample a collection of coalescing paths $\mathcal{W}$ given the environment $(\mathcal{N}_{\beta_-,\beta_+},(\omega_z)_{z\in S})$.

Theorem 6.5.7 (Constructing Howitt-Warren flows in a Brownian net) *Let $\beta \in \mathbb{R}$ and let ν be a finite measure on $[0,1]$ with $\int q^{-1}(1-q)^{-1}\nu(\mathrm{d}q) < \infty$. Let $\mathcal{N}_{\beta_-,\beta_+}$ be a Brownian net with drifts β_-,β_+ defined as in (6.39), and let S be its set of separation points. Conditional on $\mathcal{N}_{\beta_-,\beta_+}$, let $\omega := (\omega_z)_{z\in S}$ be i.i.d. $[0,1]$-valued random variables with law $\bar\nu$ defined as in (6.38). Conditional on $(\mathcal{N}_{\beta_-,\beta_+},\omega)$, let $(\alpha_z)_{z\in S}$ be independent $\{-1,+1\}$-valued random variables such that $\mathbb{P}[\alpha_z = 1 \mid (\mathcal{N}_{\beta_-,\beta_+},\omega)] = \omega_z$. Then*

$$\mathcal{W} := \{\pi \in \mathcal{N}_{\beta_-,\beta_+} : \mathrm{sgn}_\pi(z) = \alpha_z \ \forall z \in S \text{ s.t. } \pi \text{ enters } z\} \quad (6.40)$$

is distributed as a Brownian web with drift β. For any $z = (x,t) \in \mathbb{R}^2$, if z is of type $(1,2)$, then let $\pi_z^\uparrow \in \mathcal{W}(z)$ be any path in $\mathcal{W}$ entering z restricted to the time interval $[t,\infty)$; otherwise let $\pi_z^\uparrow$ be the rightmost path in $\mathcal{W}(z)$. Then

$$K_{s,t}^\uparrow(x,\cdot) := \mathbb{P}[\pi_{(x,s)}^\uparrow(t) \in \cdot \mid (\mathcal{N}_{\beta_-,\beta_+},\omega)], \qquad s \leq t, \ x \in \mathbb{R} \quad (6.41)$$

defines a version of the Howitt-Warren flow with drift β and characteristic measure ν, which satisfies condition (i′) in Remark 6.5.2.

General case: ν is any finite measure on $[0,1]$. Let $(\mu_\varepsilon)_{\varepsilon>0}$ satisfy (6.33) and let $\omega^\varepsilon := (\omega_z^\varepsilon)_{z\in \mathbb{Z}_{\mathrm{even}}^2}$ be an i.i.d. random environment with common law μ_ε. Without assuming $\int q^{-1}(1-q)^{-1}\nu(\mathrm{d}q) < \infty$, it may no longer be possible to capture the continuum random environment by a Brownian net, because either β_- or β_+ in (6.39) or both can be infinity. Instead, we follow the alternative view on the coupling between branching-coalescing and coalescing random walks in Section 6.4, where we first sample the coalescing walks and then introduce the branching points. The same approach can be applied here to encode the random environment ω in two steps.

First we construct a collection of coalescing random walks by sampling coalescing walks in the same environment ω and then average over the law of ω^ε. More precisely, from each $z = (x,t) \in \mathbb{Z}_{\mathrm{even}}^2$, the walk moves to either $(x+1,t+1)$ or $(x-1,t+1)$, represented by $\alpha_z^\varepsilon := 1$ or $\alpha_z^\varepsilon = -1$, with respective probability $\int_0^1 q\mu_\varepsilon(\mathrm{d}q)$ and $\int_0^1 (1-q)\mu_\varepsilon(\mathrm{d}q)$. Let W^ε denote this collection of coalescing random walks, which we call a *discrete web* and which has a natural dual $\widehat{W}^\varepsilon$ as illustrated in Figure 6.1. Next, we identify the law of ω^ε conditioned on W^ε.

Note that conditioned on W^ε, $(\omega_z^\varepsilon)_{z\in\mathbb{Z}_{\mathrm{even}}^2}$ are independent with conditional distribution

$$\mu_\varepsilon^{\mathrm{l}} := \frac{(1-q)\mu_\varepsilon(\mathrm{d}q)}{\int (1-q)\mu_\varepsilon(\mathrm{d}q)} \quad \text{if } \alpha_z^\varepsilon = -1; \qquad \mu_\varepsilon^{\mathrm{r}} := \frac{q\mu_\varepsilon(\mathrm{d}q)}{\int q\mu_\varepsilon(\mathrm{d}q)} \quad \text{if } \alpha_z^\varepsilon = 1. \quad (6.42)$$

Therefore conditioned on W^ε, if we sample i.i.d. random variables $(\eta_z^\varepsilon)_{\{z:\alpha_z^\varepsilon=-1\}}$ with common law μ_ε^l and i.i.d. random variables $(1-\eta_z^\varepsilon)_{\{z:\alpha_z^\varepsilon=1\}}$ with common law μ_ε^r, then $(W^\varepsilon, \eta^\varepsilon)$ provides an alternative representation for the environment ω^ε, where at each $z \in \mathbb{Z}^2_{\text{even}}$, a walk in the random environment follows the same jump as in W^ε with probability $1 - \eta_z^\varepsilon$, and jumps in the opposite direction with probability η_z^ε (see Figure 6.13(b)).

The above setup is an extension of the coupling between coalescing and branching-coalescing random walks discussed in Section 6.4, which corresponds to taking $(\eta_z^\varepsilon)_{z \in \mathbb{Z}^2_{\text{even}}}$ to be i.i.d. $\{0,1\}$-valued random variables with mean ε. As we rescale space and time by ε and ε^2 respectively, we note that W^ε converges to a Brownian web $\mathcal{W}_0$ with drift β. On the other hand, η^ε can be seen to converge to a marked Poisson point process on the set of intersection points between paths in $\mathcal{W}_0$ and paths in its dual $\widehat{\mathcal{W}}_0$, i.e., the $(1,2)$ points of $\mathcal{W}_0$. Indeed, we can regard $(\eta_z^\varepsilon)_{\{z:\alpha_z^\varepsilon=-1\}}$ as a marked point process on the set of space-time points in $\mathbb{Z}^2_{\text{even}}$ where a random walk path in W^ε is exactly one unit of distance to the left of a dual random walk path in $\widehat{W}^\varepsilon$. As noted in Section 6.4, when space-time is rescaled by $(\varepsilon, \varepsilon^2)$ and measure is rescaled by ε, the counting measure on these points of collision between forward and dual random walk paths converges to the intersection local time measure ℓ_l on points of type $(1,2)_l$ in the Brownian web. Since for every $u \in (0,1)$, by (6.33), the mark η_z^ε at each $z \in \mathbb{Z}^2_{\text{even}}$ with $\alpha_z^\varepsilon = -1$ satisfies

$$\mathbb{P}(\eta_z^\varepsilon \in (u,1]) = \frac{\int_u^1 (1-q)\mu_\varepsilon(dq)}{\int_0^1 (1-q)\mu_\varepsilon(dq)} = \frac{\varepsilon \int_u^1 q^{-1}\varepsilon^{-1}q(1-q)\mu_\varepsilon(dq)}{\int_0^1 (1-q)\mu_\varepsilon(dq)}$$

$$\sim \varepsilon \int_u^1 \frac{2}{q}\nu(dq) \quad \text{as } \varepsilon \downarrow 0,$$

it follows that $(\eta_z^\varepsilon)_{\{z:\alpha_z^\varepsilon=-1\}}$ converges to a marked Poisson point process $\mathcal{M}_l \subset \mathbb{R}^2 \times [0,1]$ with intensity measure $\ell_l(dz) \otimes \frac{2}{q}1_{\{q>0\}}\nu(dq)$. Similarly, $(1-\eta_z^\varepsilon)_{\{z:\alpha_z^\varepsilon=+1\}}$ converges to a marked Poisson point process $\mathcal{M}_r \subset \mathbb{R}^2 \times [0,1]$ with intensity measure $\ell_r(dz) \otimes \frac{2}{1-q}1_{\{q<1\}}\nu(dq)$.

The triple $(\mathcal{W}_0, \mathcal{M}_l, \mathcal{M}_r)$ then gives the desired continuum space-time random environment, wherein a random motion π must navigate along paths in $\mathcal{W}_0$, and independently at each marked $(1,2)_l$ point $(z, \eta_z) \in \mathcal{M}_l$ or marked $(1,2)_r$ point $(z, 1-\eta_z) \in \mathcal{M}_r$ it encounters, with probability η_z, π chooses orientation $-\text{sgn}_{\mathcal{W}_0}(z)$, i.e., π switches to the second outgoing path in $\mathcal{W}_0(z)$ instead of continuing along the incoming path. This description is correct when $\nu(\{0\}) = \nu(\{1\}) = 0$. However when $\nu(\{0\}) > 0$, we have excluded $\ell_l(dz) \otimes \frac{2}{q}1_{\{q=0\}}\nu(dq)$ from the intensity measure of $\mathcal{M}_l$, which has a non-negligible effect on paths sampled in the random environment. To understand this effect, we can approximate $\nu(\{0\})\delta_0(\mathrm{d}q)$ by $\nu(\{0\})\delta_h(\mathrm{d}q)$ with $h \downarrow 0$. This leads to Poisson marking $(1,2)_l$ points of $\mathcal{W}_0$ with intensity measure $\frac{2}{h}\nu(\{0\})\ell_1(\mathrm{d}z)$, and whenever a random motion π in this environment encounters such a Poisson point z, it chooses its orientation at z to be $-\mathrm{sgn}_{\mathcal{W}_0}(z)$ with probability h. The net effect is that, at each point z of a Poisson point process B_l with intensity measure $2\nu(\{0\})\ell_1(\mathrm{d}z)$, π chooses its orientation at z to be $-\mathrm{sgn}_{\mathcal{W}_0}(z)$ with probability 1. As $h \downarrow 0$, the resulting effect is that, when we sample an independent motion π' in the same random environment, then an independent copy of B_l – call it B'_l – must be sampled such that whenever π' encounters some $z \in B'_l$, it chooses its orientation at z to be $-\mathrm{sgn}_{\mathcal{W}_0}(z)$ with probability 1.

We will now give a precise formulation of how the Howitt-Warren flow can be obtained from a random environment $(\mathcal{W}_0, \mathcal{M}_l, \mathcal{M}_r)$ (cf. [92, Theorem 3.7]). As in Theorem 6.5.7, we will sample a collection of coalescing Brownian motions $\mathcal{W}$ given the environment $(\mathcal{W}_0, \mathcal{M}_l, \mathcal{M}_r)$.

Theorem 6.5.8 (Construction of Howitt-Warren flows) *Let $\beta \in \mathbb{R}$ and let ν be a finite measure on $[0,1]$. Let $\mathcal{W}_0$ be a Brownian web with drift β. Let $\mathcal{M}$ be a marked Poisson point process on $\mathbb{R}^2 \times [0,1]$ with intensity measure*

$$\ell_l(\mathrm{d}z) \otimes \frac{2}{q} 1_{\{q>0\}} \nu(\mathrm{d}q) + \ell_r(\mathrm{d}z) \otimes \frac{2}{1-q} 1_{\{q<1\}} \nu(\mathrm{d}q).$$

Conditional $(\mathcal{W}_0, \mathcal{M})$, let α_z be independent $\{-1, +1\}$-valued random variables with $\mathbb{P}[\alpha_z = +1 | (\mathcal{W}_0, \mathcal{M})] = \omega_z$ for each $(z, \omega_z) \in \mathcal{M}$, and let

$$A := \{z : (z, \omega_z) \in \mathcal{M}, \alpha_z \neq \mathrm{sgn}_{\mathcal{W}_0}(z)\}.$$

Let B be an independent Poisson point set with intensity $2\nu_l(\{0\})\ell_l + 2\nu_r(\{1\})\ell_r$. Define

$$\mathcal{W} := \lim_{\Delta_n \uparrow A \cup B} \mathrm{switch}_{\Delta_n}(\mathcal{W}_0) \qquad (6.43)$$

for any sequence of finite sets $\Delta_n \uparrow A \cup B$, where $\mathrm{switch}_{\Delta_n}(\mathcal{W}_0)$ is the set of paths obtained from $\mathcal{W}_0$ by redirecting all paths in $\mathcal{W}$ entering any $z \in \Delta_n$ in such a way that z of type $(1,2)_l$ in $\mathcal{W}$ becomes type $(1,2)_r$ in $\mathcal{W}_0$ and vice versa. Then $\mathcal{W}$ is equally distributed with $\mathcal{W}_0$, and with $\pi_z^\uparrow$ defined as in Theorem 6.5.7,

$$K_{s,t}^\uparrow(x, \cdot) := \mathbb{P}[\pi_{(x,s)}^\uparrow(t) \in \cdot | (\mathcal{W}_0, \mathcal{M})] \qquad s \leq t, \, x \in \mathbb{R}, \qquad (6.44)$$

defines a version of the Howitt-Warren flow with drift β and characteristic measure ν, which satisfies condition (i') in Remark 6.5.2.

Exercise 6.5.9 *Use Theorem 6.4.2 to show that $\mathcal{W}$ as defined in (6.43) does not depend on the choice of $\Delta_n \uparrow A \cup B$, and $\mathcal{W}$ is equally distributed with $\mathcal{W}_0$.*

6.5.3 Properties of the Howitt-Warren Flow

The construction of the continuum space-time random environment underlying the Howitt-Warren flows allows the study of almost sure properties of the flow $K^\uparrow := (K^\uparrow_{s,t}(x,\cdot))_{s<t, x\in\mathbb{R}}$ (cf. [92, Section 2]). In particular, we can study almost sure path properties of the following measure-valued process induced by the Howitt-Warren flow, called the *Howitt-Warren process*:

$$\rho_t(\mathrm{d}y) = \int K^\uparrow_{0,t}(x,\mathrm{d}y)\rho_0(\mathrm{d}x), \tag{6.45}$$

where ρ_0 can be taken to be any locally finite measure on $\mathbb{R}$ in the class

$$\mathcal{M}_g(\mathbb{R}) := \{\rho : \int e^{-cx^2}\rho(\mathrm{d}x) < \infty \text{ for all } c > 0\}, \tag{6.46}$$

where $\rho_n \in \mathcal{M}_g(\mathbb{R})$ is defined to converge to $\rho \in \mathcal{M}_g(\mathbb{R})$ if $\int f(x)\rho_n(\mathrm{d}x) \to \int f(x)\rho(\mathrm{d}x)$ for all $f \in C_c(\mathbb{R})$, which is just convergence in the vague topology, plus $\int e^{-cx^2}\rho_n(\mathrm{d}x) \to \int e^{-cx^2}\rho(\mathrm{d}x)$ for all $c > 0$.

A first consequence is that $(\rho_t)_{t\geq 0}$ is a Markov process with continuous sample path, and continuous dependence on the initial condition and starting time, which makes it a Feller process.

Theorem 6.5.10 (Howitt-Warren process) *Let $(K^\uparrow_{s,t}(x,\cdot))_{s\leq t, x\in\mathbb{R}}$ be the version of Howitt-Warren flow with drift β and characteristic measure ν, defined in Theorem 6.5.7 or 6.5.8. Let $(\rho_t)_{t\geq 0}$ be the Howitt-Warren process defined from $K^\uparrow$ as in (6.45) with $\rho_0 \in \mathcal{M}_g$. Then*

(i) *$(\rho_t)_{t\geq 0}$ is an $\mathcal{M}_g(\mathbb{R})$-valued Markov process with almost sure continuous sample paths;*
(ii) *If $(\rho_t^{\langle n\rangle})_{t\geq s_n}$ are Howitt-Warren processes defined from $K^\uparrow$ with deterministic initial condition $\rho_{s_n}^{\langle n\rangle}$ at time s_n, with $s_n \to 0$, then for any $t > 0$ and $t_n \to t$,*

$$\rho_{s_n}^{\langle n\rangle} \underset{n\to\infty}{\Longrightarrow} \rho_0 \quad \text{implies} \quad \rho_{t_n}^{\langle n\rangle} \underset{n\to\infty}{\Longrightarrow} \rho_t \quad a.s., \tag{6.47}$$

where $\Rightarrow$ denotes convergence in $\mathcal{M}_g(\mathbb{R})$.

If $\int q^{-1}(1-q)^{-1}\nu(\mathrm{d}q) < \infty$, then the above statements hold with $\mathcal{M}_g(\mathbb{R})$ replaced by $\mathcal{M}_{\mathrm{loc}}(\mathbb{R})$, the space of locally finite measures on $\mathbb{R}$ equipped with the vague topology.

We can also identify almost surely the support of ρ_t for all $t \geq 0$.

Theorem 6.5.11 (Support of Howitt-Warren process) *Let $(\rho_t)_{t\geq 0}$ be a Howitt-Warren process with drift β and characteristic measure, and the initial condition ρ_0 has compact support. Let β_-, β_+ be defined as in (6.39).*

(a) *If $-\infty < \beta_- < \beta_+ < \infty$, then a.s. for all $t > 0$, the support of ρ_t satisfies*

$$\mathrm{supp}(\rho_t) = \{\pi(t) : \pi \in \mathcal{N}_{\beta_-,\beta_+}, \pi(0) \in \mathrm{supp}(\rho_0)\}, \tag{6.48}$$

where $\mathcal{N}_{\beta_-,\beta_+}$ is the Brownian net with drift parameters β_-, β_+ as in Theorem 6.5.7.

(b) *If $\beta_- = -\infty$ and $\beta_+ < \infty$, then a.s. $\mathrm{supp}(\rho_t) = (-\infty, r_t] \cap \mathbb{R}$ for all $t > 0$, where $r_t := \sup(\mathrm{supp}(\rho_t))$. An analogue statement holds when $\beta_- > -\infty$ and $\beta_+ = \infty$.*

(c) *If $\beta_- = -\infty$ and $\beta_+ = \infty$, then a.s. $\mathrm{supp}(\rho_t) = \mathbb{R}$ for all $t > 0$.*

Theorem 6.5.11 (a) shows that when the Howitt-Warren flow can be constructed from a Brownian net, then at deterministic times, ρ_t is almost surely atomic. This result can be extended to general Howitt-Warren flows. However, almost surely there exist random times when ρ_t contains no atoms, and the only exceptions are the Arratia flow with $\nu = 0$, and the *erosion flows*, which have characteristic measures of the form $\nu = a\delta_0 + b\delta_1$ with $a, b \geq 0$ and $a + b > 0$.

Theorem 6.5.12 (Atomicness vs non-atomicness) *Let $(\rho_t)_{t \geq 0}$ be a Howitt-Warren process with drift β and characteristic measure ν.*

(a) *For each $t > 0$, ρ_t is a.s. purely atomic.*
(b) *If $\int_{(0,1)} \nu(\mathrm{d}q) > 0$, then a.s. there exists a dense set of random times $t > 0$ when ρ_t contains no atoms.*
(c) *If $\int_{(0,1)} \nu(\mathrm{d}q) = 0$, then a.s. ρ_t is purely atomic at all $t > 0$.*

We can also study ergodic properties of the Howitt-Warren process. In particular, for any Howitt-Warren process other than the measure-valued process generated by the coalescing Arratia flow, there is a unique spatially ergodic stationary law for $(\rho_t)_{t \geq 0}$, which is also the weak limit of ρ_t as $t \to \infty$ if the law of the initial condition ρ_0 is spatially ergodic with finite mean density.

Theorem 6.5.13 (Ergodic properties) *Let $(\rho_t)_{t \geq 0}$ be a Howitt-Warren process with drift $\beta \in \mathbb{R}$, characteristic measure $\nu \neq 0$, and initial law $\mathcal{L}(\rho_0)$.*

(i) *If $\mathcal{L}(\rho_0)$ is ergodic w.r.t. $T_a \rho_0(\cdot) = \rho_0(a + \cdot)$ for all $a \in \mathbb{R}$, and $\mathbb{E}[\rho_0([0,1])] = 1$, then as $t \to \infty$, $\mathcal{L}(\rho_t)$ converges weakly to a limit Λ_1 which is also ergodic with respect to T_a for all $a \in \mathbb{R}$. Furthermore, if $\mathcal{L}(\rho_0) = \Lambda_1$, then $\mathbb{E}[\rho_0([0,1])] = 1$ and $\mathcal{L}(\rho_t) = \Lambda_1$ for all $t > 0$.*
(ii) *If $\mathcal{L}(\rho_0)$ is ergodic w.r.t. T_a for all $a \in \mathbb{R}$ and $\mathbb{E}[\rho_0([0,1])] = \infty$, then as $t \to \infty$, $\mathcal{L}(\rho_t)$ has no subsequential weak limits supported on $\mathcal{M}_{\mathrm{loc}}(\mathbb{R})$, the set of locally finite measures on $\mathbb{R}$.*

Remark 6.5.14 When $v(dx) = 1_{[0,1]}(x)dx$, it is known from [64, Proposition 9(b)] that Λ_1 is the law of a random measure $\rho^* = \sum_{(x,u) \in \mathcal{P}} u\delta_x$ for a Poisson point process $\mathcal{P}$ on $\mathbb{R} \times [0, \infty)$ with intensity measure $dx \times u^{-1}e^{-u}du$.

The proof of the above results can be found in [92].

Remark 6.5.15 Howitt-Warren flows are the continuum analogues of discrete random walks in i.i.d. space-time random environments, and they share the same fluctuations on large space-time scales. In particular, for any Howitt-Warren process $(\rho_t)_{t \geq 0}$ (assuming drift $\beta = 0$ and $\rho_0(dx) = dx$ for simplicity), the rescaled current process $(\mathcal{I}(nt, \sqrt{n}x)/n^{1/4})_{t>0, x \in \mathbb{R}}$, where

$$\mathcal{I}(t,x) := \int_{\mathbb{R}} \int_R 1_{\{u<0\}} 1_{\{v>x\}} K^{\uparrow}_{0,t}(u, dv) \rho_0(du) - \int_{\mathbb{R}} \int_R 1_{\{u>0\}} 1_{\{v<x\}} K^{\uparrow}_{0,t}(u, dv) \rho_0(du),$$

converges to a universal Gaussian process as $n \to \infty$. This was shown in [106], and the same universal (Edwards-Wilkinson) fluctuations have been established earlier for random walks in i.i.d. space-time random environments (see e.g., [85]). In a different direction, the transition probabilities of a random walk in i.i.d. space-time random environments are believed to have the same universal (Tracy-Widom GUE) fluctuations as the point-to-point partition functions of a directed polymer. This was verified recently in [12] for special Beta-distributed random environments, and one expects similar results to hold for the Howitt-Warren flow $K^{\uparrow}_{0,t}(0, dx)$.

6.6 Convergence to the Brownian Web and Net

In this section, we give general convergence criteria for the Brownian web that were originally formulated in [40, 72], and simplified criteria when paths do not cross each other. We will discuss strategies for verifying these criteria and focus in particular on the convergence of coalescing random walks to the Brownian web. Lastly we will formulate a set of convergence criteria for the Brownian net and verify them for branching-coalescing simple random walks.

6.6.1 General Convergence Criteria for the Brownian Web

We give here general convergence criteria for a sequence of random variables $(X_n)_{n \in \mathbb{N}}$, taking values in the space of compact sets of paths $\mathcal{H}$, to converge in distribution to the Brownian web $\mathcal{W}$. When X_n consists of non-crossing paths, these criteria can be greatly simplified, which will be discussed in the next subsection.

First we formulate a criterion which ensures tightness for the laws of $(X_n)_{n\in\mathbb{N}}$. We then formulate criteria which ensure that any subsequential limit of $(X_n)_{n\in\mathbb{N}}$ contains a copy of the Brownian web (lower bound), but nothing more (upper bound).

Tightness: To understand the tightness criterion we will formulate, let us first see what should be the tightness criterion for a sequence of path-valued random variables $(Y_n)_{n\in\mathbb{N}}$, i.e., $Y_n \in \Pi$ with the space of paths Π defined as in Section 6.2.1. Due to the compactification of $\mathbb{R}^2$ in Section 6.2.1, it suffices to show that when restricted to any finite space-time window $\Lambda_{L,T} = [-L,L] \times [-T,T]$, the law of the random paths $(Y_n)_{n\in\mathbb{N}}$ are tight in the sense that for any $\varepsilon > 0$, there exists a modulus of continuity $\phi : [0,1] \to \mathbb{R}$, which is increasing with $\phi(\delta) \downarrow 0$ as $\delta \downarrow 0$, such that uniformly in $n \in \mathbb{N}$,

$$\mathbb{P}(\forall s \text{ with } (Y_n(s),s) \in \Lambda_{L,T} \text{ and } t \in [s, s+1], |Y_n(t) - Y_n(s)| \leq \phi(t-s)) \geq 1 - \varepsilon. \tag{6.49}$$

The modulus of continuity ϕ allows the construction of an equicontinuous, and hence compact, set of paths. To construct a ϕ that satisfies (6.49), it suffices to show that for any $\eta > 0$,

$$\lim_{\delta \downarrow 0} \limsup_{n \to \infty} \mathbb{P}(\exists s < t < s+\delta \text{ with } (Y_n(s),s) \in \Lambda_{L,T}, \text{ s.t. } |Y_n(t) - Y_n(s)| > \eta) = 0. \tag{6.50}$$

Indeed, fix a sequence $\eta_m \downarrow 0$. Then for each $m \in \mathbb{N}$, by (6.50), we can find $\delta_m > 0$ sufficiently small such that uniformly in $n \in \mathbb{N}$, the probability in (6.50) is bounded by $\varepsilon/2^m$. We can then define $\phi(h) := \eta_m$ for $h \in (\delta_{m-1}, \delta_m]$, which is easily seen to satisfy (6.49).

When we consider a sequence of random compact sets of paths $(X_n)_{n\in\mathbb{N}}$, the tightness criterion is similar, except that we need to control the modulus of continuity uniformly for all paths in X_n (cf. Exercise 6.2.2). Therefore condition (6.50) should be modified to show that for any finite $\Lambda_{L,T}$ and any $\eta > 0$,

$$\lim_{\delta \downarrow 0} \limsup_{n \to \infty} \mathbb{P}(\exists \pi \in X_n \text{ and } s < t < s+\delta \text{ with } (\pi(s),s)$$
$$\in \Lambda_{L,T}, \text{ s.t. } |\pi(t) - \pi(s)| > \eta) = 0. \tag{6.51}$$

To control the modulus of continuity of all paths in X_n simultaneously, it is convenient to divide $\Lambda_{L,T}$ into $16LT/\delta\eta$ sub-rectangles of dimension $\eta/2 \times \delta/2$, and bound the event in (6.51) by the union of events where $\Lambda_{L,T}$ in (6.51) is replaced by one of the $16LT/\delta\eta$ sub-rectangles of $\Lambda_{L,T}$. This leads to the following tightness criterion as formulated in [40, Proposition B1].

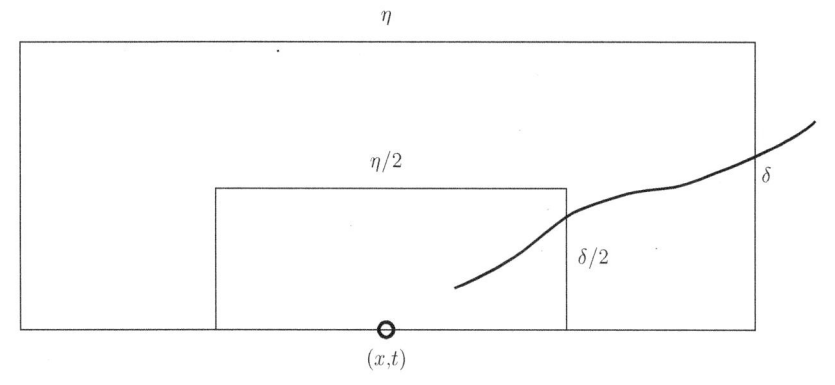

Figure 6.14. A path causing the event $A_{\delta,\eta}(x,t)$ to occur.

Proposition 6.6.1 (Tightness criterion) *The law of a sequence of $(\mathcal{H}, \mathcal{B}_\mathcal{H})$-valued random variables $(X_n)_{n\in\mathbb{N}}$ is tight if*

(T) $\quad \forall L, T \in (0,\infty), \quad \lim_{\delta \downarrow 0} \delta^{-1} \limsup_{n\to\infty} \sup_{(x,t)\in[-L,L]\times[-T,T]} \mathbb{P}(X_n \in A_{\delta,\eta}(x,t)) = 0,$

where $A_{\delta,\eta}(x,t) \in \mathcal{B}_\mathcal{H}$ consists of compact sets of paths $K \in \mathcal{H}$, such that K contains some path which intersects the rectangle $[x-\eta/4, x+\eta/4] \times [t, t+\delta/2]$, and at a later time, intersects the left or right boundary of the bigger rectangle $[x-\eta/2, x+\eta/2] \times [t, t+\delta]$ (see Figure 6.14).

Lower bound: Assuming that $(X_n)_{n\in\mathbb{N}}$ is a tight sequence of $\mathcal{H}$-valued random variables, we then need a criterion to ensure that any subsequential weak limit of $(X_n)_{n\in\mathbb{N}}$ contains almost surely a random subset which is distributed as the standard Brownian web. The following criterion serves this purpose, which is a form of convergence in finite-dimensional distributions.

(I) There exists $\pi_{n,z} \in X_n$ for each $z \in \mathbb{R}^2$, such that for any deterministic $z_1, \ldots, z_k \in \mathbb{R}^2$, $(\pi_{n,z_i})_{1\le i\le k}$ converge in distribution to coalescing Brownian motions starting at $(z_i)_{1\le i\le k}$.

If $(X_n)_{n\in\mathbb{N}}$ is a tight sequence that satisfies condition (I), then for any deterministic countable dense set $\mathcal{D} \subset \mathbb{R}^2$, we note that by going to a further subsequence if necessary, any subsequential limit $\mathcal{X}$ of $(X_n)_{n\in\mathbb{N}}$ contains a collection of coalescing Brownian motions $\{\pi_z\}_{z\in\mathcal{D}}$ starting from each $z \in \mathcal{D}$. Since $\mathcal{W} := \overline{\{\pi_z : z \in \mathcal{D}\}}$ is a standard Brownian web by Theorem 6.2.3, we obtain the desired lower bound that a.s. $\mathcal{W} \subset \mathcal{X}$.

Upper bound: Assuming that $(X_n)_{n\in\mathbb{N}}$ is a tight sequence of $\mathcal{H}$-valued random variables that satisfies condition (I), then it only remains to formulate a criterion to ensure that any subsequential weak limit $\mathcal{X}$ of $(X_n)_{n\in\mathbb{N}}$ a.s. contains no more paths than the Brownian web $\mathcal{W} \subset \mathcal{X}$. There are several approaches to this problem, depending partly on whether paths in X_n can cross each other or not.

One way is to control the expectation of the following family of counting random variables

$$\eta_{\mathcal{X}}(t,h;a,b) = |\{\pi(t+h) : \pi \in \mathcal{X}, \pi(t) \in [a,b]\}|, \qquad t \in \mathbb{R}, h > 0, a < b, \quad (6.52)$$

which considers all paths in $\mathcal{X}$ that intersect $[a,b]$ at time t and counts the number of distinct positions these paths occupy at time $t+h$. Thanks to the image set property of the Brownian web $\mathcal{W}$, which is a special case of Proposition 6.3.17 for the Brownian net, it is easily seen that if $\mathcal{X}$ contains strictly more paths than $\mathcal{W}$, than we must have $\eta_{\mathcal{X}}(t,h;a,b) > \eta_{\mathcal{W}}(t,h;a,b)$ for some rational t,h,a,b. Therefore to show $\mathcal{X} = \mathcal{W}$ a.s., it suffices to show that

$$\mathbb{E}[\eta_{\mathcal{X}}(t,h;a,b)] = \mathbb{E}[\eta_{\mathcal{W}}(t,h;a,b)] \qquad \text{for all } t \in \mathbb{R}, h > 0, a < b. \quad (6.53)$$

The following sufficient criteria have been formulated in [40] to ensure that (6.53) holds for any subsequential limit $\mathcal{X}$ of a sequence of $\mathcal{H}$-valued random variables $(X_n)_{n\in\mathbb{N}}$:

(B1') $\forall h_0 > 0$, $\displaystyle\limsup_{n\to\infty} \sup_{h > h_0} \sup_{a,t\in\mathbb{R}} \mathbb{P}[\eta_{X_n}(t,h;a,a+\varepsilon) \geq 2] \xrightarrow[\varepsilon\to 0]{} 0$,

(B2') $\forall h_0 > 0$, $\displaystyle\frac{1}{\varepsilon} \limsup_{n\to\infty} \sup_{h > h_0} \sup_{a,t\in\mathbb{R}} \mathbb{P}[X_n(t,h;a,\varepsilon) \neq X_n^-(t,h;a,\varepsilon) \cup X_n^+(t,h;a,\varepsilon)]$

$\xrightarrow[\varepsilon\to 0]{} 0$,

where $X_n(t,h;a,\varepsilon) \subset \mathbb{R}$ is the set of positions occupied at time $t+h$ by paths in X_n which intersect the interval $[a, a+\varepsilon]$ at time t, while $X_n^-(t,h;a,\varepsilon)$ (resp. $X_n^+(t,h;a,\varepsilon)$) is the subset of $X_n(t,h;a,\varepsilon)$ induced by paths in X_n which occupy the leftmost (resp. rightmost) position at time t among all paths in X_n that intersect $[a, a+\varepsilon]$ at time t. Condition (B1') ensures that for each deterministic point $z \in \mathbb{R}^2$, any subsequential limit $\mathcal{X}$ contains a.s. at most one path starting from z. Condition (B2') ensures that $\eta_{\mathcal{X}}(t,h;a,b)$ can be approximated by partitioning $[a,b]$ into equal-sized intervals and considering only paths in $\mathcal{X}$ starting at the boundaries of these intervals. Condition (B1') and (I) together ensure that paths in $\mathcal{X}$ starting at these boundary points are distributed as coalescing Brownian motions. Condition (6.53) then follows.

We thus have the following convergence criteria for the Brownian web [40, Theorem 5.1].

Theorem 6.6.2 (Convergence criteria A) *Let $(X_n)_{n \in \mathbb{N}}$ be a sequence of $(\mathcal{H}, \mathcal{B}_\mathcal{H})$-valued random variables satisfying conditions* (T), (I), (B1′), (B2′). *Then X_n converges in distribution to the standard Brownian web $\mathcal{W}$.*

Condition (B2′) turns out to be difficult to verify when paths in X_n can cross each other. This is in particular the case for non-nearest neighbor coalescing random walks on $\mathbb{Z}$, which led to the formulation in [72] of an alternative criterion in place of (B2′). The observation is that instead of $\eta_\mathcal{X}(t, h; a, b)$, we can consider the alternative family of counting random variables

$$\hat{\eta}_\mathcal{X}(t, h; a, b) = |\{\pi(t+h) \cap (a,b) : \pi \in \mathcal{X}, \pi(t) \in \mathbb{R}\}|, \qquad t \in \mathbb{R}, h > 0, a < b, \quad (6.54)$$

which considers all paths in $\mathcal{X}$ starting before or at time t and counts the number of distinct positions these paths occupy in the interval (a, b) at time $t + h$. Given $\mathcal{X} \supset \mathcal{W}$, to show that $\mathcal{X} = \mathcal{W}$ a.s., it suffices to show that

$$\mathbb{E}[\hat{\eta}_\mathcal{X}(t, h; a, b)] = \mathbb{E}[\hat{\eta}_\mathcal{W}(t, h; a, b)] \qquad \text{for all } t \in \mathbb{R}, h > 0, a < b. \quad (6.55)$$

This leads to the following alternative convergence criteria in [72, Theorem 1.4].

Theorem 6.6.3 (Convergence criteria B) *Let $(X_n)_{n \in \mathbb{N}}$ be a sequence of $(\mathcal{H}, \mathcal{B}_\mathcal{H})$-valued random variables satisfying conditions* (T), (I), (B1′) *and the following condition*:

(E) *For any subsequential weak limit $\mathcal{X}$, $\mathbb{E}[\hat{\eta}_\mathcal{X}(t, h; a, b)] = \mathbb{E}[\hat{\eta}_\mathcal{W}(t, h; a, b)]$ $\forall t \in \mathbb{R}, h > 0, a < b$.*

Then X_n converges in distribution to the standard Brownian web $\mathcal{W}$ as $n \to \infty$.

Condition (E) is actually much easier to verify than it appears. It turns out to be enough to establish the density bound

$$\limsup_{n \to \infty} \mathbb{E}[\hat{\eta}_{X_n}(t, h; a, b)] < \infty \qquad \forall t \in \mathbb{R}, h > 0, a < b, \quad (6.56)$$

which by Fatou's Lemma implies that any subsequential weak limit $\mathcal{X}$ satisfies $\mathbb{E}[\hat{\eta}_\mathcal{X}(t, \varepsilon; a, b)] < \infty$ for all $t \in \mathbb{R}$, $\varepsilon > 0$ and $a < b$. In particular,

$$\mathcal{X}^{t-}(t+\varepsilon) := \{\pi(t+\varepsilon) : \pi \in \mathcal{X}^{t-}\} \quad \text{with} \quad \mathcal{X}^{t-} := \{\pi \in \mathcal{X} : \sigma_\pi \leq t\}, \quad (6.57)$$

the set of positions at time $t + \varepsilon$ generated by paths in $\mathcal{X}$ starting before or at time t, is almost surely a locally finite subset of $\mathbb{R}$. As a random subset of $\mathbb{R}$, $\mathcal{X}^{t-}(t + \varepsilon)$ arises as the limit of $X_n^{t-}(t + \varepsilon)$,[1] and we can use Skorohod's

---

[1] In practice, $\mathcal{X}^{t-}(t+\varepsilon)$ may contain positions which arise from limits of paths in X_n that start at times $t_n \downarrow t$. Therefore we should consider instead $X_n^{(t+\delta)-}(t+\varepsilon)$ for some $\delta > 0$, so that its limit contains $\mathcal{X}^{t-}(t+\varepsilon)$.

representation theorem [9] to couple them such that almost surely, $X_n^{t^-}(t+\varepsilon)$ converges to $\mathcal{X}^{t^-}(t+\varepsilon)$ w.r.t. the Hausdorff metric on subsets of $\mathbb{R}$. For Markov processes such as coalescing random walks, we expect that the law of paths in $X_n^{t^-}$ restricted to the time interval $[t+\varepsilon,\infty)$ depends only on their positions $X_n^{t^-}(t+\varepsilon)$ at time $t+\varepsilon$, and furthermore, conditions (B1′) and (I) can be applied to these restricted paths conditioned on $X_n^{t^-}(t+\varepsilon)$. Therefore given $X_n^{t^-}(t+\varepsilon)$ converging to $\mathcal{X}^{t^-}(t+\varepsilon)$, we can apply conditions (B1′) and (I) to conclude that $\mathcal{X}^{t^-}$ restricted to the time interval $[t+\varepsilon,\infty)$ is a collection of coalescing Brownian motions starting from the locally finite set $\mathcal{X}^{t^-}(t+\varepsilon)$ at time $t+\varepsilon$, and hence

$$\mathbb{E}[\hat{\eta}_{\mathcal{X}}(t,h;a,b)] \leq \mathbb{E}[\hat{\eta}_{\mathcal{W}}(t+\varepsilon,h-\varepsilon;a,b)] = \frac{b-a}{\sqrt{\pi(h-\varepsilon)}}.$$

Sending $\varepsilon \downarrow 0$ then establishes condition (E). For non-nearest neighbor coalescing random walks, the above strategy was carried out in [72].

6.6.2 Convergence Criteria for Non-crossing Paths

When $(X_n)_{n\in\mathbb{N}}$ almost surely consists of paths that do not cross each other, i.e., X_n contains no paths π_1,π_2 with $(\pi_1(s)-\pi_2(s))(\pi_1(t)-\pi_2(t)) < 0$ for some $s<t$, tightness in fact follows from condition (I) [40, Proposition B2].

Proposition 6.6.4 (Tightness criterion for non-crossing paths) *If for each $n \in \mathbb{N}$, X_n is an $\mathcal{H}$-valued random variable consisting almost surely of paths that do not cross each other, and $(X_n)_{n\in\mathbb{N}}$ satisfies condition (I), then $(X_n)_{n\in\mathbb{N}}$ is a tight family.*

This result holds because when paths do not cross, the modulus of continuity of all paths in X_n can be controlled by the modulus of continuity of paths in X_n starting at a grid of space-time points, similar to Step (2) in the proof sketch for Theorem 6.2.3.

When X_n consists of non-crossing paths, conditions (B1′) and (B2′) can also be simplified to

(B1) $\qquad \forall h > 0, \ \limsup_{n\to\infty} \sup_{a,t\in\mathbb{R}} \mathbb{P}[\eta_{X_n}(t,h;a,a+\varepsilon) \geq 2] \xrightarrow[\varepsilon\to 0]{} 0,$

(B2) $\qquad \forall h > 0, \ \dfrac{1}{\varepsilon}\limsup_{n\to\infty} \sup_{a,t\in\mathbb{R}} \mathbb{P}[\eta_{X_n}(t,h;a,a+\varepsilon) \geq 3] \xrightarrow[\varepsilon\to 0]{} 0.$

Recall from the discussions before Theorem 6.6.2 that condition (B1) is to ensure that for each deterministic $z \in \mathbb{R}^2$, any subsequential limit $\mathcal{X}$ of $(X_n)_{n\in\mathbb{N}}$ almost surely contains at most one path starting from z. This property is easily seen to be implied by condition (I) (by the same argument as for Theorem 6.2.3 (a)) when X_n consists of non-crossing paths, which implies that $\mathcal{X}$ also consists of non-crossing paths. Therefore condition (B1) also becomes redundant, which leads to the following simplification of Theorems 6.6.2 (cf. [40, Theorem 2.2]) and 6.6.3.

Theorem 6.6.5 (Convergence criteria C) *Let $(X_n)_{n\in\mathbb{N}}$ be a sequence of $(\mathcal{H}, \mathcal{B}_\mathcal{H})$-valued random variables which a.s. consist of non-crossing paths. If $(X_n)_{n\in\mathbb{N}}$ satisfies conditions (I), and either (B2) or (E), then X_n converges in distribution to the standard Brownian web.*

Condition (B2) is often verified by applying the FKG positive correlation inequality [42], together with a bound on the distribution of the time of coalescence between two paths (see e.g., Section 6.6.3 below). However, FKG is a strong property that is not satisfied by most models. In such cases, verifying condition (B2) can be difficult. Besides checking condition (E), another alternative is to use the dual (a.k.a. wedge) characterization of the Brownian web, as noted in Remark 6.3.10, to upper bound any subsequential weak limit of $(X_n)_{n\in\mathbb{N}}$.[2]

Indeed, if $(X_n)_{n\in\mathbb{N}}$ consists of non-crossing paths and satisfies condition (I), then we can construct a collection of dual paths $\widehat{X}_n$ which almost surely do not cross paths in X_n, and the starting points of paths in $\widehat{X}_n$ become dense in $\mathbb{R}^2$ as $n \to \infty$. The tightness of $(X_n)_{n\in\mathbb{N}}$ is easily seen to imply the tightness of $(\widehat{X}_n)_{n\in\mathbb{N}}$, and any subsequential weak limit $(\mathcal{X}, \widehat{\mathcal{X}})$ of $(X_n, \widehat{X}_n)$ must satisfy the property that: for any deterministic countable dense set $\mathcal{D} \subset \mathbb{R}^2$, $\mathcal{X}(\mathcal{D})$ is distributed as a collection of coalescing Brownian motions, which by the non-crossing property a.s. uniquely determines $\widehat{\mathcal{X}}(\mathcal{D})$, which is distributed as a collection of dual coalescing Brownian motions. By the wedge characterization of the Brownian web in Remark 6.3.10, $\mathcal{X}$ equals $\overline{\mathcal{X}(\mathcal{D})}$, a standard Brownian web, if no path in $\mathcal{X}$ enters any wedge $W(\hat{\pi}_1, \hat{\pi}_2)$ of $\widehat{\mathcal{X}}(\mathcal{D})$ from outside (defined as in (6.16) with $\widehat{\mathcal{X}}(\mathcal{D})$ replacing both $\widehat{\mathcal{W}}^l$ and $\widehat{\mathcal{W}}^r$). This leads to

(U) For each $n \in \mathbb{N}$, there exists $\widehat{X}_n \in \widehat{\mathcal{H}}$ whose paths a.s. do not cross those of X_n and whose starting points are dense in $\mathbb{R}^2$ as $n \to \infty$, s.t. for any subsequential weak limit $(\mathcal{X}, \widehat{\mathcal{X}})$ of $(X_n, \widehat{X}_n)$ and any deterministic countable dense $\mathcal{D} \subset \mathbb{R}^2$, a.s. paths in $\mathcal{X}$ do not enter any wedge of $\widehat{\mathcal{X}}(\mathcal{D})$ from outside.

We then have the following convergence result.

Theorem 6.6.6 (Convergence criteria D) *Let $(X_n)_{n\in\mathbb{N}}$ be a sequence of $(\mathcal{H}, \mathcal{B}_\mathcal{H})$-valued random variables consisting of non-crossing paths and which satisfy*

---

[2] Recently a new approach to verify condition (B2) was proposed in [90], using a Lyapunov function type criterion on the gaps between three non-crossing paths, assuming that the gaps evolve jointly as a Markov process.

conditions (I) *and* (U). *Then* X_n *converges in distribution to the standard Brownian web.*

Remark 6.6.7 To verify condition (U), it suffices to show that paths in X_n do not enter wedges of $\widehat{X}_n$ from outside, and for any deterministic $z_1, z_2 \in \mathbb{R}^2$, not only do there exist paths $\hat{\pi}_{n,1}, \hat{\pi}_{n,2} \in \widehat{X}_n$ which converge to dual coalescing Brownian motions starting at z_1 and z_2 (which follows from condition (I) for $(X_n)_{n \in \mathbb{N}}$ and the non-crossing between paths in X_n and $\widehat{X}_n$), but also the time of coalescence between $\hat{\pi}_{n,1}$ and $\hat{\pi}_{n,2}$ converges to that of the coalescing Brownian motions. This ensures that the wedge $W(\hat{\pi}_{n,1}, \hat{\pi}_{n,2})$ converges and no path in the limit can enter the wedge through its bottom point. The latter can also be accomplished by showing that no limiting forward and dual paths can spend positive Lebesgue time together, as carried out in [81].

6.6.3 Convergence of Coalescing Simple Random Walks to the Brownian Web

We now illustrate how the convergence criteria in Theorems 6.6.5 and 6.6.6 for non-crossing paths can be verified for the discrete time coalescing simple random walks on $\mathbb{Z}$ (cf. [40, Theorem 6.1]).

Let X denote the collection of discrete time coalescing simple random walk paths on $\mathbb{Z}$, with one walk starting from every space-time lattice site $z \in \mathbb{Z}^2_{\text{even}}$. It is an easy exercise to show that X is a.s. precompact in the space of paths Π, and with a slight abuse of notation, we will henceforth denote the closure of X in (Π, d) also by X.

For each $\varepsilon \in (0, 1)$, let $S_\varepsilon : \mathbb{R}^2 \to \mathbb{R}^2$ denote the diffusive scaling map

$$S_\varepsilon(x, t) = (\varepsilon x, \varepsilon^2 t). \tag{6.58}$$

For a path $\pi \in \Pi$, let $S_\varepsilon \pi$ denote the path whose graph is the image of the graph of π under S_ε. For a set of paths K, define $S_\varepsilon K := \{S_\varepsilon \pi : \pi \in K\}$.

Theorem 6.6.8 *Let X be the collection of coalescing simple random walk paths on $\mathbb{Z}$ defined as above. Then as $\varepsilon \downarrow 0$, $X_\varepsilon := S_\varepsilon X$ converges in distribution to the standard Brownian web $\mathcal{W}$.*

Proof sketch. We show how the various conditions in Theorems 6.6.5 and 6.6.6 can be verified.

(I): This condition follows by Donsker's invariance principle. Indeed, coalescing random walks can be constructed from independent random walks by the same procedure as the inductive construction of coalescing Brownian motions from independent Brownian motions in the proof for Theorem 6.2.3. Furthermore, this construction is a.s. continuous w.r.t. the independent Brownian motions. Therefore (I) follows from Donsker's invariance principle for independent random walks and the Continuous Mapping Theorem.

(B2): We will verify this condition by applying the FKG inequality [42]. By translation invariance in space-time, it suffices to show that for any $t > 0$,

$$\lim_{\delta \downarrow 0} \delta^{-1} \limsup_{\varepsilon \downarrow 0} \mathbb{P}(\eta_{X_\varepsilon}(0,t;0,\delta) \geq 3) = 0.$$

Assume w.l.o.g. that $t = 1$. Formulated in terms of X, and letting $A := \delta^{-1}$ and $\sqrt{n} = \delta \varepsilon^{-1}$, it is equivalent to showing that

$$\lim_{A \to \infty} A \limsup_{n \to \infty} \mathbb{P}(\eta_X(0, A^2 n; 0, \sqrt{n}) \geq 3) = 0. \tag{6.59}$$

For $i \in \mathbb{Z}$, let π_i denote the random walk starting at $2i$ at time 0, and let $\tau_{i,j}$ denote the first meeting (coalescence) time between π_i and π_j. Then by a decomposition according to the first index $k \in \mathbb{N}$ with $\tau_{k-1,k} > A^2 n$, we have

$$\mathbb{P}(|\{\pi_i(A^2 n) : 0 \leq i \leq \sqrt{n}\}| \geq 3) = \sum_{k=1}^{\sqrt{n}-1} \mathbb{P}(\tau_{0,k-1} \leq A^2 n,$$

$$\tau_{k-1,k} > A^2 n, \tau_{k,\sqrt{n}} > A^2 n).$$

Let us restrict to the event $E_k := \{\tau_{0,k-1} \leq A^2 n, \tau_{k-1,\sqrt{n}} > A^2 n\}$ and condition on π_{k-1} and $\pi_{\sqrt{n}}$. Note that the event $\{\tau_{k-1,k} > A^2 n\}$ is increasing (while $\{\tau_{k,\sqrt{n}} > A^2 n\}$ is decreasing) w.r.t. the increments of π_k, $(\pi_k(i) - \pi_k(i-1))_{1 \leq i \leq A^2 n} \in \{\pm 1\}^{A^2 n}$, where the product space $\{\pm 1\}^{A^2 n}$ is equipped with the partial order $\prec$ such that $(a_i)_{1 \leq i \leq A^2 n} \prec (b_i)_{1 \leq i \leq A^2 n}$ if $a_i \leq b_i$ for all i. Furthermore, on the event $\{\tau_{k-1,k}, \tau_{k,\sqrt{n}} > A^2 n\}$, π_k is distributed as an independent random walk with i.i.d. increments, whose law on $\{\pm 1\}^{A^2 n}$ satisfies the FKG inequality [42]. This implies that the events $\{\tau_{k-1,k} > A^2 n\}$ and $\{\tau_{k,\sqrt{n}} > A^2 n\}$ are negatively correlated under the law of π_k with i.i.d. increments. Denoting $\mathbb{P}_k$ for probability for π_k, we then have

$$\sum_{k=1}^{\sqrt{n}-1} \mathbb{P}(\tau_{0,k-1} \leq A^2 n, \tau_{k-1,k} > A^2 n, \tau_{k,\sqrt{n}} > A^2 n)$$

$$= \sum_{k=1}^{\sqrt{n}-1} \mathbb{E}[1_{E_k} \mathbb{P}_k(\tau_{k-1,k} > A^2 n, \tau_{k,\sqrt{n}} > A^2 n)]$$

$$\leq \sum_{k=1}^{\sqrt{n}-1} \mathbb{E}[1_{E_k}\mathbb{P}_k(\tau_{k-1,k} > A^2n) \cdot \mathbb{P}_k(\tau_{k,\sqrt{n}} > A^2n)]$$

$$\leq \sum_{k=1}^{\sqrt{n}-1} \mathbb{P}(\tau_{k-1,k} > A^2n)\mathbb{P}(\tau_{k,\sqrt{n}} > A^2n)$$

$$\leq \sqrt{n}\frac{C}{\sqrt{A^2n}} \cdot \frac{C\sqrt{n}}{\sqrt{A^2n}} \leq \frac{C^2}{A^2},$$

where for the second inequality, we used the fact that $\mathbb{P}_k(\tau_{k-1,k} > A^2n)$ and $\mathbb{P}_k(\tau_{k,\sqrt{n}} > A^2n)$ are respectively functions of π_{k-1} and $\pi_{\sqrt{n}}$, which are distributed as independent random walks on the event E_k, and in the last line, we used that

$$\mathbb{P}(\tau_{0,1} > A^2n) \leq \frac{C}{A\sqrt{n}}, \tag{6.60}$$

$$\mathbb{P}(\tau_{k,\sqrt{n}} > A^2n) \leq \mathbb{P}(\exists k < i \leq \sqrt{n} : \tau_{i-1,i} \geq A^2n) \leq \frac{C\sqrt{n}}{A\sqrt{n}} = \frac{C}{A}, \tag{6.61}$$

which hold for any random walk on $\mathbb{Z}$ with finite variance [82, Proposition 32.4]. Condition (6.59) then follows.

(U): Recall from Figure 6.1 that the collection of coalescing simple random walk paths X uniquely determines a collection of dual coalescing simple random walk paths, which we denote by $\widehat{X}$. Clearly no path in X can enter any wedge of $\widehat{X}$ from outside. Furthermore, for any $z_1, z_2 \in \mathbb{R}^2$ and any choice of paths $\hat{\pi}_{\varepsilon,1}, \hat{\pi}_{\varepsilon,2} \in S_\varepsilon \widehat{X}$ which converge to dual coalescing Brownian motions starting at z_1, z_2, it is easily seen that the time of coalescence between $\hat{\pi}_{\varepsilon,1}$ and $\hat{\pi}_{\varepsilon,2}$ also converges to that between the limiting Brownian motions. Condition (U) then follows.

We will show how condition (E) is verified for general random walks in the next section. □

6.6.4 Convergence of General Coalescing Random Walks to the Brownian Web

Let X denote the collection of coalescing random walk paths on $\mathbb{Z}^2$ with one walk starting from each site in $\mathbb{Z}^2$, where the increments are i.i.d. with distribution μ with zero mean and finite variance $\sigma^2 := \sum_{x \in \mathbb{Z}} x^2 \mu(x)$, such that the walks are irreducible and aperiodic. Let $S_\varepsilon^\sigma : \mathbb{R}^2 \to \mathbb{R}^2$ be the diffusive scaling map

$$S_\varepsilon^\sigma(x,t) = (\varepsilon\sigma^{-1}x, \varepsilon^2 t), \tag{6.62}$$

and the action of S_ε^σ on paths and sets of paths is defined as S_ε in (6.58).

We have the following convergence result for general coalescing random walks on $\mathbb{Z}$, where paths may cross (see [72, Theorem 1.5] and [16, Theorem 1.2]).

Theorem 6.6.9 *Let X be the collection of coalescing random walk paths on $\mathbb{Z}$ defined as above. If the random walk increment distribution μ has zero mean, variance σ^2, and finite r-th moment $\sum_{x \in \mathbb{Z}} |x|^r \mu(x) < \infty$ for some $r > 3$, then as $\varepsilon \downarrow 0$, $S_\varepsilon^\sigma X$ converges in distribution to a standard Brownian web $\mathcal{W}$.*

Remark 6.6.10 It was pointed out in [72, Remark 4.1] that if $\sum_{x \in \mathbb{Z}} |x|^r \mu(x) = \infty$ for some $r < 3$, then tightness of the collection of coalescing random walks is lost due to the presence of arbitrarily large jumps originating from every space-time window on the diffusive scale.

Proof Sketch. For X consisting of random walk paths that can cross each other, it is not known how to verify condition (B2′) in Theorem 6.6.2. We will instead apply Theorem 6.6.3 and sketch how the conditions therein can be verified. Further details can be found in [72].

To verify condition (T), by translation invariance and reformulation in terms of the set of unscaled random walk paths X, it suffices to show that for any $\eta > 0$,

$$\lim_{\delta \downarrow 0} \delta^{-1} \limsup_{\varepsilon \downarrow 0} \mathbb{P}(X \in A_{2\delta\varepsilon^{-2}, 20\eta\varepsilon^{-1}}(0,0)) = 0, \quad (6.63)$$

where the event $E := \{X \in A_{2\delta\varepsilon^{-2}, 20\eta\varepsilon^{-1}}(0,0)\}$ (see Figure 6.14) is the event that X contains some path which intersects the rectangle $R_1 := [-\eta\varepsilon^{-1}, \eta\varepsilon^{-1}] \times [0, \delta\varepsilon^{-2}]$, and at a later time, intersects the left or right boundary of the bigger rectangle $R_2 := [-10\eta\varepsilon^{-1}, 10\eta\varepsilon^{-1}] \times [0, 2\delta\varepsilon^{-2}]$. The event E can occur either due to a random walk which starts outside the rectangle R_1 and has a jump that crosses R_1 horizontally, the probability of which can be easily shown to be negligible; or due to a random walk which starts from some $z \in R_1 \cap \mathbb{Z}^2$ and crosses the left, respectively the right, side of R_2, which events we denote respectively by E_z^- and E_z^+.

We can bound $\mathbb{P}(\cup_{z \in R_1 \cap \mathbb{Z}^2} E_z^+)$ as follows. Let $(\pi_i)_{1 \leq i \leq 4}$ be four random walks in X, starting respectively at $2.5\eta\varepsilon^{-1}$, $4.5\eta\varepsilon^{-1}$, $6.5\eta\varepsilon^{-1}$ and $8.5\eta\varepsilon^{-1}$ at time 0. Let B denote the event that each of these walks stays confined in a centered interval of size $\eta\varepsilon^{-1}$ up to time $2\delta\varepsilon^{-2}$ (see Figure 6.15). For $1 \leq i \leq 4$, let τ_i denote the first time $n \in \mathbb{N}$ when $\pi_z(n) \geq \pi_i(n)$, let τ_5 denote the first time π_z crosses the right side of R_2, and let C_i denote the event that π_z does not meet π_i before time τ_5. We

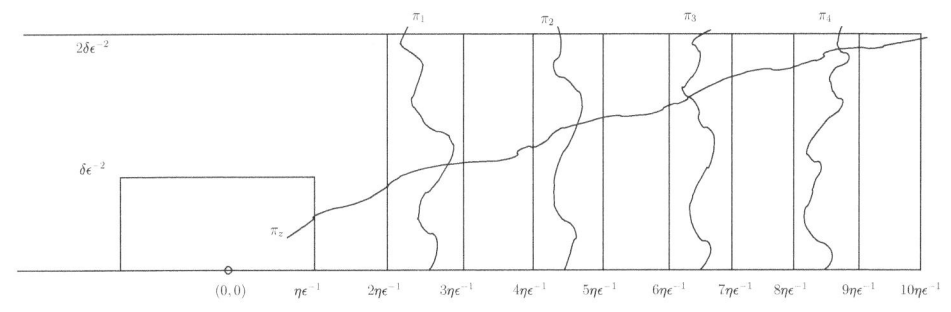

Figure 6.15. The random walks $\pi_1, \pi_2, \pi_3, \pi_4$ start respectively at $2.5\eta\epsilon^{-1}$, $4.5\eta\epsilon^{-1}$, $6.5\eta\epsilon^{-1}$ and $8.5\eta\epsilon^{-1}$ at time 0 and each stays within an interval of size $\eta\epsilon^{-1}$ up to time $2\delta\epsilon^{-2}$. The walk π_z starts from z inside the rectangle $[-\eta\epsilon^{-1}, \eta\epsilon^{-1}] \times [0, \delta\epsilon^{-2}]$ and crosses the right side of the rectangle $[-10\eta\epsilon^{-1}, 10\eta\epsilon^{-1}] \times [0, 2\delta\epsilon^{-2}]$ before time $2\delta\epsilon^{-2}$ without meeting $(\pi_i)_{1 \leq i \leq 4}$.

can then bound

$$\lim_{\delta \downarrow 0} \delta^{-1} \limsup_{\varepsilon \downarrow 0} \mathbb{P}(\cup_{z \in R_1 \cap \mathbb{Z}^2} E_z^+) = \lim_{\delta \downarrow 0} \delta^{-1} \limsup_{\varepsilon \downarrow 0} \mathbb{P}(B^c) + \lim_{\delta \downarrow 0} \delta^{-1}$$

$$\limsup_{\varepsilon \downarrow 0} \mathbb{P}(B \cap \cup_{z \in R_1 \cap \mathbb{Z}^2} E_z^+)$$

$$= \lim_{\delta \downarrow 0} \delta^{-1} \limsup_{\varepsilon \downarrow 0} \mathbb{P}(B \cap \cup_{z \in R_1 \cap \mathbb{Z}^2} E_z^+)$$

$$\leq \lim_{\delta \downarrow 0} \delta^{-1} \limsup_{\varepsilon \downarrow 0} 2\eta\delta\varepsilon^{-3} \max_{z \in R_1 \cap \mathbb{Z}^2} \mathbb{P}(B \cap E_z^+),$$

where in the second line we used Donsker's invariance principle and properties of Brownian motion. When the random walk increments are bounded by some $K < \infty$, it is easy to see that $\mathbb{P}(B \cap E_z^+) \leq C\varepsilon^4$ uniformly in $z \in R_1 \cap E_z^+$, so that the limit above equals zero.

Indeed, by successively conditioning on π_z and $(\pi_i)_{1 \leq i \leq 4}$ up to the stopping times τ_4, τ_3, τ_2 and τ_1, we note that π_z comes within distance $2K$ of π_i at time τ_i for each $1 \leq i \leq 4$, and π_z and π_i must separate by a distance of at least $\eta\varepsilon^{-1}$ before time τ_{i+1} without meeting. By the strong Markov property, these events are conditionally independent, and each event has a probability of order ε by [72, Lemma 2.4]. It then follows that $\mathbb{P}(B \cap E_z^+) \leq C\varepsilon^4$. When the random walks have unbounded increments, it is necessary to control the overshoot $\pi_z(\tau_i) - \pi_i(\tau_i)$, so that the probability of π_z overshooting more than one π_i in one jump is negligible when taken union over all starting positions $z \in R_1 \cap \mathbb{Z}^2$. This can be done when the random walk increments have finite 5th moment [72]. To relax to finite r-th moment for some $r > 3$, a multiscale argument is needed to take advantage of the coalescence and reduction of the random walks instead of the crude union bound as above. This was carried out in [16].

Condition (I) can be verified by a similar argument to that for coalescing simple random walks in Theorem 6.6.8. The complication is that random walk paths can cross without coalescing. However, for random walks with finite second moments, at the time when their paths cross, the distance between the two walks is of order one uniformly w.r.t. their starting positions, and hence it takes another time interval of order one for the two walks to coalesce. Therefore the crossing time and the coalescence time between any pair of walks are indistinguishable in the diffusive scaling limit, and forcing coalescence of the random walks at crossing times gives a good approximation, for which Donsker's invariance principle and the Continuous Mapping Theorem can be applied.

Condition (B1′) amounts to showing that

$$\lim_{\eta \downarrow 0} \mathbb{P}(\eta_X(0,n;0,\eta\sqrt{n}) \geq 2) = 0, \quad (6.64)$$

which follows from the same bounds as in (6.60)–(6.61), since on the event $\{\eta_X(0,n;0,\eta\sqrt{n}) \geq 2\}$, we must have $\tau_{i-1,i} \geq n$ for some $1 \leq i \leq \eta\sqrt{n}$, where $\tau_{i-1,i}$ is the first meeting time between the two walks in X starting respectively at $i-1$ and i at time 0.

The key to verifying condition (E) is the density bound (6.56), which formulated in terms of the set of unscaled random walks X becomes

$$\mathbb{P}(0 \in \xi^{\mathbb{Z}}(n)) \leq \frac{C}{\sqrt{n}} \quad \text{uniformly in } n \in \mathbb{N}, \quad (6.65)$$

where $\xi^A(n) := \{\pi(n) : \pi \in X, \pi(0) \in A\}$ for $A \subset \mathbb{Z}$. By translation invariance, for any $L \in \mathbb{N}$,

$$\mathbb{P}(0 \in \xi^{\mathbb{Z}}(n)) = \frac{1}{L}\mathbb{E}[|\xi^{\mathbb{Z}}(n) \cap [0,L)|]$$
$$\leq \frac{1}{L}\sum_{k \in \mathbb{Z}}\mathbb{E}[|\xi^{\mathbb{Z} \cap [kL,(k+1)L)}(n) \cap [0,L)|]$$
$$= \frac{1}{L}\sum_{k \in \mathbb{Z}}\mathbb{E}[|\xi^{\mathbb{Z} \cap [0,L)}(n) \cap [kL,(k+1)L)|]$$
$$= \frac{1}{L}\mathbb{E}[|\xi^{\mathbb{Z} \cap [0,L)}(n)|] \leq \frac{1}{L}\mathbb{E}\left[1 + \sum_{i=1}^{L-1} 1_{\{\tau_{i-1,i} > n\}}\right] \leq \frac{1}{L} + \frac{C}{\sqrt{n}},$$

where in the last inequality, we applied (6.60). Letting $L \to \infty$ then gives (6.65). The rest of the proof of condition (E) then follows the line of argument sketched after Theorem 6.6.3. □

6.6.5 Convergence to the Brownian Net

Convergence to the Brownian net so far has only been established by Sun and Swart [86] for branching-coalescing simple random walk paths with asymptotically vanishing branching probability (see Figure 6.4), and recently by Etheridge, Freeman and Straulino [34] for the genealogies of a spatial Lambda-Fleming-Viot process. We identify below the key conditions and formulate them as convergence criteria for the Brownian net, which can be applied to random sets of paths with certain non-crossing properties. Finding effective and verifiable convergence criteria for random sets of paths which do not satisfy the non-crossing condition (C) below remains a major challenge.

Given a sequence of $(\mathcal{H}, \mathcal{B}_\mathcal{H})$-valued random variables $(X_n)_{n\in\mathbb{N}}$, we first impose a non-crossing condition.

(C) There exist subsets of *non-crossing* paths $W_n^l, W_n^r \subset X_n$, such that no path $\pi \in X_n$ crosses any $l \in W_n^l$ from right to left, i.e., $\pi(s) > l(s)$ and $\pi(t) < l(t)$ for some $s < t$, and no path $\pi \in X_n$ crosses any $r \in W_n^r$ from left to right.

The second condition is an analogue of condition (I), which ensures that $(W_n^l, W_n^r)_{n\in\mathbb{N}}$ is a tight family by Proposition 6.6.4, and any subsequential weak limit contains a copy of the left-right Brownian web $(\mathcal{W}^l, \mathcal{W}^r)$ defined as in Theorem 6.3.2.

($I_\mathcal{N}$) There exist $l_{n,z} \in W_n^l$ and $r_{n,z} \in W_n^r$ for each $z \in \mathbb{R}^2$, such that for any deterministic $z_1, \ldots, z_k \in \mathbb{R}^2$, $(l_{n,z_1}, \ldots, l_{n,z_k}, r_{n,z_1}, \ldots, r_{n,z_k})$ converge in distribution to a collection of left-right coalescing Brownian motions starting at $(z_i)_{1 \leq i \leq k}$, as in Theorem 6.3.2.

The tightness of $(W_n^l, W_n^r)_{n\in\mathbb{N}}$ and condition (C) imply that $(X_n)_{n\in\mathbb{N}}$ is also a tight family, since (C) implies that almost surely, the modulus of continuity of paths in X_n (cf. (6.51)) can be bounded by the modulus of continuity of paths in $W_n^l \cup W_n^r$, whose starting points become dense in $\mathbb{R}^2$ as $n \to \infty$ by condition ($I_\mathcal{N}$).

The next condition is

(H) A.s., X_n contains all paths obtained by hopping among paths in $W_n^l \cup W_n^r$ at crossing times, defined as in the hopping construction of the Brownian net in Theorem 6.3.4.

Condition (H) ensures that any subsequential weak limit of $(X_n)_{n\in\mathbb{N}}$ contains not only a copy of the left-right Brownian web $(\mathcal{W}^l, \mathcal{W}^r)$, but also a copy of the Brownian net $\mathcal{N}$ constructed by hopping among paths in $\mathcal{W}^l \cup \mathcal{W}^r$ at crossing times.

Lastly, we formulate the analogue of condition (U) in Theorem 6.6.6, which gives an upper bound on any subsequential weak limit of $(X_n)_{n\in\mathbb{N}}$ via the dual wedge characterization of the Brownian net given in Theorem 6.3.9.

($\text{U}_\mathcal{N}$) There exist $\widehat{W}_n^l, \widehat{W}_n^r \in \widehat{\mathcal{H}}$, whose starting points are dense in $\mathbb{R}^2$ as $n \to \infty$, such that a.s. paths in W_n^l and $\widehat{W}_n^l$ (resp. paths in W_n^r and $\widehat{W}_n^r$) do not cross, and for any subsequential weak limit $(\mathcal{X}, W^l, W^r, \widehat{W}^l, \widehat{W}^r)$ of $(X_n, W_n^l, W_n^r, \widehat{W}_n^l, \widehat{W}_n^r)$ and any deterministic countable dense $\mathcal{D} \subset \mathbb{R}^2$, a.s. paths in $\mathcal{X}$ do not enter any wedge of $(\widehat{W}^l(\mathcal{D}), \widehat{W}^r(\mathcal{D}))$ from outside.

Condition (H) implies that (W^l, W^r) contains a copy of the left-right Brownian web $(\mathcal{W}^l, \mathcal{W}^r)$, and the non-crossing property implies that $(\widehat{W}^l(\mathcal{D}), \widehat{W}^r(\mathcal{D}))$ coincides with $(\widehat{\mathcal{W}}^l(\mathcal{D}), \widehat{\mathcal{W}}^r(\mathcal{D})$ for the dual left-right Brownian web. By Theorem 6.3.9, the assumption that no path in $\mathcal{X}$ enters any wedge of $(\widehat{\mathcal{W}}^l(\mathcal{D}), \widehat{\mathcal{W}}^r(\mathcal{D})$ from outside then implies that $\mathcal{X}$ is contained in the Brownian net constructed from $(\mathcal{W}^l, \mathcal{W}^r)$, which is the desired upper bound on $\mathcal{X}$.

We thus have the following convergence result.

Theorem 6.6.11 (Convergence criteria for the Brownian net) *Let $(X_n)_{n\in\mathbb{N}}$ be a sequence of $(\mathcal{H}, \mathcal{B}_\mathcal{H})$-valued random variables which satisfy conditions* (C), ($\text{I}_\mathcal{N}$), (H) *and* ($\text{U}_\mathcal{N}$) *above. Then X_n converges in distribution to the standard Brownian net $\mathcal{N}$.*

Remark 6.6.12 To verify condition ($\text{U}_\mathcal{N}$), it suffices to show that paths in X_n do not enter wedges of $(\widehat{W}_n^l, \widehat{W}_n^r)$ from outside, and when a sequence of pairs of paths in $(\widehat{W}_n^l, \widehat{W}_n^r)$ converge to a pair of dual left-right coalescing Brownian motions, the associated first meeting times between the pair also converge, so that the associated wedges converge.

The list of conditions in Theorem 6.6.11 can be verified for the collection of branching-coalescing simple random walks on $\mathbb{Z}$, which answers **Q.1** at the start of Section 6.3.

Theorem 6.6.13 *For $\varepsilon \in (0,1)$, let X_ε be the collection of branching-coalescing simple random walk paths on $\mathbb{Z}^2_{\text{even}}$ with branching probability ε, with walks starting from every site of $\mathbb{Z}^2_{\text{even}}$ (see Figure 6.4). Let S_ε be defined as in* (6.58). *Then $S_\varepsilon X_\varepsilon$ converges in distribution to the standard Brownian net $\mathcal{N}$ as $\varepsilon \downarrow 0$.*

Proof sketch. Conditions (C) and (H) hold trivially for the branching-coalescing simple random walks. Condition ($\text{I}_\mathcal{N}$) was verified in [86, Section 5] using the fact that a single pair of leftmost and rightmost random walk paths solve a discrete analogue of the SDE (6.11) for a pair of left-right coalescing Brownian motions, and furthermore, the time when the pair of discrete paths meet converges to the continuum analogue. By Remark 6.6.12, it is then easily seen that condition ($U_\mathcal{N}$) also holds, and hence Theorem 6.6.11 can be applied. □

6.7 Survey on Related Results

In this section, we survey interesting results connected to the Brownian web and net that have not been discussed so far, including alternative topologies, models whose scaling limits are connected to the Brownian web and net (including population genetic models, true self-avoiding walks, planar aggregation, drainage networks, supercritical oriented percolation), and the relation between the Brownian web and net, critical planar percolation and Tsirelson's theory of noise [95, 96].

6.7.1 Alternative Topologies

We review here several alternative choices of state spaces and topologies for the Brownian web, and compare them with the paths topology of Fontes et al. [40], as introduced in Section 6.2.1. In particular, we will review the *weak flow topology* of Norris and Turner [77], the *tube topology* of Berestycki, Garban and Sen [14] and the *marked metric measure spaces* used by Greven, Sun and Winter [52].

Another natural extension of the paths topology is to consider the space of compact sets of càdlàg paths equipped with the Hausdorff topology, where the space of càdlàg paths is equipped with the Skorohod metric after compactification of space-time as done in Figure 6.2. Such an extension has been carried out by Etheridge, Freeman and Straulino [34] in the study of scaling limits of spatial Lambda-Fleming-Viot processes.

6.7.1.1 Weak Flow Topology

In [77], Norris and Turner formulated a topology for stochastic flows, which includes the Arratia flow generated by coalescing Brownian motions.

Recall that a flow on a space E is a two-parameter family of functions $(\phi_{s,t})_{s \leq t}$ from E to E, which satisfies the flow condition $\phi_{t,u} \circ \phi_{s,t} = \phi_{s,u}$ for any $s \leq t \leq u$. It is easily seen that the Brownian web almost surely defines a flow on $\mathbb{R}$. Indeed, if $(x, s) \in \mathbb{R}^2$ is a $(1, 2)$ point in the Brownian web with one incoming and two outgoing paths, then $(\phi_{s,t}(x))_{t \geq s}$ should be defined as the continuation of the incoming Brownian path so that the flow condition is satisfied, and otherwise, $(\phi_{s,t}(x))_{t \geq s}$ can be defined to be any of the Brownian paths starting at (x, s). Note that there is no unique definition of this flow. Furthermore, it is not clear how to define a suitable topology on the space of flows in order to prove convergence of flows, as well as weak convergence of stochastic flows.

Norris and Turner addressed these issues by introducing the notion of a *weak flow*, which on $\mathbb{R}$, is a family $(\phi_{s,t})_{s \leq t}$ with the properties that

(i) For all $s \leq t$, $\phi_{s,t} \in \mathcal{D}$, the space of non-decreasing functions, so that the flow lines $(\phi_{s,t}(x))_{t \geq s}$, for $(x,s) \in \mathbb{R}^2$, do not cross.
(ii) If $\phi_{s,t}^+$ (resp. $\phi_{s,t}^-$) is the right (resp. left)-continuous version of $\phi_{s,t}$, then

$$\phi_{t,u}^- \circ \phi_{s,t}^- \leq \phi_{s,u}^- \leq \phi_{s,u}^+ \leq \phi_{t,u}^+ \circ \phi_{s,t}^+ \qquad \text{for all } s < t < u.$$

One can endow the space $\mathcal{D}$ with a metric $d_\mathcal{D}$ such that if $f, g \in \mathcal{D}$, and $f^\times, g^\times$ denote the graphs of f and g rotated clockwise by $\pi/4$ around the origin, then

$$d_\mathcal{D}(f,g) = \sum_{n=1}^{\infty} 2^{-n}(1 \wedge \sup_{|x|<n} |f^\times(x) - g^\times(x)|). \qquad (6.66)$$

Note that when f and g are distribution functions, $|f^\times - g^\times|_\infty$ is just the Lévy metric on probability measures. The space of *continuous weak flows* $C^\circ(\mathbb{R}, \mathcal{D})$ then consists of continuous maps $\phi : \{(s,t) \in \mathbb{R}^2 : s \leq t\} \to \mathcal{D}$ with $\phi_{s,s} = \text{id}$ for all $s \in \mathbb{R}$, and it is endowed with the topology of uniform convergence on bounded subsets of $\{(s,t) \in \mathbb{R}^2 : s < t\}$.

To allow for non-continuous weak flows, Norris and Turner also introduced the space of *càdlàg weak flows* $D^\circ(\mathbb{R}, \mathcal{D})$, which consists of maps ϕ from the space of bounded intervals to $\mathcal{D}$, such that $\phi_\emptyset = \text{id}$, $\phi_{(s,t)} \to \phi_\emptyset$ as $t \downarrow s$ and $\phi_{(t,u)} \to \phi_\emptyset$ as $t \uparrow u$, while $\phi_{\{t\}}$ captures the jump discontinuity of the flow at time t. One can then equip $D^\circ(\mathbb{R}, \mathcal{D})$ with a Skorohod-type topology.

It was shown in [77] that $C^\circ(\mathbb{R}, \mathcal{D})$ and $D^\circ(\mathbb{R}, \mathcal{D})$ are Polish spaces, the Arratia flow is a continuous weak flow, and a family of Poisson local disturbance flows converge in distribution to the Arratia flow. Actually [77] considered only weak flows on the circle $\mathbb{S}$. To have non-decreasing maps, a map $f : \mathbb{S} \to \mathbb{S}$ is lifted to a map from $\mathbb{R}$ to $\mathbb{R}$ by identifying $\mathbb{S}$ with $[0, 2\pi)$ and extending f outside $[0, 2\pi)$ by setting $f(x + 2\pi) = f(x) + 2\pi$. The extension to weak flows on $\mathbb{R}$ has subsequently been carried out by Ellis in [32].

Compared with the paths topology introduced in Section 6.2.1, the advantage of the weak flow topology is that it is more natural for studying stochastic flows, and it allows for discontinuity in the flow lines (paths). The limitation is that it is restricted to flows with non-crossing paths. Proving weak convergence to the Arratia flow is very simple: it suffices to verify condition (I) in Section 6.6.1, which ensures that every weak limit point contains the coalescing Brownian motions. There is no need to upper bound the limiting set of paths, unlike the paths topology of Section 6.2.1, because the weak flow topology effectively discards all paths starting from the same space-time point other than the leftmost and rightmost paths, because of the metric $d_\mathcal{D}$ on $\mathcal{D}$.

6.7.1.2 Tube Topology

Just as the paths topology of Fontes et al. [40] was inspired by a similar topology introduced by Aizenman et al. [3, 4] for two-dimensional percolation, Berestycki, Garban and Sen took inspiration from another topology for two-dimensional percolation, the *quad topology* of Schramm and Smirnov [87], and introduced in [14] the *tube topology* for sets of continuous paths in $\mathbb{R}^d$. The basic idea is that the configuration of a set of paths is captured by the set of space-time tubes these paths cross.

Let $T^* := ([T^*], \partial_0 T^*, \partial_1 T^*) := ([0,1]^d \times [0,1], [0,1]^d \times \{0\}, [0,1]^d \times \{1\})$ be the unit tube in $\mathbb{R}^{d+1}$, with $\partial_0 T^*$ and $\partial_1 T^*$ being the lower and upper faces of the tube. A tube $T = ([T], \partial_0 T, \partial_1 T)$ in $\mathbb{R}^{d+1}$ is then defined to be the image of T^* under a homeomorphism $\phi : \mathbb{R}^{d+1} \to \mathbb{R}^{d+1}$ with the property that $\partial_0 T \subset \mathbb{R}^d \times \{t_0\}$ and $\partial_1 T \subset \mathbb{R}^d \times \{t_1\}$ for some $t_0 < t_1$, and $[T] \subset \mathbb{R}^d \times [t_0, t_1]$. We call t_0 the bottom time of T, and t_1 the top time. The space of all tubes, denoted by $\mathcal{T}$, can then be equipped with the Hausdorff metric $d_{\mathcal{T}}$, with

$$d_{\mathcal{T}}(T_1, T_2) := d_{\text{Haus}}([T_1], [T_2]) + d_{\text{Haus}}(\partial_0 T_1, \partial_0 T_2) + d_{\text{Haus}}(\partial_1 T_1, \partial_1 T_2), \quad (6.67)$$

where d_{Haus} is the Hausdorff distance between subsets of $\mathbb{R}^{d+1}$. The metric space $(\mathcal{T}, d_{\mathcal{T}})$ is then separable.

Given a continuous path $\pi : [t, \infty) \to \mathbb{R}^d$ starting at time t, it is said to cross a tube T with lower face $\partial_0 T \subset \mathbb{R}^d \times \{t_0\}$ and upper face $\partial_1 T \subset \mathbb{R}^d \times \{t_1\}$ for some $t_0 < t_1$, if $t \le t_0$, $(\pi(t_0), t_0) \in \partial_0 T$, $(\pi(t_1), t_1) \in \partial_1 T$, and $(\pi(s), s) \in [T]$ for all $s \in [t_0, t_1]$. Given a set K of continuous paths in $\mathbb{R}^d$, one can then identify the set of all tubes $\text{Cr}(K) \subset \mathcal{T}$ which are crossed by some path in K. Furthermore, if K is a compact set of continuous paths w.r.t. a metric on path space defined in the same way as in Section 6.2.1, then $\text{Cr}(K)$ is in fact a closed subset of $\mathcal{T}$. Therefore a random compact set of paths can be identified with a random closed subset of $\mathcal{T}$. The state space of closed subsets of the metric space $\mathcal{T}$ can then be equipped with the Fell topology, which makes it compact (see e.g., [66, Appendix B]).

One can actually narrow down the state space further. Observe that we can define a partial order $\le$ on $\mathcal{T}$, where we denote $T_1 \le T_2$ if whenever T_2 is crossed by some path π, T_1 is also crossed by π. We denote $T_1 < T_2$ if there are open neighborhoods $U_i \subset \mathcal{T}$ around T_i, $i = 1, 2$, such that $T_1' \le T_2'$ for all $T_1' \in U_1$ and $T_2' \in U_2$. A set of tubes $S \subset \mathcal{T}$ is then called *hereditary* if $T \in S$ implies that $T' \in S$ for all $T' < T$. Note that the set of tubes $\text{Cr}(K)$ induced by a set of paths K is always hereditary. Therefore the state space for random compact sets of paths can be taken to be the space of *closed hereditary subsets* of $\mathcal{T}$, denoted by $\mathcal{H}$, equipped with the Fell topology. It can be shown that $\mathcal{H}$ is closed under the Fell topology, and hence compact. This gives the tube topology defined in [14, Section 2], and the Brownian web can be realized as a random variable taking values in $\mathcal{H}$.

The main advantages of the tube topology are: (i) tightness for a family of $\mathcal{H}$-valued random variables comes for free because the state space $\mathcal{H}$ is compact. For instance, under the tube topology, Theorem 6.6.9 on the convergence of general coalescing random walks to the Brownian web can be established under the optimal finite second moment assumption, since the higher moment assumption is only used to establish tightness under the path topology; (ii) the tube topology, being actually a weaker topology than the path topology of Fontes et al. [40] introduced in Section 6.2.1, makes it much easier to construct coalescing flows which do not satisfy the non-crossing property of the Arratia flow. In particular, the coalescing Brownian flow on the Sierpinski gasket was constructed in [14] using the tube topology, and an invariance principle was established.

To characterize a probability measure on $\mathcal{H}$, it turns out to be sufficient to determine the probability of the joint crossing of any finite collection of tubes chosen from a deterministic countable dense subset of $\mathcal{T}$. To prove the weak convergence of a sequence of probability measures on $\mathcal{H}$, it is sufficient to find a large enough set of tubes $\hat{\mathcal{T}} \subset \mathcal{T}$ such that the the probability of the joint crossing of any finite subset of $\hat{\mathcal{T}}$ converges to a limit (see [14, Proposition 2.12]). The strategy for verifying this convergence criterion for coalescing flows is similar to the proof of condition (E) in Theorem 6.6.3. First one approximates a given tube T with bottom time t_0 by tubes T_δ with bottom time $t_0 + \delta$ for $\delta > 0$, then one applies a coming down from infinity result to show that among paths which intersect the lower face of T, only finitely many remain in the lower face of T_δ at time $t_0 + \delta$, and lastly one uses condition (I) to control the joint distribution of the remaining finite collection of coalescing paths.

The tube topology is weaker than the paths topology introduced in Section 6.2.1, because the mapping from the space of compact sets of continuous paths (with the paths topology) to the space of closed hereditary sets of tubes $\mathcal{H}$ (with the Fell topology) is continuous as shown in [14, Lemma A.2]. In particular, given a set of paths K, restricting a path $\pi \in K$ to a later starting time and adding it to K has no effect under the tube topology. In a sense, the tube topology is also insensitive toward the behavior of the paths near their starting times. In particular, a sequence of paths π_n, which start at time 0 and converge to a path π uniformly on $[\delta, \infty)$ for any $\delta > 0$, will converge to π under the tube topology. But π_n may have wild oscillations in the interval $[0, 1/n)$ which prevent it from converging in the paths topology.

6.7.1.3 Marked Metric Measure Spaces

In [52], Greven, Sun and Winter treat the (dual) Brownian web as a stochastic process taking values in the space of spatially-marked metric measure spaces. The notion of a V-marked metric probability measure space was introduced by Depperschmidt, Greven and Pfaelhuber in [30], which is simply a complete separable metric space (X,d), together with a Borel probability measure μ on the product space $X \times V$ for some complete separable metric space of marks V. The probability measure μ can be regarded as a sampling measure, and each V-marked metric measure (V-mmm) space can be uniquely determined (up to isomorphism) by the joint distribution of $(v_i)_{i \in \mathbb{N}}$ and the distance matrix $(d(x_i,x_j))_{i,j \in \mathbb{N}}$, where $(x_i, v_i)_{i \in \mathbb{N}}$ are i.i.d. samples drawn from $X \times V$ with common distribution μ. A sequence of V-mmm spaces is then said to converge if the associated random vector of marks $(v_i)_{i \in \mathbb{N}}$ and the random distance matrix $(d(x_i,x_j))_{i,j \in \mathbb{N}}$ converge in finite-dimensional distribution. By truncations in the mark space V, one can also extend the notion of V-mmm spaces to the case where μ is only required to be finite on bounded sets when projected from $X \times V$ to V, which was done in [52].

To see how the Brownian web fits in the framework of V-mmm spaces, note that the coalescing Brownian motions in the (dual) Brownian web can be regarded as the space-time genealogies of a family of individuals. More precisely, if $\widehat{\mathcal{W}}$ is the dual Brownian web introduced in Section 6.2.3, then for each $z = (x,t) \in \mathbb{R}^2$, interpreted as an individual at position x at time t, $\hat{\pi}_z \in \widehat{\mathcal{W}}$ determines its spatial genealogy, with $\hat{\pi}_z(s)$ (for $s \leq t$) being the spatial location of its ancestor at time s. Coalescence of two paths $\hat{\pi}_{z_1}$ and $\hat{\pi}_{z_2}$ then signifies the merging of the two genealogy lines. If we consider all individuals indexed by $\mathbb{R}$ at a given time t, then we can measure the genealogical distance between individuals, i.e.,

$$d((x,t),(y,t)) := 2(t - \hat{\tau}_{(x,t),(y,t)}), \qquad (6.68)$$

where $\hat{\tau}_{(x,t),(y,t)}$ is the time of coalescence between $\hat{\pi}_{(x,t)}$ and $\hat{\pi}_{(y,t)} \in \widehat{\mathcal{W}}$. It is easily seen that d is in fact an ultra-metric, i.e.,

$$d((x_1,t),(x_2,t)) \leq d((x_1,t),(x_3,t)) \vee d((x_2,t),(x_3,t)) \qquad \text{for all } x_1,x_2,x_3 \in \mathbb{R}.$$

The collection of individuals indexed by $\mathbb{R}$ at time t then forms a metric space with metric d. The mark of an individual is just its spatial index, so that we can identify the metric space with the mark space $V := \mathbb{R}$, and a natural sampling measure is the Lebesgue measure on $\mathbb{R}$. There is one complication, namely that a.s. there exists $x \in \mathbb{R}$ such that multiple paths in $\widehat{\mathcal{W}}$ start from (x,t). The correct interpretation is that each such path encodes the genealogy of a distinct individual, which happens to occupy the same location. Therefore the space of population should be enriched from $\mathbb{R}$ to take into account these individuals. Such an enrichment does not affect the Lebesgue sampling measure, since it follows from Theorem 6.2.11 that a.s. there are only countably many points of multiplicity in $\widehat{\mathcal{W}}$ at a given time t. If we let (X_t, d_t, μ_t) denote the $\mathbb{R}$-mmm space induced by $\widehat{\mathcal{W}}$ on $\mathbb{R}$ at time t, then $\widehat{\mathcal{W}}$ a.s.

determines the stochastic process $(X_t, d_t, \mu_t)_{t \in \mathbb{R}}$, and conversely, it can be seen that $(X_t, d_t, \mu_t)_{t \in \mathbb{R}}$ a.s. determines $\widehat{\mathcal{W}}$, and hence the two can be identified.

We can also identify $\widehat{\mathcal{W}}$ with a single $\mathbb{R}^2$-mmm space instead of the $\mathbb{R}$-mmm space-valued process $(X_t, d_t, \mu_t)_{t \in \mathbb{R}}$. Namely, we can consider all points in $\mathbb{R}^2$ and extend the genealogical distance in (6.68) between individuals at the same time to individuals at different times:

$$d((x_1, t_1), (x_2, t_2)) := t_1 + t_2 - 2\hat{\tau}_{(x_1,t_1),(x_2,t_2)}, \qquad (6.69)$$

where $\hat{\tau}_{(x_1,t_1),(x_2,t_2)}$ is the time of coalescence between $\hat{\pi}_{(x_1,t_1)}$ and $\hat{\pi}_{(x_2,t_2)} \in \widehat{\mathcal{W}}$. We then obtain a metric space whose elements can be identified with the mark space $V := \mathbb{R}^2$ (with suitable enrichment of $\mathbb{R}^2$ to take into account points of multiplicity in $\widehat{\mathcal{W}}$), on which we equip as a sampling measure the Lebesgue measure on $\mathbb{R}^2$. We remark that one can also enlarge the mark space and let the whole genealogy line $\hat{\pi}_z \in \widehat{\mathcal{W}}$ be the mark for a point $z \in \mathbb{R}^2$.

For models arising from population genetics, V-mmm space is a natural space for the spatial genealogies, and the dual Brownian web determines the genealogies of the so-called *continuum-sites stepping-stone model*. Because V-mmm spaces are characterized via sampling, proving convergence essentially reduces to condition (I) formulated in Section 6.6.1. However, proving tightness for V-mmm space-valued random variables is a nontrivial task and can be quite involved (see e.g., [52]). Also the Brownian net cannot be characterized using V-mmm spaces, since there is no natural analogue of the metric d in (6.69) when paths can branch.

6.7.2 Other Models which Converge to the Brownian Web and Net

We review here various models which have scaling limits that are connected to the Brownian web and net. These include population genetic models such as the voter model and the spatial Fleming-Viot processes, stochastic Potts models, the true self-avoiding walks, planar aggregation models, drainage networks, and supercritical oriented percolation in dimension $1 + 1$.

6.7.2.1 Voter Model and Spatial Fleming-Viot Processes

Voter model and spatial Fleming-Viot processes are prototypical population genetic models where the spatial genealogies of the population are coalescing random walks, which converge to the Brownian web under diffusive scaling.

Arratia [1, 2] first conceived the Brownian web in studying the scaling limit of the voter model on $\mathbb{Z}$, which is an interacting particle system $(\eta_t)_{t \geq 0}$ with $\eta_t \in \Omega := \{0, 1\}^{\mathbb{Z}}$, modeling the opinions of a collection of individuals indexed by $\mathbb{Z}$. Independently for each $x \in \mathbb{Z}$, at exponential rate 1, a resampling event occurs where the voter at x picks one of its two neighbors with equal probability and changes its opinion $\eta_t(x)$ to that of the chosen neighbor. The resampling events can be represented by a Poisson point process of arrows along the edges over time, with each arrow from x to y at time t signifying the voter at x changing its opinion at time t to that of y (see Figure 6.16(a)). This is known as Harris' graphical construction. To identify $\eta_t(x)$, one just needs to trace the genealogy of where the opinion $\eta_t(x)$ comes from backward in time, i.e., follow the arrows backward in time, until an ancestor y is reached at time 0 so that $\eta_t(x) = \eta_0(y)$. The genealogy line $\hat{\pi}_{(x,t)}$ for $\eta_t(x)$ is then a continuous time random walk running backward in time. Furthermore, for multiple space-time points $((x_i, t_i))_{1 \leq i \leq k}$, their joint genealogy lines $(\hat{\pi}_{(x_i, t_i)})_{1 \leq i \leq k}$ are a collection of coalescing random walks. This is known as the *duality* between the voter model and coalescing random walks (see [54, 61] for more details). Taking the diffusive scaling limit of the joint genealogies of all voter opinions at all possible times then leads to what we now call the (dual) Brownian web.

The spatial Fleming-Viot process $(\xi_t)_{t \geq 0}$ is a measure-valued process which extends the voter model by allowing a continuum of individuals at each site $x \in \mathbb{Z}$, represented by a probability measure on the type space $[0, 1]$. Individuals in the population migrate on $\mathbb{Z}$ as independent random walks, and resampling (one individual changes its type to that of another) takes place between every pair of individuals at the same site with exponential rate 1. This model has recently been studied in [52], and the joint spatial genealogies are a collection of coalescing random walks running backward in time, where each pair of walks at the same site coalesce with exponential rate 1. It is not difficult to see that the diffusive scaling limit of these genealogies also gives the (dual) Brownian web.

It is natural to ask whether the (dual) Brownian web also determines the genealogies of a population genetic model on $\mathbb{R}$. The answer is affirmative, and such a continuum model has been studied before and is known as the *continuum-sites stepping-stone model* (CSSM) (see e.g., Donnelly et al [29]). The Brownian web effectively gives a Harris' graphical construction of the CSSM.

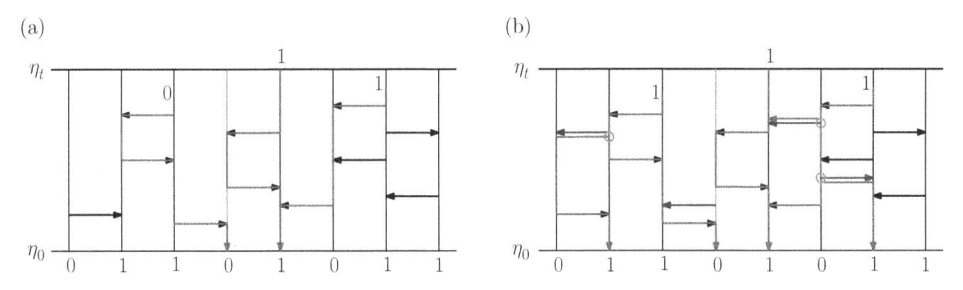

Figure 6.16. Harris' graphical construction of: (a) the voter model; (b) the biased voter model.

Convergence of the voter model as a measure-valued process on $\mathbb{R} \times \{0,1\}$ to the CSSM has been established in [6], and convergence of the genealogies of the spatial Fleming-Viot process to that of the CSSM as stochastic processes taking values in the space of $\mathbb{R}$-mmm spaces (see Section 6.7.1.3) has been established in [52]. In both results, the use of the sampling measure allows one to establish convergence under the optimal finite second moment assumption on the increments of the underlying coalescing random walks, in contrast to Theorem 6.6.9.

We remark that coalescing random walks on $\mathbb{Z}$ also model the evolution of boundaries between domains of different spins in the 0-temperature limit of the stochastic Potts model, and the Brownian web has been used to study aging in this model [37].

6.7.2.2 Biased Voter Model

We have seen that the voter model is dual to coalescing random walks. It turns out that branching-coalescing random walks is dual to the *biased voter-model*, also known as Williams-Bjerknes model [104, 83], which modifies the voter model by adding a selective bias so that type 1 is favoured over type 0. More precisely, each voter independently undergoes a second type of resampling event with rate ε, where the voter at x changes its type to that of the chosen neighbor y only if y is of type 1. The Harris' graphical construction can then be modified by adding a second independent Poisson point process of (selection) arrows along the edges over time, where a selection arrow is used only if it points to an individual of type 1 at that time (see Figure 6.16(b)). To determine $\eta_t(x)$, we then trace its genealogy backward in time by following the resampling arrows, and when a selection arrow is encountered, we follow both potential genealogies (by either ignoring or following the selection arrow). The potential genealogies then form a collection of branching-coalescing random walks with branching rate ε. It is not difficult to see that $\eta_t(x) = 1$ if and only if $\eta_0(y) = 1$ for at least one ancestor y that can be reached by the potential genealogies. This establishes the duality between the biased voter model and branching-coalescing random walks. If we consider a sequence of biased voter models with selection rate $\varepsilon \downarrow 0$, while rescaling space-time by $(\varepsilon, \varepsilon^2)$, then their genealogies converge to the Brownian net.

We remark that branching-coalescing random walks on $\mathbb{Z}$ also model evolving boundaries between domains of different spins in the low temperature limit of the stochastic Potts model, where a new domain of spins can nucleate at the boundary of two existing domains [68].

6.7.2.3 True Self-avoiding Walks and True Self-repelling Motion

Arratia first conceived what we now call the Brownian web in his unfinished manuscript [2], and the subject lay dormant until Tóth and Werner [99] discovered a surprising connection between the Brownian web and the *true self-avoiding walk* on $\mathbb{Z}$ with bond repulsion [94]. A special case of the true self-avoiding walk is defined as follows. Let $l_0(\cdot)$ be an integer-valued function defined on the edges of $\mathbb{Z}$, which can be regarded as the initial condition for the edge local time of the walk, i.e., how many times each edge has been traversed. Given the walk's position $X_n = x$ and the edge local time $l_n(\cdot)$ at time n, if $l_n(\{x-1,x\}) < l_n(\{x,x+1\})$, then with probability 1, $X_{n+1} = x - 1$; if $l_n(\{x,x+1\}) < l_n(\{x-1,x\})$, then with probability 1, $X_{n+1} = x + 1$; and if $l_n(\{x-1,x\}) = l_n(\{x,x+1\})$, then $X_{n+1} = x \pm 1$ with probability $1/2$ each. The edge local time $l_n(\cdot)$ is then updated to $l_{n+1}(\cdot)$ by adding 1 to the local time at the newly traversed edge. Such a walk is called a true self-avoiding walk because it is repelled from the more visited regions, and the laws of $(X_i)_{1 \leq i \leq n}$ are consistent as n varies, in contrast to the *self-avoiding walk*.

Interestingly, if we plot the evolution of the position of the walk together with its edge local time, then there is an almost sure coupling with a collection of forward/backward coalescing random walks. Figure 6.17 illustrates such a coupling, where the initial edge local time is given by $l_0(\{2x, 2x+1\}) = 1$ for all $x \in \mathbb{N}_0$, $l_0(\{2x-1, 2x\}) = 0$ for all $x \in \mathbb{N}$, and $l_0(\{-x-1, -x\}) = l_0(\{x, x+1\})$ for all $x \in \mathbb{Z}$. Such an l_0 corresponds to a special boundary condition for the coalescing random walks along the x-axis, as shown in Figure 6.17. The lattice-filling curve between the forward and backward coalescing random walks encodes the evolution of the position of the walk (the horizontal coordinate) and the edge local time (the vertical coordinate), and the area filled in by the curve is just the time the walk has spent. The collection of forward/backward coalescing random walks converge to the forward/backward Brownian web, albeit with the boundary condition that the path in the forward, respectively backward, web starting at $(0,0)$ is the constant

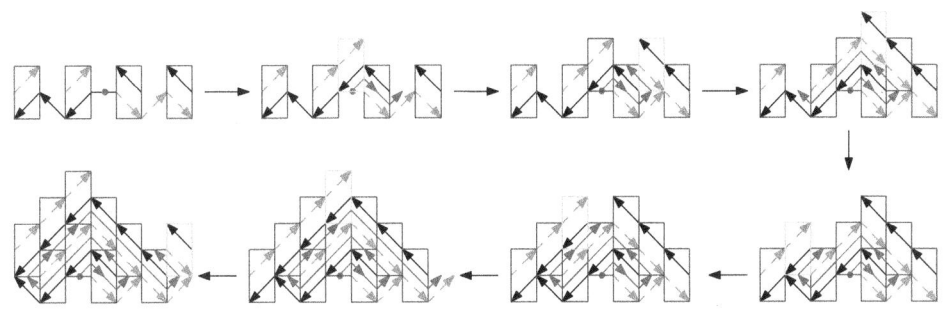

Figure 6.17. Coupling between a true self-avoiding walk (with a special initial edge local time) and a collection of forward/backward coalescing random walks (with a special boundary condition). The horizontal coordinate of the lattice-filling curve is the walk's position and the vertical coordinate is its edge local time.

path, and time for the webs now runs in the horizontal direction in Figure 6.17. It can then be shown (see [71]) that the lattice-filling curve squeezed between the coalescing random walks also converges to a space-filling curve $(X_t, L_t(X_t))_{t \geq 0}$ in $\mathbb{R} \times [0, \infty)$, squeezed between paths in the forward and backward Brownian webs. Here, X_t is the position of the so-called *true self-repelling motion*, and $(L_t(x))_{x \in \mathbb{R}}$ is its occupation time density at time t. Comparing with the discrete model, one sees that when a deterministic point $(x, y) \in \mathbb{R} \times [0, \infty)$ is first reached by the space-filling curve $(X_\tau, L_\tau(X_\tau))$ at some random time τ, $(L_\tau(a))_{a \leq x}$, respectively $(L_\tau(a))_{a \geq x}$, must equal the path in the forward, respectively backward, Brownian web starting at (x, y). Furthermore, the area under $(L_\tau(a))_{a \in \mathbb{R}}$ must be exactly equal to τ. In other words, from the realization of the forward/backward Brownian webs, one can determine when a deterministic point (x, y) is reached by the space-filling curve. By considering a deterministic countable dense set of $(x, y) \in \mathbb{R} \times [0, \infty)$, one can then determine the entire trajectory of $(X_t)_{t \geq 0}$. This is how the true self-repelling motion was constructed by Tóth and Werner in [99], which heuristically is a process with a drift given by the negative of the gradient of its occupation time density. It has the unusual scaling invariance of $(X_{at}/a^{2/3})_{t \geq 0} \stackrel{\text{dist}}{=} (X_t)_{t \geq 0}$ for any $a > 0$, and it has finite variation of order $3/2$. For further details, see [99, 26].

6.7.2.4 Planar Aggregation Models

In [76], Norris and Turner discovered an interesting connection between the Hastings-Levitov planar aggregation model and the coalescing Brownian flow. We briefly explain the model and the connection here. At time 0, we start with a unit disc denoted by K_0. A small particle P_1, which we assume to be a disc of radius $\delta > 0$ for simplicity, is attached to K_0 at a uniformly chosen random point on the boundary ∂K_0. This defines the new aggregate $K_1 = K_0 \cup P_1$. To define the next aggregate K_2, we apply a conformal map F_1 which maps K_1 back to the disc K_0 (more accurately, F_1 is a conformal map from $K_1^c \cup \{\infty\}$ to $K_0^c \cup \{\infty\}$, uniquely determined by the condition that $F_1(z) = Cz + O(1)$ for some $C > 0$ as $|z| \to \infty$). A new particle P_2 of radius δ is then attached randomly at the boundary of $K_0 = F_1(K_1)$. Reversing F_1 then defines the new aggregate $K_2 = F_1^{-1}(F_1(K_1) \cup P_2)$. The dynamics can then be iterated as illustrated in Figure 6.18. The new particles added to the aggregates $(K_n)_{n \in \mathbb{N}}$ at each step are no longer discs due to distortion by the conformal maps, and it was shown in [76] that new particles tend to pile on top of each other and form protruding fingers. Interestingly, the image of ∂K_0 under F_n forms a coalescing flow on the circle, where for $x, y \in \partial K_0 \subset \partial K_n$, the length of the arc between $F_n(x)$ and $F_n(y)$ on the unit circle is proportional to the probability that a new particle will be attached to the corresponding part of ∂K_n

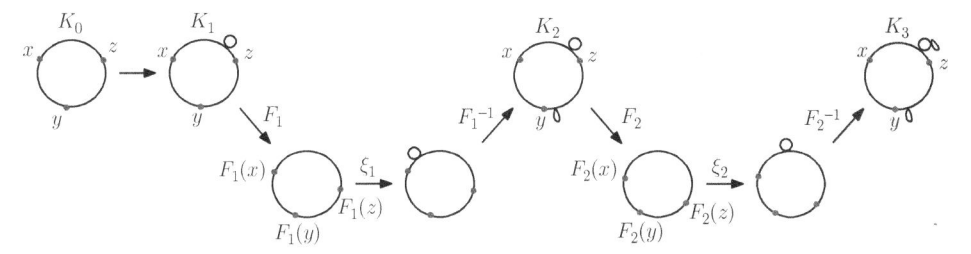

Figure 6.18. Hastings-Levitov planar aggregation model.

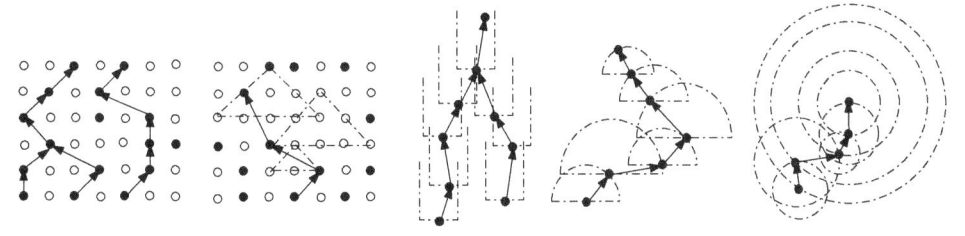

Figure 6.19. From left to right, illustration of paths in the drainage network of [50], the drainage network of [80], Poisson trees [43], the directed spanning forest [11], and the radial spanning tree [11].

between x and y. In the limit that the particle radius $\delta \downarrow 0$, while time is sped up by a factor of δ^{-3}, this coalescing flow can be seen as a localized disturbance flow, studied in [77], which converges to the coalescing Brownian flow on the circle w.r.t. the weak flow topology described in Section 6.7.1.1.

6.7.2.5 Drainage Networks and Directed Forests

Drainage networks are a class of models where coalescing paths arise naturally. First, a random subset of $\mathbb{Z}^{d+1}$ (or $\mathbb{R}^{d+1}$) is determined, which represents the water sources. Next, from each source (x, s), where $x \in \mathbb{R}^d$ and $s \in \mathbb{R}$, exactly one directed edge is drawn toward some other source (y, t) with $t > s$, representing the flow of water from (x, s) to (y, t). Examples include [50, 81], where the authors study a drainage network on $\mathbb{Z}^{d+1}$ with each vertex being a water source independently with probability $p \in (0, 1)$. From each source (x, s), a directed edge is then drawn to the closest source in the next layer $\mathbb{Z}^d \times \{s+1\}$, and ties are broken by choosing each closest source with equal probability. There are also other variants such as in [5], where the directed edge from the source $(x, s) \in \mathbb{Z}^{d+1}$ connects to the closest source in the 45° light cone rooted at (x, s), or as in [80] where the directed edge from (x, s) connects to the closest source in $\mathbb{Z}^d \times \{s, s+1, \ldots\}$ measured in ℓ_1 distance (see Figure 6.19).

There are also continuum space versions where the water sources form a homogeneous Poisson point process in $\mathbb{R}^{d+1}$. In the *Poisson trees* model considered in [43], a directed edge is drawn from each source $(x,s) \in \mathbb{R}^{d+1}$ to the source in $\{(y,t) : |y-x| \leq 1, t > s\}$ with the smallest t-coordinate. In the *directed spanning forest* model considered in [11], a directed edge is drawn from each source (x,s) to the closest source in $\mathbb{R}^d \times (s,\infty)$, measured in Euclidean distance.

Natural questions for such drainage networks include whether the directed edges form a single component, i.e., it is a tree rather than a forest. This has been shown to be the case for $d=1$ and 2 for the drainage networks on $\mathbb{Z}^{d+1}$ described above [50, 5, 80] and for the Poisson trees [43], as well as for the directed spanning forest in $\mathbb{R}^{1+1}$ [21].

The collection of paths in the drainage network, obtained by following the directed edges, can be regarded as a collection of (dependent) coalescing random walk paths. Because the dependence is in some sense local in all the models described above, it is natural to conjecture that for $d=1$, the collection of directed paths in the drainage network (after diffusive scaling) should converge to the Brownian web. This has indeed been verified for various drainage networks on $\mathbb{Z}^{1+1}$ [17, 22, 80], and for the Poisson trees on $\mathbb{R}^{1+1}$ [36]. The main difficulty in these studies lies in the dependence among the paths. The model considered in [80] is a discrete analogue of the directed spanning forest [11], and the dependence is handled using simultaneous regeneration times along multiple paths. Such arguments are not directly applicable to the directed spanning forest, and its convergence to the Brownian web remains open.

The directed spanning forest was introduced in [11] as a tool to study the *radial spanning tree*, where given a homogeneous Poisson point process $\Xi \subset \mathbb{R}^d$, a directed edge is drawn from each $x \in \Xi$ to the closest point $y \in \Xi \cup \{o\}$ that lies in the ball with radius $|x|$ and centered at the origin o. If one considers directed paths in the radial spanning forest restricted to the region (in polar coordinates) $\{(r,\theta) : r \in [an,n], |\theta + \pi/2| \leq n^{-b}\}$ for some $a \in (0,1)$ and $b \in (0,1/2)$, then after proper scaling, these paths are believed to converge to the so-called *Brownian bridge web* [46], which consists of coalescing Brownian bridges starting from $\mathbb{R} \times (-\infty, 0)$ and ending at $(0,0)$ (it can also be obtained from the Brownian web via a deterministic transformation of $\mathbb{R}^2$). For a couple of toy models, such a convergence has been established in [23, 46].

6.7.2.6 Supercritical Oriented Percolation

The oriented bond percolation model on $\mathbb{Z}^2_{\text{even}} := \{(x,t) \in \mathbb{Z}^2 : x+t \text{ is even}\}$ is defined by independently setting each oriented edge of the form $(x,t) \to (x \pm 1, t+1)$ to be either open with probability p, or closed with probability $1-p$, with retention parameter $p \in [0,1]$. The set of vertices in $\mathbb{Z}^2_{\text{even}}$ that can be reached from $z \in \mathbb{Z}^2_{\text{even}}$ by following open oriented edges is called the open cluster at z. It is known that there is a critical $p_c \in (0,1)$ such that for $p > p_c$, the open cluster at $(0,0)$

Figure 6.20. Rightmost infinite open paths in supercritical oriented percolation on $\mathbb{Z}^2_{\text{even}}$.

is infinite with positive probability, while for $p \leq p_c$, it is finite with probability one [24]. When the open cluster at $z \in \mathbb{Z}^2_{\text{even}}$ is infinite, z is called a percolation point.

In the supercritical regime $p > p_c$, the percolation points appear with a positive density, and starting from each percolation point $z = (x, t)$, we can find a rightmost path r_z among all the infinite open oriented paths starting from z (see Figure 6.20). It was shown by Durrett [24] that each $(r_z(n))_{n \geq t}$ satisfies a law of large numbers with drift $\alpha(p) > 0$, and subsequently Kuczek [58] showed that $r_z(n) - n\alpha$ satisfies a central limit theorem with variance $n\sigma(p)^2$ for some $\sigma(p) > 0$. Kuczek's central limit theorem can be easily extended to path level convergence to Brownian motion. It is then natural to ask what is the joint scaling limit of r_z for all percolation points z. Wu and Zhang [105] conjectured that the scaling limit should be the Brownian web. This may be surprising at first, because for different percolation points z_1 and z_2, r_{z_1} and r_{z_2} both depend on the infinite future, and hence could be strongly dependent on each other. However, it was shown by Sarkar and Sun in [89] that each r_z can be approximated by a thin percolation exploration cluster, such that different clusters evolve independently before they meet and quickly merge after they meet. It is then shown in [89] that after proper centering and scaling, the collection of rightmost infinity open paths from percolation points indeed converge to the Brownian web.

6.7.3 Brownian Web, Critical Percolation, and Noise

There are close parallels between the Brownian web and the scaling limit of critical planar percolation, starting from the topology. Indeed, as noted in Section 6.7.1.2, both the paths topology of [40] and the tube topology of [14] were inspired by similar topologies for planar percolation. The Brownian web is the scaling limit of discrete coalescing random walks on $\mathbb{Z}^2_{\text{even}}$ as shown in Figure 6.1, which can be seen as a dependent planar percolation model and enjoys the same duality as bond percolation on $\mathbb{Z}^2_{\text{even}}$. The scaling limit of critical planar percolation is invariant under conformal maps, while the Brownian web is invariant under diffusive scaling of $\mathbb{R}^2$, a signature of a model at criticality. The Brownian web and the scaling limit of critical planar percolation are both (and the only known examples of) two-dimensional black noise in the language of Tsirelson [95, 96], which is intimately linked to the notion of noise sensitivity and the existence of near-critical scaling limits. Indeed, the Brownian net can be regarded as a near-critical scaling limit obtained by perturbing the Brownian web. As explained in Section 6.4, such a perturbation can be carried out by Poisson marking the set of "pivotal" points (i.e., the (1,2)-points of the Brownian web) according to a natural local time measure, and then turning these points into branching points. For planar percolation, exactly the same procedure has been proposed in [19] to construct a near-critical scaling limit from the critical scaling limit, and it has been rigorously carried out in [48] and [49], where defining the natural local time measure on pivotal points alone has been a very challenging task. The fact that changing the configuration at a countable subset of $\mathbb{R}^2$, i.e., the Poisson marked pivotal points, is sufficient to alter the scaling limit indicates noise sensitivity and that the scaling limit is a black noise. Allowing the intensity of the Poisson marking of the pivotal points to vary continuously leads to dynamical evolutions of the critical scaling limit, where unusual behaviour may emerge at random "dynamical times."

In what follows, we will briefly review Tsirelson's theory of noise [95, 96], and in what sense the Brownian web and the scaling limit of critical planar percolation are both black noise. This is the key common feature that lies behind the many parallels between the two models. Such parallels will likely extend to other examples of two- or higher-dimensional black noise, once they are found. We will also briefly review results on the dynamical Brownian web.

6.7.3.1 Brownian Web, Black Noise, and Noise Sensitivity

Intuitively, the Brownian web and the scaling limit of critical planar percolation should satisfy the property that the configurations are independent when restricted to disjoint domains in $\mathbb{R}^2$. This calls for the continuous analogue of the notion of product probability spaces (or independent random variables), which led Tsirelson to define (see [96, Definition 3c1]):

Definition 6.7.1 *A (one-dimensional) continuous product of probability spaces consists of a probability space $(\Omega, \mathcal{F}, P)$ and sub-σ-fields $\mathcal{F}_{s,t} \subset \mathcal{F}$ (for all $s < t$) such that $\mathcal{F}$ is generated by $\bigcup_{s<t} \mathcal{F}_{s,t}$ and for any $r < s < t$,*

$$\mathcal{F}_{r,t} = \mathcal{F}_{r,s} \otimes \mathcal{F}_{s,t}, \tag{6.70}$$

which means that $\mathcal{F}_{r,t}$ is generated by $\mathcal{F}_{r,s} \cup \mathcal{F}_{s,t}$, and $\mathcal{F}_{r,s}$ is independent of $\mathcal{F}_{s,t}$ (i.e., $P(A \cap B) = P(A)P(B)$ for all $A \in \mathcal{F}_{r,s}$ and $B \in \mathcal{F}_{s,t}$).

We can think of $\mathcal{F}_{s,t}$ as the σ-field generated by a family of observables that only depend on what happens in the interval (s,t).

Remark 6.7.2 In dimension two or higher, we should equip $(\Omega, \mathcal{F}, P)$ with a family of sub-σ-fields $\mathcal{F}_D$, indexed by a Boolean algebra of domains $D \subset \mathbb{R}^2$ (take for example the Boolean algebra of sets generated by open rectangles), while the factorization property (6.70) becomes

$$\mathcal{F}_D = \mathcal{F}_{D_1} \otimes \mathcal{F}_{D_2} \qquad (6.71)$$

whenever $D_1 \cap D_2 = \emptyset$ and $\overline{D} = \overline{D_1 \cup D_2}$. It turns out that (6.71) cannot hold without some restrictions on the domains. The lack of a canonical choice of the family of domains in dimensions two and higher is still an issue to be resolved [98, Section 1.6].

Besides independence on disjoint domains, we also expect the Brownian web and the scaling limit of percolation to be translation invariant. This additional assumption leads to the notion of a noise (see [96, Definition 3d1]).

Definition 6.7.3 (Noise) *A noise is a continuous product of probability spaces $(\Omega, \mathcal{F}, P)$, equipped with a family of sub-σ-fields $\mathcal{F}_D$ indexed by an algebra of domains $D \in \mathcal{D}$ in $\mathbb{R}^d$, which is homogeneous in the following sense. There exists a group of isomorphisms $(T_h)_{h \in \mathbb{R}^d}$ on $(\Omega, \mathcal{F}, P)$ such that $T_{h+h'} = T_h \circ T_{h'}$ and T_h sends $\mathcal{F}_D$ to $\mathcal{F}_{D+h}$ for all $D \in \mathcal{D}$ and $h, h' \in \mathbb{R}^d$.*

In a tour de force, the scaling limit of critical planar percolation was shown to be a two-dimensional noise by Schramm and Smirnov in [87], where the sub-σ-field $\mathcal{F}_D$ (for D with piecewise smooth boundary) is generated by the indicator random variables for open crossing of quads (homeomorphic images of the unit square) contained in D. The Brownian web was shown to be a two-dimensional noise by Ellis and Feldheim in [33], where $\mathcal{F}_D$ (for open rectangles D) is generated by the Brownian web paths restricted to D. What is remarkable is that both noises are so-called *black noise* as defined by Tsirelson [95, 96], and they are the only known examples in dimension two or higher.

The notion of a noise being *black* turns out to be equivalent to the notion of a noise being *sensitive*, while a noise being *classical* (such as white noise or Poisson noise) turns out to be equivalent to a noise being *stable*. A *non-classical* noise is a noise that contains in some sense a sensitive part. To explain the underlying ideas, it is instructive to first discretize and then pass to the continuum, instead of giving directly the definition for the continuum noise.

Let us fix $\Lambda = (-1,1)^2$, and consider its discretization $\Lambda_\delta := \Lambda \cap S_\delta \mathbb{Z}^2_{\text{even}}$, where $S_\delta : (x,t) \to (\delta x, \delta^2 t)$. To each $z \in \Lambda_\delta$, we assign an i.i.d. symmetric random variable $\sigma_z \in \{+1, -1\}$, which can be defined via the coordinate map on the probability space $\Omega_\delta = \{+1, -1\}^{\Lambda_\delta}$, equipped with the discrete topology and the uniform probability measure P_δ. What noise we obtain in the continuum limit $\delta \downarrow 0$ depends crucially on the observables we choose, which will generate the σ-field $\mathcal{F}$ for the limiting noise.

If we choose our observables to be linear functions of $\sigma := (\sigma_z)_{z \in \Lambda_\delta}$, i.e., $f_\delta := \delta^{3/2} \sum_{z \in \Lambda_\delta} f(z) \sigma_z$ for continuous functions $f : \Lambda \to \mathbb{R}$, then their joint distributions converge to a nontrivial limit and the limiting noise is the classical white noise, where $\mathcal{F}$ is generated by the family of Gaussian random variables indexed by such continuous f, which we interpret as the integral of f w.r.t. the underlying white noise.

If we choose our observables to be nonlinear functions of $(\sigma_z)_{z \in \Lambda_\delta}$, then non-classical noise may appear in the limit. As illustrated in Figure 6.1, $\sigma := (\sigma_z)_{z \in \Lambda_\delta}$ uniquely determines the collection of coalescing random walks in Λ_δ. If we choose our observables to be the indicator random variables for whether there is a random walk path crossing a prescribed tube T in Λ (see Section 6.7.1.2), then their joint distributions converge to a nontrivial limit and the limiting noise is the Brownian web, where $\mathcal{F}$ is generated by the tube crossing events.

Roughly speaking, whether the limiting noise is classical or not depends on whether there are nontrivial observables that are noise sensitive as $\delta \downarrow 0$, defined as follows.

Definition 6.7.4 (Noise sensitivity) *For $\varepsilon \in (0,1)$, let $\sigma^\varepsilon := (\sigma_z^\varepsilon)_{z \in \Lambda_\delta}$ be obtained from $(\sigma_z)_{z \in \Lambda_\delta}$ by independently replacing each σ_z with an independent copy of σ_z with probability ε. A sequence of random variables $f_\delta : \Omega_\delta \to \mathbb{R}$ with $\sup_\delta E[f_\delta^2(\sigma)] < \infty$ is called noise sensitive if for each $\varepsilon > 0$,*

$$\lim_{\delta \downarrow 0} (E[f_\delta(\sigma) f_\delta(\sigma^\varepsilon)] - E[f_\delta(\sigma)]^2) = 0, \tag{6.72}$$

and the sequence is called noise stable if

$$\lim_{\varepsilon \downarrow 0} \limsup_{\delta \downarrow 0} |E[f_\delta(\sigma) f_\delta(\sigma^\varepsilon)] - E[f_\delta^2(\sigma)]| = 0. \tag{6.73}$$

There is a rich theory of noise sensitivity for functions on $(\Omega_\delta, P_\delta)$, and we refer to the lecture notes by Garban and Steif [51] for a detailed exposition and applications to percolation. For $\{0,1\}$-valued functions f_δ, noise sensitivity implies that $f_\delta(\sigma)$ and $f_\delta(\sigma^\varepsilon)$ become asymptotically independent if an arbitrarily small, but fixed, portion of σ is resampled, while noise stability implies that as $\varepsilon \downarrow 0$, the probability that $f_\delta(\sigma)$ and $f_\delta(\sigma^\varepsilon)$ coincide tends to 1 uniformly in δ close to 0.

There is a simple criterion for noise sensitivity/stability using Fourier analysis on $(\Omega_\delta, P_\delta)$. Observe that the set of functions $\chi_S := \prod_{i \in S} \sigma_i$, for all $S \subset \Lambda_\delta$, is an orthonormal basis for $L^2(\Omega_\delta, P_\delta)$. Therefore any $f_\delta \in L^2(\Omega_\delta, P_\delta)$ admits the orthogonal decomposition

$$f_\delta = \sum_{S \subset \Lambda_\delta} \hat{f}_\delta(S) \chi_S = E[f_\delta] + \sum_{k=1}^{\infty} \sum_{|S|=k} \hat{f}_\delta(S) \prod_{i \in S} \sigma_i. \quad (6.74)$$

Definition 6.7.5 (Spectral measure and energy spectrum) *The coefficients $\hat{f}_\delta(S)$ in (6.74) are called the* Fourier-Walsh *coefficients for f_δ, the measure μ_{f_δ} on $\{S : S \subset \Lambda_\delta\}$ with $\mu_{f_\delta}(S) := \hat{f}_\delta^2(S)$ is called the* spectral measure *of f_δ, and the measure E_{f_δ} on $\mathbb{N}_0$ with $E_{f_\delta}(k) = \sum_{|S|=k} \hat{f}_\delta^2(S)$ is called the* energy spectrum *of f_δ (see e.g., [51]).*

It is easily seen that

$$E[f_\delta(\sigma) f_\delta(\sigma^\varepsilon)] = E[f_\delta]^2 + \sum_{k=1}^{\infty} \varepsilon^k \sum_{|S|=k} \hat{f}_\delta^2(S) = E[f_\delta]^2 + \sum_{k=1}^{\infty} \varepsilon^k E_{f_\delta}(k), \quad (6.75)$$

and hence a sequence f_δ with $\sup_\delta E[f_\delta^2] < \infty$ is noise sensitive if and only if $E_{f_\delta}(\{1,\ldots,k\}) \to 0$ for every $k \in \mathbb{N}$, and f_δ is stable if and only if E_{f_δ} is a tight family of measures on $\mathbb{N}_0$.

As we take the continuum limit $\delta \downarrow 0$, what will happen to the spectral measure of the observables that generate the σ-field of the limiting noise? There are several possibilities: either all nontrivial square-integrable observables in the limit will have a spectral measure that is supported on $\mathcal{C}_{\text{fin}}(\Lambda) := \{S \in \Lambda : |S| < \infty\}$, in which case the limiting noise is called *classical* (and noise stable); or all nontrivial observables in the limit will have a spectral measure supported on $\mathcal{C}(\Lambda) \setminus \mathcal{C}_{\text{fin}}(\Lambda)$, with $\mathcal{C}(\Lambda)$ being the set of closed subsets of Λ, in which case the noise is called *black* (and noise sensitive); or there are nontrivial observables of both types, supported either on $\mathcal{C}_{\text{fin}}(\Lambda)$ or $\mathcal{C}(\Lambda) \setminus \mathcal{C}_{\text{fin}}(\Lambda)$, in which case the noise is called *non-classical*.

In practice, classifying a noise via discrete approximation as described above is subtle, because such approximations are not unique, and convergence of the spectral measure on $\mathcal{C}$ (equipped with Hausdorff topology) does not imply convergence of the energy spectrum. There could be observables f for the noise whose spectral measure is supported on $\mathcal{C}_{\text{fin}}$, and yet the approximating f_δ is noise sensitive (see [95, Section 5c] for the notion of *block sensitivity* which overcomes such issues). Instead, we can classify noise directly. The following definition extends [95, Definition 3d2] to general dimensions (see also [98] for an abstract formulation without associating the sub-σ-fields with domains in $\mathbb{R}^d$).

Definition 6.7.6 (Non-classical noise and black noise) *Let $(\Omega, \mathcal{F}, P)$ be a noise equipped with a family of sub-σ-fields $\mathcal{F}_D$ indexed by an algebra of domains $D \in \mathcal{D}$ in $\mathbb{R}^d$. For any $f \in L^2(\Omega, \mathcal{F}, P)$, its spectral measure μ_f is defined to be the unique positive measure on $\mathcal{C}$ (the space of compact subsets of $\mathbb{R}^d$) with the property that*

$$\mu_f(M \in \mathcal{C} : M \subset D) = E[E[f \mid \mathcal{F}_D]^2], \qquad D \in \mathcal{D}. \tag{6.76}$$

If for every $f \in L^2(\Omega, \mathcal{F}, P)$, its spectral measure μ_f is supported on $\mathcal{C}_{\text{fin}} := \{M \in \mathcal{C} : |M| < \infty\}$, then the noise is called classical; *otherwise the noise is called* non-classical. *If the spectral measure of every non-constant $f \in L^2(\Omega, \mathcal{F}, P)$ is supported on $\mathcal{C} \backslash \mathcal{C}_{\text{fin}}$, then the noise is called* black.

White noise and Poisson noise are classical, which can be seen from their chaos expansions. The scaling limit of critical planar percolation is a two-dimensional black noise as explained in [87], while the Brownian web was first shown to be a one-dimensional black noise in [95, Section 7] and [96, Section 7] (with sub-σ-fields $(\mathcal{F}_{s,t})_{s<t}$ associated with the strips $\mathbb{R} \times (s,t)$), and then shown to be a two-dimensional black noise in [33]. Interestingly, if a noise is black, then for all non-constant $f \in L^2(\Omega, \mathcal{F}, P)$, μ_f is in fact supported on $\mathcal{C}_{\text{perf}} := \{M \in \mathcal{C} : M \text{ is a perfect set}\}$, a collection of uncountable sets [95, Theorem 6d3].

The procedure of independently resampling each of a collection of ± 1-valued symmetric random variables with probability ε can in fact be carried out for noise. This allows one to define directly noise sensitivity vs stability for each $f \in L^2(\Omega, \mathcal{F}, P)$, where conditions (6.72)–(6.73) just need to be modified to remove the δ-dependence, and σ and σ^ε represent the noise and its ε-resampled version. A noise is then classical if and only if all $f \in L^2(\Omega, \mathcal{F}, P)$ are stable, and black if and only if all non-constant $f \in L^2(\Omega, \mathcal{F}, P)$ are sensitive (see [96, Section 4] and [95, Section 5] for further details). In particular, ε-resampling a black noise leads to an independent copy of the noise. To see nontrivial dependence, a different resampling procedure is needed, which is where dynamical evolution via Poisson marking of pivotal points comes in.

6.7.3.2 Dynamical Brownian Web

Because the Brownian web is a black noise, the correct way to resample the randomness underlying the Brownian web in order to see nontrivial dependence is to resample the "pivotal" points (i.e., the $(1,2)$-points of the Brownian web). This was carried out in [73] and is intimately linked to the authors' marking construction of the Brownian net in the same paper (see Section 6.4). Given a Poisson point process $M_\lambda \subset \mathbb{R}^2$ with intensity measure given by $\lambda(>0)$ times the intersection local time measure $\ell(\mathrm{d}z)$ on the $(1,2)$-points of the Brownian web (see Proposition 6.4.1), a resampled web is obtained by simply flipping the sign of each $(1,2)$ point in M_λ as defined in (6.24), instead of turning each such point into a branching point as in the construction of the Brownian net in Theorem 6.4.2. Since there is a natural coupling between M_λ for all $\lambda \geq 0$, we obtain a stochastic process in the space of compact sets of paths $(\mathcal{H}, d_\mathcal{H})$, which is in fact reversible w.r.t. the law of the Brownian web. This defines the so-called *dynamical Brownian web*. The dynamical Brownian web can also be constructed as the scaling limit of the *dynamical discrete web* introduced in [56], where *discrete web* refers to the collection of coalescing simple symmetric random walks on $\mathbb{Z}^2_{\text{even}}$ illustrated in Figure 6.1, and dynamical evolution is introduced by independently flipping at rate one the sign of the arrow from each $z \in \mathbb{Z}^2_{\text{even}}$. If space-time is rescaled by $(\varepsilon, \varepsilon^2)$, and the dynamical time is slowed down by a factor of ε, it can then be shown that the dynamical discrete web converges to the dynamical Brownian web [73].

Similar questions have been studied for percolation models [47]. There has been much study about dynamical percolation (see [84] for a survey). Furthermore, dynamical critical site percolation on the triangular lattice has been shown to converge to a continuum limit [49], the proof of which follows the same line as in the Brownian web setting (via Poisson marking of the pivotal points), except that the implementation in the percolation setting has been much more difficult.

Key questions in the study of dynamical percolation, dynamical Brownian web, and "dynamical" processes in general, are the existence of exceptional times when a given property that holds almost surely at a deterministic time fails. For dynamical percolation, where the stationary law of the process is that of the percolation model at criticality, one such property is the non-existence of an infinite cluster [84]. For the dynamical discrete web, questions on exceptional times have also been studied in [45] and [57]. In particular, it was shown in [45] that there exists a.s. a random set of dynamical times with full Lebesgue measure, when the random walk in the discrete web starting from the origin does not satisfy the Law of the Iterated Logarithm (LIL); the authors also extended this result to the dynamical Brownian web.

This is in striking contrast with the dynamical simple symmetric random walk on $\mathbb{Z}$ introduced in [15], where the sign of each increment of the walk $(X_n - X_{n-1})_{n \in \mathbb{N}}$ is independently flipped at rate one in dynamical time. In contrast with the dynamical discrete web, there is no exceptional time at which the LIL fails in this setting. The difference between the two models can be accounted for by the effect of each individual switching. In [15], a switch affects only one increment of the walk, which induces an order one perturbation in sup-norm. In contrast, changing the direction of a single arrow in the discrete web can change the path of the walk by a "macroscopic" amount. Indeed, the switching arrows encountered by the walk are comparable to the Poisson marked $(1,2)$-points encountered by a Brownian motion in the dynamical Brownian web, where each switching at a marked (1, 2)-point leads the Brownian motion to embark on an excursion from the original path, and the size of the excursion has a heavy-tailed distribution with infinite mean. This is why the random walk is more sensitive to the dynamics on the discrete web than to the dynamics on the random walk considered in [15].

6.8 Open Questions

In this section, we discuss some open questions, including how the Brownian web and net could be useful in the study of more general one-dimensional interacting particle systems, conjectures on the geometry of the Brownian net, and some other miscellaneous questions.

6.8.1 Voter Model Perturbations and Brownian Net with Killing

One further direction of research is to use the Brownian web and net to study one-dimensional interacting particle systems with migration, birth/branching, coalescence, and death. Such a particle system often arises as the spatial genealogies of a dual particle system. In suitable parameter regimes, such a particle system may converge to the Brownian net, which incorporates diffusive particle motion, branching and coalescence, and killing effect can be added by defining the so-called *Brownian net with killing*, which we will explain below. Using the convergence of the discrete particle systems to a continuum model which is amenable to analysis, one hopes to draw conclusions for the discrete particle systems, such as local survival vs extinction, as well as convergence to equilibrium.

In dimensions three and higher, a study in this spirit has been carried out by Cox, Durrett and Perkins in [18], where they studied a class of interacting particle systems on $\mathbb{Z}^d$ whose transition rates can be written as perturbations of the transition rates of a (possibly non-nearest neighbor) voter model. Their interest stemmed from the observation that several models in ecology and evolution, as well as in statistical mechanics – the spatial Lotka-Volterra model [70], the evolution of cooperation [78], nonlinear voter models [67] – can be seen as perturbations of the voter model in a certain range of their parameter space. In general, these models are difficult to study either because of a lack of monotonicity (Lotka-Volterra model) or because of the intrinsic complexity of the model (certain models of evolution of cooperation). However, by considering them as small perturbations of the voter model, it was shown in [18] that for $d \geq 3$, the properly rescaled local density of particles converges to the solution of a reaction diffusion equation, and that properties of the underlying particle system (such as coexistence of species in the Lotka-Volterra model) can be derived from the behavior of this Partial Differential Equation (PDE). Implementing an analogous program in dimensions one and two is still an open problem, although in dimension two, it is believed that the behavior of voter model perturbations should be similar to the case $d = 3$.

The genealogies of the voter model are given by coalescing random walks [61, 54], and the type of perturbations of the voter model in [18] lead to branching of the genealogies. In dimensions $d \geq 3$, because of transience of the random walk, such a collection of branching-coalescing genealogies was shown in [18] to converge (under suitable scaling and moment assumptions on the random walk and the branching mechanism) to a system of *independent* branching Brownian motions. The branching mechanism is given explicitly in terms of the "microscopic" descriptions of the system and also takes into account coalescence at the mesoscopic level. In dimension one, such a collection of branching-coalescing spatial genealogies should converge to the Brownian net under suitable scaling of parameters and space-time. One is then hopeful that in dimension one, one can also draw conclusions for the particle systems from such a convergence. One model that falls in this class is the biased annihilating branching process, introduced by Neuhauser and Sudbury in [75]. Another class is a family of one-dimensional models of competition with selection and mutation (see e.g., [7]). However, proving convergence of the particle systems to the Brownian net is a serious challenge.

6.8.1.1 Brownian Net with Killing

We briefly recall here the Brownian net with killing, which allows the modeling of the death of a particle. A discrete space analogue has been introduced in [68]. Let $b, \kappa >= 0$ be such that $b + \kappa \leq 1$, corresponding respectively to the branching and killing parameters of the system. From each point in $\mathbb{Z}^2_{\text{even}}$, an arrow is drawn from (x, t) to either $(x - 1, t + 1)$ or $(x + 1, t + 1)$ with equal probability $\frac{1}{2}(1 - b - \kappa)$; two arrows are drawn with probability b; no arrow is drawn with probability κ. Similar to the discrete net (see Figure 6.9), we can consider the set of all paths starting in $\mathbb{Z}^2_{\text{even}}$ by following arrows until the path terminates. This defines an infinite collection of coalescing random walks which branch with probability b and are killed with probability κ. This model encompasses several classical models from statistical mechanics. For $p \in (0, 1)$ and $b = p^2$, $\kappa = (1 - p)^2$, one recovers the standard one-dimensional oriented percolation model. When $b = \kappa = 0$, the trajectories are distributed as coalescing random walks.

In [74], it is shown that when the branching and killing parameters depend on some small scaling parameter ε such that $\lim_{\varepsilon \to 0} b_\varepsilon / \varepsilon = b_\infty$ and $\lim_{\varepsilon \to 0} \kappa_\varepsilon / \varepsilon^2 = \kappa_\infty$, then the collection of branching-coalescing random walks with killing introduced above converges (after proper rescaling) to a continuum object called the Brownian net with killing with parameter $(b_\infty, \kappa_\infty)$. The latter scaling of the parameters was motivated by the stochastic q-color Potts model on $\mathbb{Z}$. Indeed, in [74], it was conjectured that this model at large inverse temperature β, is asymptotically dual to a system of (continuous time) branching-coalescing random walks with killing, with branching and killing parameters

$$b_\beta = \frac{q}{2} e^{-\beta} \quad \text{and} \quad \kappa_\beta = q e^{-2\beta},$$

where q is the number of colors in the system. Thus the Brownian net could be used to construct a natural scaling limit for the stochastic Potts model at low temperature in dimension one.

The Brownian net with killing can also be directly constructed as follows. Let $\mathcal{N}_b$ be a Brownian net with branching parameter b. For every realization of the Brownian net, one can introduce a natural time-length measure on the set of points of type (p,p) in $\mathbb{R}^2$ w.r.t. the Brownian net (see Theorem 6.3.24). More precisely, for every Borel set $E \subset \mathbb{R}^2$, define the time-length measure as

$$\mathcal{L}(E) = \int_\mathbb{R} |\{x : (x,t) \in E \text{ and is of type (p,p)}\}|\, dt. \tag{6.77}$$

The Brownian net with killing with parameter (b, κ) is then obtained by killing the Brownian net paths at the points of a Poisson point process on $\mathbb{R}^2$ with intensity measure $\kappa \mathcal{L}$.

It was shown in [74] that the Brownian net with killing undergoes a phase transition as the killing parameter κ varies, similar to the percolation transition in oriented percolation. There is reason to believe [53] that the Brownian net with killing belongs to the same universality class as oriented percolation in $\mathbb{Z}^{1+1}$, in which case it may serve as a simpler model to study this universality class.

6.8.2 Fractal Structure of the Brownian Net

We explain here some open questions concerning the geometry of the Brownian net. Recall from Remark 6.3.20 the set $\mathcal{N}(*, -\infty)$ of paths started at time $-\infty$, called the *backbone* of the Brownian net $\mathcal{N}$. Let $N := \{(x,t) \in \mathbb{R}^2 : x = \pi(t) \text{ for some } \pi \in \mathcal{N}(*, -\infty)\}$ be the graph of the backbone. By the image set property (Proposition 6.3.17), every continuous function $\pi : \mathbb{R} \to \mathbb{R}$ that lies entirely inside N is a path in the backbone. In view of this, much information can be read off from the random closed set $N \subset \mathbb{R}^2$.

If we set $\xi_t := \{x \in \mathbb{R} : (x,t) \in N\}$, then $(\xi_t)_{t \in \mathbb{R}}$ is just the branching-coalescing point set in stationarity (see Section 6.3.5), which is in fact a Markov process [86, Section 1.9–1.10]. At each deterministic time t, ξ_t is distributed a Poisson point process with intensity two, which is a reversible law for this Markov process. As noted in Remark 6.3.20, by reversibility, the law of $N \subset \mathbb{R}^2$ (and likewise the collection of paths $\mathcal{N}(*,-\infty)$) is symmetric under time reversal.

It is clear that N has some sort of fractal structure. The main reason for this is the "infinite" branching rate of the Brownian particles that ξ_t consists of. Indeed, in each open time interval, a path in $\mathcal{N}(*,-\infty)$ will split infinitely often into two paths, which usually coalesce again after a very short time. But during the time that there are two paths, these too will for short moments split into two further paths, and so on, ad infinitum (see Figure 6.21).

Natural questions are: 1) is this the only way in which N has a fractal structure? 2) how can we formulate this rigorously? and 3) can we prove this? Here is an attempt to answer question 2. Let S and M be the set of all *separation points* and *meeting points* (i.e., points of types $(p,pp)_s$ and (pp,p), respectively, as defined in Section 6.2.5) that lie on N. Set $S_* := S \cup \{(*,-\infty)\}$ and $M_* := M \cup \{(*,\infty)\}$. Define $\phi : S_* \to M_*$ by

$$\phi(z) := \text{ the first meeting point of the left-most and right-most paths starting at } z, \tag{6.78}$$

where we set $\phi(z) := (*,\infty)$ if the left-most and right-most paths starting at z never meet, and for definitiveness we also define $\phi(*,-\infty) := (*,\infty)$.

By the symmetry of the backbone with respect to time reversal, we define $\phi : M_* \to S_*$ analogously, following paths in $\mathcal{N}(*,-\infty)$ backwards in time. (Note that this is different from our usual way of looking backwards. Instead of following dual paths, here we follow *forward* paths in the backbone backwards in time.)

A lot about the structure of the backbone can be understood in terms of the map ϕ. For example, separation points z such that $\phi(z) = (*,\infty)$ are $(-\infty,\infty)$-relevant separation points as defined in Section 6.4.1. By Proposition 6.4.7, these form a locally finite subset of $\mathbb{R}^2$. If we have a pair of points $z \in S$ and $z' \in M$ such that $\phi(z) = z'$ and $\phi(z') = z$, then this corresponds to a path that splits at z into two paths, which meet again at z' (see Figure 6.21). If $z = (x,t)$ and $z' = (x',t')$ form such a pair, then the left-most and right-most paths starting at z, up to their first meeting point z', enclose a compact set that we will call a *bubble*.

We conjecture that bubbles are the *only* source of the fractal structure of N. The following two conjectures make this idea precise.

Conjecture 6.8.1 (Bubble hypothesis)

(a) *Almost surely, there does not exist an infinite sequence of points $(z_k)_{k \geq 0}$ in $S_* \cup M_*$, all different from each other, such that $\phi(z_{k-1}) = z_k$ for all $k \geq 1$.*

(b) *Almost surely, there does not exist an infinite sequence of points $(z_k)_{k \geq 0}$ in $S_* \cup M_*$, all different from each other, such that $\phi(z_k) = z_{k-1}$ for all $k \geq 1$.*

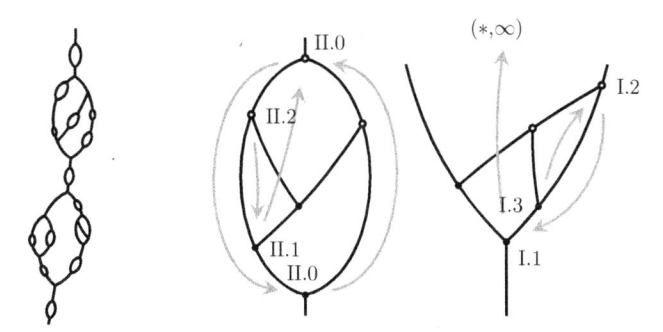

Figure 6.21. Left: schematic depiction of the fractal structure of the backbone. Middle and right: illustration of the classification of separation and meeting points on the backbone with the map ϕ (gray arrows). The leftmost picture contains two separation and two meeting points of type II.1; all others are of type II.0.

If Conjecture 6.8.1 (a) is correct, then for each separation or meeting point z that lies on the backbone, after some moment in the sequence $(z, \phi(z), \phi^2(z), \ldots)$, the same points start to repeat. Simple geometric considerations show that the limiting cycle must have length 2, i.e., there is an integer $n \geq 0$ such that $\phi^{n+2}(z) = \phi^n(z)$. Letting n denote the smallest such integer, we say that z is of *class* I.n if $\phi^n(z) = (*, \infty)$ or $(*, -\infty)$ and of *class* II.n if $\phi^n(z) \in \mathbb{R}^2$. Note that separation and meeting points of type II.0 are the bottom and top points of bubbles, as we have just defined them. Points of type II.n all lie inside bubbles, with larger n leading to a more complex left-right crossover pattern in the containing bubble (see Figure 6.21). The following conjecture says that most bubbles have a very simple internal structure.

Conjecture 6.8.2 (Bubble complexity) *Let $\mathcal{C}_{\mathrm{I},n}$, respectively $\mathcal{C}_{\mathrm{II},n}$, denote the set of meeting and separation points on the backbone N which are of class I.n, respectively II.n. Then*

(a) *The sets $\mathcal{C}_{\mathrm{II},0}$ and $\mathcal{C}_{\mathrm{II},1}$ are dense in N.*
(b) *The sets $\mathcal{C}_{\mathrm{II},n}$ with $n \geq 2$ and $\mathcal{C}_{\mathrm{I},n}$ with $n \geq 1$ are locally finite subsets of $\mathbb{R}^2$.*

As motivation for these conjectures, we state a somewhat more applied problem. Modulo a time reversal, the set $\mathcal{W}(\{0\} \times \mathbb{R})$ of all paths in the Brownian web can be interpreted as the spatial genealogies of a population living in one-dimensional space (see Section 6.7.2.1). Let M be the set of all meeting points of paths in $\mathcal{W}(\{0\} \times \mathbb{R})$, i.e.,

$$M := \{(x,t) \in \mathbb{R}^2 : t > 0, \exists \pi, \pi' \in \mathcal{W}(\{0\} \times \mathbb{R}) \text{ s.t. } \pi(t) = x = \pi'(t), \pi < \pi' \text{ on } (0,t)\}. \tag{6.79}$$

Biologically, we can interpret a point $z = (x,t)$ that is the first meeting point of paths $\pi_{(x,0)}$ and $\pi_{(x',0)}$ as the *most recent common ancestor* (MRCA) of x and x', that lived a time t in the past. It is easy to see that M is a locally finite subset of $\mathbb{R} \times (0, \infty)$.

Moving away from neutral evolution, one can interpret paths in the Brownian net as *potential* genealogies (cf. the biased voter model in Section 6.7.2.2), where in order to determine the true genealogy of an individual, one has to have information about what happens at selection events, which correspond to separation points. In this case, replacing $\mathcal{W}(\{0\} \times \mathbb{R})$ in (6.79) by $\mathcal{N}(\{0\} \times \mathbb{R})$, we can interpret the resulting set M as the set of *potential most recent common ancestors* (PMRCAs) of individuals living at time zero.

Conjecture 6.8.3 (Potential most recent common ancestors) *If we replace $\mathcal{W}(\{0\} \times \mathbb{R})$ in (6.79) by $\mathcal{N}(\{0\} \times \mathbb{R})$, then the set M is a locally finite subset of $\mathbb{R} \times (0, \infty)$.*

The definition of a potential most recent common ancestor (PMRCA) is somewhat reminiscent of the definition of a $(0, \infty)$-relevant separation point, except that in order to determine whether a meeting point is a PMRCA, we have to follow forward paths backwards in time. This is similar to the definition of the map $\phi : M_* \to S_*$ and indeed, Conjecture 6.8.3 seems to be closely related to our previous two conjectures. In particular, meeting points that lie inside a bubble (properly defined w.r.t. the finite time horizon 0) can never be PMRCAs.

6.8.3 Miscellaneous Open Questions

We collect below some other interesting questions.

(1) Is the Brownian net integrable in the sense that the branching-coalescing point set defined in (6.20) admits an explicit characterization, similar to the Pfaffian point process characterization of the coalescing point set discussed in Remark 6.2.10?

(2) Find effective criteria for general branching-coalescing particle systems, where paths can cross, to converge to the Brownian net. This would extend Theorem 6.6.11 for the non-crossing case, and pave the way for the study of voter model perturbations discussed in Section 6.8.1.

(3) Can one formulate a well-posed martingale problem for the branching-coalescing point set $(\xi_t)_{t \geq 0}$ defined in (6.20)? This may offer an alternative route to prove convergence to the Brownian net, which is based on generator convergence and avoids the paths topology.

(4) For the Howitt-Warren flow $(K_{s,t}^{\uparrow}(x, \mathrm{d}y))_{s<t, x \in \mathbb{R}}$ introduced in Section 6.5.1, which gives the transition probability kernels of a random motion in a continuum space-time random environment constructed from the Brownian

web and net, can one show that for any $x \in \mathbb{R}$, $-\log K_{0,t}^{\uparrow}(0,[tx,tx+1])$ has Tracy-Widom GUE fluctuations on the scale $t^{1/3}$ as $t \to \infty$, similar to recent results in [12] for random walks on $\mathbb{Z}$ in special i.i.d. Beta-distributed space-time random environments (cf. Remark 6.5.15)?

(5) In [35], Evans, Morris and Sen showed that coalescing stable Lévy processes on $\mathbb{R}$ with stable index $\alpha \in (1,2)$ come down from infinity, i.e., starting from everywhere on $\mathbb{R}$, the coalescing Lévy processes become locally finite on $\mathbb{R}$ at any time $t > 0$. This suggests the existence of a family of *Lévy webs*. In fact, using the tube topology reviewed in Section 6.7.1.2 and the property of coming down from infinity, it should be straightforward to construct the Lévy web w.r.t. the tube topology. For coalescing Brownian flow on the Sierpinski gasket, this was carried out in [14]. Can one construct the Lévy web in the paths topology, and what type of special points may arise in the spirit of Theorem 6.2.11?

By scaling invariance, the density of the coalescing stable Lévy flow on $\mathbb{R}$ with index $\alpha \in (1,2)$ should be $\rho_t = \frac{C}{t^{1/\alpha}}$. Identifying C would allow one to determine the sharp asymptotic rate of decay for the density of coalescing random walks on $\mathbb{Z}$ in the domain of attraction of a stable Lévy process, similar to [72, Corollary 7.1] and [88] in the Brownian case.

(6) The Brownian web has been shown by Ellis and Feldheim [33] to be a two-dimensional black noise, equipped with a family of sub-σ-fields indexed by finite unions of open rectangles in $\mathbb{R}^2$. Is there a maximal extension of the family of sub-σ-fields, indexed by a Boolean algebra of domains in $\mathbb{R}^2$ that include the rectangles, such that the Brownian web remains a noise (cf. Remark 6.7.2)? A similar question is also open for the scaling limit of critical planar percolation [98, Section 1.6].

Recently Tsirelson [97] showed by general arguments that there exists some continuous path f starting at time $-\infty$, such that $\mathcal{F}_{f_-}$ and $\mathcal{F}_{f_+}$, the σ-fields generated by Brownian web paths restricted respectively to the left or the right of the graph of f, do not jointly generate the full σ-field $\mathcal{F}$. Can we characterize the set of f for which the factorization property $\mathcal{F} = \mathcal{F}_{f_-} \vee \mathcal{F}_{f_+}$ holds? (The case $f \equiv 0$ has been treated in [33]).

(7) We expect the Brownian net with killing (see Section 6.8.1.1) to belong to the same universality class as oriented percolation (OP) on $\mathbb{Z}^{1+1}$, which corresponds to the so-called Reggeon field theory [20, 65] (see also [53] for an extensive survey on the OP universality class). The Brownian net without killing belongs to a different universality class. But is there also a field theory corresponding to the Brownian net? A positive indication is that the Brownian net has competing effects of instantaneous coalescence vs infinite effective

rate of branching, which is very much in the spirit of renormalizations in field theories.

(8) The Brownian web appears in the scaling limit of super-critical oriented percolation, as discussed in Section 6.7.2.6, and the Brownian net is expected to arise if the percolation parameter p is allowed to vary in a small interval. However, the most interesting question is: what is the scaling limit of critical (and near-critical) oriented percolation (OP) on $\mathbb{Z}^{1+1}$? The Brownian net with killing provides a simpler model to study the OP universality class. We expect that in the near-critical scaling limit, one would obtain a family of models which interpolates between the Brownian net and the scaling limit of critical oriented percolation. However, such a goal appears far out of reach at the moment, because there are no conjectures at all on what the critical scaling limit might be, and neither have critical exponents been shown to exist.

Acknowledgments

This article is based on lectures given by R. Sun in the trimester program *Disordered Systems, Random Spatial Processes and Some Applications*, at the Institut Henri Poincaré in the spring of 2015. We thank the program and the organizers for the opportunity to lecture on this topic. We thank all our collaborators and colleagues with whom we have had valuable discussions. In particular, R. Sun and E. Schertzer would like to thank Chuck Newman and Krishnamurthi Ravishankar for introducing them to this topic. R. Sun also thanks University of Warwick, Leiden University, and Academia Sinica for hospitality, where part of these notes were written. R. Sun is supported by AcRF Tier 1 grant R-146-000-185-112. J.M. Swart is sponsored by GACR grant 15-08819S.

References

[1] R. Arratia, *Coalescing Brownian motions on the line*. Ph.D. Thesis, University of Wisconsin, Madison, 1979.

[2] R. Arratia, *Coalescing Brownian motions and the voter model on $\mathbb{Z}$*. Unpublished partial manuscript.

[3] M. Aizenman, *Scaling limit for the incipient spanning clusters*, in *Mathematics of Multiscale Materials: Percolation and Composites* (K. Golden et al, eds.). New York: Springer, IMA Vol. Math. Appl. 99, (1998).

[4] M. Aizenman and A. Burchard, *Hölder regularity and dimension bounds for random curves*. Duke Math. J. 99, 419–453 (1999).

[5] S. Athreya, R. Roy and A. Sarkar, *Random directed trees and forest – drainage networks with dependence*. Electron. J. Probab. 13, 2160–2189 (2008).

[6] S. Athreya and R. Sun, *One-dimensional voter model interface revisited*. Electron. Commun. Probab. 16, 792–800 (2011).

[7] S. Athreya and J. M. Swart. *Systems of branching, annihilating, and coalescing particles*. Electron. J. Probab. 17, 1–32 (2012).

[8] D. ben-Avraham, M. A. Burschka and C. R. Doering, *Statics and dynamics of a diffusion-limited reaction: anomalous kinetics, nonequilibrium self-ordering, and a dynamic transition*. J. Stat. Phys. 60, 695–728 (1990).

[9] P. Billingsley, *Convergence of probability measures, 2nd edition*. New York: John Wiley & Sons, (1999).

[10] N. Berestycki, *Recent progress in coalescent theory*. Rio de Janeiro: Sociedade Brasileira de Matemática, Ensaios Matemáticos 16, (2009).

[11] F. Bacelli and C. Bordenave. *The radial spanning tree of a Poisson point process*. Ann. Appl. Probab. 17, 305–359 (2007).

[12] G. Barraquand and I. Corwin, *Random-walk in Beta-distributed random environment*. (2015). arXiv:1503.04117.

[13] C. Bezuidenhout and G. Grimmett, *The critical contact process dies out*. Ann. Probab. 18, 1462–1482 (1990).

[14] N. Berestycki, C. Garban and A. Sen, *Coalescing Brownian flows: a new approach*. (2013). arXiv:1307.4313. Ann. Probab. 43, 3177–3215 (2015).

[15] I. Benjamini, O. Háaggstrom, Y. Peres and J. E. Steif, *Which properties of a random sequence are dynamically sensitive?* Ann. Probab. 31, 1–34 (2003).

[16] S. Belhaouari, T. Mountford, R. Sun and G. Valle, *Convergence results and sharp estimates for the voter model interfaces*. Electron. J. Probab. 11, 768–801 (2006).

[17] C. F. Coletti, E. S. Dias and L. R. G. Fontes, *Scaling limit for a drainage network model*. J. Appl. Probab. 46, 1184–1197 (2009).

[18] J. T. Cox, R. Durrett and E. Perkins, *Voter model perturbations and reaction diffusion equations*. Astérisque 349 (2013).

[19] F. Camia, L. R. G. Fontes and C. M. Newman, *Two-dimensional scaling limits via marked nonsimple loops*. Bull. Braz. Math. Soc. (N.S.) 37, 537–559 (2006).

[20] J. L. Cardy and R. L. Sugar, *Directed percolation and Reggeon field theory*. J. Phys. A: Math. Gen. 13, L423–L427 (1980).

[21] D. Coupier and V. C. Tran, *The 2D-directed spanning forest is almost surely a tree*. Random Structures & Algorithms. 42, 59–72 (2013).

[22] C. F. Coletti and G. Valle, *Convergence to the Brownian web for a generalization of the drainage network model*. Ann. Inst. H. Poincaré Probab. Statist. 50, 899–919 (2014).

[23] C. F. Coletti and L. A. Valencia, *Scaling limit for a family of random paths with radial behavior*. (2014). arXiv:1310.6929.

[24] R. Durrett, *Oriented percolation in two dimensions*. Ann. Probab. 12, 999–1040 (1984).

[25] D. Dhar, *Theoretical studies of self-organized criticality*. Physica A: Statistical Mechanics and its Applications. 369, 29–70 (2006).

[26] L. Dumaz, *A clever (self-repelling) burglar*, Electron. J. Probab. 17, 1–17 (2012).

[27] D. Dhar, *Fragmentation of a sheet by propagating, branching and merging cracks*. Journal of Physics A: Mathematical and Theoretical. 48, 17 (2015).

[28] C. R. Doering and D. ben-Avraham, *Interparticle distribution functions and rate equations for diffusion-limited reactions*. Phys. Rev. A. 38, 3035 (1988).

[29] P. Donnelly, S. N. Evans, K. Fleischmann, T. G. Kurtz and X. Zhou, *Continuum-sites stepping-stone models, coalescing exchangeable partitions and random trees* Ann. Probab. 28, 1063–1110 (2000).

[30] A. Depperschmidt, A. Greven and P. Pfaelhuber, *Marked metric measure spaces*. Electron. Commun. Probab. 16, 174–188 (2011).

[31] M. Damron and C. L. Winter, *A non-Markovian model of rill erosion*, Networks and Heterogeneous Media. 4, 731–753 (2009).

[32] T. Ellis, *Coalescing stochastic flows driven by Poisson random measure and convergence to the Brownian web*. Ph.D. thesis, University of Cambridge, 2010.

[33] T. Ellis and O. N. Feldheim. *The Brownian web is a two-dimensional black noise*. (2012). arXiv:1203.3585. Ann. Inst. H. Poincar Probab. Statist. 52, 162–172 (2016).

[34] A. Etheridge, N. Freeman and D. Straulino, *The Brownian net and selection in the spatial Lambda-Fleming-Viot process*. (2015). arXiv:1506.01158.

[35] S. N. Evans, B. Morris and A. Sen, *Coalescing systems of non-Brownian particles*, Probab. Theory Related Fields. 156, 307–342 (2013).

[36] P. A. Ferrari, L. R. G. Fontes and X.-Y. Wu, *Two-dimensional Poisson Trees converge to the Brownian web*. Ann. Inst. H. Poincaré Probab. Statist. 41, 851–858 (2005).

[37] L. R. G. Fontes, M. Isopi, C. M. Newman and D. Stein, *Aging in 1D discrete spin models and equivalent systems*. Physical Review Letters. 87, 110201 (2001).

[38] L. R. G. Fontes, M. Isopi, C. M. Newman and K. Ravishankar, *The Brownian web*. Proc. Nat. Acad. Sciences. 99, 15888–15893 (2002).

[39] L. R. G. Fontes, M. Isopi, C. M. Newman and K. Ravishankar, *The Brownian web: characterization and convergence*. (2003). math/0304119.

[40] L. R. G. Fontes, M. Isopi, C. M. Newman and K. Ravishankar, *The Brownian web: characterization and convergence*. Ann. Probab. 32(4), 2857–2883 (2004).

[41] L. R. G. Fontes, M. Isopi, C. M. Newman and K. Ravishankar, *Coarsening, nucleation, and the marked Brownian web*. Ann. Inst. H. Poincaré Probab. Statist. 42, 37–60 (2006).

[42] C. M. Fortuin, P. Kasteleyn and J. Ginibre, *Correlation inequalities on some partially ordered sets*. Commun. Math. Phys. 22, 89–103 (1970).

[43] P. A. Ferrari, C. Landim and H. Thorisson, *Poisson trees, succession lines and coalescing random walks*. Ann. Inst. H. Poincaré Probab. Statist. 40, 141–152 (2004).

[44] L. R. G. Fontes and C. M. Newman, *The full Brownian web as scaling limit of stochastic flows*. Stoch. Dyn. 6, 213–228 (2006).

[45] L. R. G. Fontes, C. M. Newman, K. Ravishankar and E. Schertzer, *Exceptional times for the dynamical discrete web*. Stoch. Proc. Appl. 119, 2832–2858 (2009).

[46] L. R. G. Fontes, L. A. Valencia and G. Valle, *Scaling limit of the radial Poissonian web*. Electron. J. Probab. 20, paper no. 31 (2015).

[47] G. Grimmett, *Percolation*. New York: Springer-Verlag, (1999).

[48] C. Garban, G. Pete and O. Schramm, *Pivotal, cluster and interface measures for critical planar percolation*. J. Amer. Math. Soc. 26, 939–1024 (2013).

[49] C. Garban, G. Pete and O. Schramm, *The scaling limits of near-critical and dynamical percolation*. (2013). arXiv:1305.5526.

[50] S. Gangopadhyay, R. Roy and A. Sarkar, *Random Oriented Trees: A Model of Drainage Networks*. Ann. App. Probab. 14, 1242–1266 (2004).

[51] C. Garban and J. Steiff, *Noise sensitivity and percolation*, in *Probability and statistical physics in two and more dimensions* (D. Ellwood et al. eds.). Providence, RI: Amer. Math. Soc. 49–154 (2012).

[52] A. Greven, R. Sun and A. Winter, *Continuum space limit of the genealogies of interacting Fleming-Viot processes on* $\mathbb{Z}$. (2015). arXiv:1508.07169.

[53] H. Hinrichsen, *Non-equilibrium critical phenomena and phase transitions into absorbing states*. Advances in Physics. 49, 815–958 (2000).

[54] R. Holley and T. Liggett. *Ergodic theorems for weakly interacting infinite systems and the voter model*. Annals of Probab. 3, 643–663 (1975).

[55] C. Howitt and J. Warren, *Consistent families of Brownian motions and stochastic flows of kernels*. Ann. Probab. 37, 1237–1272 (2009).
[56] C. Howitt and J. Warren, *Dynamics for the Brownian web and the erosion flow*. Stochastic Processes Appl. 119, 2028–2051 (2009).
[57] D. Jenkins, *Superdiffusive and subdiffusive exceptional times in the dynamical discrete web*. Stochastic Processes Appl. 125, 3373–3400 (2015).
[58] T. Kuczek, *The central limit theorem for the right edge of supercritical oriented percolation*. Ann. Probab. 17, 1322–1332 (1989).
[59] I. Karatzas and S.E. Shreve, *Brownian motion and stochastic calculus, 2nd edition*. New York: Springer-Verlag, (1991).
[60] K. Krebs, M.P. Pfannmüller, B. Wehefritz and H. Hinrichsen, *Finite-size scaling studies of one-dimensional reaction-diffusion systems. Part I. Analytical results*. J. Stat. Phys. 78, 1429–1470 (1995).
[61] T. Liggett, *Interacting particle systems*. New York: Springer, (1975).
[62] Y. Le Jan and S. Lemaire, *Products of beta matrices and sticky flows*. Probab. Th. Relat. Fields. 130, 109–134 (2004).
[63] Y. Le Jan and O. Raimond, *Flows, Coalecence and Noise*. Annals of Probab. 32, 1247–1315 (2004).
[64] Y. Le Jan and O. Raimond, *Sticky flows on the circle and their noises*. Probab. Th. Relat. Fields. 129, 63–82 (2004).
[65] M. Moshe, *Recent developments in Reggeon field theory*. Physics Reports. 37, 255–345 (1978).
[66] I. Molchanov, *Theory of random sets*. Berlin: Springer (2005).
[67] J. Molofsky, R. Durrett, J. Dushoff, D. Griffeath and S. Levin, *Local frequency dependence and global coexistence*. Theor. Pop. Biol. 55, 270–282 (1999).
[68] Y. Mohylevskyy, C. M. Newman and K. Ravishankar, *Ergodicity and percolation for variants of one-dimensional voter models*. ALEA. 10, 485–504 (2013).
[69] R. Munasinghe, R. Rajesh, R. Tribe and O. Zaboronski. *Multi-scaling of the n-point density function for coalescing Brownian motions*. Commun. Math. Phys. 268, 717–725 (2006).
[70] C. Neuhauser and S. W. Pacala, *An explicitly spatial version of the Lotka-Volterra model with interspecific competition*. Ann. Appl. Probab. 9, 1226–1259 (2000).
[71] C. M. Newman and K. Ravishankar, *Convergence of the Tóth lattice filling curve to the Tóth-Werner plane filling curve*. ALEA Lat. Am. J. Probab. Math. Stat. 1, 333–345 (2006).
[72] C. M. Newman, K. Ravishankar and R. Sun, *Convergence of coalescing nonsimple random walks to the Brownian web*. Electron. J. Prob. 10, 21–60 (2005).
[73] C. M. Newman, K. Ravishankar and E. Schertzer, *Marking (1,2) points of the Brownian web and applications*. Ann. Inst. Henri Poincaré Probab. Statist. 46, 537–574 (2010).
[74] C. M. Newman, K. Ravishankar and E. Schertzer, *The Brownian net with killing*. Stoch. Proc. and App. 125, 1148–1194 (2015).
[75] C. Neuhauser and A. Sudbury, *The biased annihilating branching process*. Adv. Appl. Prob. 25, 24–38 (1993).
[76] J. Norris and A. Turner, *Hastings-Levitov aggregation in the small-particle limit*. Comm. Math. Phys. 316, 809–841 (2012).
[77] J. Norris and A. Turner, *Weak convergence of the localized disturbance flow to the coalescing Brownian flow*. Ann. Probab. 43, 935–970 (2015).

[78] H. Ohtsuki, H. C. Hauert, E. Lieberman and M. A. Nowak, *A simple rule for the evolution of cooperation on graphs and social networks.* Nature. 441, 502–505 (2006).

[79] V. Privman, *Nonequilibrium statistical mechanics in one dimension.* Cambridge: Cambridge University Press, (1997).

[80] R. Roy, K. Saha and A. Sarkar. *Random directed forest and the Brownian web.* Ann. Inst. H. Poincaré Probab. Statist. 52, 1106–1143 (2016).

[81] R. Roy, K. Saha and A. Sarkar, *Hack's law in a drainage network model: a Brownian web approach.* Ann. Appl. Probab. 26, 1807–1836 (2016).

[82] F. Spitzer, *Principles of random walk, 2nd edition.* New York: Springer-Verlag, (1976).

[83] D. Schwartz, *Applications of duality to a class of Markov processes.* Ann. Probab. 5, 522–532 (1977).

[84] J. Steif, *A survey of dynamical percolation. Fractal geometry and stochastics IV.* Progress in Probability. 61, 145–174 (2009).

[85] T. Seppäläinen, *Current fluctuations for stochastic particle systems with drift in one spatial dimension*, in *Ensaios Matemáticos [Mathematical Surveys]*, 18. Rio de Janeiro: Sociedade Brasileira de Matemática, (2010).

[86] R. Sun and J. M. Swart, *The Brownian net.* Ann. Probab. 36, 1153–1208 (2008).

[87] O. Schramm and S. Smirnov, *On the scaling limits of planar percolation.* Ann. Probab. 39, 1768–1814 (2011).

[88] A. Sarkar and R. Sun, *Brownian web and oriented percolation: density bounds.* RIMS Kokyuroku, No. 1805, Applications of Renormalization Group Methods in Mathematical Sciences, 90–101 (2012).

[89] A. Sarkar and R. Sun, *Brownian web in the scaling limit of supercritical oriented percolation in dimension* $1 + 1$. Electron. J. Probab. 18, Paper 21 (2013).

[90] K. Saha and A. Sarkar, *Convergence of drainage networks to the Brownian web.* (2015). arXiv:1508.06919.

[91] E. Schertzer, R. Sun and J.M. Swart, *Special points of the Brownian net.* Electron. J. Prob. 14, Paper 30, 805–864 (2009).

[92] E. Schertzer, R. Sun and J.M. Swart, *Stochastic flows in the Brownian web and net.* Mem. Amer. Math. Soc. 227, 1065 (2014).

[93] F. Soucaliuc, B. Tóth and W. Werner, *Reflection and coalescence between one-dimensional Brownian paths.* Ann. Inst. Henri Poincaré Probab. Statist. 36, 509–536 (2000).

[94] B. Tóth, *The "true" self-avoiding walk with bond repulsion on* $\mathbb{Z}$*: Limit theorems.* Ann. Probab. 23, 1523–1556 (1995).

[95] B. Tsirelson, *Scaling limit, noise, stability.* Lecture Notes in Mathematics. 1840, 1–106, (2004).

[96] B. Tsirelson, *Nonclassical stochastic flows and continuous products.* Probability Surveys. 1, 173–298 (2004).

[97] B. Tsirelson, *Random compact set meets the graph of nonrandom continuous function.* (2013). arXiv:1308.5112.

[98] B. Tsirelson, *Noise as a Boolean algebra of -fields.* Ann. Probab. 42, 311–353 (2014).

[99] B. Tóth and W. Werner, *The true self-repelling motion.* Probab. Theory Related Fields. 111, 375–452 (1998).

[100] R. Tribe, S.K. Yip and O. Zaboronski, *One dimensional annihilating and coalescing particle systems as extended Pfaan point processes.* Electron. Commun. Probab. 17, No. 40 (2012).

[101] R. Tribe and O. Zaboronski, *Pfaan formulae for one dimensional coalescing and annihilating systems*. Electron. J. Probab. 16, No. 76, 2080–2103 (2011).

[102] J. Warren, *The noise made by a Poisson snake*. Electron. J. Probab. 7, No. 21, 1–21 (2002).

[103] J. Warren, *Dyson's Brownian motions, intertwining and interlacing*. Electron. J. Probab. 12, No. 19, 573–590 (2007).

[104] T. Williams and R. Bjerknes, *A stochastic model for the spread of an abnormal clone through the basal layer of the epithelium*, in Symp. Tobacco Research Council, London, (1971).

[105] X.-Y. Wu and Y. Zhang, *A geometrical structure for an infinite oriented cluster and its uniqueness*. Ann. Probab. 36, 862–875 (2008).

[106] Jinjiong Yu, *Edwards-Wilkinson fluctuations in the Howitt-Warren flows*. Stochastic Processes Appl. 126, 948–982 (2016).

Index

1-point function 241
2-point function 243

absorbing time 68
aggregation models 346
anomalous diffusion 63
average displacement 105
Avogadro 11

Betti 17
Big Data 8
bisimilarity 26
block bundles 23
Borel-Cantelli 151
branching Brownian motion 174, 357
branching random walk 176
branching time 177
branching-coalescing point set 297
branching-coalescing random walks 285
branching scale 177
Brownian loop soups 208
Brownian net 285
Brownian web 272

category theory 2
causal reasoning 7
Cayley tree 100
Cerf 7
characteristic functions 59
characteristic polynomial 77
Chern-Simons 22
circular unitary ensemble 197
coalescing Brownian motions 270
coalescing point set 280
coalescing random walks 270
coalgebra 27
coarse graining 143
complex systems 3
computational complexity 2
conformal covariance 210

conformal dimension 239
conformal field theory 36, 206
conformal invariance 206
Conformal Loop Ensembles 207
continuity equation 45
convex functional 142
convexity 147
correlation functions 36, 211
correlation length 205
covariance 143, 153, 215, 314
critical exponents 93, 207, 363
critical point 20, 148, 205, 208
Curie-Weiss model 120, 148

diameter 77, 226
dichotomy of scales 176
diffusion coefficient 45
Dirac 24
disordered 142, 170
domain 208, 344
drainage networks 337, 347
dual 142, 274
duality 19, 146, 271, 343

Einstein 47, 143
embodied computation 14
entropy 21, 142, 171
ergodicity 10, 128
exceedances 171
extrema 166

Feynman 7
fiber bundle 33
Fick's law 45
finite graph representation 305
first arrival time 102
first passage time 69
Fleming-Viot process 335
foliation 29
formal language 10, 30

Fourier 59, 353
free energy 111, 142, 174
freezing transition 198
functional order parameter 152
functor 5

Gamma function 72
gauge group 10
Gaussian free field 167, 209
generalized random energy model 145, 152, 179
generating function 22, 213
GREM *see* generalized random energy model
Grothendieck 13
Gumbel 166, 171

Hamilton-Jacobi equation 128
Hamiltonian 103, 215
Hausdorff dimension 222, 227
heat kernel 24
Heisenberg model 114
Hodge 20
holism 3
homotopy 21
hopping construction 291
Howitt-Warren flow 312
hypergraph 16

interacting particle systems 356
intersection local time 303
inverse problem 5
Ising model 114, 211

Langevin 51
Langland 30
language equivalence 26
Laplace 61
layering operator 237
Legendre 143
log-correlated fields 172

Mac Lane, Saunders 1
magnetization 114, 145
manifold 2
marking construction 301
martingale 173, 258, 313
maximal excursion 102
mean field 114, 145
mesh construction 287
molecular density 45
molecular flux 45
Monte Carlo 83
Morse 20
multi-agent 3
multiscale decomposition 176

n-point functions 236
near-critical scaling limit 208
network theory 3
Novikov, Sergei P. 19

occupation time 103
order parameter 115, 148
order statistics 171
oriented percolation 348

paramagnet 121
partition function 97, 144, 175, 213, 322
Pearl, Judea 6, 7
percolation 206
persistent homology 5
Poisson marking 288
Popper, Karl 1
Potts model 206, 342
prime number theorem 196
probability theory 6, 56, 205
process algebra 6
pullback 32

quantum walk 103
quivers 6

random energy model 143, 175
random unitary matrices 197
random walk 206
random walk in random environment 310
random walk loop soup 208
recentering 170
reconstruction 5
recurrence 72
reductionism 3
REM, *see* random energy model
renormalizable 237
renormalization group 14
rescaling 170
Riemann hypothesis 194
Riemann zeta function 194

scaling hypothesis 205
scaling limit 205
Schramm-Loewner Evolution (SLE) 207
second moment 167
self-avoiding walk 206
self-consistency 29
self-repelling 270
self-similarity 176
semantic equivalence 26
Sherrington-Kirkpatrick model 137, 152
shock wave 129
simplicial complex 5
simulation 5
Skorohod 280
spectral dimension 88
spectrum 78
spherical model 93, 113
spin glass 152, 158
spontaneous symmetry breaking 115, 129
statistical mechanics 3, 45, 142, 173, 207, 356
statistical physics 7, 81, 205

statistics of extremes 166, 167
sticky Brownian motion 313
stochastic flow of kernels 311
subfunctor 32
subgraph 76
subleading order of the maximum 188

thermodynamic state 143
Thurston 31
tightness 323
topology 5, 76, 270
transience 72, 357
Turing 12

ultraviolet cutoff 237
undecidable 19
universality 9, 94, 167, 270
universality classes 193

variational principle 39, 137, 142, 175
voter model 270

wedge characterization 280
Weibull 166
Wiener 50
winding model 239
winding operator 237

Yang-Mills 22